FUNDAMENTALS OF AIR TRAFFIC CONTROL

FUNDAMENTALS OF AIR TRAFFIC CONTROL

FOURTH EDITION

Michael S. Nolan
Purdue University

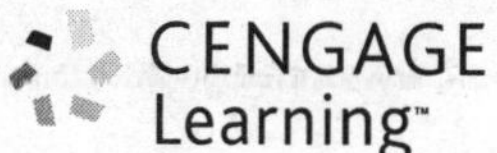

Andover • Melbourne • Mexico City • Stamford, CT • Toronto • Hong Kong • New Delhi • Seoul • Singapore • Tokyo

Fundamental of Air Traffic Control, 4e
Michael S. Nolan

ISBN-13: 978-81-315-1582-2
ISBN-10: 81-315-1582-6

Cengage Learning India Private Limited
418, F.I.E., Patparganj
Delhi 110092
India

Tel: 91-11-43641111
Fax: 91-11-43641100
Email: asia.infoindia@cengage.com

Cengage Learning is a leading provider of customized learning solutions with office locations around the globe, including Andover, Melbourne, Mexico City, Stamford (CT), Toronto, Hong Kong, New Delhi, Seoul, Singapore, and Tokyo. Locate your local office at: www.cengage.com/global

Cengage Learning Products are represented in Canada by Nelson Education, Ltd.

For product information, visit our website at **www.cengage.co.in**

Printed in India
First Indian Reprint 2011

PREFACE

This book was written after having searched long and hard for an appropriate college-level textbook on air traffic control. Various Federal Aviation Administration publications have been available for years, as have commercial introductory texts. But most of these books either describe rules and regulations or take a simplistic approach to how the air traffic control system works. No text has described not only how the ATC system works but *why* it operates the way it does. This book remedies that situation. It describes the background and history of the development of the air traffic control system, emphasizing why things are done the way they are, instead of simply repeating rules and regulations.

Throughout the text, appropriate real-life examples are used to illustrate the reasoning behind procedures used by air traffic controllers. The liberal use of figures and example phraseology assist the student in achieving an overall understanding of the air traffic control system. It is hoped that with this knowledge, future air traffic controllers will have a far better understanding of their chosen profession and can make the appropriate decisions that will lead aviation into the next century.

There are many unique features to this textbook that are not found in any other text on air traffic control. These features include the following.

History and Background

The history of the development of the air traffic control system and many of its components is included throughout the book. This history is not intended to be a dry repeat of names and dates but rather an explanation of past decisions that dramatically affected the current air traffic control system.

Illustrations and Photographs

Abundant illustrations and photographs are provided in this textbook. Air traffic control is a three-dimensional, visually oriented profession that cannot be explained simply through the use of text. These illustrations were designed to supplement the text, further explaining concepts and ideas that are difficult for the inexperienced student to visualize when simply reading about them.

Examples and Phraseology

One of the most important tasks facing a controller is the proper use of phraseology. The air traffic control system is based on understanding and usage of strange and sometimes hard to understand wording. Besides explaining the proper use of terms, the text includes numerous examples of the proper usage of phraseology.

Real-Life Examples

Throughout the text, examples found in real life are used to further explain the concepts introduced. In addition, one entire chapter is dedicated to the actual operation of the air traffic control system. "Behind the scenes" activities and coordination are described, using sample flights through actual airspace. Such examples reinforce the material introduced in earlier chapters, further clarifying and explaining some of the complicated procedures used to separate air traffic above the United States.

Chapter Material

The first four chapters of this text prepare the student for understanding the intricate procedures used in controlling air traffic. These four chapters cover fundamental topics, such as history, navigation, and phraseology. Chapters 5 through 11 detail the separation of aircraft in the ATC system. Chapter 12 is an in-depth look at the future of air traffic control, and Chapter 13 discusses employment opportunities for air traffic controllers. At the conclusion of the text is a detailed glossary of terms introduced in the book.

Collegiate Training Initiative

This textbook has been designed with the Federal Aviation Administration's College Training Initiative in mind. This program was developed to provide education to future FAA controllers. The material in this textbook provides much of the knowledge needed by tomorrow's air traffic controllers.

Acknowledgments

I would like to thank the following individuals who have made this textbook possible. Without their gracious help and assistance, it would have been impossible to complete this book: Juanita Hull, Federal Aviation Administration; James Cheesman, SRSA Corporation; Bill Goodnight, Federal Aviation Administration; and the entire staff and management of the Champaign ATCT, Lafayette ATCT, and Indianapolis ATCT and ARTCC.

The following reviewers made many helpful suggestions in development of the manuscript for this book: Roger Bacchieri, Daniel Webster College; Peter Bailey, Wilmington College; Peggy Baty, Embry-Riddle Aeronautical University; Terry S. Bowman, Southern Illinois University; Jeffry B. Burbridge, Catonsville Community College; Jonathan R. Burke, Metropolitan State College of Denver; Veronica T. Cote, Bridgewater State College; Michael Farley, Bridgewater State College; Bruce D. Hoover, Palo Alto College; Roger Matteson, Central Washington University; Patrick K. Mattson, St. Cloud State University; Mary Ozimkowski, Metropolitan State College; Keith Parkman, Embry-Riddle Aeronautical University; Martha Pearce, Arizona State University; Michael J. Polay, Embry-Riddle Aeronautical University; Robert Rogus, Mt. San Antonio College; Jose Ruiz, Southern Illinois University; Robert H. Ryder,

Delta State University; Thomas Teller, Daniel Webster College; and Henry Whitney, Mt. San Antonio College.

The air traffic control system in the United States is truly a system, which means that it takes the efforts of many individuals to make the whole thing work. I've found that publishing a textbook is much the same. Although my name is on the cover, only the concerted efforts of a diligent, professional, and very talented group of people made it possible for this book to be published. Everyone involved was just as important as everyone else, as is the case in any system. We should all be very thankful for their work and effort. I sure am.

The first edition was completed in 1990 through the efforts of a group of individuals at both Wadsworth and Del Mar Associates. Although many of them are no longer connected with this project, their legacy lives on, and I remember them fondly. They include Anne Scanlan-Rohrer, aviation editor at Wadsworth, and her editorial assistants, Leslie With, Karen Moore, and Cathie Fields. Jackie Estrada, Richard Carter, Detta Penna, Robin Witkin, and Nancy Sjoberg developed the layout and most of the illustrations for the first edition. Thanks to one and all.

The second edition was coordinated by Ruth Cottrell of Ruth Cottrell Books and Jennie Burger of Wadsworth, assisted by Barbara Britton and Charles Cox; and the third edition by Tobi Giannone of Michael Bass & Associates and Marie Carigma-Sambilay of Wadsworth, assisted by Hal Humphrey and Karen Hunt. The fourth edition was coordinated by Andy Sieverman of G&S Typesetters and Carol Benedict at Wadsworth. They were assisted by Belinda Krohmer and Jessica Reed. I can never thank all of these people enough. They continually teach me how to revise and edit this text in an attempt to keep it up to date. They are the book experts, responsible for the ultimate product you see before you. They were indispensable members of our system. Without them, this project could never have been accomplished. And of course I must thank my wife, Barbara, and my three children, Linda, Erin, and David, who gave up a lot of their time with Daddy so he could finish this revision on time. I am indebted to each and every one of you.

Mike Nolan

Michael S. Nolan (B.S., Industrial Technology, Illinois State University; M.S., Instructional Development and Educational Computing, Purdue University) is a former air traffic controller and holds licenses and certification as a commercial pilot, flight instructor, instrument instructor, tower operator, airframe and powerplant mechanic, and aviation weather observer. He has taught a variety of aviation courses at the University of Illinois and at Chanute Air Force Base as well as at Purdue University, where he currently teaches in the Aviation Technology Department.

CONTENTS

1

History of Air Traffic Control

Checkpoints

After studying this chapter, you should be able to:

1. Discuss the significance of the Airmail Act of 1925.
2. Describe how the federal government became involved in air traffic control.
3. Discuss the history of the various federal agencies involved in air traffic control.
4. Discuss the formation of organized labor unions as they pertain to air traffic control.
5. Identify the organizations currently involved in the air traffic control system.
6. Identify the various organizations that have represented air traffic controllers.
7. Identify some of the methods air traffic controllers used in the past to separate aircraft.

1903–1925: Early Aviation Developments

When the Wright brothers' experiment in flight succeeded on December 17, 1903, the world took little notice. Newspapers of that era either did not believe or belittled the accomplishments of the two brothers on that cold, blustery morning. At the start of the twentieth century, most people regarded aviation as a pastime for experimenters and daredevils. It was hard to believe that the tiny, underpowered aircraft of that era would ever evolve into a useful form of transportation. In this early period of experimentation, anyone with a mechanical aptitude could design, build, and fly an aircraft without passing any type of test or possessing any type of license. Without regulation or certification, people began to build and quite regularly crash these early flying machines. The general public was frightened by the machines and believed that only fools would fly in them. Potential investors in this new industry were fearful of risking their capital to finance an unproven and apparently dangerous industry.

In spite of this climate of fear and distrust, aviation pioneers began to demonstrate the usefulness of their primitive flying machines. As early as 1911, the first mail was carried by air. By the time the United States became involved in World War I, the airplane had demonstrated its usefulness as an observation platform and as a crude weapons delivery system.

After the war, numerous additional uses were found for the airplane. The Post Office Department began to offer routine airmail service in 1918, using U.S. Army pilots and aircraft. In 1919 the U.S. Department of Agriculture initiated experiments that would lead to the commercial use of aircraft for the application of pesticides. The first transatlantic crossing was made that year, which also saw the first experimental use of radio as a navigation aid.

Development of Airmail Service

Between 1918 and 1925, airmail service was expanded by the Post Office Department until full transcontinental service was finally achieved. Until 1923 most of the mail was flown during daylight hours, since a safe, reliable form of nighttime navigation had not been developed. In 1921 the first experimental night flight was conducted, using bonfires located along the navigation route. These bonfires were replaced in 1923, when a 72-mile stretch of airway between Dayton and Columbus, Ohio, was experimentally lit with electric and gas arc lighting. The experiment proved successful, and airway lighting was soon introduced across the country. By 1924 the portion of the transcontinental airway between Cheyenne, Wyoming, and Chicago, Illinois, was lit, and routine night flights were being conducted along this section of the airway.

By 1925 the commercial potential for aviation had been established, and the Post Office Department found itself under pressure to expand airmail service at a faster rate than was possible for a government agency. Since aviation appeared to be a commercially viable industry, it was felt that airmail service could now be handled by private airline companies. A resolution to permit private contracting, introduced by Congressman Clyde Kelly of Pennsylvania, was signed into law on February 2, 1925, and became known as the **Airmail Act of 1925**. The Airmail Act authorized the postmaster general to contract with private individuals and corporations for the purpose of transporting airmail. Between 1925 and 1927, airmail contracts were bid to private corporations, and commercial aviation became a reality.

After this act was signed into law, many companies that had been sitting on the sidelines earnestly jumped into the aviation field. Boeing, Douglas, and Pratt and Whitney were just a few of the companies that bid to supply aircraft and engines to the budding airmail industry. Even the great Henry Ford entered the picture, producing the famous Ford Trimotor and operating an air cargo airline between Detroit and Chicago.

The Morrow Report

With this increase in air activity came an increased desire for some type of national regulation of the industry. Prior to this time, individual states had reserved the right to test and certify pilots, but many were hesitant to exercise this authority. The aviation industry was still fragile, and public sentiment favored federal government regulation to unify the industry through a common set of rules, procedures, and certification. It was felt that government regulations were needed if the aviation industry were to grow and prosper. Without this regulation, the public's trust could not be gained.

A joint congressional committee recommended the formation of an advisory board composed of prominent businessmen to recommend the possible extent of federal involvement in the aviation industry. In 1924 President Calvin Coolidge appointed Dwight Morrow to head this board and make recommendations as to future government policy. The Morrow board presented its final report to the president on December 2, 1925. The **Morrow Report** recommended that military and civilian aviation operate separately, with the Department of Commerce to be given the responsibility for the promotion and the regulation of the civilian aviation industry.

1925–1934

Air Commerce Act

President Coolidge endorsed the findings of the Morrow Report and passed it along to Congress. He requested that the board's recommendations be implemented as soon as possible. After the inevitable discussion and negotiations, Congress approved, and President Coolidge signed the **Air Commerce Act** into law on May 20, 1926.

As Senator Hiram Bingham, who introduced the Air Commerce Act into the Senate, explained, the purpose of the act was "not so much to regulate as to promote civilian aviation." The Air Commerce Act made it the duty of the secretary of commerce (at that time Herbert Hoover) to encourage the growth of the aviation industry through the establishment of airways and navigation aids. In addition, the secretary was authorized to regulate the industry as necessary to elevate the public's perception of aviation as a safe mode of transportation. To this end, Hoover instituted a program to license pilots and mechanics and to regulate the use of these airways. These responsibilities were delegated to the Aeronautics Branch of the Department of Commerce, which was headed by the new director of aeronautics, Clarence M. Young.

In May 1927 Charles Lindbergh captured the attention of the nation with his daring flight across the Atlantic. During that same year, the first ground-to-air experimental radio was installed in an aircraft. In 1928 the first seven airmail-route radio stations were installed. Many of today's airlines were born in this era. Colonial Airlines (American), Western Express (TWA), Northwest Airlines, and United Airlines were all formed during this exciting period of air transportation growth.

Evolution of Air Traffic Control

Prior to the early 1930s, there was little need for an organized system of air traffic control in the United States. Almost all of the aerial traffic in this country was conducted in daylight with clear flying conditions. Advances in aircraft control and navigation had yet to be made that would permit flight at night or during periods of restricted visibility. The practice of "see and be seen" became the principal method of traffic avoidance. This meant that pilots could fly only in conditions that would permit them to see other aircraft and alter their flight path in time to avoid them.

According to this principle, pilots were required to fly clear of any clouds and only in areas where the visibility was at least 3 miles. These rules have been only slightly modified since that time and are now known as **visual flight rules (VFR).** (A discussion of the current version of these flight rules is presented in Chapter 3.) Since the aircraft used by the airlines in the 1930s were relatively slow and the pilots could readily see and avoid other aircraft, the establishment of an organized air traffic control system was not deemed necessary. But by the late 1930s the capability of aircraft to fly at night and in marginal weather conditions had improved tremendously. Instrumentation had been designed that would permit pilots to control the aircraft without visual reference to the natural horizon. In addition, a system of ground-based radio **navigation aids (navaids)** was being constructed to permit pilots to navigate without ground reference. When this equipment was installed, pilots were able to take off, cruise en route, and land in weather conditions that would not permit them to see and avoid other aircraft.

Because all of these aircraft eventually had to land at an airport, it was inevitable that the airspace within the immediate vicinity of busy airports became congested, and some form of local air traffic control would soon be needed. The problem of airspace congestion was compounded by the fact that the airports of that era only remotely resembled the airports of today. An airport in the 1930s rarely had designated runways. It usually consisted of a large, rectangular plot of land covered with either sod or cinders.

After flying over the airport to observe the wind direction, local traffic, and runway conditions, the pilots themselves decided in which direction they wished to land. During the approach and landing, the pilots were kept busy try-

Figure 1-1. The first air traffic controller, Archie W. League, shown in his winter clothing at the St. Louis Lambert Municipal Airport in 1929.

ing to spot other aircraft, decide who had priority, and maneuver their planes behind the others, allotting sufficient time for a previous plane to land, brake to a stop, and taxi clear of the runway prior to their arrival. In addition, pilots needed to constantly scan the airport surface area to detect aircraft taxiing for takeoff. To decrease ground roll distance, pilots usually maneuvered their aircraft to land or take off into the wind. On windy days this forced most of the pilots to land and take off in the same direction. But on calm days, aircraft could be seen landing and taking off in every direction. It was immediately apparent that some form of air traffic control would have to be initiated around airports or accidents would begin to occur at an increasing rate.

Air Traffic Controllers The earliest method of regulating takeoffs and landings required an air traffic controller to stand in a prominent location on the airfield and use colored flags to communicate with the pilots. If the controller waved a green flag, it meant that the pilots were to proceed with their planned takeoff or landing. But if the controller waved a red flag, the pilots were to hold their position until the controller had determined that it was safe to continue. At that time, the controller would wave the green flag, advising the pilots that they could proceed. The first airport to hire this type of air traffic controller was the St. Louis (Missouri) airport. In 1929, St. Louis hired Archie League as the nation's first air traffic controller (see Figure 1-1). Before taking on this

FAA

Figure 1-2. Archie League standing next to a spotlight while guiding down an aircraft during IFR weather.

role, League had been a barnstormer, a mechanic, and the operator of a flying circus.

League controlled air traffic from a wheelbarrow on which he had mounted a beach umbrella. In the morning, he would pack the wheelbarrow with a beach chair, water, a note pad, a pair of colored flags, and his lunch. He would wheel his equipment out to the approach end of the runway, where he would use the flags to advise the pilots to either continue their approach or hold until the traffic was clear. At the end of the day, League would repack his equipment into the wheelbarrow and return to the terminal. He performed these tasks both winter and summer, beginning a 36-year career in air traffic control (see Figure 1-2). Other large cities soon saw the advantages of this system and began to employ air traffic controllers at their own airports.

Although workable, this early, crude form of air traffic control had many obvious drawbacks. Since the controller usually stood near the approach end of the runway, he was far more likely to attract the attention of departing rather than arriving aircraft. Pilots inbound for landing found it difficult to determine which direction to land and to see the air traffic controller's location. And if more than one aircraft was inbound, it became difficult, if not impossible, for the air traffic controller to give different instructions to each plane. It was also difficult for the controller to determine whether the pilots had actually received and understood the intended instructions. And it was impossible to use this system of communication at night or during stormy weather. Fortunately, at that time few aircraft flew during such weather conditions.

Light Guns In an attempt to rectify some of these problems, the controller's colored flags were soon replaced by **light guns.** A light gun is a device that permits the controller to direct a narrow beam of high-intensity colored light to a specific aircraft (see Figure 1-3). Light guns were equipped with a gunsight that let the controller accurately aim the beam of light at one particular aircraft. The gun was also equipped with different-colored lenses to permit the controller to easily change the color of the light.

The controller operated the light gun either from a glassed-in room on top of a hangar, called a **control tower,** or from a portable light gun station located near the arrival end of the runway.

The light gun signals used by the early controllers resembled the colored-flag system. A red light advised the pilots to hold their aircraft, whereas a green light advised them to proceed. Most of the busy airports soon built control towers and installed these light guns. The control towers were usually placed on top of one of the highest structures at the airport. Controllers working in the tower now had an unobstructed view of the airport and the surrounding airspace. They no longer had to stand out next to the runway, exposed to the elements.

Light guns are still used today at most control towers. They are used to communicate when either the radios in the control tower or the aircraft are inoperative or when an aircraft is not radio equipped. The light gun code has not changed significantly since the 1930s. The official light gun signals in use today are listed in Table 1-1.

FAA

Figure 1-3. Using a light gun signal.

Table 1-1. Light Gun Signals

Color and Type of Signal	***Meaning with Respect to Aircraft on the Ground***	***Meaning with Respect to Aircraft in Flight***
Steady green	Cleared for takeoff	Cleared to land
Flashing green	Cleared to taxi	Return for landing (to be followed by a steady green at the proper time)
Steady red	Stop	Give way to other aircraft and continue circling
Flashing red	Taxi clear of runway in use	Airport unsafe, do not land
Flashing white	Return to starting point on airport	Not applicable
Alternating red and green	Exercise extreme caution	Exercise extreme caution

Although the light gun was an improvement over the colored-flag system of air traffic control, a number of important deficiencies still remained. When inbound to the airport, the pilots were usually busy flying their aircraft and were unable to maintain a constant lookout for changing light gun signals. As a result, the controller might not be able to transmit critical instructions to pilots who were performing some other task and whose attention was diverted. The light guns were also useless in bad weather, since airborne particles of dust and moisture easily diffused and refracted the light beam. Furthermore, the controller could never be quite sure whether the pilot had received or properly interpreted the light gun signal. The controller could give instructions to the pilots, but the pilots had no means for communicating to the air traffic controller. It was apparent that a reliable, two-way communications system would have to be developed.

Radio Communication The modern system of air traffic control was born at the Cleveland (Ohio) airport. The city of Cleveland constructed a control tower on top of an old hangar and equipped this facility with radio transmitting and receiving equipment. The communications transmitters were 15-watt radios that permitted voice communication with pilots over a distance of approximately 15 miles. Using this primitive radio equipment, the air traffic controller could communicate directly with the pilots of properly equipped aircraft. In addition, the pilots could respond to these instructions or initiate communication with the controllers. This system permitted the controllers to issue traffic instructions, weather information, and airport landing conditions to the pilots of radio-equipped aircraft. This voice communication could be maintained night and day, in good weather or in bad.

The control tower was also equipped with light guns to permit controllers to communicate with aircraft that were not radio equipped. The light guns were also used for backup communications in case the radio equipment in either the control tower or the aircraft malfunctioned. By being located on top of the highest structure at the airport, the controllers had an unobstructed view of the airport surface area and the approaches to the landing area. This permitted the controllers to issue instructions that would properly sequence aircraft inbound for landing with those attempting to depart. Most of the busy airports around the country followed Cleveland's example and constructed radio-equipped air traffic control towers (see Figure 1-4).

Despite the dramatic safety improvement that these towers offered, their effectiveness was limited, since the primitive radio equipment was heavy, clumsy to use, unreliable, and relatively expensive. The airlines were hesitant to install this equipment on planes since it would replace valuable, revenue-producing space on the aircraft. Furthermore, most of the small aircraft in use during this era were manufactured with electrical systems that provided insufficient power to operate the radios, and the owners were often unable to afford the expensive electrical system modifications that would permit them to benefit from this advance in air traffic control technology.

Michael Nolan

Figure 1-4. The control tower at Indianapolis's Stout Field.

Air traffic controllers and pilots were also severely handicapped by the lack of a standardized set of rules or phrases to be used when communicating with each other. Some pilots contacted the control tower when they were 5 to 10 miles away from the airport, whereas other pilots neglected to contact the controllers until they were almost ready to land. Even though the air traffic controllers were federally certified, they were still airport employees, and pilots had no legal obligation to contact them. And if radio contact was established, there was little agreement on the phraseology that should be used. Many pilots simply did not understand the instructions that were being transmitted to them.

Despite these serious limitations, this early form of air traffic control worked remarkably well. Radio permitted the controllers to pass along valuable information and control instructions to the pilots of properly equipped aircraft, and the pilots could acknowledge receipt of these instructions and make accurate position reports to the controllers.

Instrument Flying At the same time that control towers were being constructed, aircraft designers were beginning to produce a new generation of faster, higher flying transport aircraft specifically designed for airline operation. These aircraft were equipped with advanced instrumentation and radio navigation equipment that would permit their pilots to fly in weather conditions that had made navigation impossible just 10 years previously. Using these instruments and the ground-based radio navigation aids installed by the federal government, the airlines began to routinely conduct flights of hundreds of miles through cloud and fog with no outside reference. These flight conditions, where aircraft control and navigation are conducted solely by reference to cockpit instrumentation, are known as **instrument meteorological conditions** (IMC).

Pilots of properly equipped aircraft could now fly in conditions where in-flight visibility might be measured in feet instead of miles. Pilots were able to land when visibilities were less than 2 miles. Certainly in these flight conditions, the "see and be seen" concept of aircraft separation was inadequate. In addition, as the airlines introduced into service newer airliners such as the DC-2, DC-3, and Boeing 247, the wide disparity in performance between these aircraft and the older planes in service became more readily apparent. This mix of aircraft with different cruising airspeeds and flight characteristics increased the complexity involved in separating aircraft and made it much more difficult for the air traffic controller to properly and safely sequence traffic inbound for landing. The only reason that midair collisions occurred infrequently was that few aircraft were flying in reduced visibility conditions at the same altitude, on the same route, at the same time.

By the early 1930s the airspace around major airports had become increasingly crowded; people living around these airports felt that the risk of midair collisions had increased and feared that colliding aircraft might crash into their neighborhoods. These residents began to pressure the states and the localities around the major airports to pass laws restricting air travel over their jurisdictions. It was apparent that utter chaos in the aviation industry would result if every state enacted legislation restricting or banning flight over certain areas. These restrictions would retard the growth of the airline industry and might jeopardize its very existence.

1934–1945

Establishment of the Bureau of Air Commerce

In response to this threat to the nation's interstate commerce, in 1934 Congress created the **Bureau of Air Commerce** (part of the Department of Commerce) as the agency responsible for the regulation of traffic along the nation's airways. This act made the federal government responsible for the licensing of pilots, the establishment of airways and navigation aids, and the actual separation and safety of the aircraft using these airways. In 1936 the Bureau of Air Commerce established rules to be followed by pilots when flying on the airways in instrument meteorological conditions. These rules are known as **instrument flight rules (IFR)**.

Because most of the major airports had already constructed and staffed air traffic control towers, the most pressing need facing the aviation industry

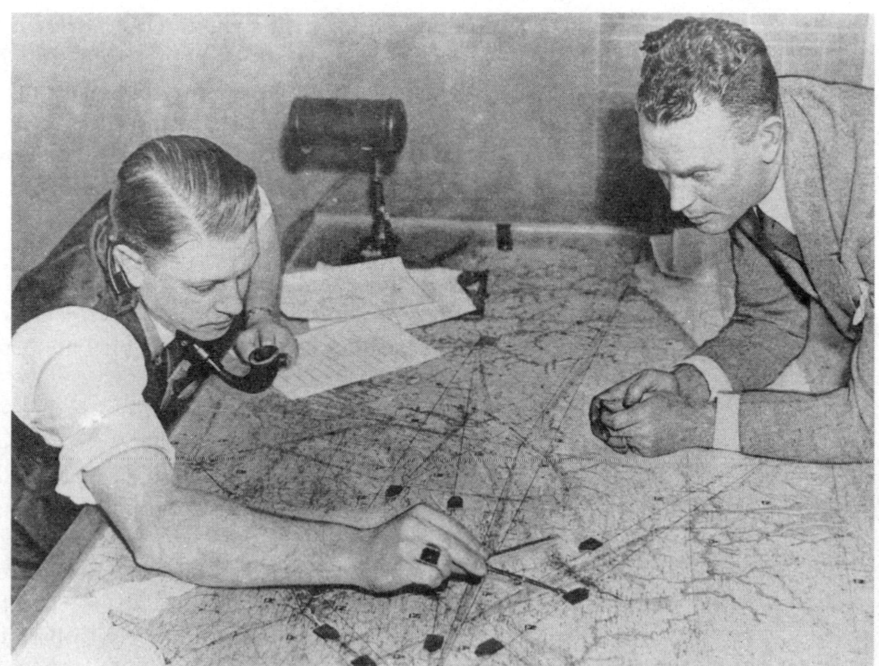

FAA

Figure 1-5. Controllers at the Newark ATCU separating en route traffic using maps and shrimp boats.

was for the separation of aircraft flying between airports. The airways in the eastern United States had become increasingly congested, and during periods of IFR weather, near misses began to occur with increasing frequency. Some form of traffic control on these busy airways would have to be established as soon as possible.

As a result of Depression-era budget restrictions, the Department of Commerce was unable to quickly form an air traffic control system and requested that the major airlines themselves take the initiative and develop a number of **airway traffic control units** (ATCUs) that would separate aircraft operating on the federal airways. The federal government promised that it would take possession of and operate these facilities at a later date. On December 1, 1935, TWA, American, Eastern, and United Airlines formed the first experimental airway traffic control unit at the Newark (New Jersey) airport (see Figure 1-5). The responsibility of the ATCU was to separate traffic operating on the airway during periods of IFR weather. During VFR weather, pilots flying the airways would still separate themselves using the see-and-avoid principles of air traffic control.

En Route Air Traffic Control

Following the operational success of the first ATCU, the four airlines were encouraged by the Department of Commerce to open additional units in Chicago,

Cleveland, Pittsburgh, and Oakland. These ATCUs were opened a short time thereafter and were staffed by employees of the airlines. By mutual agreement, each of these ATCUs assumed responsibility for separating IFR traffic within a selected area of airspace. It was not mandatory that every pilot contact the ATCUs, however. Military and noncommercial civilian aircraft were not required by law to contact the ATCUs. Fortunately, most of the IFR-certified aircraft were operated by the airlines.

Because of the technical limitations of 1930s radio equipment, all communication between the pilots and controllers was accomplished through an intermediary, either an airline dispatcher or a radio operator. Whenever pilots planned to fly in poor weather conditions, they filed an instrument flight plan with an airline dispatch office. This flight plan included the type of aircraft to be flown, the names of departure and arrival airports, the estimated departure time and time en route, the airline flight number, the requested route of flight, the aircraft's cruising airspeed, and the requested cruising altitude.

The airline dispatcher on duty forwarded this information by telephone to the ATCU with responsibility for the departure airport. The air traffic controllers on duty determined whether the route and altitude requested by the pilot might conflict with other aircraft and modified the flight plan as necessary to ensure the safe separation of aircraft. The controllers then issued an air traffic control clearance to the dispatcher that was to be relayed to the pilot. This clearance included the departure time, route of flight, and cruising altitude. The dispatcher relayed this information to the pilot, either in person or by radio.

The air traffic controllers in the ATCU wrote the appropriate flight plan information on a chalkboard and on a note card. This card was then attached to a brass holder that was called a **shrimp boat** by the controllers because of its resemblance to a small fishing boat. These shrimp boats would be moved along an airway map, indicating the approximate positions of the aircraft as they flew toward their destinations.

As each plane progressed through the ATCU's airspace, the pilots would transmit their position to an airline company radio operator, who then relayed this information to the ATCU controller by telephone or telegraph. The ATCU controllers updated the aircrafts' information on their blackboard and note cards and continued to move the shrimp boats along the map indicating each plane's approximate position. If a controller detected a potential conflict between two aircraft, appropriate route or altitude changes would be issued to one or both aircraft. These instructions were telephoned to the airline radio station nearest the last reported positions of the aircraft. The airline radio operator would then try to relay this information to the pilots. If the radio operator was unable to contact the aircraft, the ATCU controller would telephone other airline radio stations and ask that they try to make contact with the aircraft. Because of the problems inherent with the frequencies used by radio transmitters and receivers of that era, the controller might be required to telephone a number of radio operators before one could be found who could establish contact with the desired aircraft. Under certain adverse weather conditions, the radio operators might not be able to make contact with a particular aircraft for hours at a time.

Normally, three air traffic controllers were on duty in the ATCU at any one time. Each controller was assigned different job responsibilities. The "A" controller was responsible for the safe separation of all the participating aircraft operating within the ATCU's area of jurisdiction. Using the information provided on the flight plan and gathered through position reports, the "A" controller determined whether potential conflicts existed and undertook corrective action. The "A" controller communicated with the airline radio operators by telephone and issued the clearances that were then relayed to the pilots. As position reports were obtained from the pilots, the "B" controller was responsible for moving the shrimp boats across the airway map. In addition, the "B" controller received updated weather reports and was responsible for disseminating this information to the pilots. Using position reports obtained from the pilots, the estimated airspeeds from the flight plans, and the estimated winds aloft, the "C" controller calculated the future location of each aircraft. The "C" controller wrote this information on the blackboard and on the note cards attached to the shrimp boats. During periods of reduced staffing (such as evenings, weekends, and holidays), there might be two or possibly only one controller staffing the ATCU, and the responsibilities were divided evenly.

During periods of good weather, when pilots could legally fly under the existing VFR flight rules, the ATCU controllers exercised **passive control** of the aircraft. Passive control means that the controllers would track and update the flight path of each aircraft and would advise the pilot of the presence of other aircraft only if they were predicted to be within about 15 minutes' flying time of each other. The controllers would not issue any instructions to try to separate these aircraft unless either pilot requested this service. Since the weather was VFR, it was assumed that the pilots could see and avoid each other.

Whenever adverse weather conditions existed and the pilots were unable to operate in VFR conditions, the controllers began to exercise **active control** of air traffic along the airways. Active air traffic control assumes that the pilots cannot see and avoid each other, and the controllers must issue instructions to ensure that all participating aircraft remain safely separated (see Figure 1-6).

Although the air traffic control units were successful in accomplishing the initial objective of separating aircraft along the busiest sections of the airways, a number of problems were immediately apparent. Many airline companies were operating in the United States, but only four of them were chosen to operate the ATCUs. Any pilot who wished to participate in the air traffic control system was required to file a flight plan and receive a clearance from the controllers at the ATCU. The fact that the controllers were all employees of the four airline companies led to numerous complaints of favoritism and of unjustified holding of competing and privately owned aircraft. In addition, the legal authority of the ATCUs and their controllers was questionable. Pilots were not required by law to file flight plans until August of 1936. An additional problem was that few established or standardized procedures existed for the separation of aircraft operating along the airways.

FAA

Figure 1-6. Controllers at the St. Louis Center in 1939 separating en route aircraft using flight progress strips.

There was also little agreement as to how the transfer of control would occur when the aircraft entered the local area around the airport. Since the air traffic control towers were operated by the cities, whose controllers did not even have to be federally certified, little agreement or coordination occurred between the towers and the ATCUs.

On June 6, 1937, the Department of Commerce began to acquire the ATCUs from the airlines and staff them with federally certified controllers. The federal government renamed these facilities **airway traffic control stations (ATCSs).** Many of the ATCU employees transferred from the airlines to the government. In most cases, these employees took a considerable pay cut to do so. With the acquisition of the ATCSs, the Department of Commerce began to implement standardized air traffic control procedures. In May 1938 the Department of Commerce also became the licensing authority for all civilian air traffic controllers, both those employed in the ATCSs and those operating the air traffic control towers.

Copeland Committee

On May 6, 1935, a TWA airliner crashed outside of Kansas City, killing five persons including Senator Bronson M. Cutting of New Mexico. This accident and a number of other factors prompted Congress to commission a report on air traffic safety and the operation of the Bureau of Air Commerce. The Senate appointed Royal S. Copeland, the chairman of the Commerce Committee, to head the commission. The preliminary report issued by this committee (known as the Copeland Committee) was released on June 30, 1936. The report was a scathing (and in retrospect very biased) indictment of the Bureau of Air Commerce. As a subordinate bureau in the Department of Commerce, the Bureau of Air Commerce had become enmeshed in politics and had found it difficult to improve the airway system in the midst of the Depression. The report blamed the bureau for providing insufficient funding and maintenance of airway navaids. At the same time, a Bureau of Air Commerce accident report placed the blame for the crash on the pilots of the TWA aircraft. The controversy that ensued harshly pointed out the problems in the nation's air traffic control system. Both Congress and President Franklin D. Roosevelt decided that something needed to be done.

Civil Aeronautics Act of 1938

In a move to eliminate the Bureau of Air Commerce, on June 23, 1938, Congress passed the Civil Aeronautics Act, which in turn created the **Civil Aeronautics Authority (CAA)**. The CAA became the only independent authority in the U.S. government at that time. One of the Copeland Commission findings was that the Bureau of Air Commerce had been assigned contradictory responsibilities. On the one hand, it was supposed to promote aviation, yet on the other hand, it was supposed to regulate it. The bureau was responsible for operating many components of the air traffic control system but was also responsible for investigating accidents that might be caused by deficiencies in the air traffic control (ATC) system itself.

To try to solve some of these problems, the Civil Aeronautics Act divided the functions of the CAA into three groups. A five-person Civil Aeronautics Authority was given the responsibility of issuing airline route certificates and determining airline fares. The members of the authority were appointed by the president and could be removed only for cause. An independent Air Safety Board was also established to investigate aviation accidents and make safety recommendations. Finally, a CAA administrator, to be appointed by the president, was charged with fostering aviation, maintaining the airways, and controlling air traffic. The administrator was subject to dismissal by the president for any reason. To perform these various tasks, the CAA absorbed most of the employees of the Bureau of Air Commerce.

The Civil Aeronautics Act also provided for CAA certification of air traffic controllers who worked in the air traffic control towers. Most of the tower controllers were still employed by the municipalities that owned and operated the airports. At the same time, the CAA began a program to slowly take possession of the control towers and their controller work force but was hampered by budget restrictions imposed by Congress. By 1941, as America approached the beginning of World War II, the CAA was finally able to absorb the employees of most of the municipally operated air traffic control towers. On July 1 of that year, the CAA established the Air Traffic Control Division, which was given responsibility for operating the control towers and the newly named **airway traffic control centers (ATCCs).**

Within a short time, it became obvious that the CAA organization was unwieldy and defective, since the responsibilities of the three dominant groups overlapped. For example, it was the Air Safety Board's responsibility to determine the cause of an accident, the five-man authority's job to make recommendations, and the administrator's role to attempt to implement these recommendations. No one group had either ultimate authority or responsibility for aviation safety. Institutional paralysis began to set in. Within a short time, dissension among the three groups diminished the effectiveness of the CAA as a whole. Public disagreement among the groups threatened to destroy the CAA. President Roosevelt directed the Bureau of the Budget to undertake a study of the CAA structure and make recommendations for reorganization.

1940 Reorganization of the CAA

In 1940, under the authority conferred by the Reorganization Act of 1939, the president chose to reorganize the CAA in accordance with the recommendations made by the Bureau of the Budget.

The functions of the Air Safety Board and the five-person authority were combined into a new organization known as the **Civil Aeronautics Board (CAB).** The CAB was placed administratively within the Department of Commerce but exercised its duties with little outside influence. The Office of the Administrator was placed under the auspices of the Department of Commerce and was renamed the **Civil Aeronautics Administration (CAA).** The independent status of the Civil Aeronautics Authority turned out to be extremely short lived.

After the reorganization, **civil air regulations (CARs)** were mandated by the CAA to give legal authority to its controllers when separating aircraft on the nation's airways. Under the CARs, by December 1, 1941, pilots wishing to fly IFR would have to be certified by the Department of Commerce and their aircraft would have to carry federally mandated equipment. Every civilian pilot flying IFR would be required to file a flight plan and follow the instructions issued by the controllers manning the control towers and the ATCCs. Eventually, the old system of airline and municipal control would be dismantled and a new era of federal control of the nation's airspace would begin.

By November 1941 the CAA had established 23 ATCCs and controlled 100 percent of the civilian traffic operating on the federal airways during instrument meteorological conditions. The CAA had also installed sufficient navigation equipment to create federal airways that connected all of the major cities and busy airports across the United States. The airspace used by these airways was known as **controlled airspace.** This meant that during periods of IFR weather, CAA controllers would be responsible for the separation of civilian aircraft in these areas. Much of the rest of the nation's airspace was not a part of the federal airway system; this space was designated as **uncontrolled airspace.** This meant that the CAA would not regulate or separate aircraft operating in these areas. Pilots were legally free to fly in uncontrolled airspace during IFR weather conditions, but the CAA would not be responsible for separating their aircraft. Even today, the federal government declines responsibility for the separation and safety of aircraft flying in uncontrolled airspace. A complete description of these and other airspace categories is provided in Chapter 3.

The War Years

The beginning of World War II brought about lasting and important changes in the structure of the CAA and aviation as a whole. In 1939 aviation was ranked as the 46th largest industry in the country. By 1943 it had become the largest industry in the world. Its importance in world commerce and the conduct of the war cannot be overstated.

Aviation suddenly captured the interest of both the American people and the military establishment: For the first time, countless thousands of individuals flew either commercially or on military transports. The military services used aircraft both for fighting the war and for transporting troops and materials. World War II also caused a tremendous increase in the CAA's operating budget. It was this increased funding that enabled the CAA to take over the operation of most of the nation's control towers. Additional funds were granted to develop emergency landing areas for military aircraft. Unfortunately, research and development on future civilian air traffic control and navigational systems virtually halted in order to pursue the immediate needs of the war effort.

The war also had some harmful effects on civilian aviation. By emergency order, civilian flying was all but banned along the American coasts, and inland flyers were required to comply with additional restrictions. Aircraft fuel was either rationed or cut off completely. Military aircraft soon made up the bulk of the flights operating on the nation's airways. In 1940 only 30 percent of IFR flights were military; by 1943 this figure had increased to 85 percent. The U.S. Army, which was the arm of the military most involved in aviation, began to draft CAA controllers to operate their own ATC facilities. The Army also began to requisition civilian airliners to use as military transport aircraft.

Since the vast majority of the flights around the country were now military, the Army expressed a desire to appropriate and operate the air traffic control system for the duration of the war. CAA officials adamantly opposed this move, surmising that it would be difficult, if not impossible, to wrest control of the system away from the Army at the conclusion of the war. In an attempt to defuse and define the situation, on December 12, 1941, President Roosevelt signed Executive Order 8974, which stated:

> In the administration of the statutes relating to civil aviation the Secretary of Commerce is directed to exercise his control and jurisdiction over civil aviation in accordance with requirements for the successful prosecution of the war, as may be required by the Secretary of War.
>
> The Secretary of War is authorized and directed to take possession and assure control of the civil aviation system, or systems, or any part thereof, to the extent necessary for the successful prosecution of the war.

Various factions in the U.S. Army recommended that the secretary of war invoke the provisions of paragraph 2 of the executive order to completely militarize the CAA. CAA administrators insisted, however, that they were best equipped to operate a system to separate both civilian and military aircraft. The

CAA opinion prevailed, and an uneasy truce resulted, although many factions in the Army continued to press for a full military takeover of the ATC system.

The Interdepartmental Air Traffic Control Board (IATCB) was formed on April 7, 1941, to coordinate activities between the CAA and the various military services. This board remained in existence until 1946. During the war, the CAA established *approach control* facilities at the busiest air traffic control towers. Instead of simply handling takeoffs and landings, the approach controllers assumed responsibility for separating arriving or departing aircraft out to a distance of about 15 to 20 miles. This reduced the burden on the controllers working at the air traffic control centers and allowed tower controllers to more easily sequence arriving and departing aircraft into the local traffic pattern.

During 1942, the CAA established **interstate airway communication stations (INSACSs)** that were strategically placed to offer flight advisory services to aircraft operating along the federal airways. The INSACs were staffed by air traffic controllers who communicated directly with pilots by radio and passed along weather information and instructions from the controllers working at the ATCCs. As these INSACSs were completed, it was no longer necessary to utilize airline radio operators to relay instructions from the controller to the pilots. The INSACSs were able to accept position reports from the pilots and relay these reports to the centers. Eventually, the INSACSs began to offer preflight weather briefings by telephone and to file flight plans for pilots. The INSACSs were the precursors to the **flight service stations (FSSs)** operated by the federal government today.

Civilian Versus Military Air Traffic Control

The U.S. Army was still dissatisfied with the decision authorizing the CAA as the primary agency responsible for both civilian and military air traffic control. On May 13, 1943, in a bid to reduce the CAA's control of military traffic, the Army began to staff the ATCCs with military air traffic controllers. These controllers assumed the responsibility for tracking Army aircraft but did not usurp the CAA controllers when separating these aircraft. The Army was not satisfied with this limited control, however, and in January 1944 set up 23 flight control centers of its own. These military air traffic control centers paralleled those of the CAA but were responsible for separating military aircraft.

As World War II drew to a close in the mid-1940s, the pent-up growth of civil aviation around the world exploded, creating a need for further air traffic control equipment and personnel. In 1945 the Provisional International Civil Aviation Organization (PICAO) was formed in an attempt to coordinate this growth. This organization was superseded by the **International Civil Aviation Organization (ICAO)** in 1947. ICAO eventually accepted the U.S. navigation and communication system as the worldwide standard for air traffic control. In addition, ICAO selected English to be the common language of air traffic control worldwide (see Figure 1-7).

In an attempt to plan for the nation's growth in aviation, the **Air Coordinating Committee (ACC)** was established on March 27, 1945. By interdepartmental agreement, the ACC was staffed by members of the State Department,

Figure 1-7. Controllers in the tower at the Chicago Midway Airport in 1946.

CAA, War Department, Post Office Department, and Bureau of the Budget. The ACC's primary responsibilities were to coordinate with ICAO and to make recommendations on technical, economic, and industrial matters relating to aviation. In 1945 the ACC absorbed the responsibilities of the Interdepartmental Air Traffic Control Board. The ACC received formal status when President Harry Truman signed Executive Order 9781.

1945–1955

RTCA Special Committee 31 Report

On March 1, 1947, the ACC requested that the **Radio Technical Commission for Aeronautics (RTCA)** form a task force to try to predict the future needs of the nation's air traffic control system. The RTCA was composed of members from the State Department, War Department, Coast Guard, Federal Communications Commission, and CAA. The RTCA formed **Special Committee 31 (SC-31)**, which completed its final report on May 12, 1948.

The SC-31 report recommended that a common air traffic control system be developed that would serve the needs of both military and civilian pilots.

The committee avoided recommending the development of any unusual technical equipment and insisted that any additions to the air traffic control system meet the following requirements:

1. The new system must permit aircraft to be flown safely.
2. It must improve the flow of air traffic.
3. Any airborne equipment must be both simple and lightweight.
4. Any new system must impose a minimum burden on the pilot or ground personnel.
5. The installed equipment must require a minimum of funding from either taxpayers, airlines, or private pilots.

To meet these requirements, the report recommended that a common navigation system be developed around the newly designed **VHF omnidirectional range (VOR)** and **distance measuring equipment (DME).** The report also recommended that **airport surveillance radar (ASR)** be installed at busy airports and at air traffic control centers. It was recommended that a lightweight **transponder** be developed that could be installed on aircraft and provide altitude and identification information to ground-based radar. Finally, the report recommended that the newly designed **instrument landing system (ILS)** be installed in conjunction with **precision approach radar (PAR)** to improve the capabilities of aircraft attempting to land in poor weather conditions. (A detailed explanation of all of this equipment is provided in the following chapters.) In 1948, the **Air Navigation Development Board (ANDB)** was formed to oversee the implementation of the ATC system described in SC-31.

Although the SC-31 report was hailed as a milestone in air traffic control development, a number of immediate problems surfaced. With the federal budgetary restraints due to expenditures during the war still in place, Congress found itself both unable and unwilling to appropriate the funds necessary to expand the needed air traffic control services. At the same time, the military began to pursue the development of an incompatible navigation system known as **tactical air navigation (TACAN).** In addition, the military services preferred the sole utilization of precision approach radar instead of the instrument landing system. In general, however, the SC-31 report laid the groundwork for the development of the next-generation air traffic control system.

Air Traffic Congestion

As the 1940s drew to a close, the air traffic control system, rooted to procedures developed in the early 1930s, could no longer handle the tremendous numbers of aircraft using it. Even then, center controllers were still tracking and separating aircraft by writing their approximate positions on paper strips and moving shrimp boats along a map. Because of the inaccuracies inherent in this system, controllers were required to separate aircraft by at least 10 minutes. At the typical cruising speeds of aircraft in use in the 1940s, this meant that aircraft were being separated by between 50 and 100 miles. This procedure resulted in an excessive amount of airspace being assigned to each aircraft.

Journal of Air Traffic Control

Figure 1-8. Controllers separating en route aircraft at the New York Center in 1949.

There was simply not adequate airspace available to allocate every aircraft 50 to 100 miles. Controllers had no option on busy days other than to hold aircraft in flight or delay their departure until sufficient airspace was available (see Figure 1-8).

Whenever traffic delays became excessive and VFR flight conditions existed, many pilots chose to operate under VFR flight rules. These rules required that pilots see and avoid other aircraft within their immediate vicinity. But during extended periods of marginal weather, every pilot who wished to fly on a federal airway needed a clearance from air traffic control. This application of separation to every aircraft extended the air traffic control system beyond its capacity and forced air traffic controllers to limit access to the nation's airways. Most pilots chose to accept this delay, but others departed and operated IFR in uncontrolled airspace. Since the CAA did not regulate flights in uncontrolled airspace, these pilots would not experience any delays. But no separation service was offered to these pilots, and they could only hope that no other aircraft were within their immediate vicinity.

When certain areas, such as New York or Chicago, experienced poor weather conditions, the resulting delays rippled throughout the ATC system, affecting traffic as far as 1,000 miles away. As more and more airlines began to operate, and with over 5,000 private aircraft being added to the general aviation fleet every year, the air traffic control system was rapidly becoming critically overloaded. This overload reached epic proportions in the middle of the 1950s. When inclement weather approached the New York City area on September 15,

1954, air traffic controllers were confronted with a record number of pilots filing instrument flight plans. Airliners and private aircraft along the eastern seaboard of the United States were delayed for hours before they could be allowed to depart toward their destinations. On that day, called "Black Wednesday" by New York City residents, over 45,000 airline passengers and hundreds of private aircraft were substantially delayed because of traffic congestion.

As air traffic continued to increase during the decade, even clear weather in the major metropolitan areas of the country could not totally prevent delays. As the military began to introduce jet-powered fighter aircraft, and the airlines introduced into service bigger, faster, and higher flying airliners, many pilots realized that it would soon be impossible to avoid other aircraft using the "see and be seen" VFR rules of traffic avoidance. Two pilots approaching each other at 500 miles per hour might have less than 10 seconds to locate one another, evaluate the potential traffic conflict, and take corrective action. Subsequently, even during periods of VFR weather, an increasing number of pilots chose to file IFR flight plans to ensure air traffic separation. This practice decreased the chance of a midair collision between two aircraft operating on IFR flight plans but did not eliminate the risk of collision with an aircraft operating under VFR flight rules. When operating in VFR conditions, pilots on IFR flight plans were still required to see and avoid other aircraft that could be flying VFR.

As the demand on the air traffic control system increased, both pilots and controllers realized that it would only be a matter of time until the ATC system became completely saturated and traffic came to a standstill around major airports and airways.

1955–1965

Implementation of Radar

Even though the SC-31 report of 1948 recommended the installation of radar to assist controllers to separate aircraft, it was not until the late 1950s that a civilian radar system was installed by the CAA. Although radar had been developed and perfected during World War II, early radars were designed to detect incoming aircraft and direct interceptors to their location. This was a much different task than trying to use radar to separate aircraft. In 1949 the U.S. Air Force had begun to develop a computerized radar defense system known as **SAGE,** an acronym for **semiautomated ground environment.** This system used multiple radar sites to display all the aircraft within a designated area on a radar screen, making it possible for military controllers to vector air defense fighters toward invading enemy aircraft. By the early 1950s, millions of dollars had been spent by the Department of Defense to develop this system. But by the time the SAGE system became operational, the Air Force recognized that it might not be needed since intercontinental ballistic missiles were beginning to replace the long-range bombers that the SAGE system was designed to detect.

The Air Force recommended to the president that since the SAGE system was already operational, and since Air Force controllers were already experi-

FAA

Figure 1-9. Controllers using radar and shrimp boats to control traffic circa 1960.

enced in operating the equipment, military controllers be authorized to use SAGE to separate high-altitude aircraft. At the very least, they recommended that the CAA purchase and use the SAGE system for air traffic control. The CAA in turn recommended that the air traffic control centers be equipped with new radar systems expressly designed for locating, tracking, and separating high-altitude aircraft. The CAA felt that the SAGE system was not satisfactorily suited to the separation of aircraft and felt that a new radar system should be designed strictly for that purpose. The inevitable squabbling over which system should be installed ended with the president authorizing the CAA to develop a nationwide civilian system of air traffic control radar (see Figure 1-9).

In 1956 the first **air route surveillance radar** was purchased by the CAA for use in the air traffic control centers. Also that year the first air traffic control computer was installed at Indianapolis Center. Research and development began on a secondary radar system that would use a transponder in each aircraft to display the aircraft's identification and altitude on the controller's radar screen. In 1957 this system was experimentally implemented and is known as the air traffic control radar beacon system (ATCRBS). This system is fully described in Chapter 8.

Budget Cutbacks

During the 1950s, the CAA's budget requests were routinely reduced because of more politically pressing problems. From 1950 to 1954, the CAA's budget decreased from $187 million to $116 million per year. The budget for airway facilities development and acquisition was slashed from $37 million a year to just $7 million a year.

It soon became obvious that the CAA could not effectively improve the nation's air traffic control system with these shrinking appropriations. In fact, it would prove to be almost impossible to maintain the current, outmoded air traffic control system at such a reduced funding level. As appropriations were cut, research and development on advanced radar and computer systems had to be postponed, and even routine maintenance on navigation and air traffic control equipment was reduced to a minimum.

The budget cuts of the 1950s made it impossible for the CAA to implement many of the programs designed to increase the safety and efficiency of the air traffic control system. The implementation of additional air route surveillance radars for the air traffic control centers was delayed, along with the establishment of additional air traffic control towers. The CAA did, however, recommend that many of the INSACSs be closed. Since the centers had eventually been equipped with remote radio transmitters and receivers that permitted direct pilot-to-controller communication, the INSACSs were no longer needed to relay clearances and position reports. This move to reduce expenditures by the CAA during an extremely meager funding period was immediately met with opposition from both Congress and the affected local communities. Although the INSACSs were not particularly large, in a small city where one might be located the loss of even a few federal jobs was perceived as a tremendous blow. Congressional pressure forced the CAA to withdraw the INSACSs closing plan. This further depleted the funds available to the CAA for air traffic control modernization.

The congressional short-sightedness that had reduced the CAA's research and development program forced the CAA to delay implementing many of the SC-31 recommendations. Because of insufficient funding, by 1954 only 47 percent of the "interim" SC-31 plan had been completed. The target date for completion of the entire program had been 1953. Now, at the current rate of progress, it was calculated that the SC-31 recommendations would be completed in the year 2011! Soon the nation would regret this lack of legislative foresight.

The Question of Airway Safety

On Saturday, June 30, 1956, one of a series of major aircraft accidents occurred that made the American public regret the low priority Congress had given air traffic control funding. This midair collision finally persuaded legislators to embark on a massive ATC modernization plan. At 9:01 A.M. TWA Flight 2, a Lockheed Super Constellation, departed the Los Angeles airport en route to Kansas City, St. Louis, and Washington, D.C. At 9:04 A.M., United Airlines Flight 718 left Los Angeles for Chicago. Both aircraft were initially assigned the same route, with the TWA aircraft climbing to 19,000 feet and the United aircraft climbing to 21,000 feet. At about 10:00 A.M. the TWA flight encountered some air turbulence and the pilots requested a higher cruising altitude. The air traffic controller assigned TWA Flight 2 a cruising altitude of 20,000 feet. Since the established federal airway was not the most direct route from Los Angeles to Chicago, the pilots of both aircraft eventually requested that they be allowed to alter their route of flight, leaving the airway and entering uncontrolled airspace, to provide a scenic view of the Grand Canyon to their passengers.

During VFR weather conditions, it was not uncommon for pilots to request permission to deviate from the airways and take "short cuts" to their destination. Air traffic controllers routinely approved this route deviation, and the pilots realized that the controllers would no longer provide separation services. It then became necessary for the pilots to operate under VFR flight rules and provide their own separation from other aircraft. When the pilots of both the TWA and United aircraft failed to make any additional position reports, an intensive ground and air search was begun. The wreckage of the two aircraft was eventually found scattered in a remote gorge in the Grand Canyon. The aircraft had collided in midair, killing over 120 people. There were no survivors.

The American public was both shocked and outraged to hear that two modern, sophisticated aircraft flying in near perfect weather conditions had collided in midair. During the ensuing investigation, the CAA publicly denied responsibility for the collision, since the pilots of both aircraft had requested permission to fly in uncontrolled airspace and were responsible for their own safety and separation. But it soon became obvious that the CAA did not have enough airways, airspace, or controllers to be able to offer positive separation to every aircraft trying to fly across the country. Additional airways had not been developed due to insufficient funding during the previous decade. The CAA had not been able to purchase sufficient radio navigation aids or hire sufficient air traffic controllers to properly separate air traffic over much of the continental United States.

In reality, it was not the CAA's fault, but that of Congress, which had refused to appropriate sufficient funds to operate the ATC system. Shortly after the investigation of this accident, the CAA requested a $250 million appropriation from Congress to upgrade the airway system. These funds were used to purchase sufficient radar surveillance equipment to permit air traffic controllers to monitor and separate all the traffic operating above 18,000 feet. The CAA also requested funding to almost double the available navigation aids and to open 40 new control towers. In addition, the CAA was permitted to hire 1,400 more air traffic controllers. In 1956 the first 23 air route surveillance radars (ARSRs) were ordered. The CAA stated that an additional 50 to 60 ARSRs were needed to provide radar coverage over the entire continental United States but agreed to use as many Air Force defense radars as possible to minimize the acquisition cost. Congress immediately approved this funding request. Eventually, radar surveillance and positive control of high-altitude aircraft would be implemented and would greatly improve the airway safety record in the coming decades.

Creation of the Federal Aviation Agency

In 1957 the CAA announced a plan to have a scaled-down version of the SC-31 recommendations implemented by 1962 — 9 years later than the original estimate for the full implementation. Unfortunately, by this time many of the recommendations made in SC-31 were already obsolete. When the report had first been commissioned in 1948, turbine-powered airliners were only a gleam in some engineer's eye. No one predicted that by 1957 one British and two American jet aircraft would be nearing production. The current, underfunded air traffic control system was not ready for this influx of high-flying passenger aircraft.

An independent Airways Modernization Board (AMB) was formed in 1957 to coordinate civilian-military aviation electronics research and development. The AMB immediately began to conduct research on air traffic control computers, transponders, and advanced radar equipment. In 1958 the AMB opened its own research and development facilities separate from those of the CAA. Located in Atlantic City, New Jersey, this complex became known as the **National Aviation Facilities Experimental Center (NAFEC).** As research activity rapidly increased and funding was diverted to NAFEC, the CAA was eventually forced to close its own Technical Evaluation and Development Center in Indianapolis.

Unfortunately, this sudden increase in funding did little to immediately improve the capabilities and operation of the ATC system. Woefully underfunded for the previous two decades, the system could not be brought up to speed in such a short time. The air traffic controllers had been frustrated by their working conditions and compensation for years, and this new promise of future equipment and increased funding rang hollow in their ears. They had seen the recommendations of SC-31 dragged out for years and had no reason to believe that it might be any different this time around.

For these and many other reasons, air traffic controllers began to leave the profession at an ever-increasing rate. Some left because of overwork, and others left because of the generally low pay. In some centers as many as 30 per cent of the controllers tendered their resignation in 1957 alone. This high attrition rate made it even harder for the CAA to increase the controller complement. Where previously the CAA had to hire about 50 controllers per month, it now had to try to hire over 400 quality recruits per month just to keep up with the replacement of controllers who were leaving the profession. By 1957, because of a lack of experienced controllers, most center controllers were working 6 days a week, 8 hours a day with no breaks. Often they might be required to work 60 to 70 days in a row with little time off. In response to this pressure, a group of discontented controllers formed the **Air Traffic Control Association (ATCA)** to assert controllers' demands for increased pay and improved working conditions. By 1960 this group boasted over 9,000 members.

In an attempt to improve safety, in 1957 the CAA designated all of the airspace above 24,000 feet as controlled airspace. This milestone development improved the separation of high-altitude airline flights operating IFR but did nothing to separate them from military or private aircraft that chose to fly in this airspace under VFR flight rules. As the CAA debated the wisdom of requiring IFR flight plans of any aircraft operating above 24,000 feet, a number of accidents occurred that would force the hand of the CAA.

On April 21, 1958, an Air Force jet collided with a United Airlines DC-7 at 21,000 feet near Las Vegas. The fighter had been descending toward Nellis Air Force Base under VFR conditions. Both of the fighter pilots and 47 persons on the DC-7 died in that accident. In less than a month, another accident occurred near Brunswick, Maryland. An Air National Guard jet operating VFR collided with a Capital Airlines turboprop, killing 12 people. Although many congressmen called for the immediate implementation of positive control of every aircraft operating at high altitude, it was apparent that the CAA did not have the capacity to do so. Positive control requires that every aircraft be actively separated by the air traffic control system. To positively separate every aircraft operating above 18,000 feet would mean an immediate and substantial influx of personnel and equipment.

In 1956 President Dwight D. Eisenhower selected Edward Curtis, a Kodak vice president and Army Air Corps major general, to direct a long-range study of the nation's aviation system. Curtis and his staff took this responsibility seriously and prepared an extensive report for the president in 1957. The committee recommended that in the interim the Airways Modernization Board be given the task of consolidating the government's efforts in aviation research and development. The eventual goal of the AMB was to design a common air traffic control system that would serve the needs of both military and civilian aircraft. Curtis also recommended the permanent formation of an independent agency that would absorb the functions of the CAA, and eventually those of the AMB. This new agency would be known as the **Federal Aviation Agency (FAA).**

It took 2 years to shepherd the appropriate legislation through Congress, but with the backing of Senators Mike Monroney of Oklahoma and Warren Magnuson of Washington, the Federal Aviation Agency was created by act of Congress and began operation on December 31, 1958. This new federal agency was administered by a cabinet-level officer who was appointed by and directly responsible to the president of the United States. No longer would the air traffic control system be handicapped because of bureaucratic infighting within the Department of Commerce. The new FAA would receive its funding directly from Congress and the FAA administrator would report personally to the president.

The employees of the newly created FAA faced an enormous task. The air traffic control system had been undermanned and underfunded for over 20 years. Although impressive safety measures had been taken to separate high-flying airliners, the low altitudes in the immediate vicinity of the major terminals were still congested, a major source of traffic delays. High-speed military aircraft and low-speed private aircraft were flying in VFR conditions within the vicinity of the nation's major airports while an ever-increasing number of commercial airline flights were attempting to use the same airspace. During the 1950s, few innovative procedures had been developed and sufficient radar surveillance equipment had not been installed by the CAA. Numerous near collisions were being reported each year by both pilots and controllers. A major accident around an airport seemed inevitable.

The New York City Disaster

Just as the new FAA was in the process of getting organized, disaster struck. On December 16, 1960, a United Airlines DC-8 and a TWA Super Constellation collided over New York City. One hundred twenty-eight people on board the two aircraft died, as did eight people on the ground. The accident pointed out many of the difficulties the FAA still faced. The ensuing investigation revealed that both of the aircraft were on IFR flight plans and both were in contact with the appropriate air traffic control facilities. The inquiry also revealed that the United Airlines aircraft had experienced partial navigation receiver failure, but the pilots had not informed the air traffic controller of the malfunction. The United aircraft had been cleared to enter a holding pattern, pending clearance to land at New York's Idlewild airport (now John F. Kennedy International Airport). The pilots of this aircraft were then advised to contact the tower controllers at Idlewild.

At the same time, the TWA Super Constellation had been placed in a holding pattern awaiting clearance to land at New York's La Guardia airport. The pilots of the TWA aircraft were in communication with the tower controllers at La Guardia. Both of the aircraft had been assigned the same altitude but were assigned to two different holding patterns that were safely separated. The pilots of the United Airlines aircraft entered their assigned holding pattern at an excessive airspeed and flew outside the confines of their designated holding pattern airspace. As a result, they strayed into the airspace reserved for the TWA flight and eventually collided with it. The investigation determined that proper procedures had been applied by the controllers in the two control towers and placed the blame for the accident on the pilots of the United aircraft.

The investigators determined that the United pilots had used improper procedures while entering the holding pattern and should have advised the controllers of their navigation receiver problems. The accident report did mention, however, that had the control towers been equipped with surveillance radar, the air traffic controllers might have detected the impending collision and issued corrective instructions to one or both of the aircraft. This realization hastened the installation of radar equipment at busy airports and led to the eventual establishment of the New York approach control facility (known as the Common IFR Room) that would be equipped with radar and would be responsible for the separation of all aircraft inbound to the New York metropolitan area.

Project Beacon

The air traffic control system in the United States had been constructed haphazardly in response to situations instead of in anticipation of them. In an attempt to rectify this, President Kennedy issued an order on March 8, 1961, requesting the FAA to "conduct a scientific, engineering overview of our aviation facilities and related research and development and to prepare a practicable long-range plan to insure efficient and safe control of all air traffic within the United States." A task force was created by the Federal Aviation Agency to carry out the wishes of the president and to report its findings to the FAA administrator. This task force was to investigate the current air traffic control system and make recommendations for improving it. The task force was known as **Project Beacon.**

After close to a year of investigating the current and planned improvements to the nation's air traffic control system, the Project Beacon task force issued its final report. The report stated that although the FAA had many projects in development, no overall direction or coordination seemed to guide these projects. Much of the FAA's research was still based on the SC-31 report, which was almost 20 years old. The task force found that much of the research and development work was focusing on technically advanced equipment, while very little work was being done on short-range problems that needed immediate attention. FAA research personnel were working to develop advanced air traffic control computers and three-dimensional radar, while controllers in the field were complaining that the current radar system was unsuitable.

The Project Beacon task force essentially agreed with the controllers that before an advanced computerized air traffic control system was developed, the current radar equipment would have to be modernized and improved. The task force report recommended that the FAA install sufficient radar surveillance equipment across the country to permit air traffic controllers to maintain continuous radar monitoring of aircraft from takeoff to landing. The Project Beacon report also stressed the use of secondary radar and transponders to assist controllers in identifying each aircraft and determining its altitude.

The task force also recommended that computer processing equipment be installed at air traffic facilities to assist controllers with their clerical duties and to help them more readily interpret radar information. In the early 1960s controllers were hand printing flight progress strips and passing along information to other controllers using teletypes and party-line telephone circuits. The task force recommended that the FAA develop a computerized flight information system that would automatically print out flight progress strips and continuously distribute updated flight information. This system would be designed to permit air traffic controllers to communicate with each other and pass along essential flight information without using the telephone. The system was ultimately developed by the FAA and is known as the **flight data processing (FDP)** system.

The Project Beacon task force also recommended that a computer-driven display system be developed that would show the aircraft's identification, altitude, and airspeed directly on the radar scope. Placing this information directly on the radar display would help eliminate the confusion when controllers try to identify each blip on the radar screen and correlate this information with that contained on the plastic shrimp boats and on the flight progress strips. The computer would be designed so that it could be custom programmed to compute and predict the future flight path of each aircraft and advise the controller if two aircraft were going to approach too close to each other or descend too close to the ground. In addition, when an aircraft neared the boundary of a controller's area of responsibility, the computer would ensure that the aircraft's identification would begin to flash on the next controller's radar screen; with the push of a button, the next controller would take responsibility for the aircraft's separation (see Figure 1-10).

Although a common system was envisioned, because of the different requirements of the center and terminal controllers, two distinctly different computerized radar systems were eventually developed. The system used in the air route traffic control centers (ARTCCs) is called **radar data processing (RDP)**,

FAA

Figure 1-10. Controllers in the tower cab at the Washington Dulles airport in 1963.

whereas the system used in the control towers and approach controls is called the **automated radar terminal system (ARTS)**. These two systems are described in detail in Chapter 8.

Controller Unionization

In the early 1960s, labor unrest began to appear again within the FAA. Air traffic controllers, who had endured years of low salaries and unpleasant working conditions, began to earnestly embrace professional associations to represent them. The Air Traffic Control Association was the first of many that represented various groups of air traffic controllers. These professional associations differed from trade unions in that they spoke out on the controllers' behalf and lobbied for ATC system improvements but refrained from collective bargaining and other typical union activities.

In 1961 President Kennedy signed Executive Order 10988, which gave trade unions the right to represent air traffic controllers. The executive order made no distinction between professional associations and trade unions. Within a matter of years, various associations began to organize the controller work force. These organizations included ATCA, the National Association of Government Employees (NAGE), and the National Association of Air Traffic Specialists (NAATS).

The FAA faced a distinctive problem in that it was one of the few federal agencies whose operation was vital to the well-being of the country but whose work force was permitted to become unionized. Although strikes were illegal for federal employees, it was felt that the possibility existed for disruptive labor activity if this situation was allowed to endure. The FAA administrator at the time, Najeeb Halaby, proposed that this problem be solved through the formation of a semimilitary organization known as the Federal Air Service (FAS). Under this plan, every controller would be a member of the FAS, which would be a group similar to the U.S. Coast Guard. Although remaining technically civilian, this group of federal employees would be subject to military induction during times of national crisis. The new organization would be considered vital to the national defense and as such would not be permitted to be unionized under the president's executive order.

This proposal drew immediate and vociferous opposition from both controllers and members of Congress. After years of attempting to pry aviation control away from the military, they saw the FAS as an opening that might lead to further military control of aviation. Congressional hearings determined that this organization was unnecessary and actually more expensive than the current system. By 1963 the Federal Air Service concept was no longer being seriously considered. Although highly disliked, the concept did attempt to define the role of the FAA in a national emergency. In 1964 President Johnson signed Executive Order 11161, which directed the FAA administrator and the secretary of defense to draw up plans whereby the FAA would be absorbed by the Defense Department in times of national emergency.

Public discussion did little to pacify the working controllers. Frustration was reaching a peak, and the controllers were becoming increasingly militant. Most felt that FAA management had been inattentive to their concerns. The controllers were generally dissatisfied with both of the major associations (ATCA and NAGE) that were attempting to represent them. Many felt that ATCA would not effectively press the issues that were important to them. NAGE, on the other hand, represented many different types of government employees, which disappointed many controllers who were still proud to be part of an elite government group. A group of New York–area controllers eventually formed an association in an attempt to better represent their special interests. After an accidental meeting between one of the group and well-known attorney F. Lee Bailey, he helped them create a new national controllers' organization, the **Professional Air Traffic Controllers Organization (PATCO).** PATCO was run by controllers with membership limited to controllers. In a short time, PATCO became one of the most militant and vocal controller organizations. It would play a large part in future FAA-labor relations.

1965 to the Present

Department of Transportation

During the Johnson administration in the 1960s, the consensus in the federal government was that as the government became more involved in transportation issues, every transportation function of the government should reside in one cabinet-level agency. This arrangement would theoretically make it easier for overall transportation policy to be developed and implemented. It had become apparent during the construction of major airports around the country that no one form of transportation was completely independent of the others. For instance, in many cities, modern and expensive airports had been constructed but were wasting away for want of decent ground access to the airport. Millions of dollars were being wasted on federally sponsored projects because of a lack of overall direction.

This cabinet-level coordinating agency became a reality on April 1, 1967, when the **Department of Transportation (DOT)** was created by Congress. The Federal Aviation Agency was merged into the DOT, with its stature and administrator downgraded. The initials FAA now stood for **Federal Aviation Administration,** a part (albeit the largest part) of the Department of Transportation. The FAA administrator would no longer report directly to the president but would instead report to the secretary of transportation. New programs and budget requests would have to be approved by the DOT, which would then include these requests in the DOT's overall budget and submit it to the president.

The **National Transportation Safety Board (NTSB)** was also created on this date. The NTSB was charged with investigating and determining the cause of transportation accidents and making recommendations to the secretary of transportation. The Civil Aeronautics Board was merged into the DOT, with its responsibilities limited to the regulation of commercial airline routes and fares.

With this new organization in place, the FAA administrator would have to learn to operate within the growing bureaucracy of the Department of Transportation. No longer could the administrator make direct appeals to the president or Congress. The FAA became part of a larger organization that included the Federal Highway Administration, the Federal Railroad Administration, the Coast Guard, and the Saint Lawrence Seaway Commission.

Continued Labor Unrest

As the 1960s drew to a close, airports around the country were becoming increasingly congested, and delays were skyrocketing. A number of midair collisions around major airports had shaken the public's confidence in the air traffic control system. Because of increased defense spending to fund the conflict in Vietnam, the FAA's budget was constantly being reduced. The equipment recommended by the Project Beacon task force was being installed, but at a much slower pace than originally planned. The air traffic controllers working in the towers and the centers were becoming increasingly irritated with delays in equipment acquisition and blamed FAA mismanagement for their problems. The FAA was forced to spend most of its shrinking appropriations on simply maintaining the current air traffic control system, not improving it.

The FAA had to stretch out major equipment procurement programs and even temporarily closed the controller training school in Oklahoma City. This was a critical time for training in the FAA. Many of the controllers who had been hired in the 1940s were retiring, while few new controllers were being hired to replace them. As air traffic continued to increase, the FAA's management relationship with the controllers continued to worsen. PATCO spokesmen asserted that both the FAA and the DOT were unnecessarily delaying the installation of sufficient air traffic control equipment. The union charged that the FAA was not hiring enough new air traffic controllers to properly and safely operate the nation's air traffic control system.

On July 3, 1968, PATCO flexed its muscles by announcing "Operation Air Safety," which ordered member controllers to strictly adhere to established separation standards to aircraft. The resultant delay of traffic was the first of many official and unofficial "slowdowns" that the union was to initiate. In 1969 the U.S. Civil Service Commission ruled that PATCO was no longer a professional association but was in fact a trade union. The controllers' disaffection with the FAA reached a critical point on March 25, 1970, when the newly designated union orchestrated a controller sick-out. To protest many of the FAA actions that they felt were unfair, over 2,000 controllers around the country did not report to work as scheduled and informed management that they were ill. Management personnel were required to assume many of the duties of the missing controllers. With traffic around the country delayed for hours, the union and FAA management came to an agreement that returned most of the controllers to work. After fierce negotiations and court battles, the FAA agreed to hire back most of the "sick" controllers, and the union agreed never to sponsor an illegal strike.

During the controller sick-out, it became apparent that the ATC system was operating nearly at capacity. It had become necessary to reroute or delay hundreds of IFR flights in order to reduce the traffic congestion over the busiest areas of the country. The FAA requested, and Congress appropriated, additional funds to accelerate the installation of many of the automated systems recommended by the Project Beacon task force. Even with this additional funding, the FAA was still years behind the planned schedule for automation. In addition, the FAA reopened the training academy in Oklahoma City and began to hire air traffic controllers at an increasing rate. Salaries were increased to help attract and retain controllers. During the 1970s, the FAA made steady progress toward the goal of automating many of the functions of the ATC system. But it would prove to be impossible to make up for the lack of funding the FAA had experienced in the 1960s.

Airline Deregulation

In 1978 the **Airline Deregulation Act** was enacted by the Carter administration. This act greatly reduced the influence of the Civil Aeronautics Board and provided for its eventual dissolution. Prior to ratification of the Deregulation Act, the airlines were required to petition the CAB for any route addition, deletion, or change. In addition, the fare structure for airline flights had been highly regulated by the CAB. With the passage of the Deregulation Act, the airlines were free to determine their own fares and route structures without government approval. This forced the airline industry to compete for passengers as never before. New airlines were formed in record time and soon began to operate in direct competition with older, more established airlines. In response to this competitive threat, the established airlines reevaluated their markets and began to overhaul their route structures, all without the government approval that had been needed prior to deregulation. As a result of this new competition among airlines, fares were reduced to all-time lows while record numbers of travelers chose to travel by air.

The Deregulation Act of 1978 disrupted this nation's air traffic control system in ways never foreseen by its architects. Air traffic activity increased at rates that had been impossible to predict. Airport activity increased faster than new controllers could be trained or new equipment could be moved to adapt to the changing traffic flows. Many of the major airlines adopted "hub-and-spoke' route systems that threatened to overwhelm the air traffic control system at some airports. A hub-and-spoke system eliminates many nonstop direct routes and designates one airport to act as the "hub" airport for the region. Flights depart other airports in the area and converge on the hub airport at approximately the same time. Most of the passengers disembark from their flights and are shuttled to another aircraft that will fly them to their destination. When all of the passengers have boarded the proper aircraft, the airliners depart. Since all of the aircraft arrive or depart at approximately the same time, this creates a tremendous but transient strain on the ATC system.

Many of the airlines chose to locate their hubs at airports that were ill-equipped to handle this tremendous growth in traffic. The affected ATC facilities were required to have sufficient controllers and ATC equipment in place to handle the enormous but momentary peaks in air traffic throughout the day. The FAA was unable to quickly adapt the air traffic control system to meet this tremendous increase in air traffic. Controllers could not be trained nor could equipment be installed quickly enough to meet the new demands on the ATC system (see Figure 1-11).

Controllers' Strike of 1981

During the 1970s FAA management had caved in to a number of PATCO demands. The union had requested and received sole representation rights for controllers in the towers and centers, an early retirement program, a medical disability program, airline familiarization flights, and a number of changes in the compensation rules. The one item that PATCO wanted but was never able to get was the release of air traffic controllers from the Civil Service System. As long as controllers were considered Civil Service employees, any change in controller working conditions also affected federal employees nationwide. PATCO looked to the Postal Service, which was an independent agency born of the illegal strike by postal workers in the early 1970s, as a role model.

After a bitter internal fight, a more militant faction of controllers took charge of PATCO in early 1981. This group advocated a showdown with the FAA to finally force the issue. Despite being warned by the FAA, Congress, and the president, PATCO staged an illegal strike on August 3, 1981. Two days later, after the controllers disregarded a presidential ultimatum to return to work, their employment was terminated by the FAA. Over 10,000 controllers participated in this illegal job action and were fired.

Figure 1-11. Controllers in the Long Island (New York) tower in 1986.

In the wake of the strike and the mass firings, the FAA was faced with the enormous task of hiring and training enough controllers to replace those who had been fired. Since it takes at least 3 years in normal conditions to train a new controller, temporary flight restrictions were necessary to reduce the workload on the controllers and management personnel who were now staffing the ATC system. A system of airport reservations was established by the FAA to reduce the flow of air traffic into major airports. In addition, the FAA implemented an advanced form of **flow control** that restricted aircraft departures until it was determined that sufficient airspace was available for each aircraft.

Flow control techniques had been experimented with on a limited basis and were perfected when the FAA realized that PATCO was considering engaging in an illegal strike. Flow control procedures required that the specialist at the FAA's **Central Flow Control Facility (CFCF)** calculate optimal airport acceptance rates and attempt to match the inbound flow of aircraft to that acceptance rate. This procedure substitutes ground delays for airborne holding and reduces airspace congestion around busy airports.

In the aftermath of the strike, the FAA claimed that staffing levels would be restored within 2 years. In fact, it took close to 10 years before overall staffing levels returned to normal. During that time, the effects of deregulation became apparent, with airlines choosing to operate from large, congested hub airports. Hub-and-spoke procedures required that the airlines concentrate arrivals and departures within limited time frames.

Delays began to increase, as did controller workload at most air traffic control facilities. New ATC equipment was slow to become operational, while existing equipment became difficult to maintain and operate. The nation's air traffic controllers (most of whom were hired to replace striking PATCO controllers) again became dissatisfied with their relations with the FAA and in 1987 voted to form a new union: the **National Air Traffic Controllers Association (NATCA).** Although NATCA leadership has promised never to condone an illegal strike, it actively pressures Congress and the FAA to hire more controllers and to accelerate the installation of advanced air traffic control systems.

ATC Modernization Programs

The FAA developed a billion dollar, multiyear air traffic control modernization program that was to be known as the **advanced automation system (AAS).** This system was to use new computers and controller workstations that would provide controllers with advanced displays and traffic management tools. This program also involved the consolidation of many local approach controls into the existing air route traffic control centers. These new facilities, which were to be known as area control facilities, were to be equipped with the new AASs. A similar system would also be installed in the remaining approach control facilities and would be known as the terminal advanced automation system (TAAS).

Design criteria were established and a prime vendor for the program was selected, but due to a variety of problems, including poor project management and oversight, ill-defined system architecture, and rapidly changing technology, the program dragged on with little progress. In the mid-1990s, the program was canceled during a massive restructuring.

There are many reasons for this program's lack of progress, but various studies by the **Government Accounting Office (GAO)** and others indicated that the government procurement process in general, and FAA procedures in particular, made it difficult to develop and establish a long-term modernization program in an area of rapidly changing technology. By the time the FAA had defined the standards, developed specifications, and awarded contracts, much of the technology under consideration was very often obsolete. Many individuals and organizations, both in and out of government, have questioned the FAA's ability to handle such a huge modernization project. Various proposals for reorganizing the FAA have been suggested. Some involve simply relieving the FAA of much of the paperwork and red tape inherent in any governmental organization. Others propose removing the FAA from the DOT and making it an autonomous federal entity such as Amtrak. Others propose converting the FAA into a private corporation. Under all of these proposals, the air traffic control functions might be removed from the FAA, but air safety and regulation would remain in the federal government. Other national governments, most recently Canada and New Zealand, are experimenting with converting to such a system.

The FAA has already conducted some programs similar to these proposals in some respects. In recent years, many of its training functions have been contracted out to private industry and educational institutions. The operation of many low-activity control towers has also been contracted out to private industry. In most of these cases, the FAA provides operational funding for these towers to approved contractors who then operate the towers much more cheaply than the FAA could. Similar programs have been proposed by Congress and constituent groups. But each proposal has its advantages and disadvantages, as well as its own constituent groups, both pro and con. The legal implications of an ATC-related accident have not really been tested in a court of law. Politically, it is difficult to close, sell, or transfer federal facilities, move federal employees, or transfer federal employees to private contractors. Such proposals raise as many or more political considerations as operational questions. It may be years before an organizational solution to the FAA's problems can be implemented.

Air Traffic Control Equipment

Since some of the critical air traffic control equipment still in use by the FAA is over 25 years old and is difficult to operate and maintain, the FAA has developed an interim modernization plan that it hopes can be implemented rapidly. These new system upgrades use primarily modified **commercial off the shelf (COTS)** systems that are readily available. By the mid-1990s, the FAA had placed an order for new displays at the ARTCCs and new computer processing systems. The displays, known as **display system replacements (DSRs)**, will provide the controllers with programmable color displays similar to high-resolution computer screens. The **standard terminal automation replacement system (STARS)** program will receive and process target reports, weather, and other nontarget messages from both terminal and en route digital sensors. STARS will automatically track primary and secondary surveillance targets and will provide aircraft position information to the **enhanced traffic management system (ETMS)** being developed by the FAA.

STARS will integrate many of the computer functions of the ARTCCs and adjacent terminal control facilities. It will be used to effect handoffs, will include interfacility communications, and can support the processing and transmission of radar, beacon, weather, and air traffic data to the displays located within the terminal control facility and local and remote towers.

While developing and installing this interim system, the FAA has also committed to a long-term upgrade of the nation's **communication, navigation, surveillance (CNS)** and **air traffic management (ATM)** systems. This long-term upgrade will bring the U.S. air traffic control system into compliance with future ICAO standards for navigation and air traffic management. This integrated system, recently agreed to by ICAO nations, is known as the **future air navigation system (FANS)**.

The current CNS system in the United States is composed primarily of ground-based electronic systems. Air traffic management functions are provided primarily by air traffic controllers monitoring this equipment, making decisions, and communicating these instructions via voice radio systems. The future CNS/ATM system as envisioned by the ICAO and the FAA places much of the CNS function with space-based satellite systems and utilizes high-capacity ground-based computers to provide controllers with many ATM tools. When implemented, this combination should permit pilots to fly more direct routes

Table 1-2. Present and Future CNS/ATM Systems

	Present	Future
Communications	Domestic VHF/UHF Oceanic HF Limited data link (mode-C, ACARS)	Satellite voice Satellite or mode-S data link
Navigation	Ground-based transmitters (VOR, NDB, ILS)	Satellite based
Surveillance	Radar or position reporting	Automatic Dependent Surveillance (satellite)
Air traffic management	Controller interprets data, then makes decisions	Computer makes short-term aircraft separation decisions; controller manages airspace

from their departure airport to their destination with minimal route or altitude changes. The controllers would play a more passive role, intervening only when needed. This new concept of air traffic control, known as free flight, may soon become the new model for air traffic control in the next century. Details of the required CNS/ATM programs and improvements leading to free flight are described in Table 1-2 and discussed in detail in Chapter 12.

Staffing Problems

The FAA faces a shortage of controllers just past the turn of the century. By act of Congress, controllers cannot be older than 30 years of age when first employed by the FAA. Depending on the type of position they have held during their careers, most controllers can retire after 20 to 30 years of service.

Virtually the entire active controller work force in the FAA was hired between 1981 and 1990. Over 10,000 controllers were hired to replace the fired PATCO strikers, and most of the controllers who did not go out on strike have either retired or moved into management or supervisory positions. Thus, the FAA has a comparatively young and healthy work force with little turnover, and therefore few new employees. Between the years 2002 and 2012, however, the majority of the controller work force is slated to be eligible for retirement. Their departure will leave the FAA with a tremendous challenge: how to screen, train, and certify the needed controllers.

One recent program that attempts to meet these needs is the **collegiate training initiative (CTI)**. Under this program, selected colleges and universities will include the basic skills training needed by the FAA as part of their aviation curriculum. Graduates of these programs will, upon completion of other FAA requirements, be eligible for hiring and placement in the FAA system. The FAA selected 11 schools for the program in 1997, with more planned for the future.

All these changes, both in progress and proposed, indicate a bold, new, exciting, yet uncertain future for air traffic control. Everyone agrees that air travel will increase, aircraft will get larger, and airspace will become more congested. New equipment and procedures are badly needed. Many factors, such as politics, incidents, and possibly accidents, will have a dramatic influence on the air traffic control system. A blueprint for change is there, but can and will the changes be made?

KEY TERMS

active control
advanced automation system (AAS)
Air Commerce Act
Air Coordinating Committee (ACC)
Air Navigation Development Board (ANDB)
air route surveillance radar
Air Traffic Control Association (ATCA)
air traffic controller
air traffic management (ATM)
Airline Deregulation Act
Airmail Act of 1925
airport surveillance radar (ASR)
airway traffic control centers (ATCCs)
airway traffic control stations (ATCSs)
airway traffic control units (ATCUs)
automated radar terminal system (ARTS)
Bureau of Air Commerce
Central Flow Control Facility (CFCF)
Civil Aeronautics Administration (CAA)
Civil Aeronautics Authority (CAA)
Civil Aeronautics Board (CAB)
civil air regulations (CARs)
collegiate training initiative (CTI)
commercial off the shelf (COTS)
communication, navigation, surveillance (CNS)
control tower
controlled airspace
Department of Transportation (DOT)
display system replacements (DSRs)
distance measuring equipment (DME)
enhanced traffic management system (ETMS)
Federal Aviation Administration (FAA)
Federal Aviation Agency (FAA)
flight data processing (FDP)
flight service stations (FSSs)
flow control
future air navigation system (FANS)
Government Accounting Office (GAO)
instrument flight rules (IFR)
instrument landing system (ILS)
instrument meteorological conditions (IMC)
International Civil Aviation Organization (ICAO)
interstate airway communication stations (INSACSs)
light guns
Morrow Report
National Air Traffic Controllers Association (NATCA)
National Aviation Facilities Experimental Center (NAFEC)
National Transportation Safety Board (NTSB)
navigation aids (navaids)
passive control
precision approach radar (PAR)
Professional Air Traffic Controllers Organization (PATCO)
Project Beacon
radar data processing (RDP)
Radio Technical Commission for Aeronautics (RTCA)
semiautomated ground environment (SAGE)
shrimp boat
Special Committee 31 (SC-31)
standard terminal automation replacement system (STARS)
tactical air navigation (TACAN)
transponder
uncontrolled airspace
VHF omnidirectional range (VOR)
visual flight rules (VFR)

REVIEW QUESTIONS

1. How did the federal government become involved in air traffic control?
2. How did airmail affect air traffic control?
3. What are VFR and IFR?
4. Who was the first air traffic controller?
5. What were and where were the first air traffic control facilities?
6. What is the history of the Federal Aviation Administration?
7. What is the history of labor organizations in the air traffic control profession?
8. What was the primary cause of the FAA's failure to upgrade the nation's air traffic control system?
9. What is meant by CNS/ATM?
10. How will the CNS/ATM system have to be upgraded if free flight is to become a reality?

2

Navigation Systems

Checkpoints

After studying this chapter, you should be able to:

1. Identify the en route navigation aids in use today.
2. Identify the approach navigation aids in use today.
3. Briefly explain the operating principles behind each of these navigation aids.
4. Properly interpret an instrument approach procedures chart.
5. Properly interpret an instrument en route navigation chart.
6. Properly interpret a VFR sectional chart.
7. Determine whether an instrument approach procedure is a precision or a nonprecision procedure.

One of the first scheduled airline flights in the United States occurred just prior to World War I. The St. Petersburg–Tampa Airboat Lines was established to provide regular passenger service between the two Florida cities. For 3 months, during the winter of 1914, the airline flourished. But when spring arrived, the tourists departed north, and with the lack of passengers, the airline folded. No other serious attempts at starting airline service were made during World War I.

At the conclusion of the war, the federal government disposed of many of its military aircraft, selling them to private individuals as surplus property. This enormous influx of inexpensive aircraft helped establish the aviation industry in the United States. Some airline companies were formed after the war using these surplus aircraft, but they proved to be as short-lived as the St. Petersburg–Tampa line. The available war surplus aircraft were expensive to operate and maintain, forcing the airlines to charge passengers high fares. Only the wealthy could afford to fly at these high prices, and they were accustomed to traveling in luxury, not in war surplus aircraft. Trying to lure passengers using these aircraft thus proved to be nearly impossible, and most of the fledgling airline companies folded.

In 1916, in the midst of World War I, Congress had authorized the Post Office Department to institute the nation's first official airmail service. The war delayed the implementation of this policy until 1918. The first flight, from New York City to Washington, D.C., was finally conducted on May 15 of that year, using U.S. Army aircraft. Airmail service soon proved to be commercially successful, and within 3 months the Post Office Department began to transport the mail using its own aircraft and pilots. Additional routes were soon added, and the Post Office Department eventually came to provide airmail service from coast to coast.

Within a few years, in an attempt to stabilize the fledgling airline industry, the Post Office Department began to contract airmail routes to the few remaining airline companies still struggling to survive. Airmail contracts proved to be a lifesaver to these airlines, since they could now transport mail while conducting passenger flights and use the airmail payments as a subsidy to reduce fares and attract more passengers. The resultant increase in revenue permitted the airlines to dispose of their war surplus aircraft and invest in larger and more luxurious aircraft specifically designed to carry passengers. But this merging of passenger and airmail service complicated airline scheduling and operations. When carrying only airmail, airlines could delay flights because of poor weather conditions or darkness. But delays were unacceptable when carrying fare-paying passengers. Passengers demanded that the airlines fly consistent schedules with as few delays as possible. If the airlines hoped to lure passengers away from their main competitor, the railroads, they would have to offer fast, timely flights with few or no delays. Methods that would permit flying during poor weather or at night would have to be developed if the airlines were to survive and prosper.

Visual Navigation

Initially, because they lacked flight instruments or navigation systems, airline pilots were limited to daylight flying during good weather conditions. The pilots were forced to use outside visual references to control their aircraft's attitude, relying on the natural horizon as a reference. They would note any changes in the flight attitude of their aircraft and make the necessary control adjustments that would keep their aircraft in level flight.

Pilotage

Pilots navigated from airport to airport using either **pilotage** or **deduced reckoning** (commonly called *dead reckoning*). Pilotage required that the pilot use a map of the surrounding area as a reference. The pilot would draw a line on the map, extending from the departure to the destination airport, and note any prominent landmarks that would be passed while in flight. As the aircraft passed these landmarks, the pilot would note any deviation from the planned flight path and adjust the aircraft's heading to return to the preplanned course.

Since the winds at the aircraft's cruising altitude usually caused the aircraft to drift either left or right of course, the pilot was forced to constantly alter the aircraft's heading to counteract these **crosswinds.** This change in heading is known as the **crosswind correction angle,** or **wind correction angle.** The resultant path that the aircraft flies over the ground is known as the **ground track,** or the **course.**

Aeronautical Charts

The maps used by pilots in the early 1920s were common road maps available at automobile service stations. These maps were unsuitable for aerial navigation since they lacked the necessary landmark information needed to accurately navigate from one airport to the next. It soon became apparent that pilots needed a specialized chart expressly designed for use in aeronautical navigation. The U.S. government then developed and began to print such air navigation charts, known as **sectional charts.**

Sectional charts are aeronautical charts scaled 1:500,000, or about 8 statute miles to the inch. Sectional charts are still used today and depict the relevant information needed by pilots to navigate accurately and safely. This information includes cities, highways, railroads, airport locations, terrain features, and distinctive objects (see Figure 2-1). Sectional charts also depict navigation aids, federal airways, and air traffic control facilities. With very little change over the years, sectional charts are still being printed by the **National Ocean Service** (part of the U.S. Department of Commerce) and are primarily used by pilots flying under VFR rules (see Figure 2-2). In addition, pilots flying IFR usually carry appropriate sectional charts in case of navigational equipment failure. If IFR pilots should encounter any electronic navigation problems during flight, they may be able to continue under VFR conditions using sectional charts for **visual navigation.**

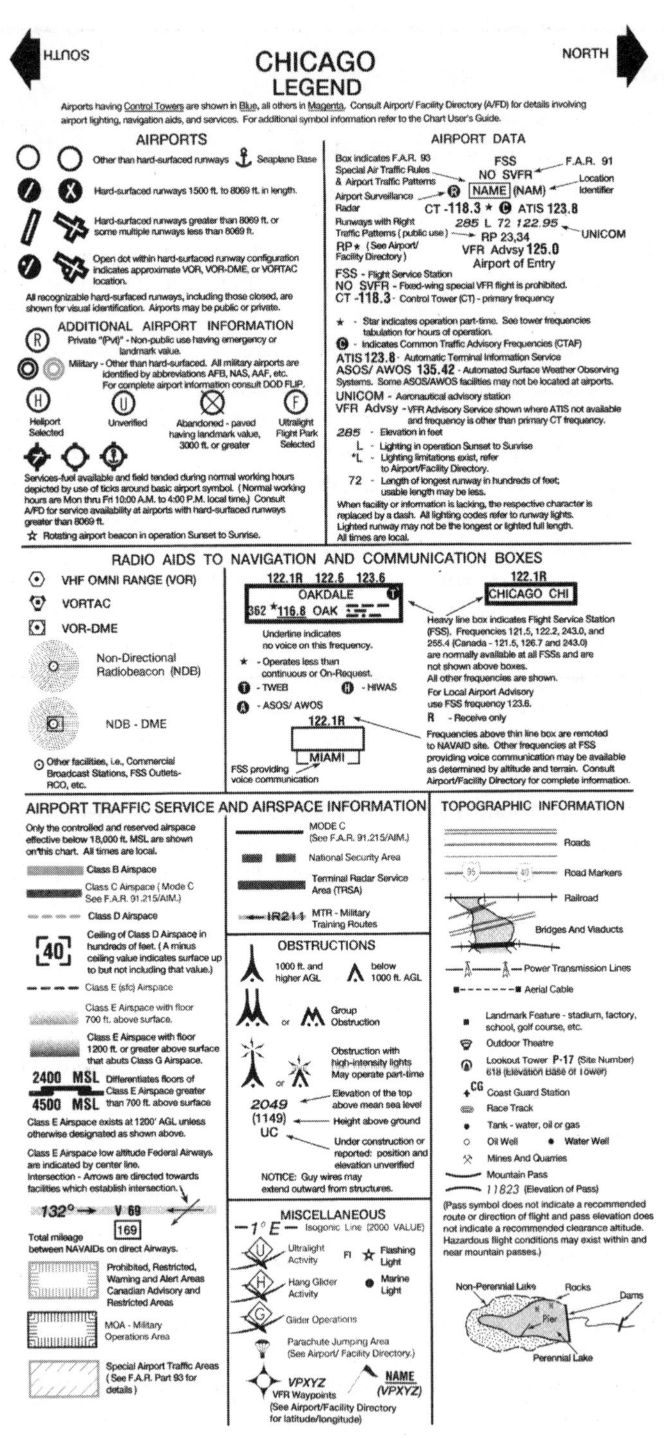

Figure 2-1. Example of a legend for a sectional chart.

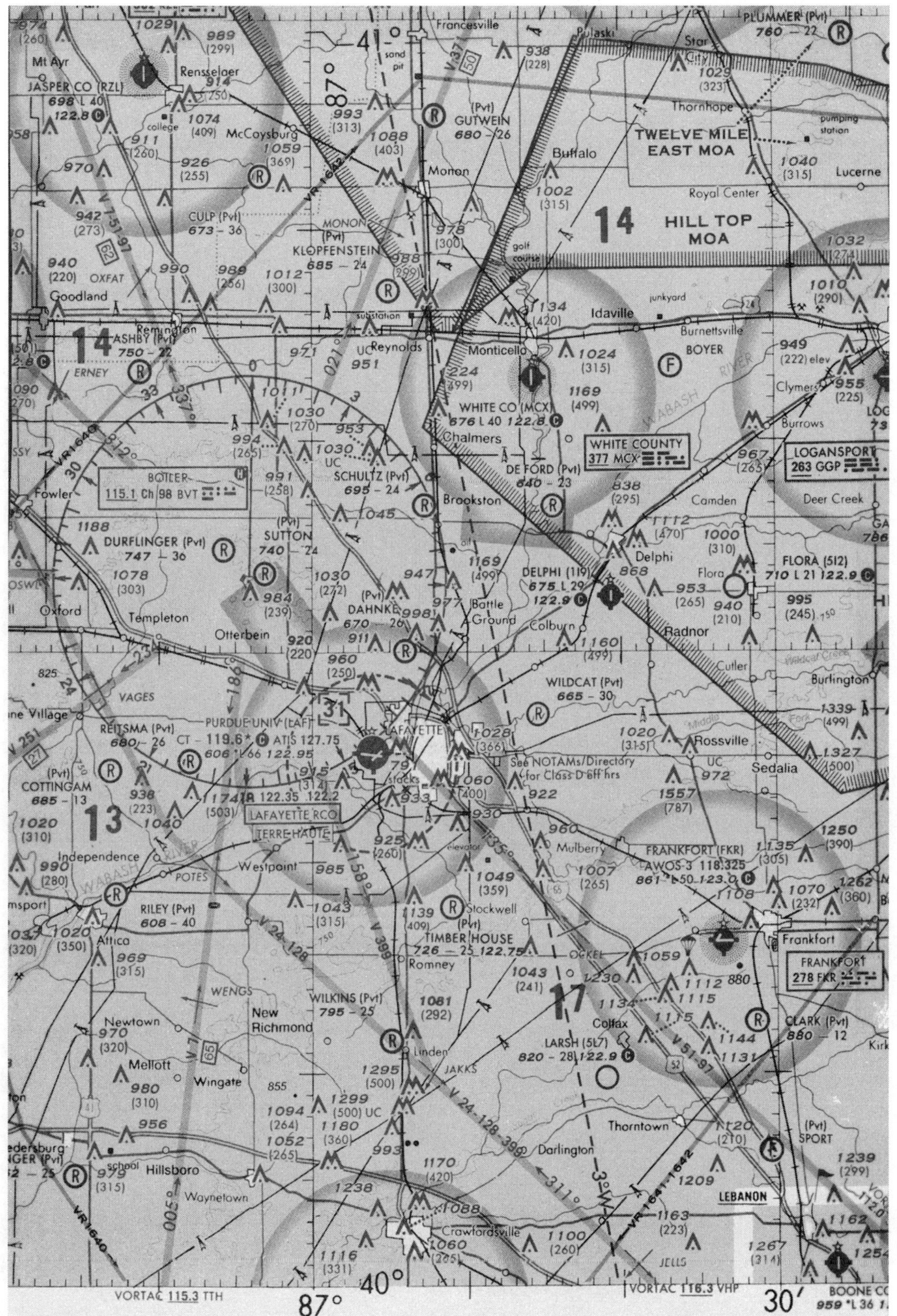

Figure 2-2. Sample sectional chart.

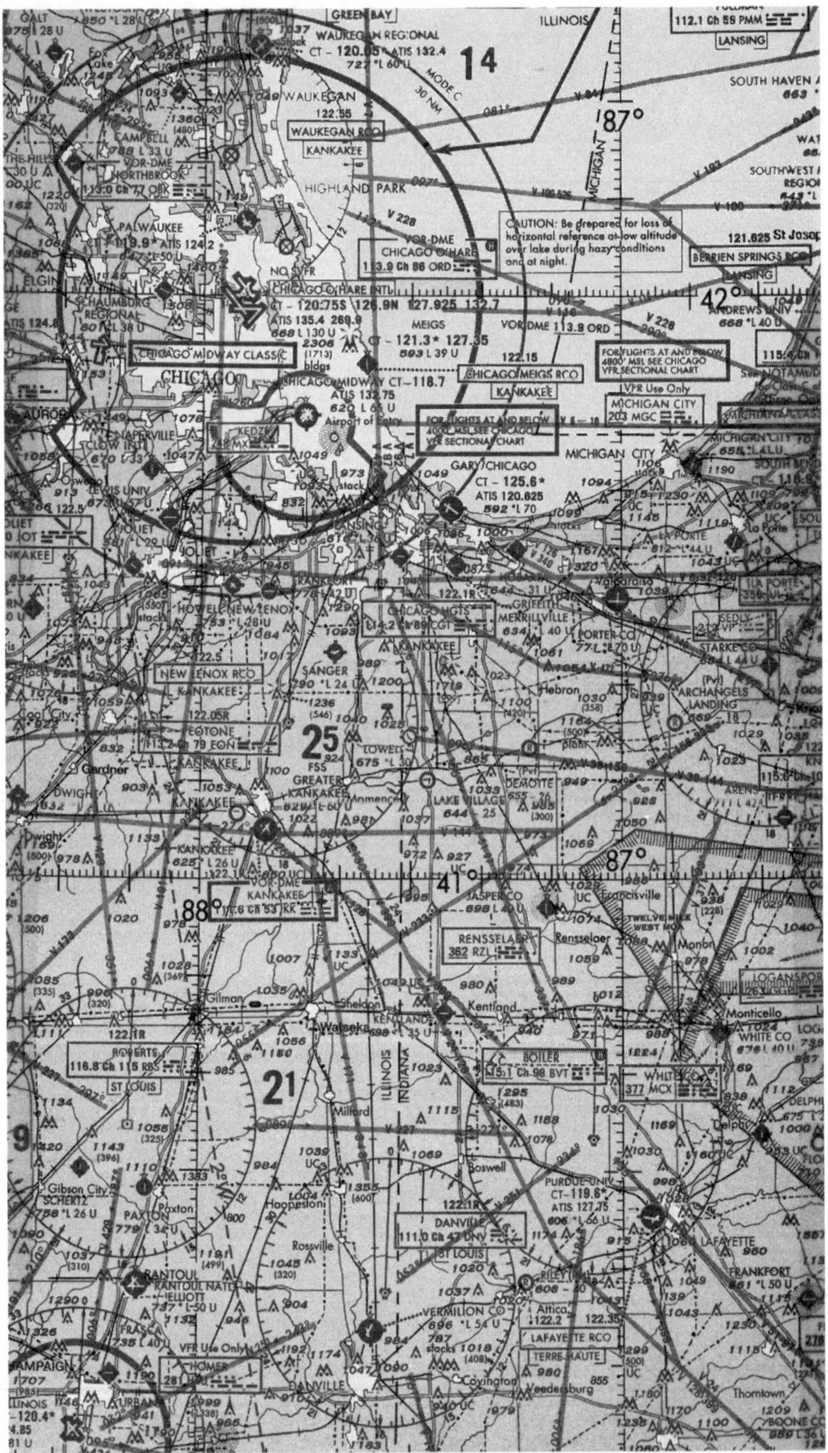

Figure 2-3. Sample world aeronautical chart.

Some pilots carry **world aeronautical charts (WACs)** instead of sectionals during IFR flights (see Figure 2-3). WACs are similar to sectionals but are scaled 1:1,000,000, or about 16 miles to the inch. They present less-detailed information to the pilot but cover a larger area than a sectional chart.

Dead Reckoning

When flying using VFR rules, most pilots use dead reckoning, in combination with pilotage, to navigate to their destination. With dead reckoning, the pilot uses the forecast winds at the planned cruising altitude and applies trigonometry to deduce the proper heading that the aircraft should fly to counteract the crosswind. Properly calculated, this method of navigation is very accurate; however, it is hampered by the fact that the winds-aloft information is a forecast, not a reflection of the actual winds. To verify that dead reckoning has calculated the proper heading, the pilot must still visually check the accuracy of the deduced heading by using a sectional chart.

Flight Planning

The first step in planning a flight using both dead reckoning and pilotage is to determine the **true course** that will lead the aircraft to the destination airport. This is accomplished by drawing a line from the departure airport to the destination on the sectional chart. The pilot then determines the angle of this course in reference to **true north,** using a device called a plotter. The pilot obtains the forecast wind speed and direction at the chosen cruising altitude and, using either a mechanical or an electronic computer, calculates the **true heading** that the aircraft must fly.

The deduced true heading is the direction that the aircraft must be aimed in order to track to the desired destination. If there is not wind at the aircraft's cruising altitude, the true heading and the true course will be exactly the same. But if the pilot encounters a crosswind, he or she must angle the aircraft into the wind to remain on course. The angular difference between the aircraft's heading and the true course is the crosswind correction, or wind correction angle.

Aircraft Instrumentation

Magnetic Compass

Aeronautical charts cannot be properly used by pilots unless they have accurate aircraft heading information. All of these charts are oriented with respect to true north. Unfortunately, the only instrument aboard most aircraft that actually indicates heading is a **magnetic compass,** which usually points toward **magnetic north** (see Figure 2-4).

The angular difference between true north and magnetic north is known as variation (see Figure 2-5). The variation depends on the aircraft's current location. In different areas of the United States, the variation may range from 0° to as much as 20°. To properly use the magnetic compass when navigating, the pilot must add the variation to or subtract it from the aircraft's true heading to

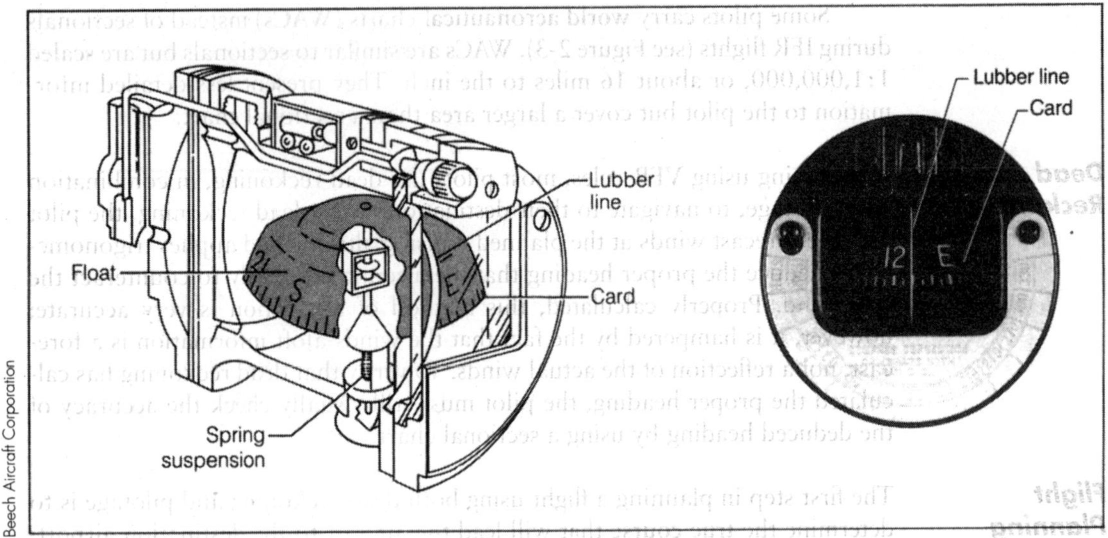

Figure 2-4. The magnetic compass.

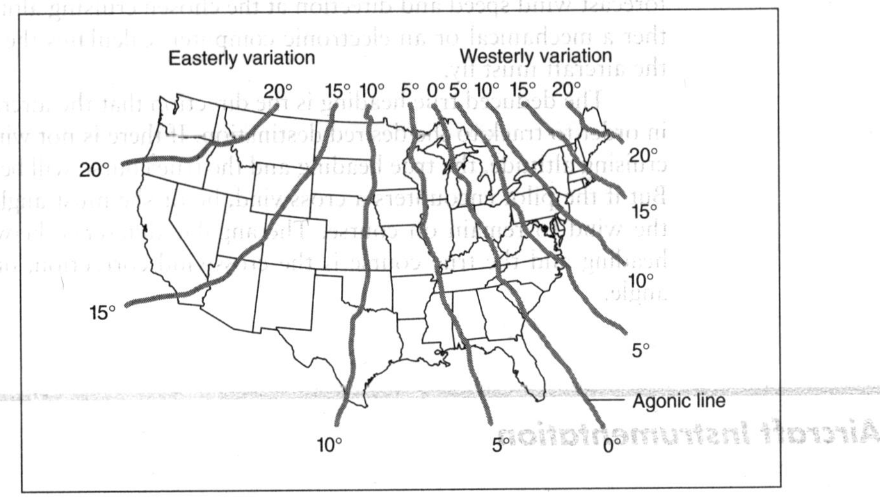

Figure 2-5. Variation chart.

determine the **magnetic heading** that must be flown. The pilot may then fly this heading using the aircraft's magnetic compass.

Although the magnetic compass is a relatively reliable instrument, it is subject to various inaccuracies. One of these inaccuracies is known as **deviation.** Deviation is caused by the stray magnetic fields of electrical equipment or metallic structures within the aircraft. Since all aircraft contain some stray magnetic fields, every plane is required to be equipped with a **compass deviation card** that lists the inaccuracies and the correction that must be applied when interpreting the magnetic compass.

A few other conditions can cause the magnetic compass to indicate inaccurately. During changes in airspeed or while the aircraft is turning, the magnetic compass will not indicate correctly. These particular inaccuracies are known as *acceleration* and *turning errors*. In general, the only time that the magnetic compass can be accurately interpreted is when the aircraft is in straight and level, unaccelerated flight. In addition, placement of a metal or magnetized object (such as a flashlight, clipboard, or screwdriver) near the compass will alter the local magnetic field and cause magnetic compass errors.

Heading Indicator

Many of the problems inherent in the magnetic compass can be alleviated by using a **heading indicator** (see Figure 2-6). Because the heading indicator is a gyroscopic instrument, it is not subject to the same problems that affect the magnetic compass. The heading indicator is initially set by the pilot while on the ground. When properly set, it accurately reflects the aircraft's magnetic heading during flight. As the aircraft turns, the heading indicator rotates, constantly displaying the correct heading.

The heading indicator is not subject to acceleration or turning errors, and it is immune to stray magnetic fields. It has, however, a few inherent problems. Since it is unable to sense magnetic fields, it must be properly adjusted by the pilot before being used. If the pilot sets the indicator incorrectly, it will not accurately reflect the aircraft's magnetic heading. In addition, since the heading indicator is subject to internal bearing friction and will slowly drift and begin to indicate inaccurately, the pilot must constantly check its accuracy and reset it as necessary during the flight. It is also possible, though highly unlikely, that the heading indicator will fail mechanically, not indicating the proper heading even when properly set. The heading indicator is also subject to precession and should be periodically reset during the flight.

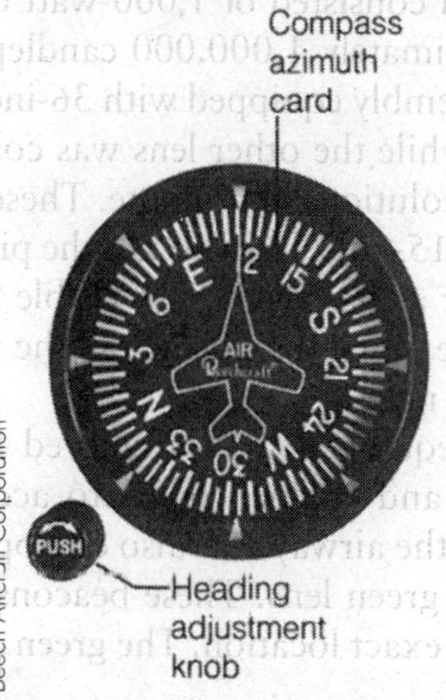

Figure 2-6. A heading indicator.

VFR Navigation

In theory, using dead reckoning alone, the pilot should be able to fly the computed heading and arrive over the airport at the calculated time. But in reality, because of imprecise winds-aloft forecasts, most pilots use a combination of pilotage and dead reckoning. The proper heading and time must still be deduced and used, but en route navigation checkpoints are established and marked on the appropriate sectional charts. As the pilot flies toward the destination, he or she makes periodic checks to determine whether the aircraft is still on course. If it has deviated from the planned route, the pilot will adjust the aircraft's heading to return to and remain on the desired flight path. As archaic and old fashioned as this may seem, it is still the primary method of navigation for most VFR pilots today.

Although visual navigation works quite well during daylight hours, at night or in marginal weather conditions it is almost impossible for pilots to see objects on the ground and make an accurate determination of their aircraft's position. Sparsely populated areas of the country may not offer sufficient ground references to permit the pilot to determine the aircraft's location. If and when the pilot finally arrives at the destination airport, he or she may find it difficult to actually locate the runway in the dark and land the aircraft. The solution, of course, is to have both airport and airway lighting.

In the 1920s, airports were illuminated through the use of **airport boundary lighting,** which consisted of steady-burning 40-watt lights on wooden stakes every 300 feet around the perimeter of the airport. Eventually, these lights were equipped with lenses to concentrate the light beam and were mounted on orange-colored steel cones so that they could also be clearly seen during daylight hours. With the outline of the airport now quite visible, the pilots were able to safely land and take off at night.

As noted in Chapter 1, the first airway lighting was also instituted in the 1920s. At equal intervals along the airway, rotating beacon lights were installed that delineated the airway's center line (see Figure 2-7). These rotating beacons were installed on steel towers and consisted of 1,000-watt electric lamps that produced a white light of approximately 1,000,000 candlepower. Each lamp was housed in a rotating drum assembly equipped with 36-inch-diameter lenses at each end. One lens was clear while the other lens was colored. The beacon rotated at a speed of about six revolutions per minute. These rotating beacons were installed along the airway at 15-mile intervals. As the pilots flew along the airway, the beacons would appear as flashes of light visible from a distance of over 40 miles. To visually navigate along the airway, all the pilot needed to do was to fly from one beacon to the next.

Each rotating beacon was equipped with a colored lens that uniquely identified that particular beacon and enabled pilots to accurately determine their position. Each airport along the airway was also equipped with a rotating beacon having one clear and one green lens. These beacons were designed to help pilots determine the airport's exact location. The green and white rotating

Michael Nolan

Figure 2-7. A rotating beacon.

beacons are still used at civilian airports today. Other color combinations are used to differentiate other types of airports. The assigned colors for rotating beacons are as follows:

White and green	Land airport
Green and green*	Land airport
White and yellow	Water airport
Yellow and yellow*	Water airport
Green, white, and white	Military airport
Green, yellow, and white	Lighted heliport

* Green or yellow rotating beacons are used to prevent confusion when another airport with a similarly colored rotating beacon is located nearby.

Instrument Flying

Lighting the airports and airways proved to be a tremendous advance in nighttime navigation but still required that pilots fly in weather conditions that would permit them to see the rotating beacons. If a pilot flew in or above a cloud layer, or if the flight visibility diminished to less than 15 miles, he or she would be unable to see the rotating beacons and navigate to the destination airport. As advances were made in aircraft design and instrumentation, it soon became possible for pilots to control their aircraft using just cockpit instrumentation, flying their aircraft without visual reference to the natural horizon. The new cockpit instruments were based on gyroscopic principles and included the **artificial horizon** (now called the **attitude indicator**), the heading indicator, and the turn and bank indicator (now called the **turn coordinator**).

The attitude indicator (Figure 2-8) mimics the movements of the natural horizon, providing the pilot with accurate aircraft attitude information. Using the attitude indicator, pilots can determine whether their aircraft is banked and whether its nose is pointed up or down. This allows them to adjust their aircraft's flight attitude and keep the aircraft upright and under control.

The heading indicator, as described previously in this chapter, provides a more stable and accurate indication of the aircraft's flight direction than does the magnetic compass. The turn coordinator (Figure 2-9) is utilized by the pilot to indicate the direction and the rate of turn.

These instruments, used in conjunction with the already existing **altimeter** (Figure 2-10) and **airspeed indicator** (Figure 2-11), made it possible for pilots to accurately control their aircraft without reference to the natural horizon.

Unfortunately, the federal airway system had not kept pace with these instrumentation developments and until the late 1920s was still based on sectional charts and rotating beacons, which required that pilots have at least 15 to 20 miles of visibility to navigate at night. There was no provision for navigation when the visibility dropped below these values.

Beech Aircraft Corporation

Figure 2-8. The attitude indicator.

Figure 2-9. The turn coordinator.

Figure 2-10. The altimeter.

Figure 2-11. The airspeed indicator.

Electronic Navigation

Four-Course Radio Range

In an early attempt to remedy this situation, the federal government began to install the four-course radio range in the late 1920s. This new radio device was placed at intervals along each federal airway and permitted the pilot to navigate without using outside visual references. Four-course radio ranges soon became the U.S. and international standard for aviation navigation and were widely used in the United States until the 1950s.

The four-course radio range used a 1,500-watt transmitter that operated on a frequency between 190 and 565 kHz. The transmitting antenna consisted of two single-wire vertical loops strung out on five wooden masts. These wires were attached to the masts to form two figure-eight patterns (see Figure 2-12).

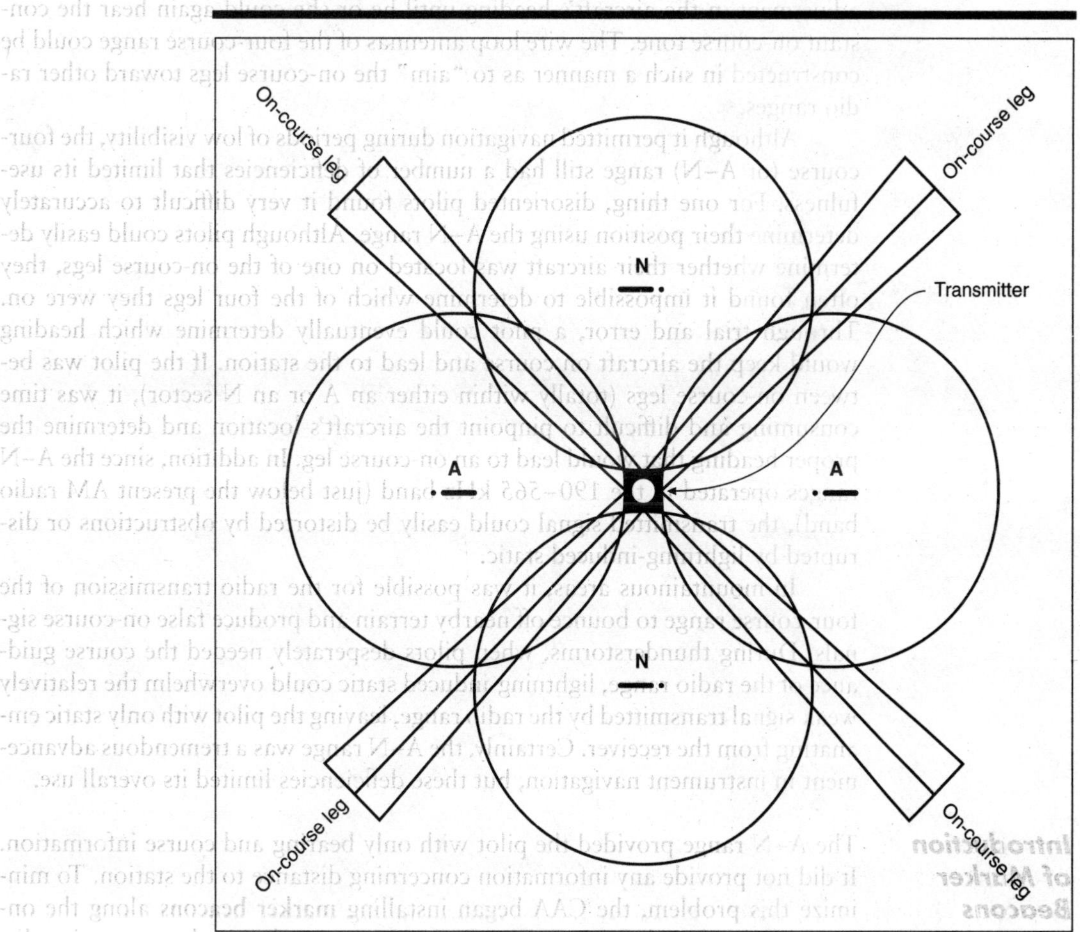

Figure 2-12. Field pattern of four-course radio range.

This arrangement produced two separate radio transmission patterns that overlapped slightly. One loop constantly transmitted the Morse code for the letter A (dot-dash); the other transmitted the Morse code for the letter N (dash-dot). Any pilot wishing to use a four-course range simply tuned the receiver on the aircraft to the proper frequency and listened to the transmitted signal through earphones. If the aircraft was located somewhere within the A sector, the pilot would hear Morse code for the letter A (dot-dash) constantly being repeated. If the aircraft was in the N sector, the pilot would hear the Morse code for the letter N (dash-dot). If the aircraft was located where the two transmissions overlapped (the on-course line of position), the dot-dash and the dash-dot would be of equal strength and would produce a constant tone in the pilot's headset.

When navigating to a four-course radio range, all the pilot needed to do was to head the aircraft toward one of these on-course "legs," then proceed along it to the radio range. If the aircraft drifted off course, the pilot would begin to hear the individual A or N Morse code becoming dominant, requiring an adjustment in the aircraft's heading until he or she could again hear the constant on-course tone. The wire loop antennas of the four-course range could be constructed in such a manner as to "aim" the on-course legs toward other radio ranges.

Although it permitted navigation during periods of low visibility, the four-course (or A–N) range still had a number of deficiencies that limited its usefulness. For one thing, disoriented pilots found it very difficult to accurately determine their position using the A–N range. Although pilots could easily determine whether their aircraft was located on one of the on-course legs, they often found it impossible to determine which of the four legs they were on. Through trial and error, a pilot could eventually determine which heading would keep the aircraft on course and lead to the station. If the pilot was between on-course legs (totally within either an A or an N sector), it was time consuming and difficult to pinpoint the aircraft's location and determine the proper heading that would lead to an on-course leg. In addition, since the A–N ranges operated in the 190–565 kHz band (just below the present AM radio band), the transmitted signal could easily be distorted by obstructions or disrupted by lightning-induced static.

In mountainous areas, it was possible for the radio transmission of the four-course range to bounce off nearby terrain and produce false on-course signals. During thunderstorms, when pilots desperately needed the course guidance of the radio range, lightning-induced static could overwhelm the relatively weak signal transmitted by the radio range, leaving the pilot with only static emanating from the receiver. Certainly, the A–N range was a tremendous advancement in instrument navigation, but these deficiencies limited its overall use.

Introduction of Marker Beacons

The A–N range provided the pilot with only bearing and course information. It did not provide any information concerning distance to the station. To minimize this problem, the CAA began installing **marker beacons** along the on-course legs. These low-powered radio beacons were designed to transmit a distinctive tone and code that could be received by the aircraft as it passed directly overhead. The pilots could use the code to identify which beacon was crossed and use this information to accurately determine their aircraft's position along the on-course leg. But whenever the aircraft was between marker beacons or no longer on one of the on-course legs, the marker beacons were useless in helping determine the aircraft's location.

Nondirectional Beacons

While the CAA was developing and installing A–N radio ranges, the **nondirectional radio beacon (NDB)** was also being developed. The NDB transmits a uniform signal omnidirectionally from the transmitter, utilizing the low- and medium-frequency band (190–540 kHz). The receiver on the aircraft (known as a **direction finder,** or DF) was originally equipped with a looptype antenna that the pilot rotated manually. When the antenna was rotated so that the plane of the loop was perpendicular to the transmitted signal, the "null" position was reached, and the pilot would be able to hear the transmitted signal. Using the magnetic compass and the NDB receiver, the pilot could then determine the aircraft's bearing from the nondirectional beacon. This bearing could be plotted on a chart as line of position. Plotting lines of position from two NDBs permitted the pilot to pinpoint the aircraft's exact location. If the pilot wished to fly toward the NDB, he or she would turn until the NDB station was located directly ahead of the aircraft. If the winds aloft caused the aircraft to drift off course, the pilot would readjust the aircraft's heading, keeping the NDB directly ahead of the aircraft. This method of navigation is called **homing.**

Automatic Direction Finder

Trying to manually manipulate the DF antenna while flying the aircraft proved to be a cumbersome method of navigation and usually provided the pilot with relatively inaccurate position information. As advances were made in aircraft electronics, the manually operated NDB receiver was soon replaced by the **automatic direction finder (ADF),** which could electronically determine the bearing to the NDB and display this information to the pilot (see Figure 2-13). Using ADF equipment in conjunction with the aircraft's heading indicator, the

Figure 2-13. ADF receiver and indicator.

pilot could easily determine the aircraft's relative bearing from the station and use this information to determine the proper heading that would lead to the beacon.

The development of the automatic direction finder hastened pilot acceptance of the NDB as a navigation aid. The first NDB was installed in the United States in 1924. By 1964, 272 high-powered NDBs had been installed throughout the country. A series of federal airways utilizing NDBs for en route navigation were soon developed. Because these airways were designated by a color and a number (for example, RED-64 or GREEN-32), they were soon referred to simply as **colored airways.**

Compass Locators

In addition to their role as en route navigation aids, NDBs were located at airports or along instrument-approach paths to assist pilots who were conducting such approaches. NDBs along the final approach are known as **compass locators.** In 1965 the federal government began to decommission the high-powered NDBs used for en route navigation. Due to their extremely low cost and ease of installation, however, low-power units continued to serve smaller airports as instrument-approach aids.

Visual Aural Range

In 1937 the Radio Development Section of the Bureau of Air Commerce demonstrated an improved radio range at its research center in Indianapolis. This new radio range, called the **visual aural range (VAR),** was an improvement over the old A–N range in two major areas. The VAR was designed to operate in the **very high frequency** (VHF) band located around 63 mHz. This frequency band was chosen since transmitters operating on VHF frequencies are rarely affected by static caused by lightning. VHF transmissions are also **line of sight,** which means that they do not follow the curvature of the earth. One significant advantage of using VHF frequencies is that although they can easily be blocked by terrain and obstructions, they are seldom reflected by them. The use of VHF frequencies would thus minimize the reflection problem that plagued the A–N ranges.

The VAR also solved the orientation problem inherent with the A–N range by transmitting four radio signals instead of two. While retaining the Morse-coded A and N signals, the VAR also transmitted overlapping "blue" and "yellow" signals perpendicular to the A and N signals (see Figure 2-14). An instrument on board the aircraft would indicate whether it was in the blue or the yellow sector. The pilot was still required to listen to the VAR to determine whether the aircraft was within the A or the N sector, however. The addition of the overlapping color signals gave each sector a unique identification that enabled pilots to accurately determine their aircraft's location.

The first operational VAR was installed at Matawan, New Jersey, in 1944. By 1948 a total of 68 VARs had been commissioned by the CAA and were located along federal airways. The VAR never gained wide acceptance, however, since it was soon replaced by an improved radio range that emitted an infinite number of courses instead of just four. This new navigation aid was called the

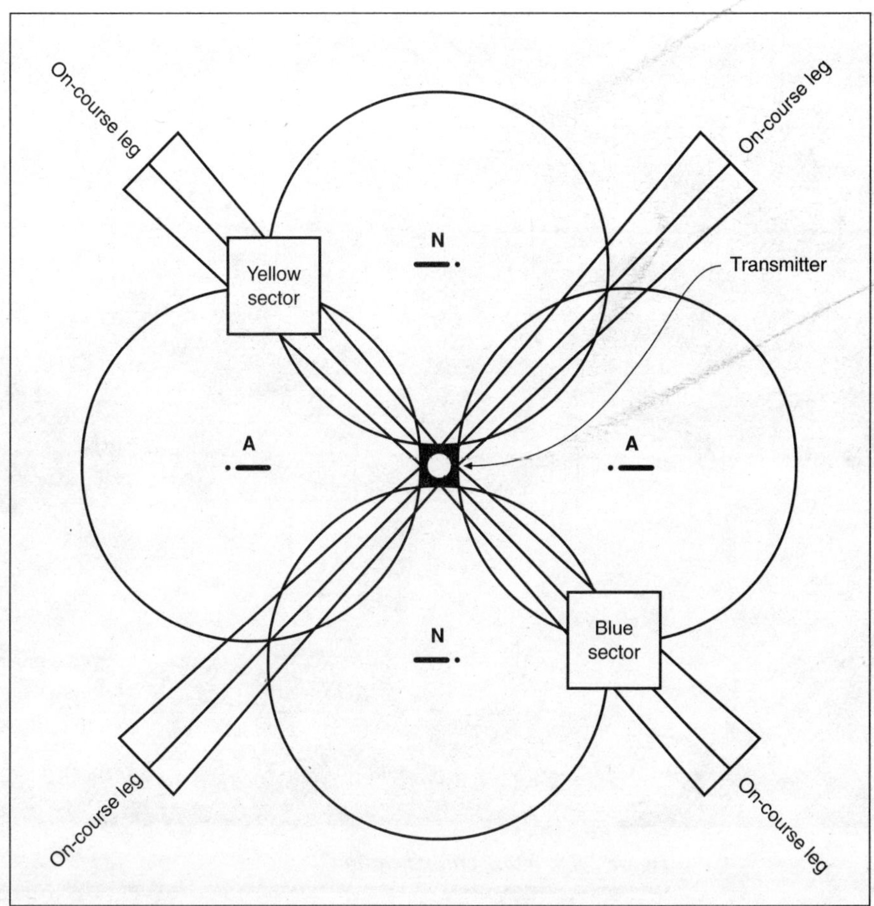

Figure 2-14. VAR operation.

VHF omnidirectional range (VOR). In the early 1950s, as VORs were being installed around the country, the CAA began to decommission the VARs, with the last being retired from service in 1960.

VHF Omnidirectional Range (VOR)

Research on a radio range that would offer pilots more than four courses and transmit in the static-free VHF radio spectrum had started in 1937. The Washington Institute of Technology delivered the first operable VHF omnidirectional range (VOR) to the CAA in 1944. This experimental VOR operated on a frequency of 125 mHz. After extensive testing and development by the CAA, three prototype VORs were installed at Patuxent River, Maryland; Philipsburg, Pennsylvania; and Ogden, Utah. After operational testing at these three sites, the CAA adopted the VOR as the national civil navigation standard in 1946.

Wadsworth, Inc. photo by Paul Bowen Photography, Inc.

Figure 2-15. VOR ground station.

The VOR was also selected as the international civil navigation standard in 1949 by the International Civil Aviation Organization (see Figure 2-15).

VOR Operation The VOR offered a number of improvements over the old A–N and VAR methods. The VOR transmits an infinite number of navigation courses, selectable by the pilot, instead of just four. The VOR is also relatively immune to the reflections and static inherent in the operation of the A–N ranges.

Each VOR is assigned a frequency between 108.10 and 117.90 mHz. The VOR transmission is modulated with two signals: a *reference-phase signal* that is constant in all directions and a *variable-phase signal* whose phase varies with azimuth. The variable-phase signal is modulated so that at magnetic north the reference and variable signals are precisely in phase with each other. In any other direction, the VOR is designed so that the two signals are no longer in phase.

The VOR receiver on board the aircraft measures the phase difference between the two signals to determine the azimuth angle of the aircraft in relation

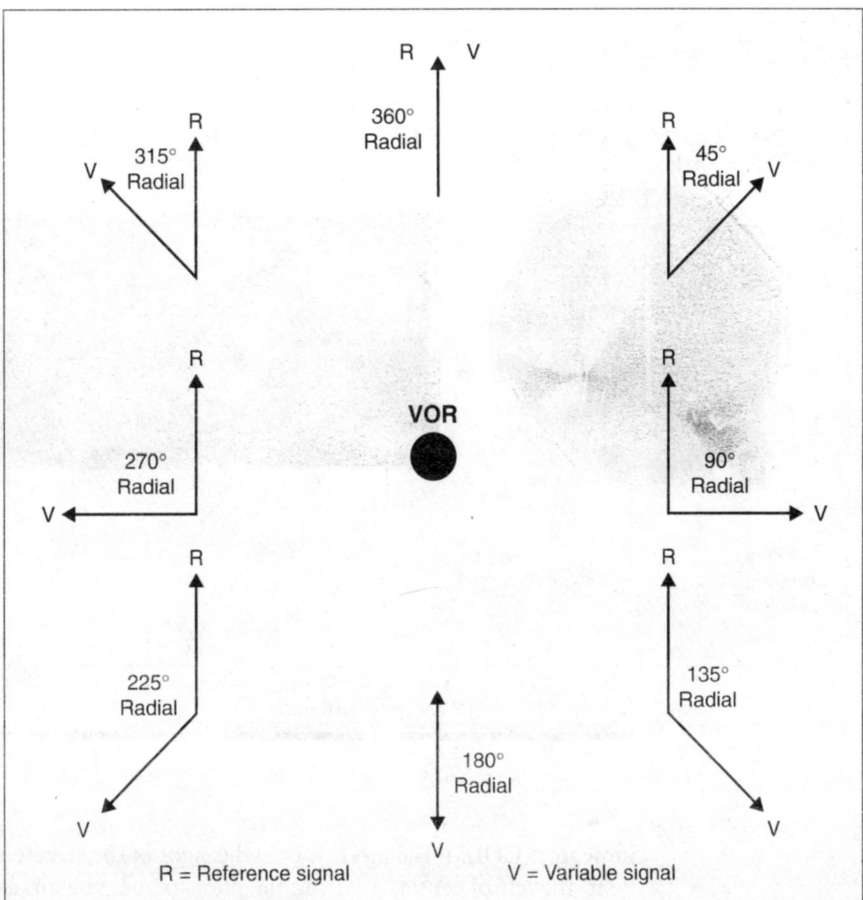

Figure 2-16. VOR operation.

to the VOR transmitter. When the aircraft is directly east of the VOR, the variable signal will lag the reference signal by 90°. An aircraft located directly east of the VOR is said to be on the 90° **radial** of the VOR (see Figure 2-16). An aircraft directly south of the VOR will receive the variable signal lagging the reference signal by 180° and will be on the 180° radial. An aircraft located on the 359° radial (north of the VOR) will receive the variable signal lagging the reference signal by 359°.

The radial to be flown by the pilot is selected on the aircraft's VOR indicator (see Figure 2-17) using the **omni bearing selector (OBS)**. After selecting the appropriate VOR frequency, the indicator in the cockpit will inform the pilot whether the selected course will lead to the station or away from it (known as the To-From flag). The VOR indicator will also display any lateral deviation from the selected course, using a vertical pointer known as the **course deviation**

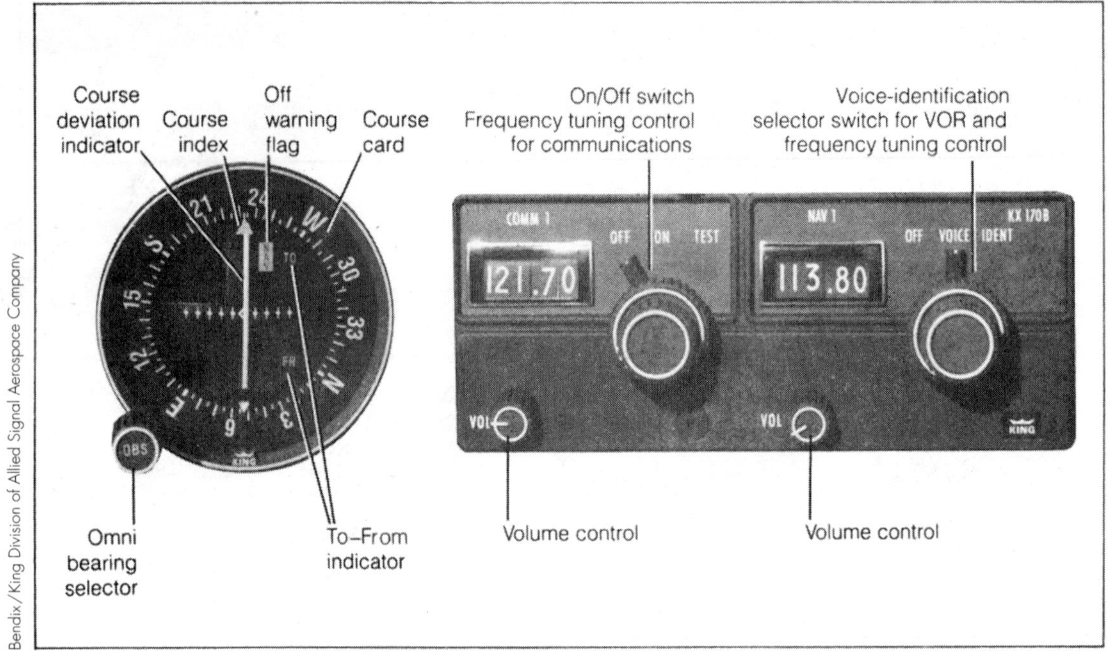

Figure 2-17. VOR receiver and indicator.

indicator (CDI). If the aircraft is to the right of the selected course, the CDI will be to the left of center, advising the pilot to alter the aircraft's course to the left in order to return to the selected radial. If the aircraft is left of course, the CDI will be right of center. If the aircraft is precisely located on the radial selected by the pilot, the CDI will be centered.

VORs used for en route navigation have an output power of 200 watts and are assigned a frequency between 112.00 and 117.90 mHz. This signal permits en route VOR reception up to a distance of 200 miles. Terminal VORs (used solely for instrument approaches) have an effective radiated power of 50 watts and are assigned a frequency between 108.10 and 111.80 mHz. Terminal VORs can be received up to a distance of about 25 nautical miles. Since VOR transmissions are line of sight, these reception distances vary depending on the receiving aircraft's altitude (see Figure 2-18).

VOR Categories A number of difficulties were encountered as soon as the CAA began to install VORs along the federal airways. Since VHF transmissions are line of sight, low-flying aircraft were unable to receive the VOR signal if they were "below the horizon." This limitation forced the CAA to place the VORs no farther than 80 miles from each other to ensure adequate reception for aircraft operating at low altitudes. Because only a limited number of fre-

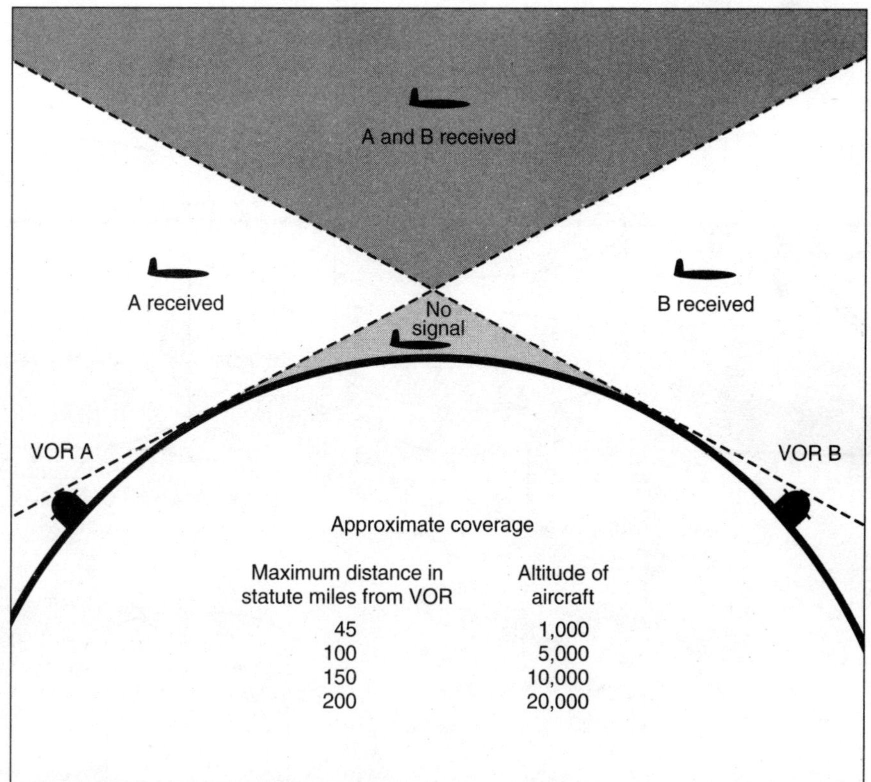

Figure 2-18. VOR reception distances.

quencies can be assigned to VORs, some would have to be assigned the same operating frequency, which could cause interference problems for aircraft operating at very high altitudes, as they might receive the signals being broadcast from two or more VORs operating on the same frequency. The resulting interference would render the navigation signal unusable.

The CAA responded by designating every VOR as a terminal, low-, or high-altitude VOR. **Terminal VORs (TVORs)** are low powered and are usable up to a distance of 2.5 nautical miles. TVORs are not to be used for en route navigation but are reserved for local navigation and instrument approaches. **Low-altitude VORs** guarantee interference-free reception to aircraft operating up to 40 nautical miles away. This interference-free zone is guaranteed only at or below 18,000 feet. Low-altitude VORs cannot be used by aircraft operating above 18,000 feet or farther than 40 miles away, as there is no guarantee that another VOR operating on the same frequency will not cause interference. **High-altitude VORs** are used by aircraft operating between 18,000 and 60,000 feet, at ranges up to 200 nautical miles. These limitations imposed upon VORs are known as **service volumes** (see Figures 2-19 and 2-20).

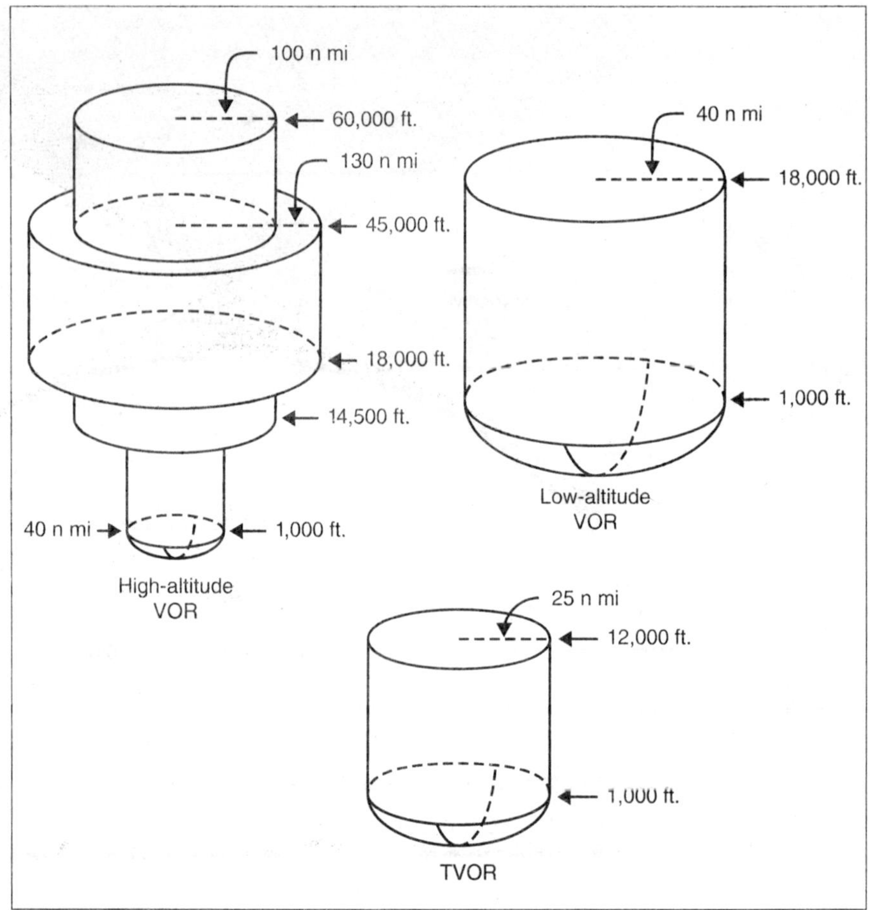

Figure 2-19. VOR service volumes.

Unusable Radials Testing of the VOR found that a clear zone of several thousand square feet around the VOR was necessary for proper operation. Any obstruction within this area could blank out or reflect some of the signal from the VOR and cause incorrect course information to be transmitted to the aircraft. Tall buildings located thousands of feet from the VOR transmitter could even distort the transmitted signal. In an attempt to solve this problem, the CAA developed the **Doppler VOR (DVOR).**

Although the operating principles of this VOR differ radically from a conventional VOR, the information available to the pilot is exactly the same. The VOR receiver on board the aircraft is unable to differentiate between Doppler or conventional VOR transmissions. Doppler VOR is less sensitive to reflections from buildings or terrain. In locations unsuitable for conventional VOR installation, a Doppler VOR might be necessary (see Figure 2-21).

VOR/VORTAC/TACAN NAVAIDs

Normal Usable Altitudes and Radius Distances

Class	*Altitudes*	*Distance (nautical miles)*
T	12,000 and below	25
L	Below 18,000	40
H	Below 14,500	40
H	14,500 — 17,999	100
H	18,000 — FL 450	130
H	Above FL 450	100

L/MF Radio Beacon (RBN)

Usable Radius Distances for All Altitudes

Class	*Power (watts)*	*Distance (nautical miles)*
CL	Under 25	15
MH	Under 50	25
H	50 — 1,999	50
HH	2,000 or more	75

ILS

Usable Height and Distance*

Height (feet) above transmitter	*Distance (nautical miles) from transmitter*
4,500	10 (for glideslope)
4,500	18 (for localizer)

Figure 2-20. VOR service volume chart.

If the DVOR fails to correct the reflections or blanking, the affected radials must be listed as unusable, which means that although the pilot may be able to receive these radials, they are *not* accurate and should not be used. Unusable VOR radials are published in the Airport Facility Directory (see Figure 2-22).

Minimum Altitudes

The CAA faced an enormous task when trying to determine where VORs should be located. CAA planners had to consider potential obstructions, terrain, and the position of other VORs operating on the same frequency to determine that a suitable interference-free signal could be received by any aircraft operating along a VOR-equipped airway. The airways must be constantly flight checked. After these checks, a **minimum en route altitude** (MEA) is designated

FAA

Figure 2-21. A Doppler VOR station.

for each airway. Aircraft operating at or above the MEA are guaranteed clearance above any obstruction located along or near the airway.

Along some airways, if they differ from the MEA, **minimum obstruction clearance altitudes (MOCAs)** are also designated (see Figure 2-23). MOCAs are lower than MEAs and are designed to provide obstacle clearance only. In case of an emergency, the pilot may safely descend to the MOCA and will still be guaranteed obstacle clearance. Pilots flying at the MOCA altitude are also guaranteed proper VOR reception as long as they are within 22 nautical miles of the VOR.

Airways

When VOR airways are designated, their identifying numbers are preceded by the letter V if they are low altitude airways, or the letter J if they are high altitude airways.

Aircraft Positioning Methods

The VOR provides only bearing information to the pilot (known as rho), not distance from the station (known as theta). There are only two ways for a pilot using the VOR to accurately determine an aircraft's position: using either rho–rho or rho–theta position determination. Rho–rho position determination requires that the pilot obtain bearing information from two different VORs. Using airborne VOR equipment, the pilot can plot a line of position from each VOR. These two lines of position (or radials) are then plotted on a navigation chart, with the aircraft being located at the intersection of the two radials (see Figure 2-24).

ILLINOIS **19**

CHICAGO–O HARE INTL (ORD) 14 NW UTC–6(–5DT) N41°58.78′ W87°54.27′ **CHICAGO**
668 B S4 **FUEL** 100LL, JET A1 OX 1, 2, 3, LRA ARFF Index E **COPTER**
RWY 14R–32L: H13000X200 (ASPH–CONC–GRVD) S–100, D–185, DT–350 HIRL CL **H–3G, L–23A, A**
RWY 14R: ALSF2. TDZL. Vent. **RWY 32L:** MALSR. TDZL. **IAP**
RWY 09R–27L: H10141X150 (ASPH–CONC–GRVD) S–100, D–185, DT–350 HIRL CL
RWY 09R: MALSR. TDZL. **RWY 27L:** MALSR. TDZL. Pole.
RWY 14L–32R: H10003X150 (ASPH–GRVD) S–100, D–185, DT–350 HIRL CL
RWY 14L: ALSF2. TDZL. **RWY 32R:** MALSR. TDZL. Light.
RWY 04R–22L: H8071X150 (ASPH–GRVD) S–100, D–200, DT–350 HIRL CL
RWY 04R: MALSR. TDZL. **RWY 22L:** MALSR. TDZL. Light standard.
RWY 09L–27R: H7967X150 (ASPH–CONC–GRVD) S–100, D–210, DT–350 HIRL CL
RWY 09L: MALSR. TDZL. **RWY 27R:** MALSR. TDZL. Antenna.
RWY 04L–22R: H7500X150 (ASPH–GRVD) S–100, D–185, DT–350 HIRL CL
RWY 04L: LDIN. Light standard. **RWY 22R:** MALSR.
RWY 18–36: H5341X150 (ASPH) S–60, D–100, DT–150 MIRL
RWY 18: Tree.

AIRPORT REMARKS: Special Air Traffic Rules—Part 93; High Density Arpt, prior reservations req; SEE AIM. Attended continuously. PAEW near various twys. Due to obstructed vision; O'Hare twr is unable to provide ATC svc on Twy Y between Twy T and United Airlines hangar. Lgt poles bordering the Signature Aviation ramp. Arpt nighttime. noise abatement procedures are in effect from 0400 to 1300Z‡; ctc arpt manager 312–686–2200. Rwy 18 CLOSED tkf. Rwy 14R–32L possible low level turbulence on AER 14R. Unlighted 867 ft MSL crane 1 mile NW append Rwy 14L. Rwys 09R–27L and 04L–22R irregular surface intermittent grooves. Birds on and invof arpt. Pyrotechnics and bird cannons in use for bird control. No run–ups permitted on the Rwy 14R or Rwy 14L pads. During periods of cold weather; the apch end of Rwy 32R may not be visible from the twr due to steam plume from arpt heating plant. Be aware of duplicate alpha-numeric twy designators and terminal gate designations involving the letters H, K, L and M. Acft with wingspan 170 ft; other than B–747 require 48 hrs PPR—call 312–686–2255. Ldg fee. Flight Notification Service (ADCUS) available. NOTE: See Land and Hold Short Operations Section.

WEATHER DATA SOURCES: ASOS (312) 462–0118. LLWAS. TDWR.

COMMUNICATIONS: ATIS 135.4 (773) 601–7769 **UNICOM** 122.95
KANKAKEE FSS (IKK) LC 626–8266. TF 1–800–WX–BRIEF. NOTAM FILE ORD.
Ⓡ **APP CON** 119.0 (360°–179°) 128.45 125.7 (180°–359°) 121.15 124.35
O'HARE TOWER 126.9 (N) 120.75 (S) **GND CON** 121.75 (outbound) 121.9 (inbound)
CLNC DEL 121.6 119.25 **PRE–TAXI CLNC** 121.6 119.25 **GND METERING** 121.675
Ⓡ **DEP CON** 125.0 (340°–159°) 125.4 (220°–339°) 127.4 (160°–219°)

AIRSPACE: CLASS B: See VFR Terminal Area Chart.

RADIO AIDS TO NAVIGATION: NOTAM FILE ORD.
(L) VORW/DME 113.9 ORD Chan 86 N41°59.26′ W87°54.29′ at fld. 650/2E. **HIWAS.**
SEE SPECIAL NOTICES.
VOR unusable:
140°–155° byd 15 NM all altitudes
100°–140° and 155°–060° byd 35 NM blo 2500′.
DEANA NDB (H–SAB/LOM) 350 ME N41°58.07′ W88°01.59′ 083° 5.5 NM to fld. Unmonitored.
General outlook only on TWEB 0400–1100Z‡. TWEB OTS indef.
LEAMA NDB (LOM) 368 OH N42°04.16′ W87°59.45′ 144° 6.6 NM to fld.
ROAMY NDB (LOM) 394 OR N42°03.36′ W88°00.47′ 135° 6.5 NM to fld.
TAFFS NDB (LOM) 414 IA N41°59.06′ W87°47.34′ 267° 5.2 NM to fld. Unmonitored.
ILS 110.75 I–IDN Rwy 32R.
ILS 110.5 I–JAV Rwy 09L.
ILS 110.1 I–LQQ Rwy 22L. LOC OTS indef.
ILS/DME 111.1 I–TSL Chan 48 Rwy 27L. MM OTS indef.
ILS/DME 109.75 I–ORD Chan 34(Y) Rwy 14R. LOM ROAMY NDB.
ILS/DME 111.1 I–MED Chan 48 Rwy 09R. LOM DEANA NDB.
ILS 111.3 I–RXZ Rwy 22R.
ILS/DME 108.95 I–RVG Chan 26Y Rwy 32L.
ILS 110.5 I–IAC Rwy 27R. LOM TAFFS NDB.
ILS 110.1 I–FJU Rwy 04R.
ILS/DME 110.9 I–OHA Chan 46 Rwy 14L. LOM LEAMA NDB.
ILS 111.3 I–HNA Rwy 04L. LOC only.

COMM/NAVAID REMARKS: CLASS B frequency 133.5 for VFR acft req tfc advisories or to transition CLASS B South of O'Hare. CLASS B frequency 120.55 for VFR acft req tfc advisories or to transition CLASS B North of O'Hare. If unable to ctc Chicago App/Dep Control 133.1; call 312–686–0681 to obtain IFR CLNC.

Figure 2-22. Unusable radials listed in the Airport Facility Directory (gray screen).

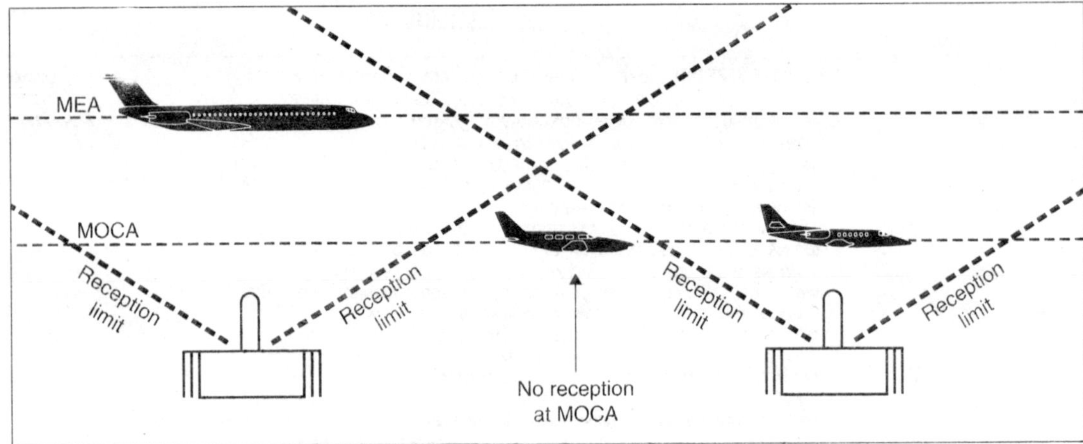

Figure 2-23. Minimum en route and minimum obstruction clearance altitudes.

The rho–rho method of position determination requires that the aircraft be within the service volume of both VOR transmitters. These two stations should also be at approximately right angles to each other. Since the VOR receiver on the aircraft can legally have an accuracy of ±6°, this in effect makes each radial 12° wide. The aircraft's location will be somewhere within the area defined by the limits of the VOR receivers' accuracy. If the two radials do not bisect each other at approximately right angles, the area defined by the two radials becomes much larger, thereby making the position determination less accurate (see Figure 2-25).

DME Position Determination

If a pilot wishes to determine an aircraft's location using just one station, rho–theta position determination techniques must be used. The pilot must determine on which radial the aircraft is located (rho), then utilize distance measuring equipment (DME) to determine the aircraft's distance (theta) from the VOR transmitter. Rho–theta position determination requires specialized DME equipment both on the aircraft and at the VOR transmitter.

The DME system uses the principle of elapsed time measurement as the basis for distance measurement. The DME system consists of an **interrogator** located on board the aircraft and a **transponder** located at the ground station. At regularly spaced intervals, the interrogator transmits a coded pulse on a frequency of around 1,000 mHz (see Figure 2-26).

When the ground-based DME transponder receives this pulse, it triggers a coded reply that is transmitted on a different frequency. When the interrogator receives this pulse, the elapsed **range time** is electronically calculated. Range time is the interval of time between the transmission of an interrogation and the receipt of the reply to that interrogation. The approximate range time for a signal to travel 1 nautical mile and return is 12.36 microseconds. The DME equipment on board the aircraft measures the elapsed time between interrogator

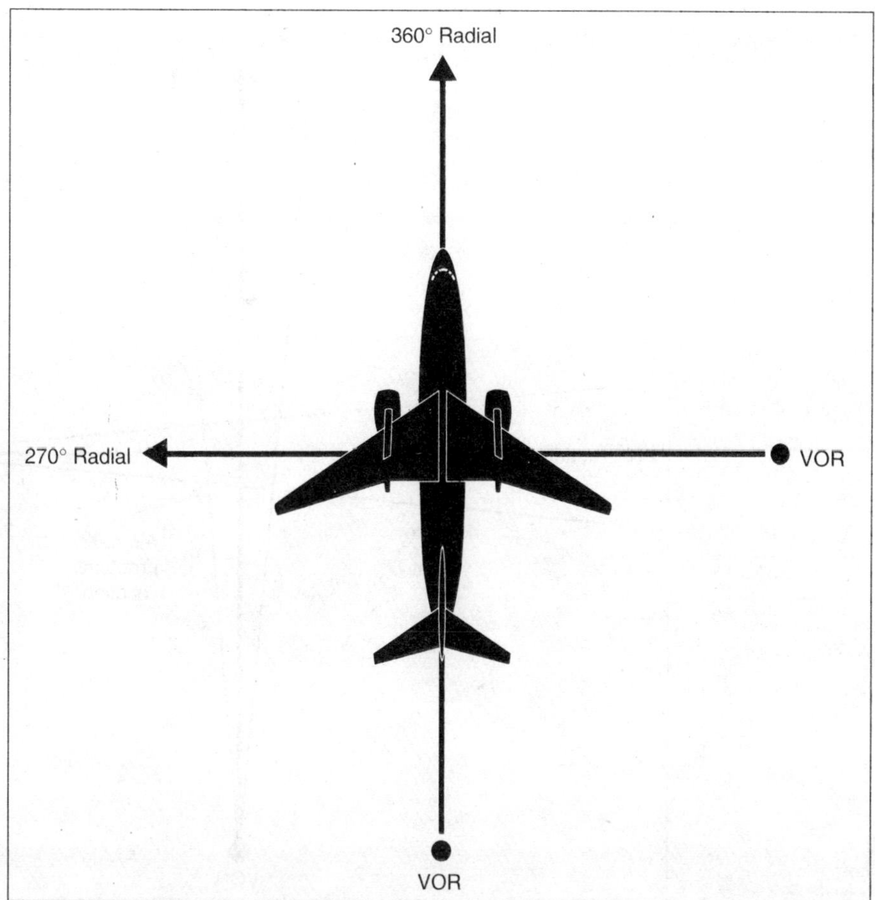

Figure 2-24. Plotting aircraft position using two VORs.

transmission and reception of that signal. This time is divided by 12.36 microseconds, providing the distance the aircraft is from the ground station. This determination is known as the line of sight or **slant range distance.**

Slant range is the actual distance between the aircraft and the ground-based DME transponder. As the aircraft's altitude increases, the difference between slant range and ground distance increases. For instance, if an aircraft is 5.0 ground miles from the DME station, at an altitude of 6,000 feet, the DME indicator on board the aircraft will indicate approximately 5.1 nautical miles from the station. But if the aircraft is directly over the DME station, at an altitude of 30,000 feet, the DME indicator will also indicate about 5.1 nautical miles (see Figure 2 27).

The difference between slant range and ground distance is most pronounced when aircraft are operating at high altitudes fairly close to the DME

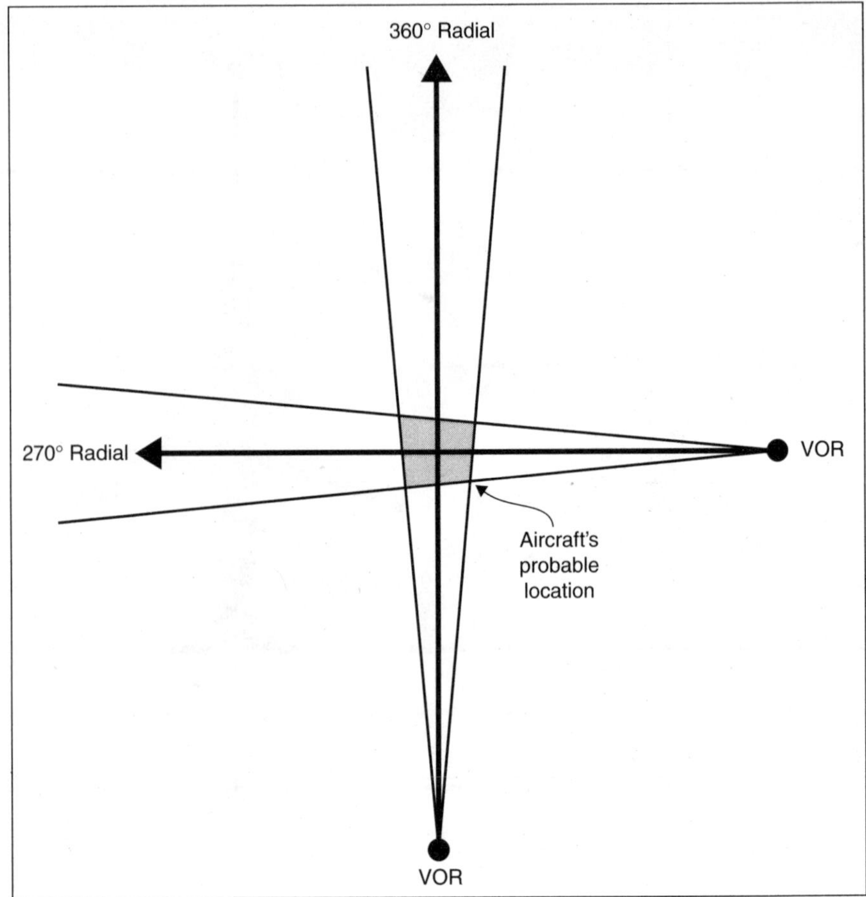

Figure 2-25. Actual location of an aircraft using two VORs for position determination.

ground station. This difference has been taken into consideration by the FAA when determining holding-pattern sizes, intersection locations, and airway positioning.

Tactical Air Navigation (TACAN)

The VOR-DME system has deficiencies that make it unusable for certain military operations. A conventional VOR transmitter is fairly large and needs an extensive clear zone around it to minimize reflections. In addition, since all of the DME interrogators on board aircraft transmit at the same frequency when interrogating a station, a DME ground station can become saturated from too many aircraft within its vicinity interrogating at the same time. If this happens, the interrogator signals may interfere with one another and cause inaccurate DME distances to be displayed in the cockpit.

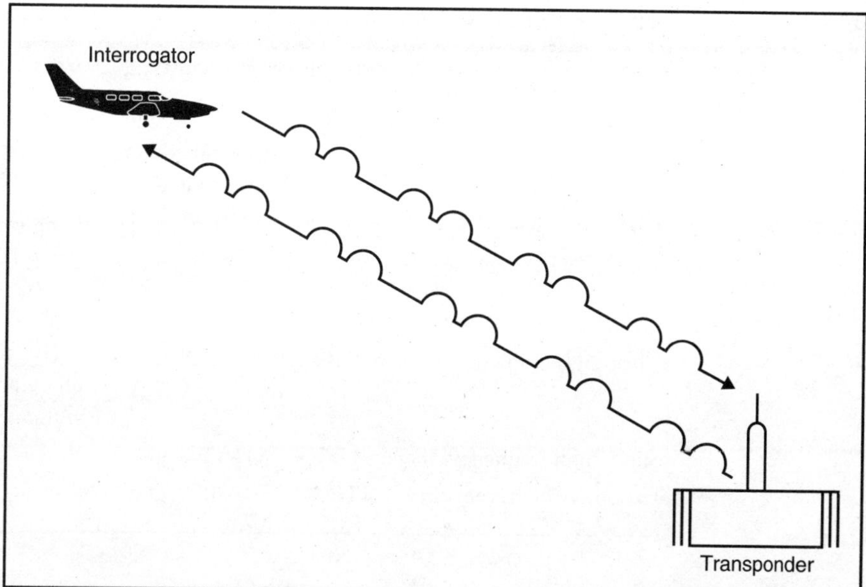

Figure 2-26. DME operation.

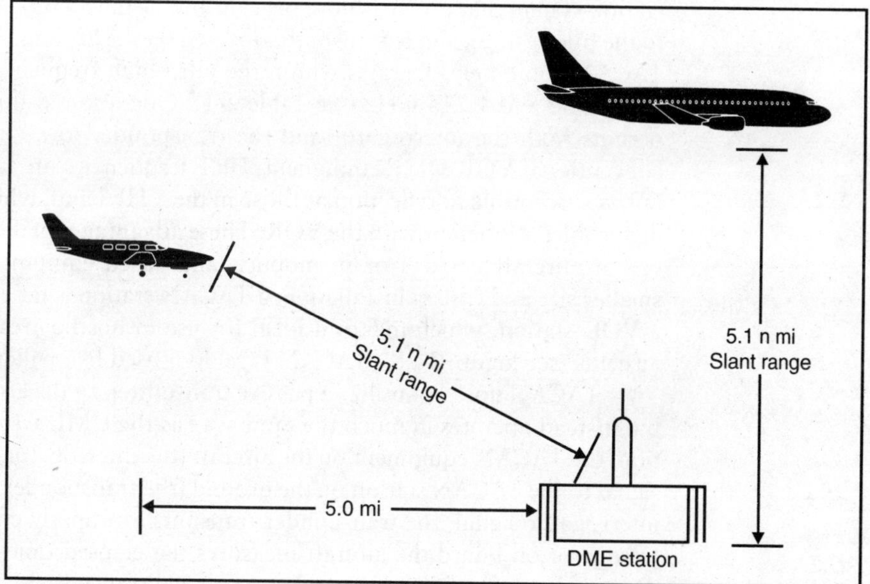

Figure 2-27. DME slant range measurement.

Table 2-1. Radio Frequency Allocation

Name	Abbreviation	Frequency	Uses
Very low frequency	VLF	3–30 kHz	Naval communication
Low frequency	LF	30–300 kHz	LORAN, NDB
Medium frequency	MF	300–3,000 kHz	NDB
High frequency	HF	3–30 mHz	Long-range communications
Very high frequency	VHF	30–300 mHz	VOR, localizers, marker beacons, civil communications
Ultra-high frequency	UHF	300–3,000 mHz	DME, TACAN, MLS, glide slope, military communications
Super high frequency	SHF	3–30 gHz	Radar
Extremely high frequency	EHF	30–300 gHz	

After an extensive evaluation of the civilian VOR-DME system, the Department of Defense chose to develop an alternative navigation system known as tactical air navigation (TACAN). TACAN is a polar coordinate–based navigation system that provides both bearing and distance (rho–theta) information to the pilot using a single transmitter located on the ground. This ground-based TACAN equipment operates within the **ultra-high frequency** (UHF) band between 960 and 1,215 mHz (see Table 2-1). Operation in this frequency range permits both the interrogator and the transponder to be much smaller than conventional VOR-DME equipment. UHF frequencies are line of sight but are not as susceptible to reflection as those in the VHF band, which reduces the siting problems inherent with the VOR. These advantages make TACAN ideal for use on aircraft carriers or in mobile, land-based equipment. Because of its smaller size and ease of installation, a TACAN station is far easier to move than a VOR station, which makes it ideal for use in hostile areas or at temporary airfields (see Figure 2-28). TACAN is seldom used by civilian aircraft.

TACAN does not utilize a passive transmitter on the ground like the VOR but instead operates in much the same way as the DME system. During operation, the TACAN equipment on the aircraft (the interrogator) transmits a coded signal to the TACAN station on the ground (the transponder). On receipt of the interrogator signal, the transponder transmits a properly coded reply. The interrogator on board the aircraft measures the elapsed time and calculates the distance between the aircraft and the TACAN transmitter. (This is done in the same manner as with civilian DME equipment.) The interrogator on board the aircraft also decodes the signal and determines the aircraft's azimuth from the TACAN ground station. The airborne equipment can then display both bearing and distance information to the pilot, using a display system similar to civilian VOR-DME indicators.

E-Systems, Montek Division

Figure 2-28. A mobile TACAN ground station.

VORTAC

While the military was developing TACAN, the CAA was developing and implementing the civilian VOR-DME system. Congress expressed concern over the increased expense of developing, operating, and maintaining two separate navigation systems when both would provide pilots with the same navigational information. The CAA recommended adoption of VOR-DME as the civil navigation standard, since system implementation had already begun and VOR-DME receivers were readily available at a lower cost than TACAN equipment. In addition, the CAA believed that the VOR-DME system was more flexible, since VOR and DME equipment could be purchased separately. The CAA preferred a system that would permit the pilot to purchase just VOR equipment; DME equipment could be installed in each aircraft at a later date if the pilot felt that the expense was justified. In addition, since the CAA had previously recommended that pilots install VOR equipment and many pilots had already made this expensive investment, the CAA felt that it would be unfair to require aircraft owners to remove their VOR equipment and install even more expensive TACAN receivers.

The Department of Defense, however, believed that TACAN was better suited to military operations because of its smaller size and portability. After years of negotiations, the CAA and the Department of Defense eventually agreed that civilian aircraft would be permitted to use ground-based TACAN transponders to provide distance information while still using VOR ground stations for azimuth information. Military aircraft, however, would be equipped solely with TACAN equipment and would be dependent on it for both azimuth and distance information.

E-Systems, Montek Division

Figure 2-29. A VORTAC ground station.

The military and the CAA agreed to place VORs and TACANs at the same locations utilizing common physical structures. This combined navigation aid would henceforth be known as **VORTAC.**

The VORTAC system was chosen by Congress to become the nation's new en route navigation standard, providing both distance and bearing information to military and civilian aircraft. TACAN frequencies would be paired with the appropriate VOR frequencies to simplify pilot operation. To use VORTAC, all that civilian pilots needed to do was select the appropriate VOR frequency, and the DME interrogator would automatically tune itself to the proper TACAN UHF frequency. Military pilots using TACAN were required to select an appropriate channel number, and their receiver automatically tuned itself to the proper frequency.

Most of the VORs across the United States were soon co-located with TACANs and became VORTACs (see Figure 2-29). In locations where the military had no need for TACAN but civilian aircraft still needed some form of navigation, a VOR with civilian DME equipment was installed. (These VOR-DME facilities cannot be used by military aircraft unless they are VOR equipped.) In addition, some locations justify installation of a VOR station but not a DME station. In this case a VOR is installed without associated DME or TACAN equipment. Such facilities can be used for azimuth information by aircraft equipped with VOR.

Area Navigation

To navigate the airways using the VORTAC system, pilots are required to fly from VORTAC to VORTAC until they reach the destination airport. Because of airport locations and VORTAC placement restrictions, it is seldom possible to navigate in a straight line from the departure to the destination airport. This navigation restriction forces pilots to fly a longer distance than necessary. It also creates congestion in the air traffic control system, since every aircraft operating under an IFR flight plan is forced to navigate along a limited number of airways. In an attempt to alleviate this congestion, a number of systems have been developed to permit pilots to bypass the airway system and navigate directly to the destination airport. These various systems are collectively referred to as **area navigation, or RNAV.**

Doppler Radar

One of the first area-navigation systems adopted for use was **Doppler radar.** The Doppler radar system is composed of a radar transmitter, a receiver, a signal processor, and display unit, all installed on board the aircraft. The Doppler system constantly transmits a radar signal straight down from the aircraft at a precise frequency. After the radar signal has reflected off the ground back to the receiver, the signal processor compares the frequency of the transmitted signal with that returned to the aircraft.

If the aircraft were not moving at all, no detectable change would be notable in the frequency of the transmitted radar signal. But when the aircraft is moving in any direction, either longitudinally or laterally, the radar frequency will change as it reflects off the earth's surface. This phenomenon is called the **Doppler effect,** and the change in frequency is known as a **frequency shift.** The signal processor on board the aircraft measures the frequency shift and uses this information to calculate the aircraft's ground speed and true course. This information is then displayed in a manner that permits pilots to navigate to their destination.

The Doppler radar system measures only the aircraft's relative motion over the earth's surface; it cannot actually determine an aircraft's location. For the system to operate correctly, the pilot must input the starting position of the aircraft into the doppler system before takeoff. Any error in this input will cause the system to inaccurately calculate the current position of the aircraft. This is the primary disadvantage of the Doppler radar system.

Since the Doppler system is self-contained within the aircraft, it operates without using any ground-based navigation stations (such as VORTAC or NDB) and can be used where navigational aids are sparse or nonexistent. This characteristic makes Doppler radar ideal over long stretches of desert or ocean. Doppler radar is no longer one of the most accurate RNAV systems, however, and is rapidly being replaced for primary navigation by the systems described next; if installed, it is usually utilized as a backup navigational system.

Course-Line Computers

The **course-line computer** (CLC) was developed to permit pilots to use existing VORTAC stations to fly directly from one airport to another. Using rho–theta navigation principles, the course-line computer can determine the aircraft's position utilizing any VORTAC or VOR-DME station. Upon receiving the azimuth and distance information from a VORTAC station, the CLC mathematically calculates the bearing and distance from the aircraft to any desired location and produces navigation instructions that lead the pilot to that point.

The CLC accomplishes this task by electronically creating a **phantom VORTAC** station (known as a **waypoint** at the desired destination and then providing bearing and distance information to that station using the aircraft's VOR and DME indicators (see Figure 2-30). During flight, the pilot selects an appropriate VORTAC station and electronically "moves" it to the desired location. The CLC then constantly obtains position information from the VORTAC, calculates the bearing and distance to the waypoint, and displays the course guidance information to the pilot using the CDI (course deviation indicator) on the aircraft. Distance to the waypoint is constantly displayed on the DME indicator.

The primary limitation to CLC-based area navigation is that the waypoint must be located within the service volume of an actual VORTAC station. If the aircraft is not in a position to receive an accurate navigation signal from an existing VORTAC, the CLC cannot determine the aircraft's present location or compute the bearing and distance to the waypoint. This limitation forces the pilot to electronically create a sufficient number of waypoints along the planned route of flight to permit a straight course to be flown.

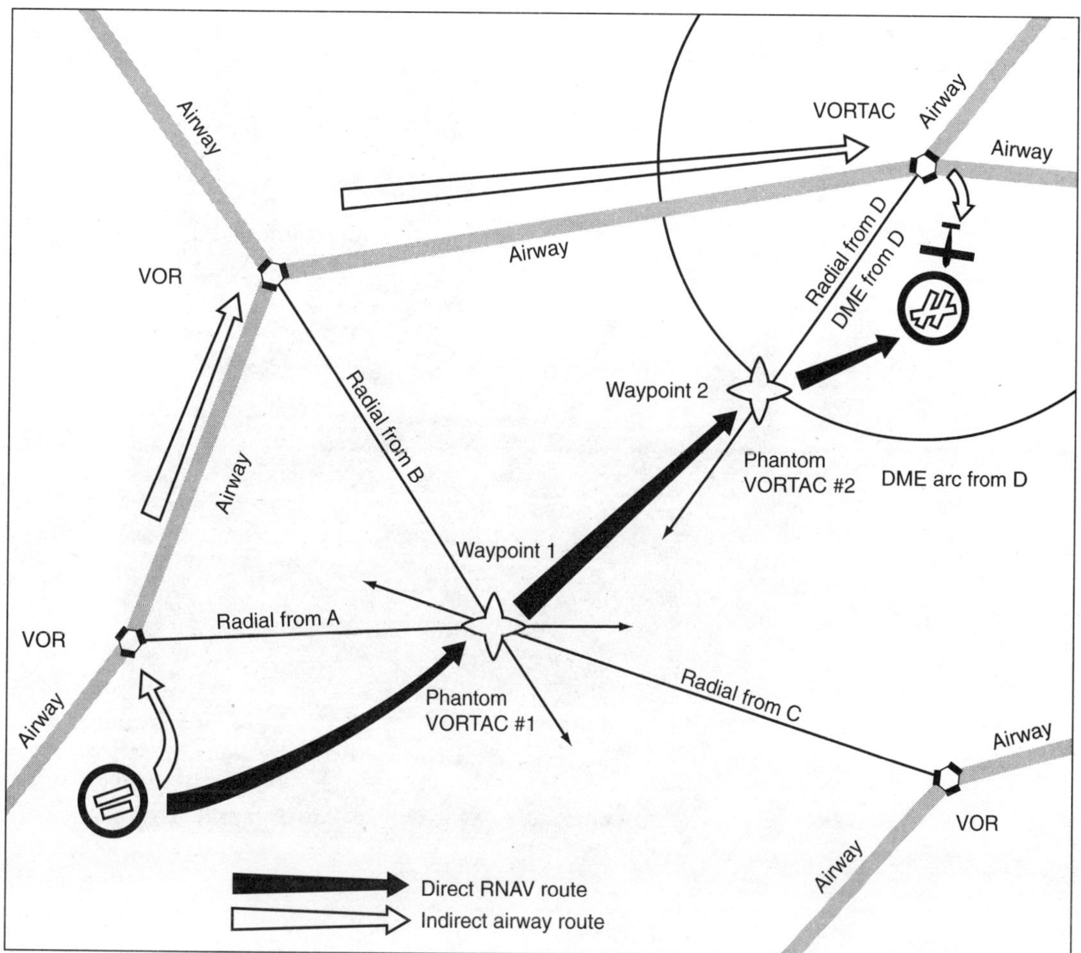

Figure 2-30. CLC(course-line computer)-based RNAV.

During the entire flight, the aircraft must be within reception distance of one of the selected VORTACs, and the pilot must locate waypoints within each VORTAC's service volume. If the aircraft strays outside the service volume, the CLC will be unable to receive sufficient information to provide course guidance to the waypoint. This reduces the CLC's effectiveness over sparsely settled terrain. CLC-based RNAV can be used over most of the continental United States, however.

CLC-based RNAV can also be used as a navigational aid when approaching airports. Upon arriving within the vicinity of the destination airport, the pilot can electronically move a VORTAC and place it at the center of the destination airport, simplifying instrument approach procedures. VFR pilots

Bendix/King Division of Allied Signal Aerospace Company

Figure 2-31. A typical light aircraft RNAV system.

can use CLCs to assist in navigating to airports that are not served by VORs or NDBs. The course-line computer is one of the most common area-navigation systems in use today and is normally simply called RNAV (see Figure 2-31).

LORAN

The **long-range navigation (LORAN)** system was initially developed as a maritime navigation system. Since LORAN stations provide coverage primarily over the Atlantic and Pacific Oceans, where aviation navigation aids are virtually nonexistent, LORAN was eventually adapted for aviation use. LORAN differs from most aviation navigation systems in that it is a **hyperbolic navigation system,** rather than a rho–theta navigation system such as VORTAC. When using LORAN, the pilot plots multiple hyperbolic lines of position to determine the aircraft's position.

The **LORAN-A** system consists of a **master station** and a **slave station** installed about 500 nautical miles apart. At precise intervals, the master station transmits a coded pulse in the 1700–2000 kHz band. When the slave station receives this pulse, it transmits another coded pulse on the same frequency. The LORAN receiver on board the aircraft measures the time delay and displays this information on an indicator. The pilot can then use this time-delay information to plot a line of position on which the aircraft is located. After plotting the first line of position (LOP), the pilot repeats this procedure using a second pair of stations. The second LOP will intersect with the first one, defining the aircraft's exact location.

LORAN-A was never designed to be used by high-speed aircraft. Since a significant amount of time can elapse between the plotting of the first and the second LOPs, there were always inherent inaccuracies whenever an aircraft's position was being determined using LORAN-A.

LORAN-C

LORAN-A was operated by the U.S. Coast Guard and was decommissioned in the 1990s. LORAN-B was a replacement system that was developed but never made operational. LORAN-D is a short-range military version used for pinpoint navigation. **LORAN-C** is the current civilian version of LORAN and was again designed to be used primarily for maritime navigation. LORAN-C operates on the same general principles as LORAN-A but uses a computer to quickly and accurately plot multiple lines of position (see Figure 2-32). Since

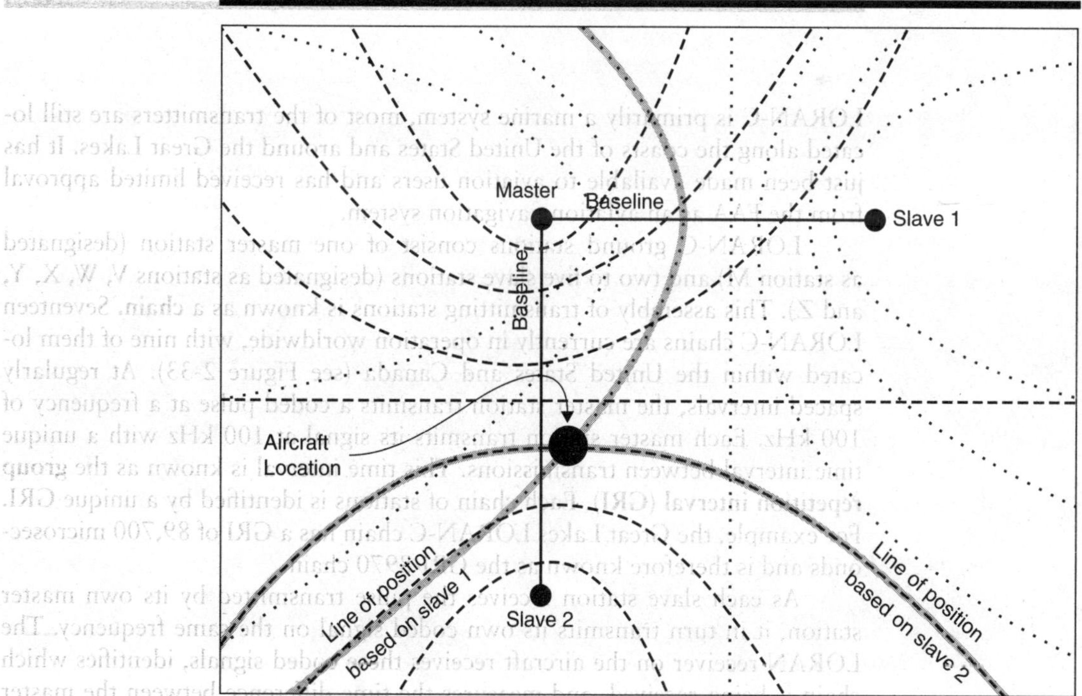

Figure 2-32. LORAN-C line-of-position plotting.

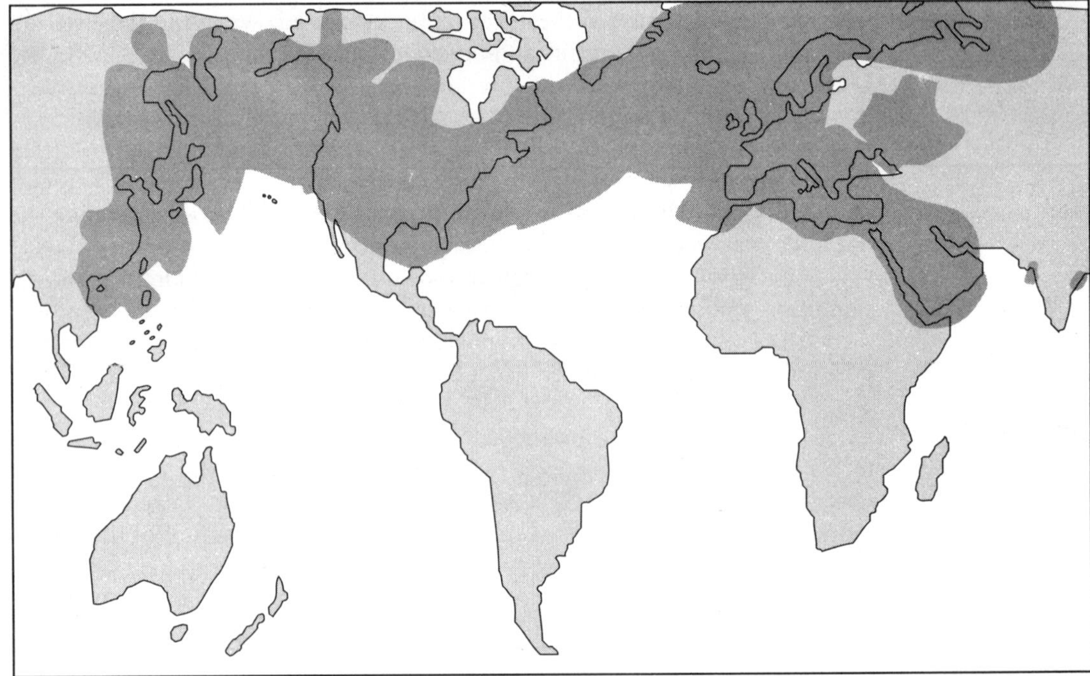

Figure 2-33. Darker areas indicate LORAN-C worldwide coverage.

LORAN-C is primarily a marine system, most of the transmitters are still located along the coasts of the United States and around the Great Lakes. It has just been made available to aviation users and has received limited approval from the FAA as an aviation navigation system.

LORAN-C ground stations consist of one master station (designated as station M) and two to five slave stations (designated as stations V, W, X, Y, and Z). This assembly of transmitting stations is known as a **chain.** Seventeen LORAN-C chains are currently in operation worldwide, with nine of them located within the United States and Canada (see Figure 2-33). At regularly spaced intervals, the master station transmits a coded pulse at a frequency of 100 kHz. Each master station transmits its signal at 100 kHz with a unique time interval between transmissions. This time interval is known as the **group repetition interval (GRI).** Each chain of stations is identified by a unique GRI. For example, the Great Lakes LORAN-C chain has a GRI of 89,700 microseconds and is therefore known as the GRI-8970 chain.

As each slave station receives the pulse transmitted by its own master station, it in turn transmits its own coded signal on the same frequency. The LORAN receiver on the aircraft receives these coded signals, identifies which chain is being received, and measures the time difference between the master and each of the slave-station transmissions. The computer in the receiver uses these time differences to plot multiple lines of position. The LORAN receiver can then plot up to five LOPs from each chain of stations. The LORAN receiver on board the aircraft (Figure 2-34) then electronically determines the intersection of these LOPs and displays the aircraft's position to the pilots as latitude-longitude coordinates or as a bearing and distance from any preselected location.

Since all of the LORAN ground stations operate at the same frequency (100 kHz), the airborne receiver can use the transmissions from other LORAN chains to confirm its initial position determination. As the aircraft continues along its flight, the LORAN receiver constantly calculates the aircraft's new position and uses this information to compute the aircraft's course and ground speed. Using this information, the pilot can program the LORAN-C receiver to guide the aircraft to the desired destination.

The LORAN-C receiver displays course guidance and distance information in a number of different formats, all of which provide the same essential information to the pilot. This information includes ground speed, ground track, course to be flown, distance to the destination airport, and estimated time of arrival.

LORAN-C is a fairly accurate navigation system but has a number of important limitations. The radio frequencies used by LORAN are in the low-frequency (LF) band and are not line of sight, which makes it possible for an aircraft to receive the LORAN signal at a distance of up to 1,500 miles from the transmitter. This is usually beneficial but can sometimes prove to be a disad-

II Morrow, Inc.

Figure 2-34. A typical LORAN-C receiver.

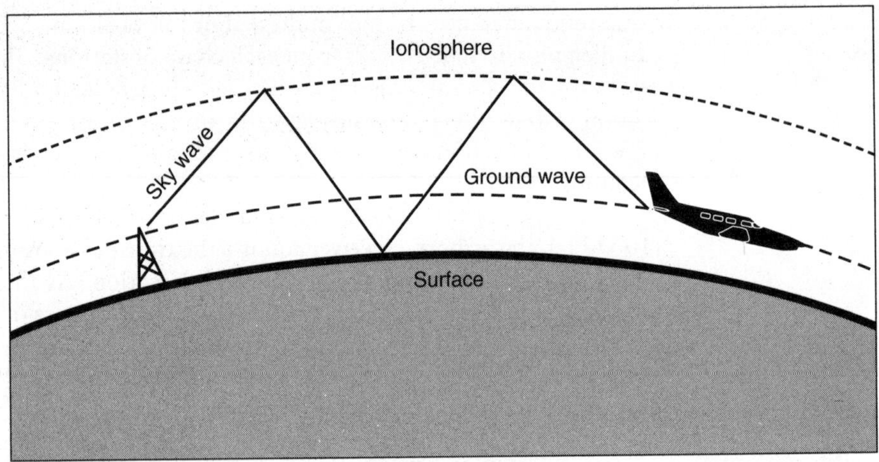

Figure 2-35. LORAN-C ground wave versus sky wave.

vantage. During certain atmospheric conditions (usually at twilight), an aircraft might receive two or more distinct signals from each master and slave station. The first signal is the **ground wave,** which is the signal the LORAN receiver is designed to utilize. Under certain conditions, the LORAN transmission may also travel into space and reflect off the ionosphere and return to the aircraft. This secondary signal takes longer to reach the aircraft and can confuse the LORAN receiver, since it now receives two pulses from every transmitter (see Figure 2-35). This condition makes it impossible for the LORAN receiver to accurately determine time delays or plot lines of position. Under these circumstances the receiver is designed to ignore the transmissions from the affected chain and must be switched to an alternate chain of stations.

Global Navigation Satellite System

The newest and most accurate form of en route navigation has been accepted by the United States and ICAO as the navigation system for the twenty-first century. This space-based system, using satellites owned and operated by the U.S. government and possibly Russian satellites, will be known as the **Global Navigation Satellite System (GNSS).**

The GNSS — known in the United States as the Global Positioning Satellite (GPS) system, and in Russia as GLONASS — will provide worldwide navigation service using a system of at least 21 operational satellites in orbit. Aircraft equipped with relatively low-cost receivers will be able to pinpoint their location and their altitude using GNSS.

GNSS operates similarly to LORAN and other hyperbolic navigation systems. The only difference is that the transmitters are located on orbiting satellites instead of fixed ground locations. GNSS also operates at much higher frequencies, which permits lateral positional accuracy of plus or minus 100 meters anywhere in the world.

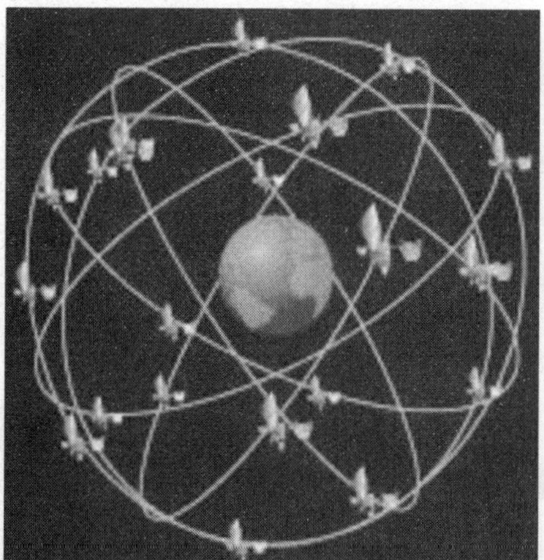

Figure 2-36. Global Positioning Satellite system.

Global Positioning Satellite

GPS is a space-based positioning, velocity, and time system composed of 24 satellites (21 operational plus 3 spares) in six orbital planes. The satellites operate in circular orbits arranged so that a minimum of five satellites are in view at any one time to users worldwide (see Figure 2-36).

GPS operations is based on the concept of ranging and triangulation from a group of satellites in space that act as precise reference points. A GPS receiver measures distance from the satellite using the travel time of a radio signal. Each satellite transmits a specific code, called a course/acquisition (CA) code, that contains information on the satellite's position, the GPS system time, its clock error, and the health and accuracy of the transmitted data. GPS satellites have very accurate atomic clocks to calculate signal travel time. The GPS receiver matches each satellite's CA code with an identical code contained in the receiver's database. By shifting its copy of the satellite's code in a matching process, and by comparing this shift with its internal clock, the receiver can calculate how long it took the signal to travel from the satellite to the receiver. The distance derived from this method is called a pseudo-range, because it is not a direct measurement of distance but a measurement based on time. Pseudo-range is subject to several error sources, such as an ionospheric delay or time disparities between the atomic clocks in the satellites and the GPS receiver, which the receiver can correct.

In addition to knowing the distance to a satellite, a receiver needs to know the satellite's exact position in space; this is known as its ephemeris. Each satellite transmits ephemeris information about its exact orbital location. The GPS receiver uses this information to precisely establish the position of the satellite.

Using the calculated pseudo-range and the position information supplied by the satellite, the GPS receiver mathematically determines its position by triangulation (see Figure 2-37). The GPS receiver needs at least three satellites with timing corrections from a fourth satellite to yield an unaided, unique, and

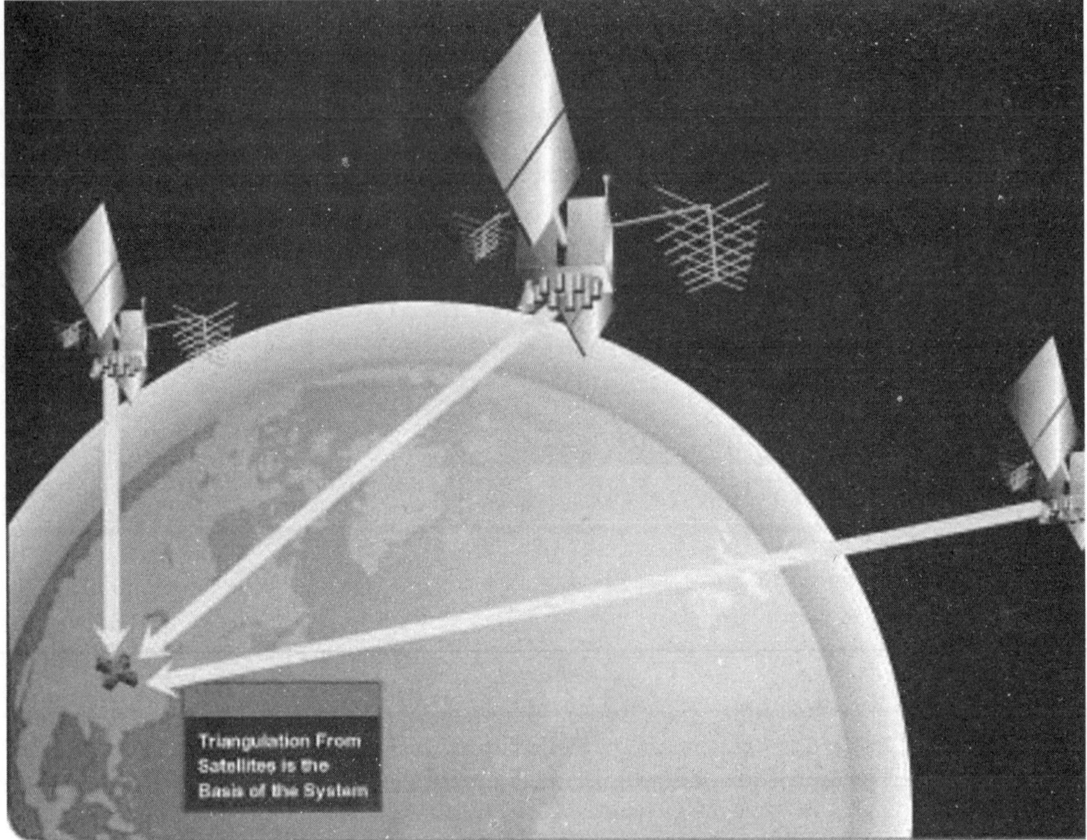

Figure 2-37. Satellite triangulation as used by the GPS system.

true three-dimensional position (latitude, longitude, and altitude). The GPS receiver computes navigational values such as distance and bearing to a waypoint, ground speed, estimated time en route, estimated time of arrival, and winds aloft by using the aircraft's known latitude/longitude, measuring relative movement, and referencing these to a database built into the receiver. The receiver uses data from the best four satellites, automatically adding signals from a new one as it drops signals from another, to continually calculate its position.

The GPS receiver (see Figure 2-38) verifies the integrity of the signals received from the GPS constellation through receiver autonomous integrity monitoring (RAIM). RAIM is an independent means for the receiver to determine whether a satellite is providing corrupted information. At least one satellite, in addition to those required for navigation, must be in view for the receiver to perform the RAIM function; therefore, for RAIM to work correctly, five satellites must be in view of the receiver.

Garmin, Inc.

Figure 2-38. Example of a high-end, panel-mounted GPS unit.

Due to its accuracy and worldwide availability, GNSS has been designated by ICAO as the future navigation system that will meet all civil aviation needs, including departure terminal, oceanic, en route, nonprecision approach, precision approach, and surface navigation. The transition from Earth-based to satellite systems has begun and should be complete sometime around the year 2010.

Approved GPS Uses GNSS has already been approved for some use in the National Airspace System. GPS is currently approved as the primary means of navigation over ocean areas by all properly equipped aircraft and has also been approved by the FAA as a supplemental navigation aid for en route flight and nonprecision approaches.

The GPS system does not provide sufficient accuracy, availability, and integrity to be used as a primary means of navigation for all aviation operations, however. Even when PPS becomes available to civilian users, some form of augmentation of the basic GPS system will still be needed for GPS to become the primary means for providing instrument approaches. GPS augmentation is generically known as **differential GPS (DGPS)**. The feasibility of DGPS is well proven and is already providing accurate GPS signals for many nonaviation applications, such as agriculture, trucking, and rail car tracking. The primary challenge facing aviation users is developing an ultra-reliable system that provides accurate positioning information to rapidly moving aircraft.

Two types of augmentation technologies are currently being developed by the FAA: **wide area augmentation system (WAAS)** and **local area augmentation system (LAAS)**. Wide area augmentation is currently under development and is a geographically expansive augmentation to the basic GPS service. WAAS can improve the accuracy, integrity, and availability of the basic GPS signals over an area as large as the continental United States. When fully implemented, WAAS will allow GPS to be used as a primary means of navigation for en route travel and nonprecision approaches in the United States as well as for making Category I precision approaches. The wide area of coverage for this system includes the entire United States and some outlying areas. Other countries have plans for augmenting the GNSS signal over their respective territories.

WAAS is based on a network of approximately 35 ground reference stations (see Figure 2-39), which receive signals from GPS satellites. The location

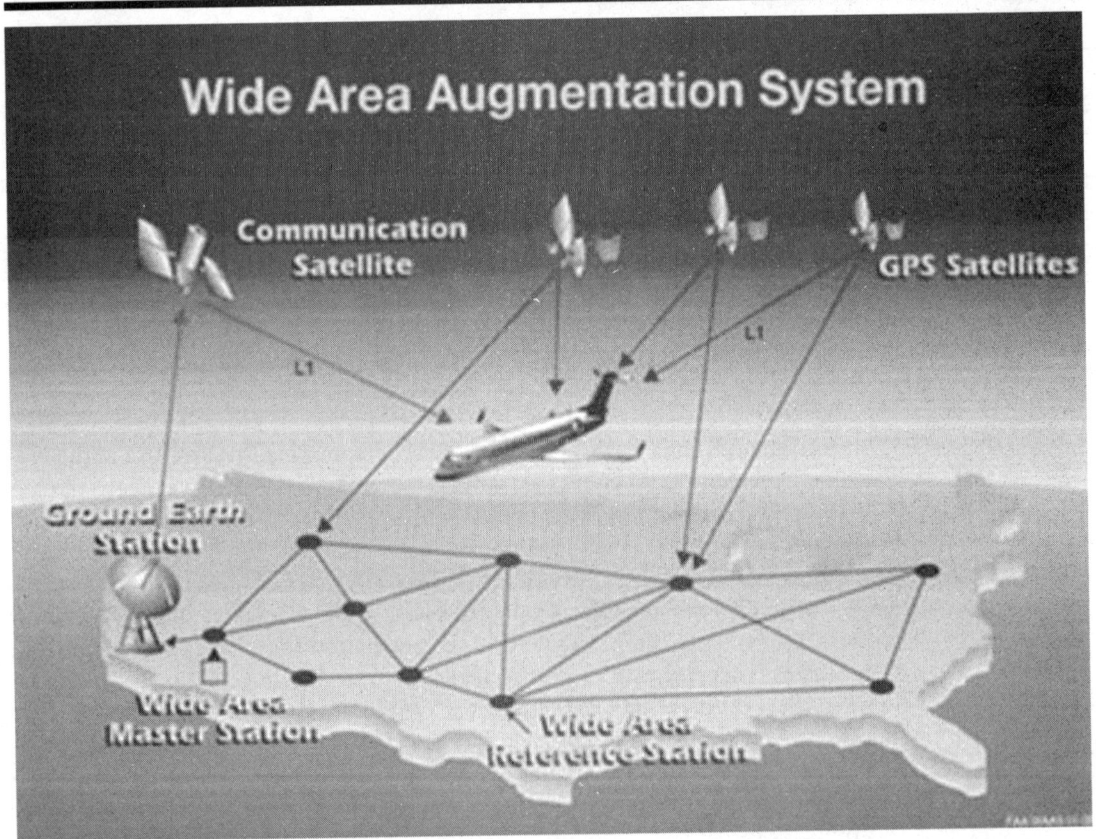

Figure 2-39. Wide area augmentation system.

of each of these reference stations is precisely surveyed, enabling each to determine any error in the GPS signal being received at its own location. This information is then passed to a master control station, which calculates correction algorithms and assesses the integrity of the system. These data are then put into a message format and sent for uplink to a geostationary communications satellite in orbit above the United States. The corrective information calculated by the master control station is forwarded by this satellite to GPS receivers onboard the aircraft, which then make the needed adjustments.

Local area augmentation system (LAAS) is intended to complement WAAS to support Category II and III precision approach applications. LAAS will also provide Category I capability at locations where WAAS cannot, as well as provide a signal that could be used for surface navigation in and around the airport area. LAAS uses a technique of determining a single, highly accurate correction that accounts for all expected common errors between a local reference and users. LAAS will operate much like WAAS but will broadcast navigation information directly from a ground station to aircraft located within a 30-nautical-mile area centered on the LAAS station. This service area would typically encompass a specific airport or airports within close proximity. Coupled with the existing aircraft receivers and display equipment, LAAS will be able to provide ILS look-alike functions and will permit GPS to emulate existing airborne-based ILS display and autopilot systems.

Although the service volume of LAAS is much smaller than WAAS, the local area nature of LAAS will provide greater accuracy than WAAS. LAAS should, therefore, be able to provide precision approach quality signals beyond the capability of WAAS. LAAS should theoretically meet Category I, II, and III instrument approach minima for all eligible runways in the local service area. Additionally, the LAAS signal should provide new services, such as airport surface navigation and collision avoidance during periods of low visibility. LAAS will operate independently of WAAS while being fully compatible with it. This will enable LAAS to provide a satellite-based, independent backup to WAAS within the United States.

GPS Implementation Plans IFR operations in oceanic areas can be currently conducted utilizing GPS if appropriate aircraft avionics are installed. GPS domestic en route and terminal IFR operations are also permitted as soon as the approved GPS receivers are installed on the aircraft. In addition, the avionics and ground-based transmitters necessary to receive all the facilities (such as VORTAC) appropriate for the route to the destination airport and any required alternate airport must also be installed and operational.

The FAA has developed the GPS approach overlay program, which is a three-phase plan to permit pilots to use GPS for existing instrument approach. This plan will eventually transition the U.S. terminal navigation system from a ground-based to a space-based GPS system. IFR pilots may use GPS in accordance with any phase of the overlay program as soon as the proper avionics systems are installed and the appropriate requirements are met. This general approval by the FAA to use GPS to fly instrument approaches is limited to U.S.

airspace, however. The use of GPS in any other airspace must be authorized by the appropriate aviation/air traffic control authority.

Inertial Navigation System

The **inertial navigation system** (INS) is similar to Doppler radar systems in that it precisely measures any change in an aircraft's direction of flight and uses this information to determine position, ground speed, and the course to be flown to the destination airport. An INS contains accelerometers that can measure the slightest change in an aircraft's speed or direction of flight. At the beginning of each flight, the pilot is required to program the aircraft's exact location into the INS computer (see Figure 2-40). Using the information obtained from the accelerometers, the INS computer on board the aircraft determines the aircraft's speed and direction of flight. Using this information, the INS can calculate the course to be flown and the estimated time of arrival. This information is then displayed to the pilot or directed to the aircraft's autopilot.

When used correctly, the INS is highly accurate. INS information may be accurate to within ±.25 mile after a transoceanic flight in excess of 14 hours. Since the INS is independent of ground-based radio navigation stations, it can be used by aircraft anywhere around the world. But as with Doppler radar, pilots must be careful to correctly enter the aircraft's initial starting position into the navigation computer prior to depature. Since every subsequent position de-

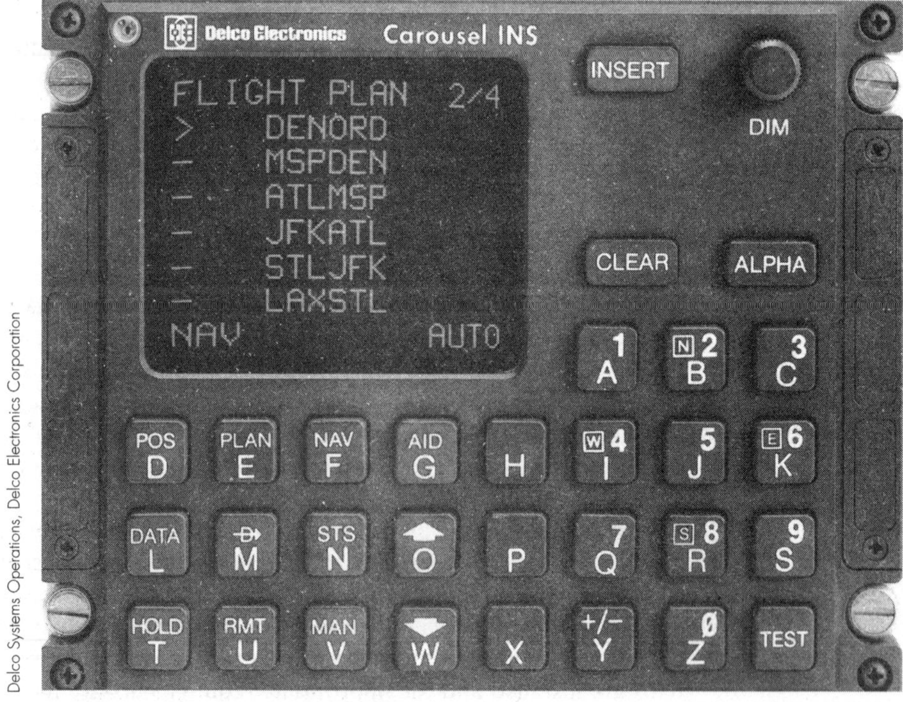

Delco Systems Operations, Delco Electronics Corporation

Figure 2-40. An inertial navigation system.

termination will be made based on this initial programming, any input errors will render all subsequent navigation information invalid.

In an attempt to reduce the risk of pilots' erroneously programming the INS computer, most manufacturers have designed their inertial navigation systems to interconnect with other navigation systems on board the aircraft (such as LORAN or VORTAC). Using the information available from these systems, the INS can continuously examine its own calculations and determine their validity. If a gross discrepancy is noted by the INS, the pilot will be alerted.

The INS is certified by the FAA as a primary means of en route navigation. The INS is fairly expensive and is normally found only on large, expensive commercial aircraft or business jets.

Instrument Approach Procedures

The navigation systems discussed to this point are those primarily utilized for en route navigation between airports. If, upon arrival at the destination airport, the pilot can see the airport and safely perform an approach to the runway and land, the pilot may use either a **visual approach** or a **contact approach.** The specific differences between these two approach procedures will be covered in Chapter 6. In general, a visual approach can be conducted if the visibility is greater than 3 miles. Visual approaches can be initiated by either the pilot or the controller. Contact approaches may be conducted whenever the visibility is greater than 1 mile. Only the pilot can initiate this type of approach. In this chapter, both types of approaches will be generically referred to as visual approaches.

During a visual approach, the pilot accepts the responsibility for navigating to the airport and avoiding any obstacles within the local area. When visual approaches are being conducted, air traffic controllers are still responsible for separating aircraft that are using them from aircraft operating on IFR flight plans.

If the weather conditions at the destination airport are such that the pilot is unable to, or chooses not to, conduct a visual approach, he or she must conduct an **instrument approach procedure (IAP).** During the conduct of an instrument approach, the pilot must follow a specified procedure that provides course guidance and obstacle clearance. This procedure guides the pilot to the destination airport where he or she can then make a safe landing.

Instrument approach procedures are designed and published by the U.S. government and are made available to pilots and private corporations. When requested by a sponsoring agency, specially trained FAA personnel accurately determine the routes and altitudes that aircraft will fly when approaching the airport under marginal weather conditions. These specialists use the procedures contained in the **terminal instrument approach procedures (TERPS)** manual published by the FAA. The TERPS manual specifies the criteria that must be met before the FAA can certify an instrument approach procedure. TERPS specialists ensure that pilots complying with a published instrument approach procedure will avoid every obstacle in the vicinity of the approach path and will still be able to safely land at the completion of the approach.

When the FAA specialists are finished designing an instrument approach procedure, specially trained FAA pilots conduct **flight checks** in specially instrumented aircraft to ensure that the approach procedure actually meets TERPS criteria. After this flight check, the FAA publishes the instrument approach and permits pilots to utilize these procedures (see Figure 2-41). These instrument approach procedures are actually considered Federal Aviation Regulations (FARs), FAR 91 mandates that pilots comply with these procedures when conducting an instrument approach.

The National Ocean Service (NOS) and Jeppessen Incorporated (a privately owned company) use the TERPS information to publish **instrument approach procedure charts** (sometimes called **approach plates**) that graphically depict the transition from the airway structure to the actual instrument approach procedure. Each publisher uses the same information when designing its charts but presents this information differently. NOS charts are primarily used by the FAA, Department of Defense, and general aviation pilots. Jeppessen (or JEPP) charts are primarily used by airline, corporate, and some general aviation pilots.

Segments of an Instrument Approach Procedure

An instrument approach procedure essentially consists of four components: the initial approach, intermediate approach, final approach, and missed approach segments. A detailed description of each segment is provided in the TERPS manual, available from the U.S. Government Printing Office.

Initial Approach Segment The **initial approach segment** is designed to transition the aircraft from the en route airway structure to the intermediate approach segment. The initial approach segment begins at one of the **initial approach fixes (IAFs)** located along the federal airways. This segment is usually defined as a heading or a radial to fly from the IAF to the intermediate approach segment. The initial approach segment specifies the minimum allowable altitude that may be flown along that route. There is usually one initial approach segment for every airway that pilots might be using as they approach the airport. The initial approach segment terminates when it joins the intermediate approach segment.

Intermediate Approach Segment The **intermediate approach segment** is designed to permit the pilot to descend to an intermediate altitude and align the aircraft in order to make an easy transition to the final approach segment. The intermediate approach segment terminates at the **final approach fix (FAF)**, which is designated on the approach chart with a maltese cross for nonprecision approaches and a lightning bolt for precision approaches. There is usually only one intermediate approach segment for every approach. It is not ordinarily identified as such on an approach chart. The intermediate segment may simply consist of a course to fly that leads to the final approach fix or it may be part of a procedure turn.

DEPARTMENT OF TRANSPORTATION - FEDERAL AVIATION ADMINISTRATION

NDB **STANDARD INSTRUMENT APPROACH PROCEDURE**

FLIGHT STANDARDS SERVICE - FAR PART 97. 27

Bearings, headings, courses and radials are magnetic. Elevations and altitudes are in feet, MSL, except HAT, HAA, and RA. Altitudes are minimum altitudes unless otherwise indicated. Ceilings are in feet above airport elevation. Distances are in nautical miles unless otherwise indicated, except visibilities which are in statute miles or in feet RVR.

TERMINAL ROUTES

FROM	TO	COURSE AND DISTANCE	ALT
BVT VORTAC	LA LOM	173/7.8	2300
OKK VORTAC	LA LOM	263/45.9	2600
OCKEL INT	LA LOM	301/17.6	2300
POTES INT	LA LOM	024/6.7	2300
VAGES INT (IAF)	LA LOM (NoPT)	150/1.5 (Hdg) & 100/7.8 (LA Brg 280)	2300
STAKS INT (IAF)	SLANT INT (NoPT)	053/6.5 (DNV R-053)	2300
SLANT INT	LA LOM	100/9.8 (LA Brg 280)	2300

MISSED APPROACH

MAP: 4.7 miles after LA LOM.
Climbing left turn to 2300 direct BVT VORTAC and hold; or when directed by ATC, climbing right turn to 2300 direct LA LOM.

ADDITIONAL FLIGHT DATA

Hold NW, RT, 143 inbound.
FAS Obst: 830 Lighted Tower 402637/865947.

1. PT R SIDE OF CRS 280 OUTBND 2300 FT WITHIN 10 MI. OF LA LOM (IAF)
2.
3. FAC 100 FAF LA LOM DIST FAF TO: MAP 4.7 THLD 4.7
4. MIN ALT LA LOM 2300
5. MSA FROM LA LOM 2600

VAR 1W YR 85

MINIMUMS

TAKEOFF: [ ] STD [X] SEE FAA FORM 8260-15 FOR THIS AIRPORT ALTERNATE: [X] NA

CAT. →	A			B			C			D			E		
	MDA	VIS	HAT/HAA	MDA	VIS	HAT/HAA	MDA	VIS	HAT/HAA	MDA	VIS	HAT/HAA	MDA	VIS	HAT/HAA
S-10	1140	3/4	539	1140	3/4	539	1140	1	539	1140	1½	539			
Circling	1160	1	554	1160	1	554	1160	1½	554	1260	2	654			

NOTES

CITY AND STATE	ELEVATION AIRPORT 606	TDZE 601	FACILITY IDENT.	PROC. NO.-AMDT. NO.-EFFECTIVE DATE	SUP. / AMDT. 10 / DATED 28 AUG 86
Lafayette, IN	Purdue University		LA	14 JAN 88 NDB Rwy 10, Amdt 11	

FAA Form 8260-5 (3-76) SUPERSEDES PREVIOUS EDITION

Figure 2-41. FAA form 8260: written description of a standard instrument approach procedure (NDB runway 10 approach at Lafayette, Indiana).

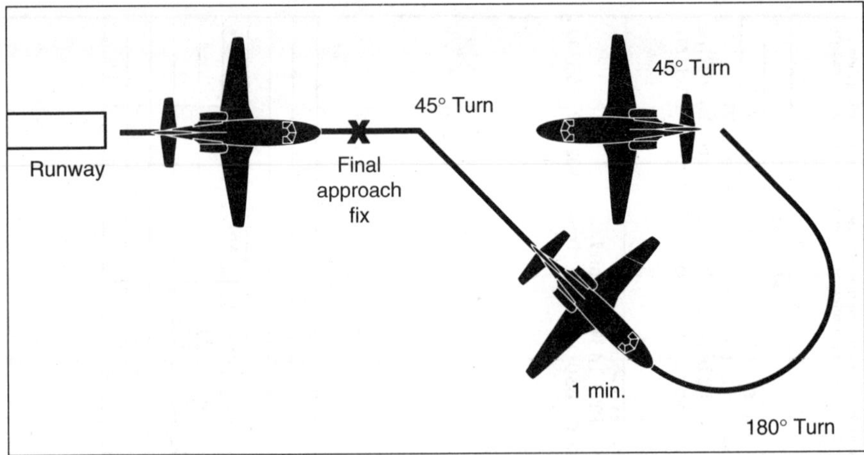

Figure 2-42. Procedure turn.

Procedure turns are necessary whenever the heading of the initial approach segment is nearly opposite that of the intermediate segment. A **procedure turn** is a maneuver performed in a designated area of airspace where the pilot turns the aircraft around and tracks inbound on the intermediate approach segment (see Figure 2-42). Typically, the airspace reserved for a procedure turn includes all of the airspace on one side of the approach course within a distance of about 10 nautical miles from the final approach fix. The pilot is authorized to use all of this airspace when reversing course from an initial to the intermediate approach segment.

Final Approach Segment The **final approach segment** is used to navigate the aircraft to the runway and properly position it to permit a safe landing. This segment begins at the final approach fix and ends at the **missed approach point (MAP).** The final approach segment guides the aircraft to the desired runway utilizing a navigation aid located either at the airport or nearby. The navigation aid can be one of two general types: precision or nonprecision. A precision approach aid provides the pilot with both lateral and vertical course guidance to the approach end of the runway. A **nonprecision approach** aid provides only lateral guidance to the pilot.

Nonprecision Approach During a nonprecision approach, upon reaching the final approach fix the pilot descends to a predetermined **minimum descent altitude (MDA)** published on the instrument approach chart (see Figure 2-43). The pilot maintains this altitude while tracking along the final approach segment toward the missed approach point. If the runway or runway environment is sighted prior to reaching the MAP and the pilot feels that a safe landing can be made, he or she is legally authorized to continue the approach and land. If the

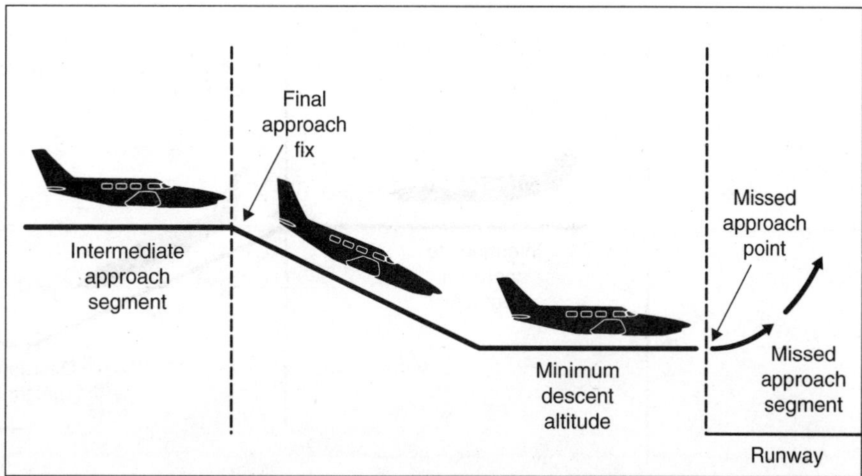

Figure 2-43. Nonprecision approach procedure using a minimum descent altitude.

runway is not in sight prior to reaching the MAP, or if a safe landing cannot be accomplished, the pilot must transition to the **missed approach segment,** which usually leads back to an initial approach fix. This is called **missed approach procedure.**

Precision Approach During the conduct of a precision approach, the pilot descends while tracking along the final approach segment. The precision approach aid provides an electronic descent path for the pilot known as a **glide path.** When the designated altitude (known as the **decision height,** or **DH**) has been reached, the pilot must determine whether a safe landing can be made (see Figure 2-44). If, in the pilot's opinion, it is safe to land, he or she is legally authorized to continue the descent and land. But if the pilot determines that it is not safe to continue, a transition to the missed approach segment must be made and the missed approach procedure must be conducted.

Because of the accuracy of precision approach aids, the pilot is usually authorized to descend to a lower altitude before making a decision about landing. This makes a precision approach much more valuable to the pilot during periods of marginal weather. Since precision approach aids are usually more expensive to purchase, install, and operate than nonprecision aids, they are normally reserved for use at airports that experience a significant amount of marginal weather conditions.

Terminal Arrival Area Criteria Most existing VOR, NDB, or instrument landing system (ILS) approach procedures require that the aircraft transition from the en route airway structure to the instrument approach procedure using specified ground tracks defined by ground-based navigation aids. The advent of

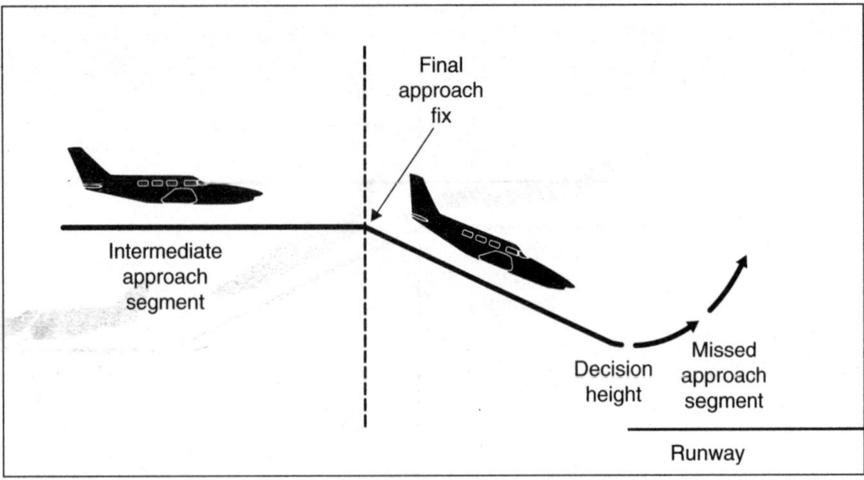

Figure 2-44. Precision approach procedure using a decision height.

GNSS, coupled with the concept of free flight, means that many aircraft will no longer fly these routes. Requiring a transition from free flight to a fixed route structure as the aircraft nears the airport will reimpose a traffic constraint, nullifying any operational advantage offered by GNSS/free flight.

In response to the introduction of GPS navigation systems in the United States, the FAA has begun to establish new, standardized instrument approach configurations for use at each airport. These new criteria are defined in FAA Order 8260, entitled Terminal Arrival Area (TAA) Design Criteria. These standardized approach procedures will be established for all new GPS approach procedures. Most likely, any remaining non-GPS instrument approach procedure retained by the FAA will be converted similarly.

Instead of defining specific routes and altitudes that the aircraft must use while transitioning to the GNSS-based instrument approach, TAA criteria define one final approach, one missed approach, and three initial approach fixes in addition to three airspace areas (see Figure 2-45). These fixes are arranged in a T-shaped configuration.

In the example shown in Figure 2-46, aircraft approaching the airport from the southwest would be located in airspace A and would navigate direct to the initial approach fix (in this example, called Alpha). Aircraft approaching from the southeast would navigate direct to Charlie, whereas aircraft approaching in a clockwise arc from west to east (airspace B), would navigate directly to Bravo.

To provide obstacle and terrain clearance, each airspace area extends 30 nautical miles out from its associated initial approach fix. A minimum altitude is established for each area that provides a minimum of 1,000 feet of obstacle clearance. If a single altitude cannot be provided for a particular sector, step-

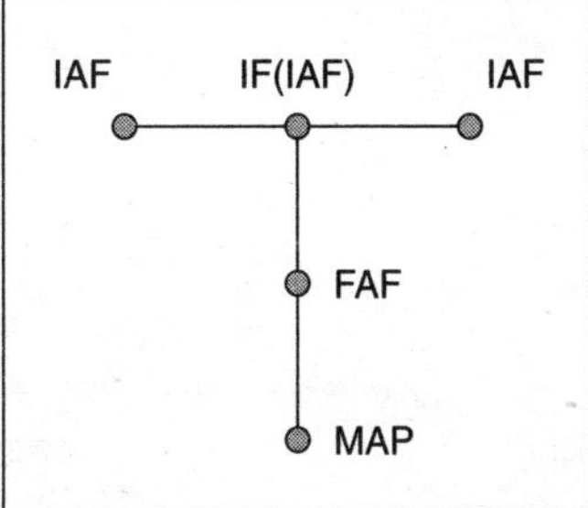

Figure 2-45. Basic TAA configuration.

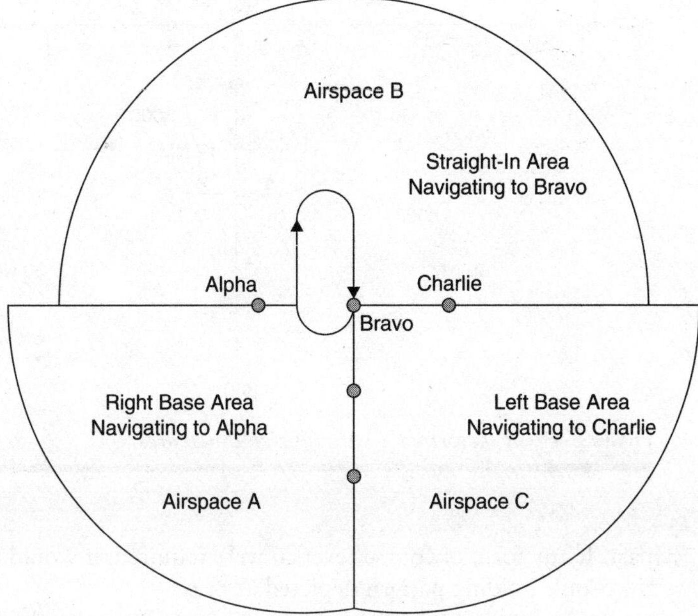

Figure 2-46. Standard TAA with airspace areas.

down areas will be developed. Instead of being fixed points in space, as they are in current instrument approach procedures, the step-down areas will exist as arcs centered on the appropriate initial approach fix (see Figure 2-47).

Aircraft initiate the instrument approach entering via either airspace A or C, proceed direct to the appropriate fix (either Alpha or Charlie), and then turn 90° and fly direct to Bravo. Aircraft entering airspace B would simply proceed direct to Bravo. These three flight segments are the initial approach segments. The length of these segments will be between 3 and 10 miles, depending on the speed of the aircraft that typically use this approach. Any aircraft required to hold would do so at Bravo. Holding at the final approach fix would no longer be a common procedure. Procedure turns would also become a thing of the

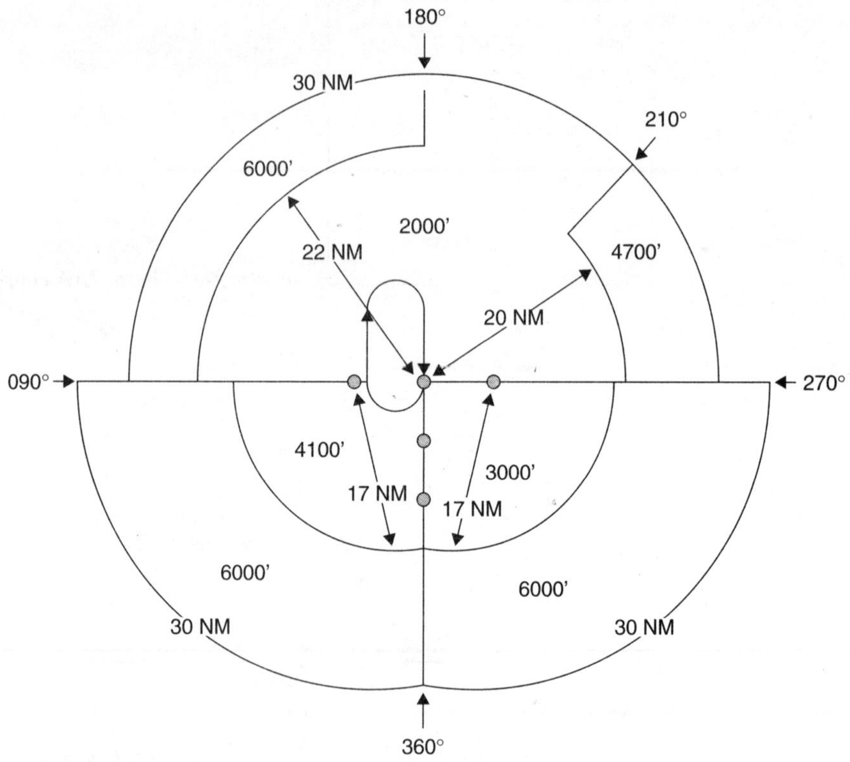

Figure 2-47. A sectorized TAA with step-down arcs.

past. If any form of course reversal were required, it would be accomplished at the 5-mile holding pattern depicted at Bravo.

The intermediate segment exists from Bravo to the final approach fix. An altitude will be established that provides the aircraft with 1,000 feet of vertical clearance from all obstructions during this narrow segment. This segment terminates at the final approach fix. At that point the pilot would fly the approach to the missed approach point just like any other approach. The approach could be a nonprecision GPS approach, or if some form of vertical guidance were provided, the pilot would conduct a precision instrument approach. In most procedures, the missed approach procedure will require the aircraft to turn either right or left, returning to either Alpha or Bravo where the approach will recommence.

The plan is that this new configuration will become the new standard for every existing and newly developed instrument approach procedure. This will provide a standardized configuration that permits direct flight to the beginning of every approach, thereby increasing efficiency and no longer tying aircraft to fixed, inflexible routes and altitudes while still providing safe separation and transition to the instrument approach (see Figure 2-48).

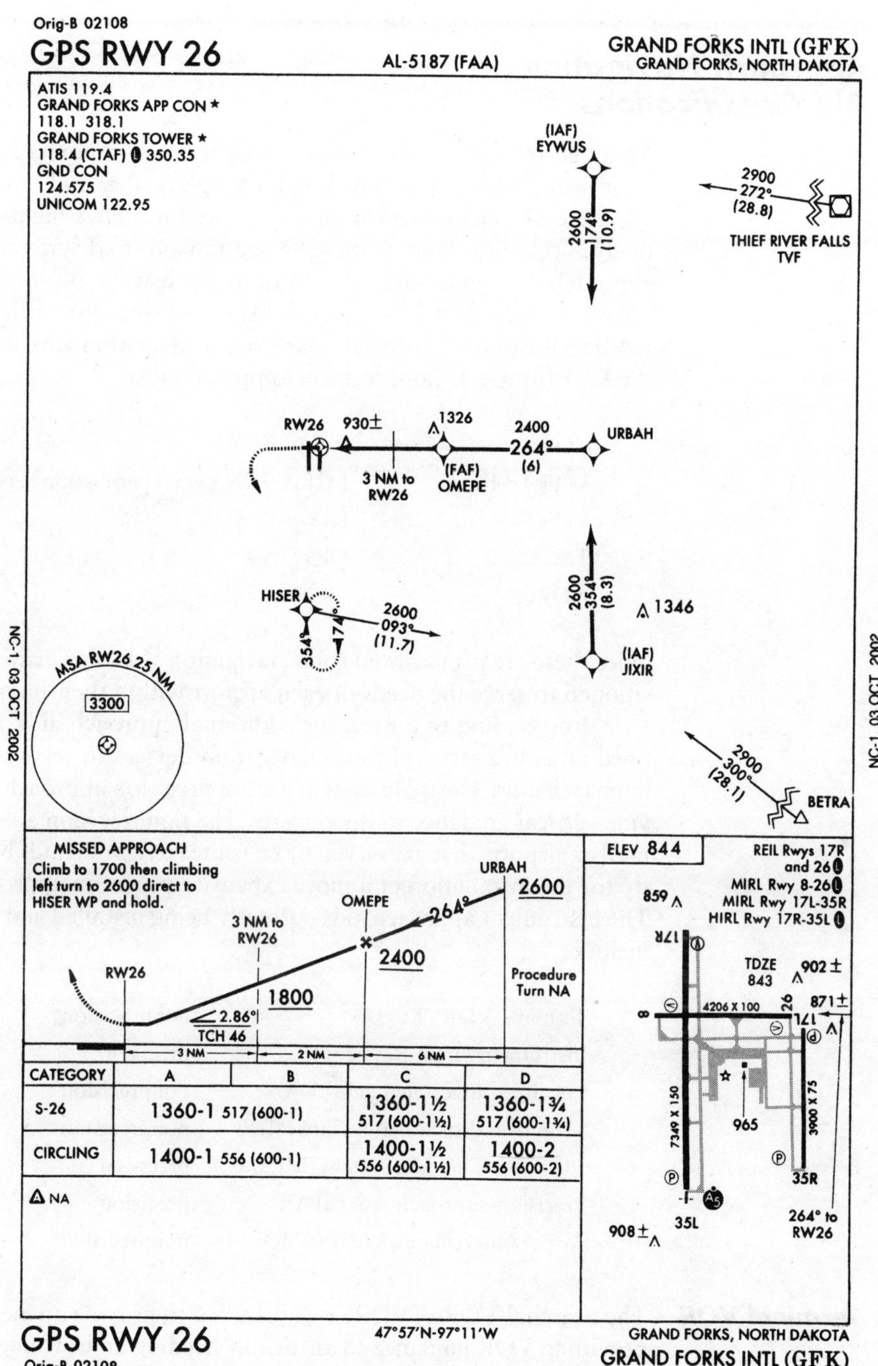

Figure 2-48. GPS runway 26 approach at Grand Forks, North Dakota.

Approach Navigation Aid Classifications

Many en route navigation aids can be used as nonprecision approach aids if their transmitted signal is of a high enough quality and can be safely used during the entire instrument approach procedure. FAA flight-check aircraft routinely check the quality of these navigation aids to determine their suitability as approach navigation aids. An en route navigation aid used for an instrument approach is classified as a nonprecision aid since no vertical guidance is provided to the pilot. The following en route navigation aids have been certified by the FAA for use as nonprecision approach aids:

VOR	RNAV (CLC)
VOR-DME	LORAN-C (special permission needed)
VORTAC	INS
TACAN	GPS
NDB	

Since these are primarily en route navigation aids, they may not be properly positioned to serve the needs of each airport within their immediate vicinity.

Responding to a need for additional approach aids, the FAA has developed an entire series of radio navigation devices to serve solely as instrument approach aids. The main ones in use are precision approach aids since they provide vertical guidance to the runway. The nonprecision aids are designed to be used at airports that are served by en route navigation aids but that do not qualify for the installation of a more expensive precision approach navigation aid. The instrument approach aids currently being installed and utilized by the FAA include:

Terminal VOR (TVOR)	nonprecision
Instrument landing system (ILS)	precision
Localizer directional aid (LDA)	nonprecision
Simplified directional facility (SDF)	nonprecision
Microwave landing system (MLS)	precision
Precision approach radar (PAR)	precision
Airport surveillance radar (ASR)	nonprecision

Terminal VOR

The terminal VOR (TVOR) was designed to provide an inexpensive method of providing VOR guidance to an airport needing an instrument approach. A terminal VOR is a low-powered version of a standard VOR that is usable to a distance of 25 nautical miles. The TVOR does *not* provide distance information unless a civilian DME is co-located at the facility. Terminal VORs are not normally combined with TACANs.

Instrument Landing System

The instrument landing system (ILS) is designed to provide the pilot with an approach path that is perfectly aligned with the runway centerline. An ILS provides both lateral and vertical guidance to the pilot (see Figure 2-49).

The ILS system is equipped with three different types of transmitters: the localizer, the glide slope, and two or three marker beacons.

Localizer The **localizer** system consists of a transmitter building, localizer antenna, and monitoring equipment (see Figure 2-50). Typically, the localizer antenna is located about 1,000 feet beyond the departure end of the runway being served by the ILS. The transmitter building is about 300 feet to one side of the localizer antenna. On older installations, the monitoring equipment is mounted on wooden posts a short distance in front of the antenna array. In newer installations, it is an integral part of the antenna. The localizer operates within the VHF band between 108.10 and 111.95 mHz (see Figure 2-51).

The localizer provides the pilot with lateral course guidance information only. The antenna radiates a signal that is aligned with the runway centerline and is modulated with two different tones: 90 Hz and 150 Hz. The final approach course is produced as a result of the radiation patterns emanating from the antenna array. The array is situated such that the 150 Hz tone is predominant on the right side of the runway while the 90 Hz tone predominates on the left (see Figure 2-52).

Along the centerline, both tones are of equal strength. An aircraft to the right of the centerline will receive predominantly the 150 Hz tone. The airborne receiver will detect this condition and move the vertical needle of the ILS indicator to the left of center, thereby advising the pilot to alter the aircraft's course to the left. If the aircraft is to the left of course, the 90 Hz tone will dominate and the vertical needle of the ILS indicator will move to the right, advising the pilot to alter the aircraft's course to the right. If the aircraft is established on course, the 90 and the 150 Hz tones will be of equal strength and the vertical needle on the ILS indicator will be centered (see Figure 2-53).

The localizer signal is transmitted along a fairly narrow path extending 35° to the left and right of the runway centerline and out from the transmitter to a distance of 10 nautical miles. Between 10 and 25 nautical miles from the runway, the localizer is certified to be accurate within a range of only 10° either side of the extended centerline. The localizer signal is approximately 7° high.

The ILS receiver on the aircraft is designed so that when the vertical needle on the indicator is fully displaced on one side or the other (known as **full-scale deflection**), the aircraft is 3° off course. Since the localizer is an angular device the on-course beam narrows as the aircraft approaches the antenna. Ten miles from the end of the runway, a full-scale deflection indicates that the aircraft is about a half mile off course. When crossing the approach end of the runway, 3° off course translates to approximately 300 feet.

The localizer is one of the most precise and sensitive navigation aids available for instrument approaches. Unfortunately, the localizer signal can be easily reflected off of terrain, buildings, aircraft, vehicles, and power lines, thereby creating **course scalloping** or **false courses** (see Figure 2-54). When an ILS is

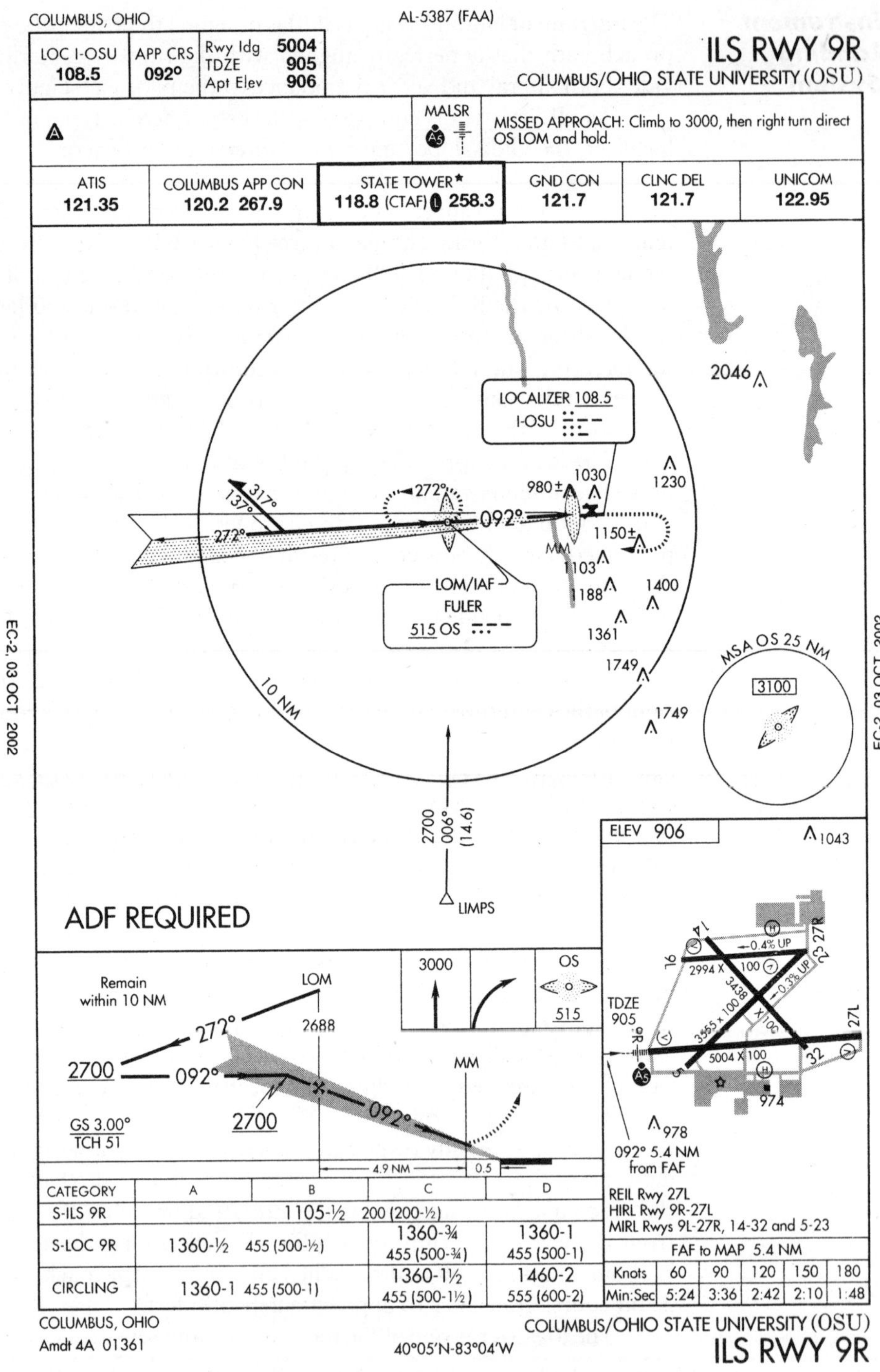

CATEGORY	A	B	C	D
S-ILS 9R	1105-½ 200 (200-½)			
S-LOC 9R	1360-½ 455 (500-½)		1360-¾ 455 (500-¾)	1360-1 455 (500-1)
CIRCLING	1360-1 455 (500-1)		1360-1½ 455 (500-1½)	1460-2 555 (600-2)

FAF to MAP 5.4 NM					
Knots	60	90	120	150	180
Min:Sec	5:24	3:36	2:42	2:10	1:48

Figure 2-49. ILS 9R approach at Columbus–Ohio State University Airport.

Michael Nolan

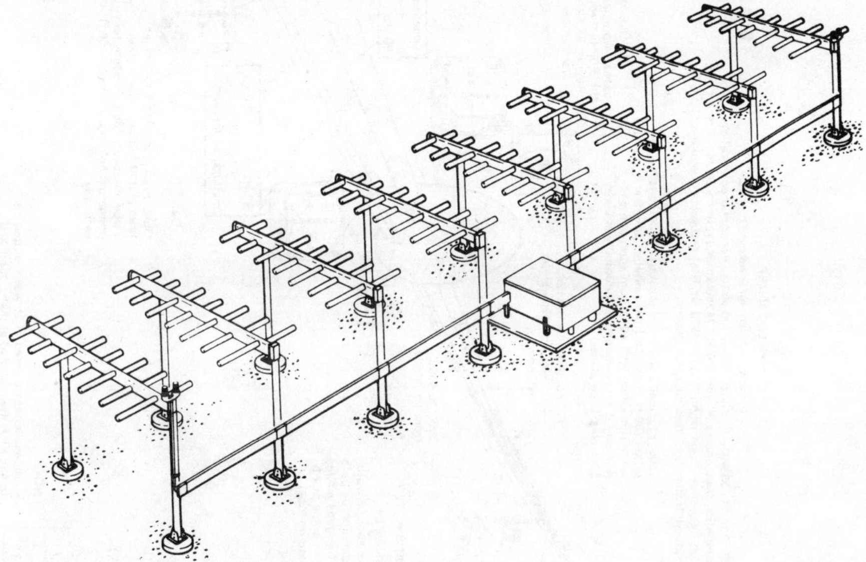

Figure 2-50. An ILS localizer antenna installation.

initially being installed, the localizer radiation pattern is carefully studied to ensure that nearby buildings and power lines will not unduly interfere with the accuracy of the transmission. However, a strong signal reflected off of a nearby object will create a change in the radiation pattern of the localizer and artificially "move" the localizer centerline to the left or right.

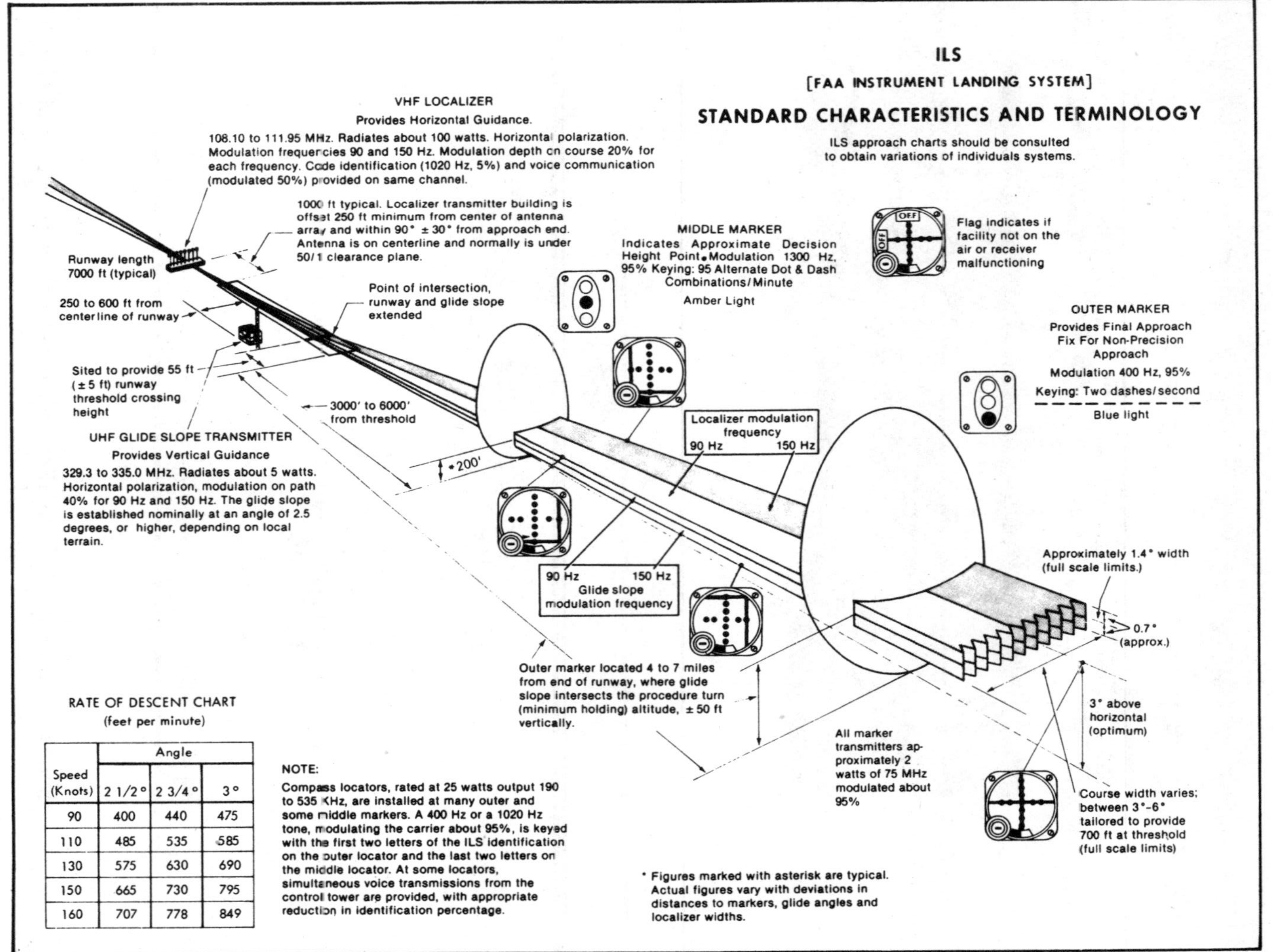

Speed (Knots)	Angle		
	2 1/2°	2 3/4°	3°
90	400	440	475
110	485	535	585
130	575	630	690
150	665	730	795
160	707	778	849

Figure 2-51. Graphic depiction of an instrument landing system.

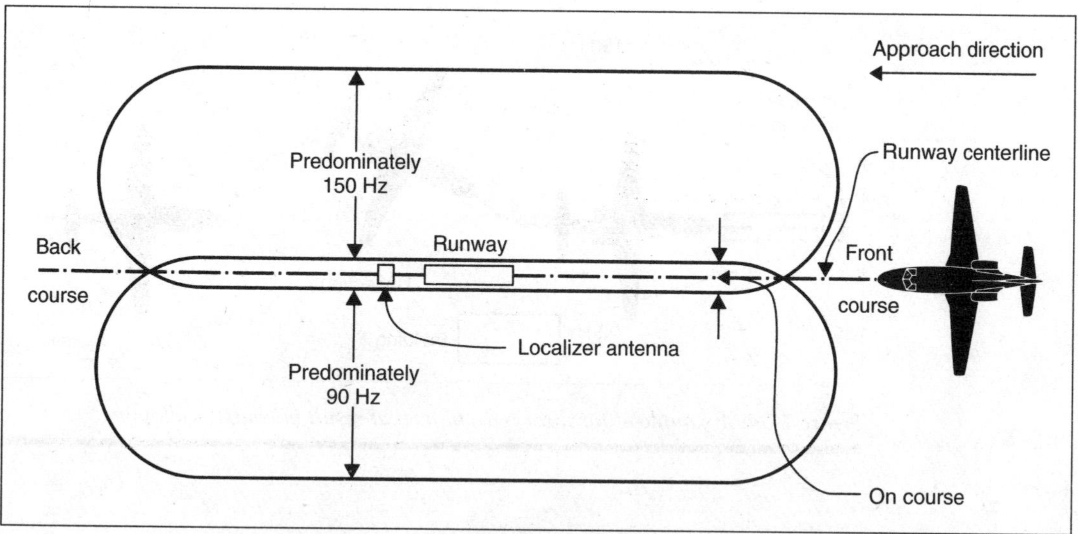

Figure 2-52. ILS approach procedure and the localizer radiation pattern.

Bendix/King Division of Allied Signal Aerospace Company

Figure 2-53. A typical light aircraft ILS receiver.

If this situation is encountered, FAA technicians attempt to solve the problem by installing a different type of localizer antenna. In some cases, even this remedy will not solve the reflection problem and the localizer must be relocated. It may be possible to move the antenna off the runway centerline or to redirect the final approach course somewhat. If either of these modifications is

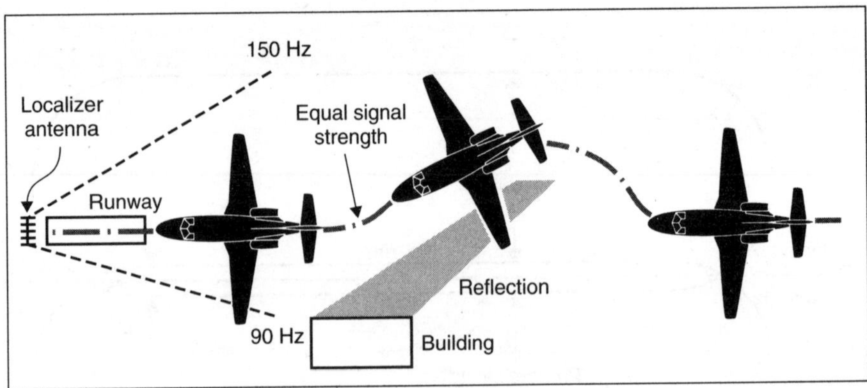

Figure 2-54. Example of localizer reflections that result in course scalloping.

necessary, the localizer is no longer considered to be part of an ILS and is called a **localizer directional aid (LDA)** (see Figure 2-55).

This name change is necessary to alert pilots that the localizer is not aligned with the runway centerline. When a localizer is offset in this manner, vertical guidance is not normally provided, making an LDA-based instrument approach a nonprecision approach. The conversion of a localizer to an LDA is done only as a last resort, since an LDA procedure requires pilots to make a low-altitude turn to line up with the runway just prior to landing.

Airway facility technicians employed by the FAA are responsible for ensuring that reflections from terrain, buildings, and power lines do not disturb the localizer transmission. It is the air traffic controller's responsibility to ensure that aircraft and vehicles do not interfere with the localizer transmission whenever ILS approaches are in progress. To prevent any inadvertent reflections, **localizer critical areas** have been established for every localizer antenna. Each localizer installation is unique and may not have the same critical area, but in general the standard localizer critical area is shaped as in Figure 2-56.

Other than aircraft landing and exiting the runway, aircraft conducting the missed approach procedure, or aircraft using the runway for departure and flying over the localizer antenna, no vehicles or aircraft are allowed within the localizer critical area when ILS approaches are in progress. When weather conditions are extremely poor (such as visibility below 1/2 mile or ceilings below 200 feet), no aircraft or vehicles are allowed in the critical area for any reason when an aircraft is inside the final approach fix during an ILS approach. The exact criteria to be followed concerning localizer critical areas will be covered in Chapter 6 and can be found in the *Air Traffic Control Handbook*.

The localizer transmission is radiated in a pattern that can be received from both the approach and the departure ends of the runway. The front course of the localizer is the transmission that serves as the primary instrument approach. The **localizer back course** is a mirror image of the front course, serving

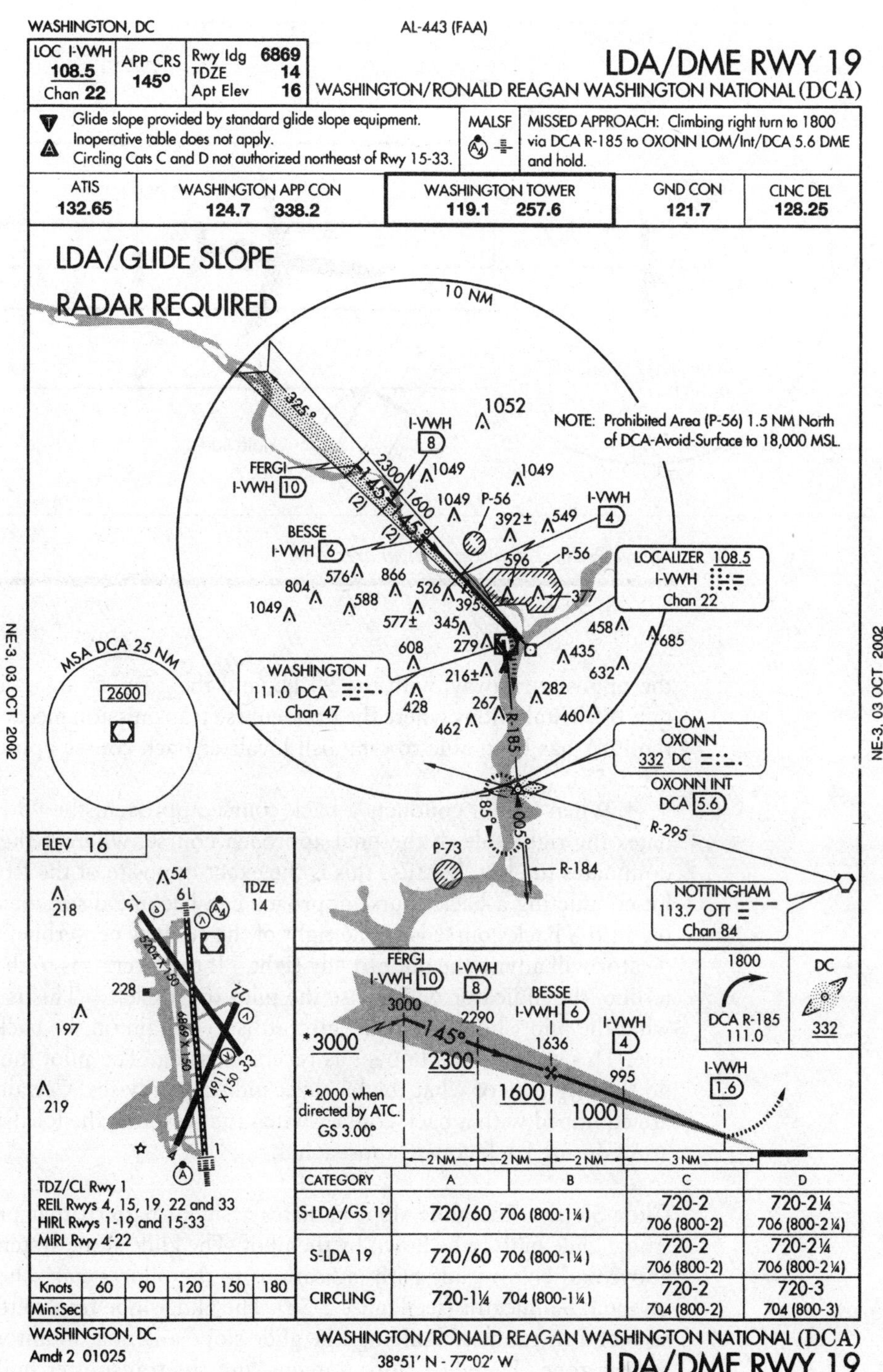

CATEGORY	A	B	C	D
S-LDA/GS 19	720/60	706 (800-1¼)	720-2 706 (800-2)	720-2¼ 706 (800-2¼)
S-LDA 19	720/60	706 (800-1¼)	720-2 706 (800-2)	720-2¼ 706 (800-2¼)
CIRCLING	720-1¼	704 (800-1¼)	720-2 704 (800-2)	720-3 704 (800-3)

Knots	60	90	120	150	180
Min:Sec					

Figure 2-55. LDA runway 18 approach at Ronald Reagan Washington National Airport.

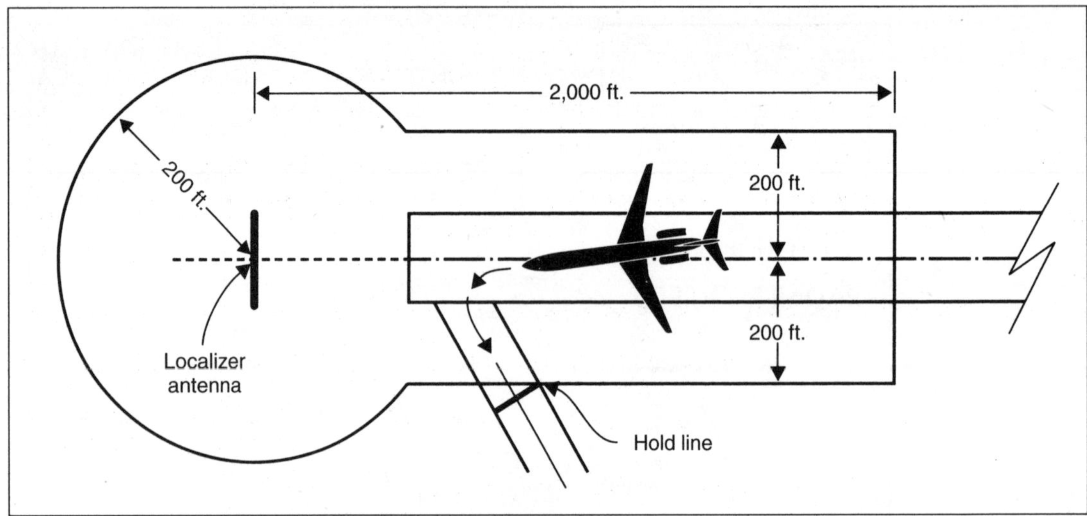

Figure 2-56. Localizer critical area.

the opposite runway, with the 90 Hz and the 150 Hz areas reversed. At certain ILS installations where the back course transmission meets TERPS criteria, the FAA has been able to establish **localizer back course approaches** (see Figure 2-57).

When a pilot conducts a back course approach, the 90 Hz signal dominates the right side of the final approach course, whereas the 150 Hz signal dominates the left. Because this is the exact opposite of the front course, a pilot conducting a back course approach can become disoriented. If an aircraft on an ILS back course is to the right of the runway centerline, the localizer indicator will advise the pilot to "fly right." If the aircraft is to the left of the centerline, the indicator will advise the pilot to "fly left." This is the opposite to what the pilot should do if the aircraft is to remain on the back course centerline. This condition is known as **reverse sensing.** The pilot must remember to do the opposite of what the localizer indicator advises. Certain ILS indicators are equipped with a back course switch that reverses the localizer needle operation during back course approaches.

Glide Slope The **glide slope** radiates a signal pattern that provides an electronic glide path to be flown by the pilot. The glide slope system provides both above and below glide path indications to the pilot, using a horizontal needle on the ILS indicator (see Figure 2-58). The glide slope transmitting system consists of a transmitter building, the glide slope antenna, monitor antennas, and a **clear zone.** The glide slope antenna and the transmitter building are about 500 feet from the runway centerline and about 1,000 feet from the approach end of the runway.

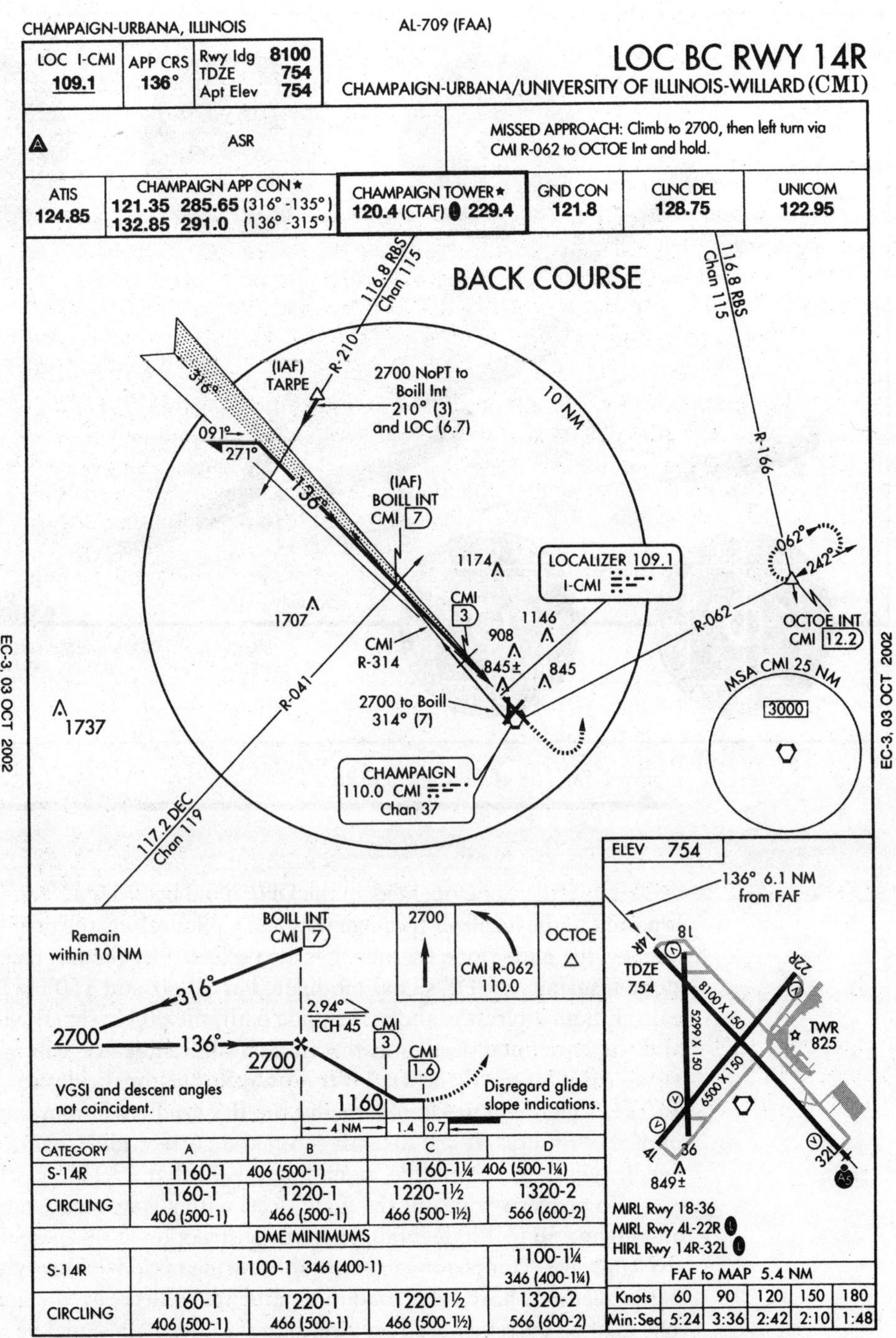

CATEGORY	A	B	C	D
S-14R	1160-1	406 (500-1)	1160-1¼ 406 (500-1¼)	
CIRCLING	1160-1 406 (500-1)	1220-1 466 (500-1)	1220-1½ 466 (500-1½)	1320-2 566 (600-2)
DME MINIMUMS				
S-14R	1100-1 346 (400-1)			1100-1¼ 346 (400-1¼)
CIRCLING	1160-1 406 (500-1)	1220-1 466 (500-1)	1220-1½ 466 (500-1½)	1320-2 566 (600-2)

FAF to MAP 5.4 NM					
Knots	60	90	120	150	180
Min:Sec	5:24	3:36	2:42	2:10	1:48

Figure 2-57. Localizer back course runway 14R approach at the University of Illinois Willard Airport.

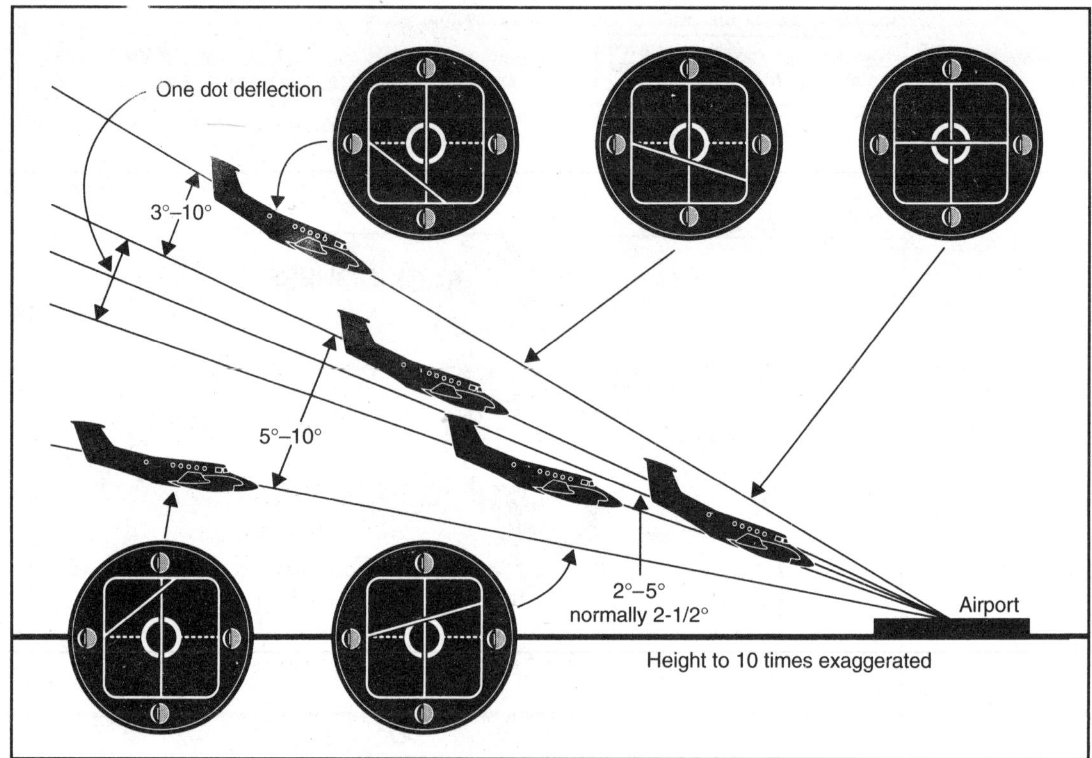

Figure 2-58. ILS glide slope indications.

The glide slope operates in the UHF band between 329 and 335 mHz and is paired to the localizer frequency. When a pilot selects the proper localizer frequency, the glide slope frequency is also selected by the ILS receiver. The glide slope transmits a UHF signal modulated at 90 Hz and 150 Hz just like the localizer. If an aircraft is above the glide path, the 90 Hz signal will predominate and the horizontal needle on the aircraft's ILS indicator will move down, advising the pilot to "fly down." If the aircraft is below the desired glide path, the 150 Hz signal will predominate and the ILS needle will move up, advising the pilot to "fly up." If the aircraft is on the correct glide path, the horizontal needle will be in the middle, signalling ON GLIDE PATH.

To properly transmit the glide slope with a single antenna would require an antenna 50 to 100 feet tall. Since an obstruction at this height next to an active runway is completely unacceptable during periods of low visibility, a number of methods have been tried in an attempt to decrease the antenna's height. The method currently used requires that the glide path signal be reflected off the ground directly in front of the antenna. This area is known as the *glide slope reflecting area.* This solution reduces the height of the antenna mast to about

30 feet (see Figure 2-59). Since the glide slope antenna is highly directional, there is no back course to a glide slope.

A number of other problems are inherent to the current glide slope method of transmission. One such problem is the creation of **false glide paths.** At most glide slope installations, the desired glide path angle is about 3° above horizontal. But when the glide slope bounces off the ground, a number of additional glide paths are also created. Fortunately, none of these false glide paths are lower than 3°, but many exist above this level. The first false glide path usually occurs at an angle of about 9°.

To ensure that the correct glide path is used by pilots conducting an ILS approach, it is imperative that the aircraft be allowed to transition from the airway structure to the ILS at an altitude that will place the aircraft below any of the false glide paths.

Any obstruction directly in front of or to the side of the glide slope transmitter might reflect some of the signal and cause glide slope scalloping. To prevent this, an area directly in front of the glide slope antenna has been designated the **glide slope critical area** (see Figure 2-60). When aircraft are conducting ILS approaches, this area *must* be kept clear of aircraft, vehicles, deep snow, or any objects that may interfere with the correct operation of the glide slope transmitter.

Another of the problems inherent in bouncing the glide slope is that extensive site preparation is needed to ensure that the ground in the reflecting area will properly reflect the glide slope at the desired approach angle. If the area in front of the antenna does not offer the proper reflectivity, it may have to be resurfaced, which is usually expensive. Many factors can temporarily change the reflectivity of this zone. Water-soaked ground, excessive snow, or extremely long grass can all cause the glide slope to reflect at the wrong angle. To ensure proper glide slope operation, receivers called **glide slope monitors** are located within the clear zone. If these monitors detect that the glide slope radiation pattern is no longer within established tolerances, the glide slope transmitter is automatically shut down.

Marker Beacons Marker beacons are located at known distances along the final approach course of the ILS to provide position information to pilots conducting the approach. Marker beacons transmit a cone-shaped signal on a frequency of 75 mHz, uniquely coded to identify each type of beacon.

Outer marker (OM) beacons are usually located on the ground about 5 miles from the approach end of the runway (see Figure 2-61). When a properly equipped aircraft flies over an outer marker, a blue light flashes and a 400 Hz series of continuous dashes is emitted from the marker beacon receiver on board the aircraft (see Figure 2-62).

The **middle marker (MM)** is usually about 3,000 feet (or half a mile) from the approach end of the runway and causes an amber light to flash and a series of 1,300 Hz dots and dashes to be heard in the cockpit. The middle marker is usually located such that an aircraft properly positioned on the glide slope will

Michael Nolan

Figure 2-59. An ILS glide slope antenna installation.

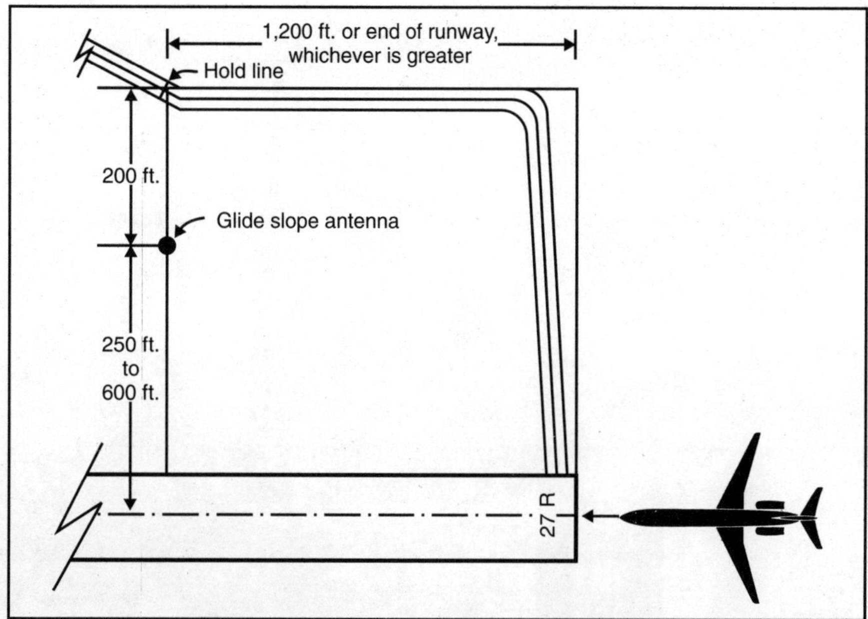

Figure 2-60. Glide slope critical area.

overfly it at approximately 200 feet. This is the normal decision height for a Category I ILS approach.

If a Category II ILS has been installed, an **inner marker (IM)** is placed approximately 1,000 feet from the end of the runway. The inner marker is located at the point where an aircraft on the glide slope passes through an altitude of 100 feet. This is the decision height for a Category II ILS approach. The inner marker causes a white light to flash and a 3,000 Hz series of continuous dots to be heard in the cockpit.

Compass Locators At many ILS installations, a low-powered nondirectional beacon (NDB) may be co-located with either the outer or the middle marker. Such nondirectional beacons assist the pilot when transitioning from the airway structure to the ILS. An NDB used for this purpose has a transmitter power of less than 15 watts and is known as a compass locator. Combining a compass locator with an **outer marker (OM)** creates a facility known as a **locator outer marker (LOM).** When co-located with a middle marker, the facility is known as a **locator middle marker (LMM).** Since the increased use of radar in the terminal environment has diminished the need for compass locators, the FAA has begun to decommission the few existing locator middle markers and will install locator outer markers only where operationally necessary.

Figure 2-61. A marker beacon transmitter.

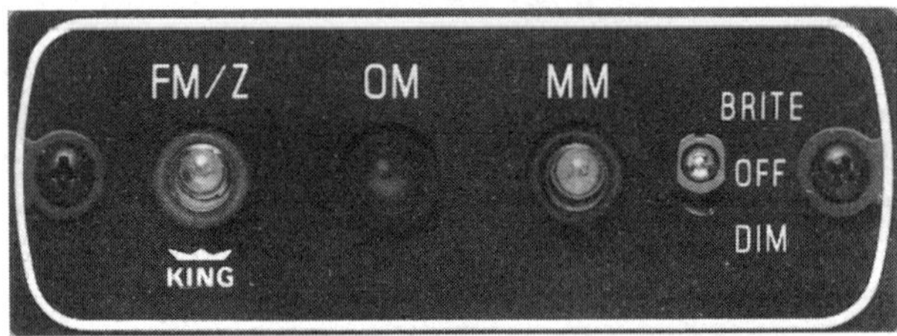

Figure 2-62. A typical light aircraft marker beacon receiver.

ILS Distance Measuring Equipment In rare instances, distance measuring equipment may be installed at the localizer site to provide distance information to an aircraft conducting an ILS approach. DME is usually used when the local terrain precludes the installation of 75 mHz outer or middle markers. The proper DME frequency is automatically selected when the pilot tunes in the appropriate localizer frequency. DME-equipped aircraft can then use this distance information in place of the marker beacons.

ILS Categories ILS systems are currently classified into one of three categories, each category being defined in terms of minimum visibility and decision

Table 2-2. ILS Categories

ILS Category	Decision Height	Visibility or RVR
I	200 feet	½ mile or 2,400 feet
II	100 feet	1,200 feet†
IIIa	*	700 feet†
IIIb	*	150 feet†
IIIc	*	‡

* No decision height specified. Visibility is the only limiting factor.
† No fractions of miles authorized when determining visibility. The runway served by the ILS must have operable RVR equipment.
‡ No ceiling or visibility specified. Aircraft must be equipped with automatic landing equipment.

height altitudes (see Table 2-2). Minimum visibility is measured in fractions of a mile when measured by human observes or in hundreds of feet when measured by a runway visual range.

The standard ILS is a Category I, which provides accurate guidance information in visibilities as low as 1/2 mile and ceilings as low as 200 feet. These minima are representative of a standard ILS installation.

With a slight change in the ground equipment, an ILS installation may be certified as a Category II, which permits a properly rated pilot to utilize the ILS in visibilities as low as 1,200 feet or ceilings as low as 100 feet (see Figure 2-63). The additional equipment required for a Category II installation includes more stringent localizer and glide slope monitoring equipment, an inner marker, and additional approach lighting. Pilots and aircraft must be certified to use a Category II ILS and its associated minima. Those pilots not certified to Category II minima may still use a Category II ILS down to Category I minima.

In those locations that qualify, a Category III ILS may be installed (see Figure 2-64). A Category III ILS installation is much more expensive since it requires completely redesigned localizer and glide slope equipment. Category III ILS approaches are of three types: IIIa, IIIb, or IIIc. Category IIIc approaches may be conducted when the ceiling or visibility is zero! Aircraft conducting Category III approaches must be equipped with **autoland** devices that automatically land the aircraft. Category III installations are rarely justified for use in this country. Few airports need this type of approach and few aircraft are equipped to utilize them.

Runway Visual Range **Runway visual range (RVR)** equipment measures the visibility along the runway being utilized for instrument approaches. The RVR system consists of a transmissometer projector, a transmissometer detector, a data converter, and a remote digital display.

In a typical RVR installation, the **transmissometer projector** and the **transmissometer detector** are located to the side of the runway, approximately

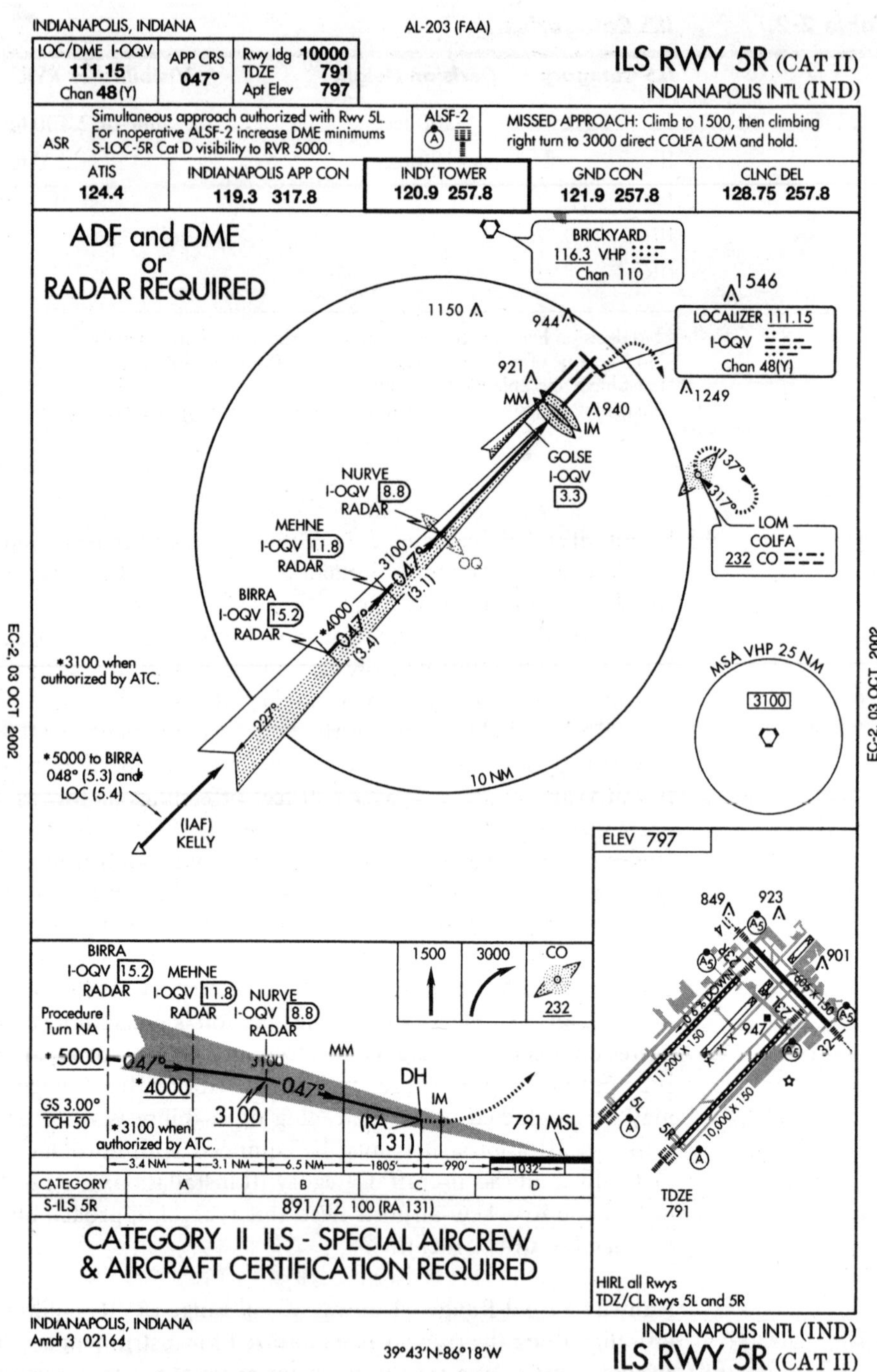

Figure 2-63. Category II ILS runway 5R approach at Indianapolis, Indiana.

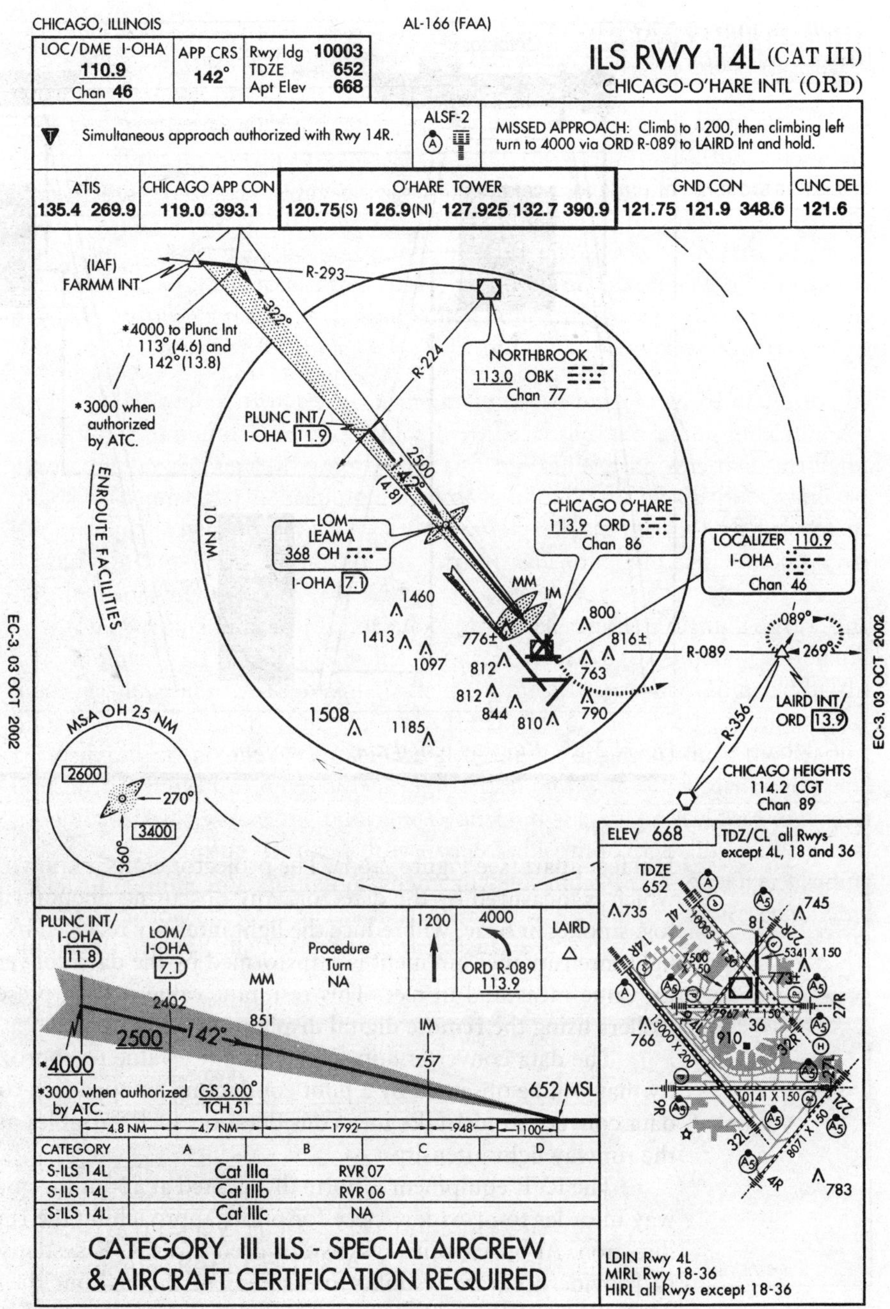

CATEGORY	A	B	C	D
S-ILS 14L	Cat IIIa	RVR 07		
S-ILS 14L	Cat IIIb	RVR 06		
S-ILS 14L	Cat IIIc	NA		

Figure 2-64. Category III ILS runway 14L approach at Chicago O'Hare International Airport.

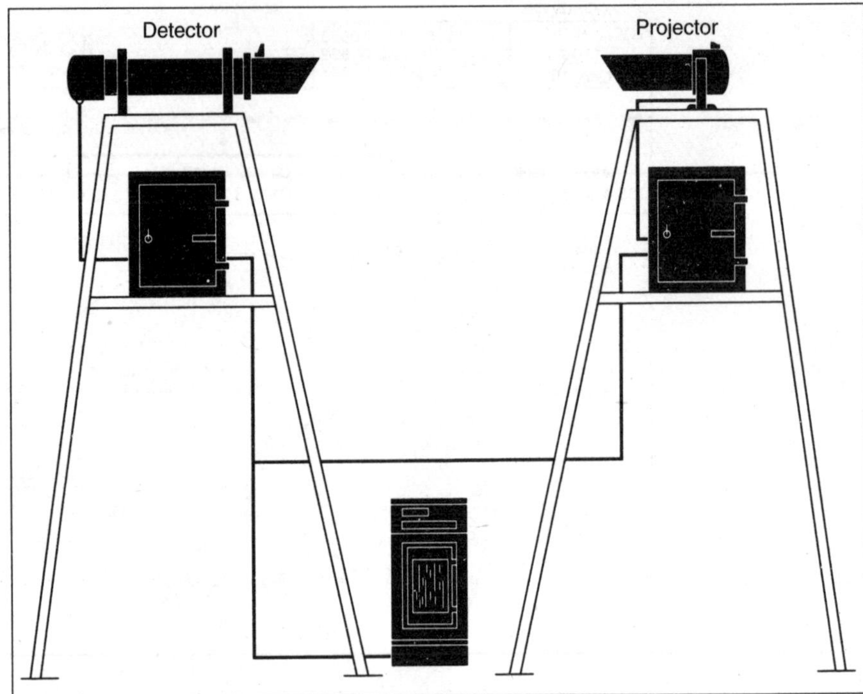

Figure 2-65. Runway visual range operation.

500 feet apart (see Figure 2-65). The projector emits a known intensity of light, which is measured by the detector. Any obscuring phenomenon, such as rain, fog, smoke, or haze, will reduce the light intensity received by the detector. The light intensity measurement is transformed by the **data converter** into a visibility value measured in feet. This resultant value is then presented to the controllers using the **remote digital display.**

The data converter adjusts the visibility value to approximate the visibility that will be observed by a pilot conducting an approach to the runway. The data converter must take into consideration such variables as time of day and the runway light intensity.

The RVR equipment is normally located at about the midpoint of the runway in order to provide service for pilots approaching the runway from either direction. At busier runways, two or even three RVR systems may by installed to provide accurate visibility measurement throughout the runway's length. These three RVR installations are called the touchdown, midpoint, and rollout RVRs.

Simplified Directional Facility

At some locations where the installation of an expensive ILS cannot be justified but where the existing navigation aids are unsuitable for the development of an instrument approach, a **simplified directional facility (SDF)** may be installed.

An SDF provides course guidance similar to but less accurate than the localizer component of an ILS. An SDF transmitter broadcasts in the same frequency range as the ILS (108.10 to 111.95 mHz), with a signal modulated at 90 Hz and 150 Hz. An SDF approach does not provide glide path information. Marker beacons and compass locators may be utilized as part of an SDF approach.

The SDF final approach course may not be aligned with the runway and is wider than an ILS localizer. SDFs are usually much cheaper and easier to install and maintain than an ILS and are well suited for smaller airports or for use as a secondary approach at an airport already equipped with an ILS.

Runway and Approach Lighting

When night flying was first introduced, most airports consisted of an open area covered with either turf or cinders. Pilots could land in whatever direction they chose. Rotating beacons provided the pilots with the general location of the airport but did not provide sufficient visual cues to permit them to actually locate the cinder area and land. This problem was solved through the introduction of airport boundary lighting, previously described in this chapter.

In the late 1930s, because of the increased weight of aircraft that were being introduced into service, most of the airports began to construct concrete runways to replace the cinder landing surfaces. These runways were usually about 1 mile long and about 100 feet wide. Since each airport had only two or three runways, airport boundary lighting no longer satisfactorily assisted the pilot in locating the runway at night. A different type of lighting needed to be developed.

Runway Lighting

Runway Edge Lights Many different types of runway lighting systems were examined, including runway floodlights and neon lights. After numerous experiments by both civilian and military aviation authorities, it was eventually agreed that **runway edge lights** should be the standard type of runway lighting.

Runway edge lights are placed on either side of the runway, spaced approximately 200 feet apart, outlining the edges of the runway. These lights are usually placed on short metal poles to elevate them from any obstruction such as long grass or drifting snow. Runway lights are white and are usually covered with a **Fresnel lens** (see Figure 2-66). Fresnel lenses are designed to focus the emitted light, concentrating it along and slightly above the horizontal plane of the runway's surface.

The lights installed on the last 2,000 feet of runways used for instrument approaches utilize lenses that are half white and half amber. These lights appear amber to a landing pilot, warning that the far end of the runway is fast approaching. The ends of the runway are clearly designated through the use of **runway threshold lights,** which are similar to runway lights but utilize red and green split lenses. As the pilot approaches the runway to land, the threshold

Michael Nolan

Figure 2-66. A typical runway Fresnel light installation.

lights on the near end of the runway appear green, while those on the far end of the runway appear red.

Runway light systems are normally operated from the control tower and are turned on during nighttime hours and during daylight whenever the visibility is less than 2 miles or at the pilot's request. Whenever the control tower is not in operation, the lights are either left on or are operated using **pilot-controlled lighting (PCL)** systems. PCL systems permit pilots to switch on the lights by pressing their microphone switch a number of times in rapid succession, producing an audible click on the control tower frequency. The number of clicks controls both the operation and the intensity of the runway lighting system. For more information about pilot-controlled lighting systems, consult the *Aeronautical Information Manual.*

Runway Light Intensity Runway light systems are classified according to the brightness they are capable of producing. **Low-intensity runway lighting (LIRL)** is the least expensive to install and is typically equipped with 15-watt bulbs that operate on one intensity level. This intensity level is known as step one.

The standard type of lighting for a runway used for instrument approaches is **medium-intensity runway lighting (MIRL).** Medium-intensity lights are similar in construction to low-intensity lights but are usually equipped with 40-watt bulbs. MIRL can be operated on three intensity levels: step one, step two, and step three. When operated on step one, medium-intensity lights produce the same light level as low-intensity lights (15 watts). When functioning on step two, they operate at about 25 watts, and on step three they operate at the maximum-allowable 40-watt level. During normal operation, medium-intensity lights are usually set to step one. This intensity is increased whenever the pilot requests or when the visibility drops below 3 miles.

Runways that are heavily utilized during periods of low visibility may be equipped with **high-intensity runway lighting (HIRL).** High-intensity runway lights operate on five steps ranging from 15 watts to 200 watts. High-intensity lights are operated on step one until the visibility begins to decrease below 5 miles. At that point, higher intensities are used, with step five being reserved for periods when the visibility is less than 1 mile.

Embedded-in-Runway Lighting Runways that are used extensively during periods of low visibility may be equipped with an assortment of embedded runway lights that provide the pilot additional visual cues when landing. These systems include touchdown zone lighting, runway centerline lighting, and taxiway turnoff lighting.

During periods of very low visibility, the runway edge lighting does not provide the pilot with sufficient visual cues to properly land the aircraft. In the 1950s, when ILS was initially being installed, various pilot groups complained that landing in these conditions was like landing in a black hole. They reported that during the last few seconds of the approach, as they were raising the aircraft's nose for landing, the runway edge lights were too far apart to provide an accurate altitude reference. In an attempt to provide additional visual cues during this critical phase of landing, a new supplemental lighting system was developed, known as **touchdown zone lighting.** Touchdown zone lights are embedded in the runway and extend from the landing threshold to a point 3,000 feet down the runway. Touchdown zone lights use 100- to 200-watt bulbs and are placed in sets of three, on both sides of the runway centerline. Touchdown zone light intensities are stepped in conjunction with the runway edge lights.

In conditions of reduced visibility, runway edge lights do not provide sufficient directional guidance information to enable pilots to accurately steer their aircraft along the center of the runway. To assist the pilot, many airports have installed **runway centerline lights.** Centerline lights are similar to touch-down zone lights but are placed along the entire centerline, at 75-foot intervals. Runway centerline lights are bidirectional: in the first part of the runway the lights are white, while the last 1,000 feet of centerline lights are red; in the 2,000 feet preceding the red lights, the centerline lights alternate red and white to warn pilots that the runway end is approaching. Runway centerline lights are also varied in intensity in proportion to the setting chosen for the runway edge lights.

When visibility is reduced, many pilots find it difficult to identify the intersecting taxiways for exiting the runway. Runway utilization rates are reduced as pilots taxi slowly, trying to find the proper turnoff. To reduce this taxi time, some airports have installed **taxiway turnoff lights,** which are similar to centerline lights but are used to delineate the path that the pilot should use for exiting the runway. Taxiway turnoff lights are inset into the runway's surface and are spaced at 50-foot intervals. These lights are colored green and extend from the runway centerline to the proper intersecting taxiway.

Large airports may have a myriad of taxiways, runways, and vehicular paths that all look similar to a pilot unfamiliar with the airport. To assist these pilots, **taxiway edge lighting** systems have been developed. Taxiway edge lights are similar to runway edge lights but operate at reduced wattage and are equipped with blue lenses. Taxiway centerline lights are green. Taxiway lights may operate at different intensity levels and are usually operated from the control tower.

Approach Lighting Systems

One of the most complex tasks facing pilots occurs near the end of an instrument approach, when they make the transition from instrument to visual flying. During this transition, they must locate the runway and properly maneuver the aircraft for landing within seconds. In conditions of low visibility, a pilot may be able to see only about 2,000 feet ahead of the aircraft. In today's modern jets, this distance can be covered in less than 20 seconds. Within this short time, the pilot must locate the runway, determine the aircraft's position, make any necessary adjustments in flight attitude, and then land the aircraft. Without some form of visual assistance, this task is virtually impossible to perform safely in so short a time.

These problems were noted as early as 1932 by officials from the airlines and the Bureau of Air Commerce. Experiments were conducted as early as 1935 in an attempt to simplify the transition from instrument to visual flight during an approach. These experiments led to the construction of a number of different types of **approach lighting systems.** Approach lights are placed along the extended centerline of the runway and usually extend from the runway threshold out to a point where the pilot might make the transition from instrument to visual flying. Approach lighting systems are designed to provide the pilot with visual cues that will permit accurate aircraft control during the final approach and landing phase of the flight.

Experimental Systems The first experimental approach lighting system consisted of three high-intensity incandescent lights placed approximately 500 feet apart along the extended centerline of a runway at the Newark (New Jersey) airport. Later experimental systems installed at the airport included neon bar lights and 1,500-foot rows of incandescent lights.

Additional experiments were conducted at the Indianapolis (Indiana) and Nantucket (Massachusetts) airports. In 1945 the CAA, Army Air Corps, and Navy Department agreed to join efforts to establish the **Landing Aids Experiment Station (LAES)** at the Naval Air Station at Arcata, California, where most of the pioneering research in approach lighting was conducted.

Opinions differed about the requirements and the configurations for approach lighting systems. The military services preferred a system that did not lie along the extended centerline of the runway. Military officials felt that the area directly below the aircraft should remain clear of obstructions during the final phase of the approach. They preferred approach lights to be placed to the left, to the right, or on both sides of the aircraft.

The CAA, on the other hand, preferred to place the approach lights directly under the aircraft. Although this system created a slight obstruction problem, it did not require pilots to look out their side window to see the approach lights. An approach lighting system along the centerline of the runway would permit pilots to concentrate directly ahead of the aircraft, which would simplify runway detection and make it easier to note any changes in aircraft altitude or attitude.

By 1953 each organization had selected a different system as its standard. The CAA selected a centerline system, known as Configuration A. The Air Force and the Navy chose systems known, respectively, as Configuration B and Configuration C. In 1958, after years of discussion, the Department of Defense agreed to cease building any additional Configuration B and C systems and to use Configuration A approach lighting in all new installations.

The CAA configuration consisted of a series of high-intensity white lamps placed five abreast, extending from the runway threshold out to a distance of 2,400 to 3,000 feet (see Figure 2-67). These light bars were spaced 100 feet apart. At a point 1,000 feet from the end of the runway, a triple set of light bars provided the pilot with both roll guidance and a definite, unmistakable distance indication. The threshold of the runway was delineated with a series of four red light bars and a continuous line of green threshold lights.

Michael Nolan

Figure 2-67. A typical approach light installation.

Michael Nolan

Figure 2-68. A typical high-intensity approach light.

To provide for identification of the approach lighting system, a high-intensity strobe light was placed on each of the light bars that extended beyond the 1,000-foot mark (see Figure 2-68). These strobe lights flashed in sequence, at a rate of two times per second, and appeared to the pilot as a moving ball of light leading to the runway. These **sequenced flashing lights (SFL)** are also referred to by the slang name "the rabbit." This combination approach lighting system became the standard for runways equipped with Category I ILS and is known as approach lighting system type 1, or **ALSF-1.**

When Category II and III ILSs were being developed, it was realized that an improved approach lighting system was necessary. During Category II approaches, the pilot may be required to transition to visual references during the last 15 seconds of flight. Category III approaches permit the pilot even less time to make this transition. In response, the FAA developed an improved approach lighting system known as approach lighting system type 2, or **ALSF-2** (see Figure 2-69).

ALSF-2 is similar to ALSF-1 but includes additional lighting during the last 1,000 feet. A supplemental set of white light bars is located 500 feet from the runway threshold to provide the pilot with an additional distance indication. Red light bars are also placed on both sides of the centerline, providing pilots with aircraft roll guidance during the last 1,000 feet.

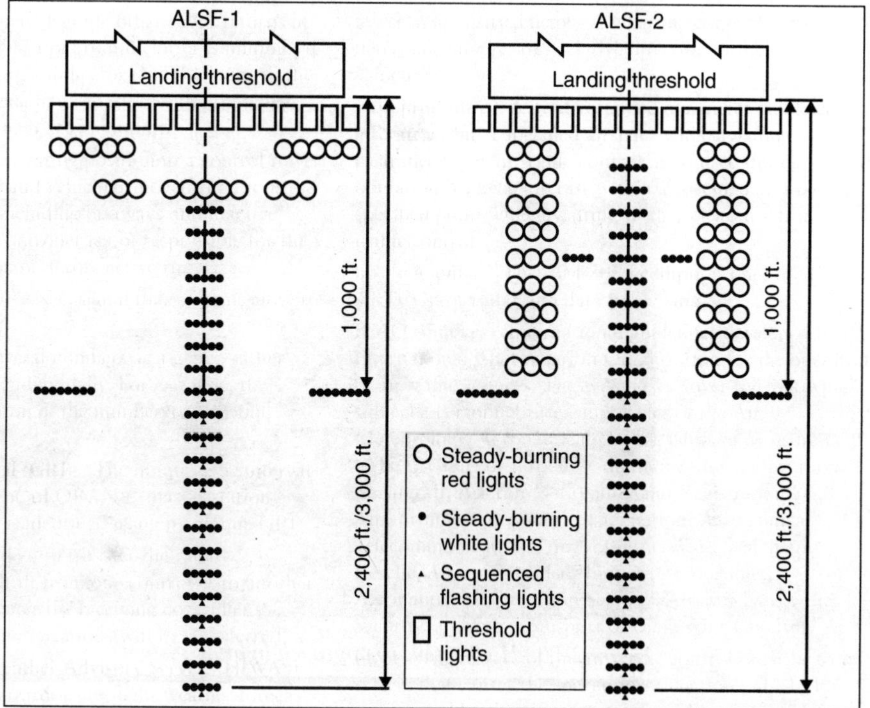

Figure 2-69. ALSF-1 and ALSF-2 installations.

ALSF-2 approach lighting systems are wired such that the additional lights can be switched off whenever Category I ILS approaches are being conducted and appears as an ALSF-1 systems. The system is operated in the ALSF-2 configuration only when the pilot requests or when the visibility decreases below ¾ mile.

Both the ALSF-1 and ALSF-2 systems are similar to high-intensity runway lighting systems in that they can be set to one of five intensity steps. Step five, the brightest, is used only during periods of extremely low visibility.

Simplified Approach Lighting Systems Both the ALSF-1 and the ALSF-2 systems are expensive to install, operate, and maintain. This expense can be justified only at airports that use this type of equipment routinely. At most airports, a smaller, less expensive system can provide pilots with the same benefits as these larger systems.

Some runways are located such that identification of the extended runway centerline is difficult. If extensive instrument approaches are not being conducted to that runway, a full approach lighting system may be economically unfeasible. It is usually more practical to simply install the sequenced flashing

FAA

Figure 2-70. High-intensity approach lighting system.

lights and let them guide pilots to the runway end. When installed in this manner, SFLs are usually spaced 200 feet apart and are known as **runway alignment indicator lights (RAILs).**

At some locations, a full-length (3,000-foot) approach lighting system is unnecessary. For many, the FAA has chosen to install a version of ALSF-1 that is only 1,200 feet long. This system utilizes the same high-intensity white approach lights as the ALSF-1 system, but they are spaced at 200-foot intervals. This is known as the **simplified short approach lighting system (SSALS)** (see Figure 2-70).

In most of these locations, runway alignment indicator lights are also installed out to a distance of 2,400 feet. In this configuration, the system is known as the **simplified short approach lighting system with RAIL (SSALR)** (see Figure 2-71). Most ALSF-1 and ALSF-2 systems are wired such that they can operate as SSALR systems during periods when low-visibility approaches are not being conducted.

The FAA has begun to place a smaller approach lighting system at most airports that do not routinely conduct a large number of low-visibility approaches. This system is designed to include most of the important components available in the ALSF and SSALR systems but reduces the installation, operating, and maintenance expenses.

This system, known as the **medium-intensity approach lighting system with RAIL (MALSR)**, operates with only three steps of intensity, using medium-

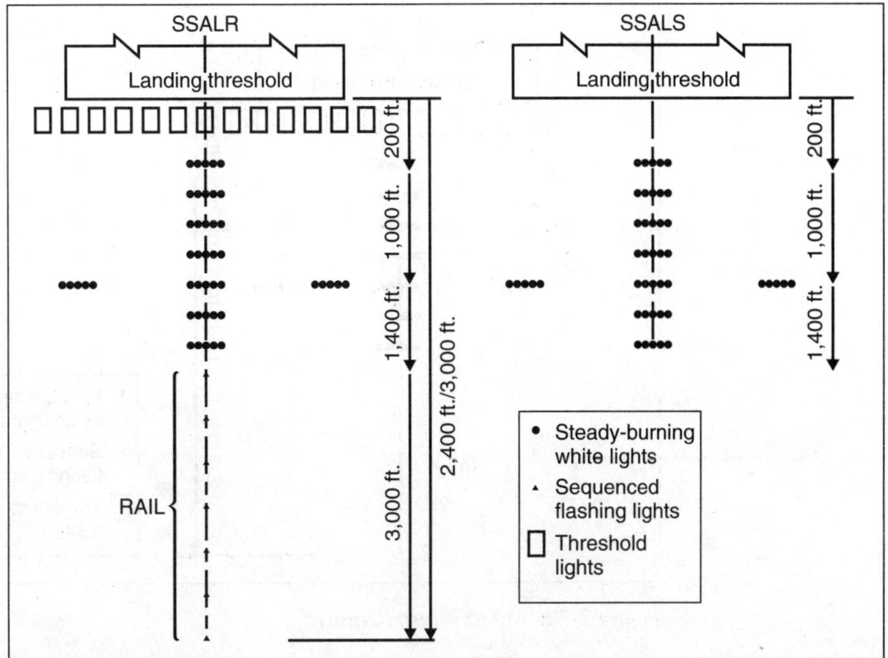

Figure 2-71. SSALR and SSALS installations.

intensity white lamps. MALSR systems extend 2,400 feet from the runway threshold, with the light bars spaced at 200-foot intervals. MALSR systems operate on step one through step three, with step three being equivalent in intensity to step three on an ALSF system (see Figure 2-72).

VFR Approach Lighting Systems

At airports in densely populated areas, it may be extremely difficult for a pilot flying VFR to identify the location of the runways. Thus, it may be necessary to provide pilots with a positive means of locating the runways. If the area is particularly noise sensitive and pilots are required to fly a specific flight path to the runway, it may also be necessary to delineate the approach flight path. Two types of identifier lights have been developed for these purposes: runway end identifier lights and omnidirectional approach lights.

Runway End Identifier Lights Runway end identifier lights (REILs) provide pilots with rapid and unmistakable identification of the end of the runway. REIL units are located on both sides of the approach end and are synchronized to flash together two times per second. Each light is unidirectional, is pointed approximately 15° away from the centerline, and flashes with an intensity of 600 watts (see Figures 2-73) and (2-74). Some REIL units are single step, whereas others may be three step and connected to the runway light-intensity controller.

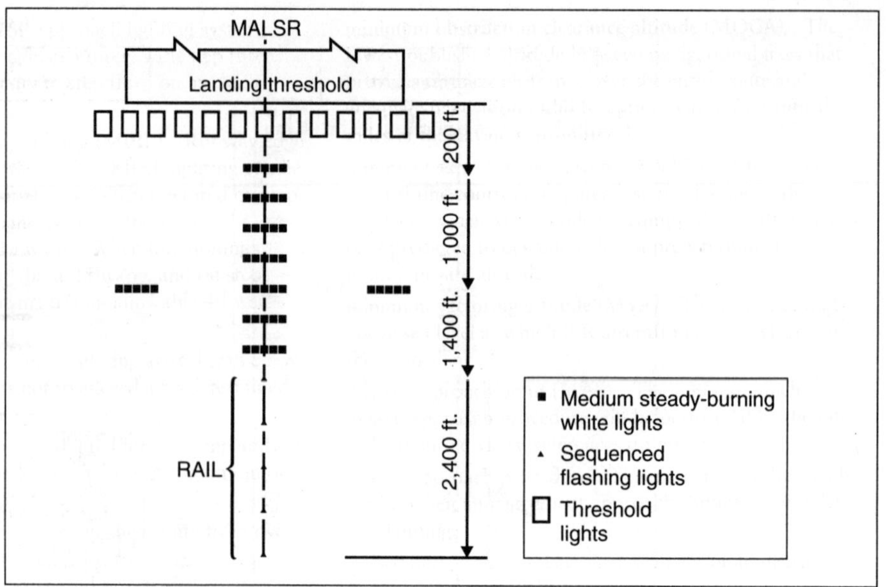

Figure 2-72. MALSR installation.

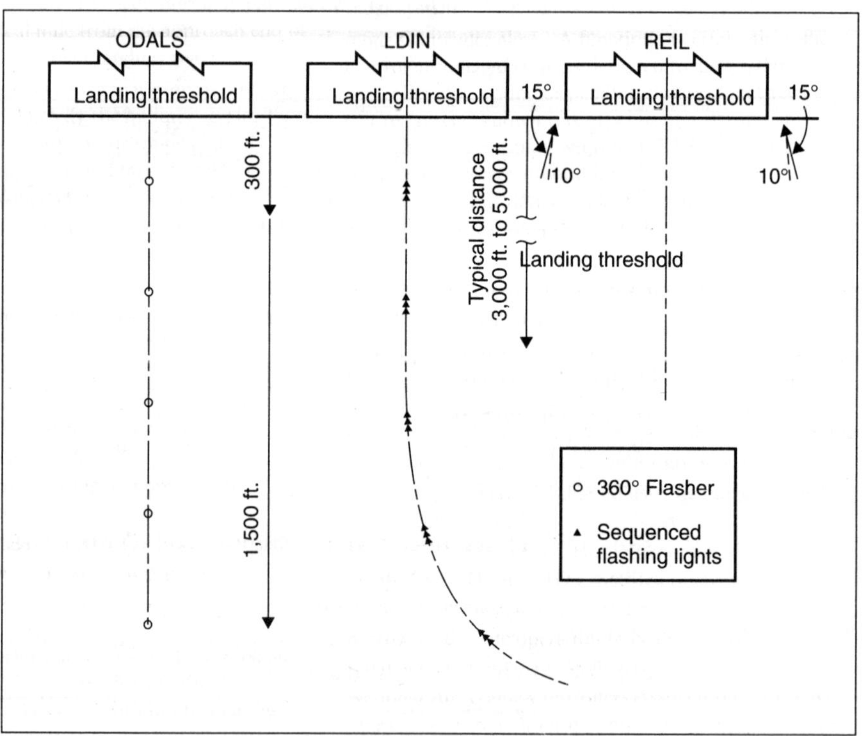

Figure 2-73. ODALS, LDIN, and REIL installations.

Figure 2-74. A typical runway end identifier light installation.

Omnidirectional Approach Lighting System **Omnidirectional approach lighting systems (ODALSs)** are used to delineate the flight path that should be used by a pilot approaching a specific runway. The lights are installed in groups, are omnidirectional, and flash in sequence. ODALSs may be installed directly in front of the runway or may be placed miles from the airport, under the proper flight path. ODALSs are also being used experimentally to delineate VFR flight paths in congested areas.

Vertical Guidance Systems The previously described systems were primarily designed to provide lateral guidance to the pilot, with vertical guidance being provided by an electronic glide path. At night, or during periods of reduced VFR visibility, pilots are deprived of many of the visual cues used to determine the proper glide path. Without these cues, pilots may be unable to correctly orient their aircraft during the final approach phase and may misjudge their distance, glide angle, or rate of descent. Any miscalculation of one of these factors may cause the pilot to incorrectly approach the runway and collide with obstructions in the approach path or land at an excessive speed and roll off the end of the runway. Since it is financially impractical to install an electronic glide path at every runway across the United States, an inexpensive method of glide path indication was necessary.

After extensive evaluation at the National Aviation Facilities Experimental Center (NAFEC) in Atlantic City, New Jersey, in 1960 the FAA introduced the **visual approach slope indicator (VASI)** system. VASI lights are designed to

Michael Nolan

Figure 2-75. A typical VASI light installation.

be installed on runways with or without ILS approaches and can provide pilots with accurate glide path information as far as 20 miles from the runway. The VASI system utilizes either two or three light units arranged to provide the pilot with a visual glide path. These light units are next to the runway, with the first located approximately 700 feet and the second approximately 1,200 feet from the approach end (see Figure 2-75).

Each VASI unit provides a narrow beam of light filtered such that the upper portion (above the glide path) of the beam is white and the lower portion (below the glide path) is red. Pilots looking at a VASI light know that the aircraft is too high if they see a white light and too low if they see a red light. The two VASI units are installed such that a pilot on the desired glide path is above the near VASI (the white beam) but below the far VASI (the red beam). A pilot who is too high will see the white light from both units, whereas the pilot who is too low will see the red beams from both (see Figure 2-76).

The glide path provided by the standard two-light VASI system is of insufficient altitude for large aircraft (such as DC-10s and 747s) conducting approaches to the runway. At airports frequented by these types of aircraft, a third light bar is installed farther down the runway. Pilots of these aircraft utilize the middle and far VASI units, and pilots of small aircraft utilize the near and middle VASI units.

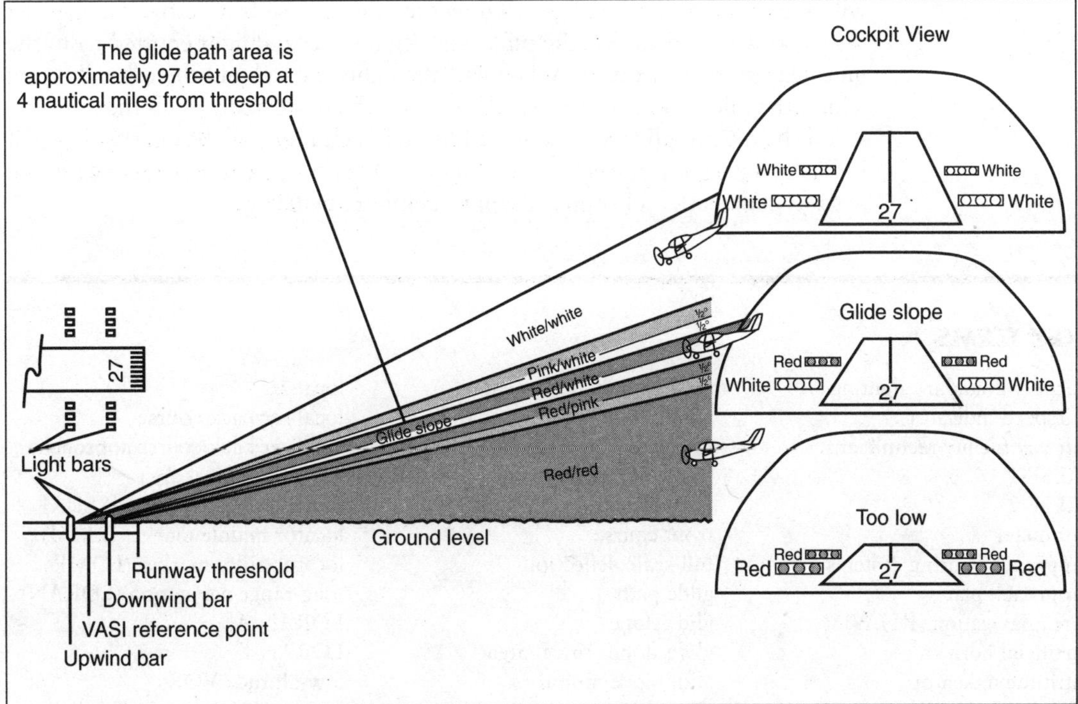

Figure 2-76. VASI operation.

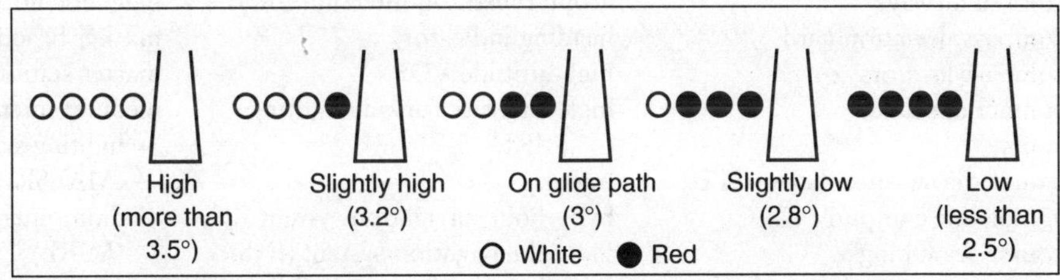

Figure 2-77. PAPI operation.

Precision Approach Path Indicator The VASI system is highly effective but can be difficult to use since the pilot must constantly observe light units that are separated by up to 1,000 feet. A similar system, the **precision approach path indicator (PAPI),** has been developed that remedies this situation (see Figure 2-77).

PAPI units are similar to VASI units but are installed in a single row. Each light unit emits a white and a red beam but at progressively higher angles. If the pilot is more than half a degree above the desired flight path, all the light units will appear to emit white light. But as the pilot descends to a lower angle, the system is designed so that the pilot will begin to see red light emitted from the unit nearest to the runway. When half the lights are red and the other half are white, the pilot is on the desired glide path, which is usually 3°. If the pilot descends below this glide path angle, additional light units will be observed as red. If all the light units appear red, the pilot is in excess of half a degree below the desired glide path and should begin to climb immediately.

KEY TERMS

airport boundary lighting
airspeed indicator
airway facility technicians
ALSF-1
ALSF-2
altimeter
approach lighting systems
approach plates
area navigation (RNAV)
artificial horizon
attitude indicator
autoland
automatic direction finder (ADF)
chain
clear zone
colored airways
compass deviation card
compass locators
contact approach
course
course deviation indicator (CDI)
course-line computer (CLC)
course scalloping
crosswind correction angle
crosswinds
data converter
decision height (DH)
deduced (dead) reckoning
deviation
differential GPS (DGPS)
direction finder (DF)
Doppler effect
Doppler radar
Doppler VOR (DVOR)
false courses
false glide paths
final approach fix (FAF)
final approach segment
flight check
four-course radio range
frequency shift
Fresnel lens
front course
full-scale deflection
glide path
glide slope
glide slope critical area
glide slope monitors
Global Navigation Satellite System (GNSS)
ground track
ground wave
group repetition interval (GRI)
heading indicator
high-altitude VORs
high-intensity runway lighting (HIRL)
homing
hyperbolic navigation system
inertial navigation system (INS)
initial approach fixes (IAFs)
initial approach segment
inner marker (IM)
instrument approach procedure (IAP)
instrument approach procedure charts
intermediate approach segment
interrogator
Landing Aids Experiment Station (LAES)
line of sight
local area augmentation system (LAAS)
localizer
localizer back course
localizer back course approaches
localizer critical area
localizer directional aid (LDA)
locator middle marker (LMM)
locator outer marker (LOM)
long-range navigation (LORAN)
LORAN-A
LORAN-C
low-altitude VORs
low-intensity runway lighting (LIRL)
magnetic compass
magnetic heading
magnetic north
marker beacons
master station
medium-intensity approach lighting system with RAIL (MALSR)
medium-intensity runway lighting (MIRL)
middle marker (MM)
minimum descent altitude (MDA)
minimum en route altitude (MEA)
minimum obstruction clearance altitude (MOCA)
missed approach point (MAP)
missed approach procedure
missed approach segment
National Ocean Service
nondirectional radio beacon (NDB)
nonprecision approach
omni bearing selector (OBS)

omnidirectional approach lighting system (ODALS)
outer marker (OM)
phantom VORTAC
pilot-controlled lighting (PCL)
pilotage
precision approach
precision approach path indicator (PAPI)
procedure turn
radial
range time
remote digital display
reverse sensing
runway alignment indicator lights (RAILs)
runway centerline lights
runway edge lights
runway end identifier lights (REILs)
runway threshold lights
runway visual range (RVR)
sectional charts
sequenced flashing lights (SFL)
service volumes
simplified directional facility (SDF)
simplified short approach lighting system (SSALS)
simplified short approach lighting system with RAIL (SSALR)
slant range distance
slave station
taxiway edge lighting
taxiway turnoff lights
terminal instrument approach procedures (TERPS)
terminal VORs (TVORs)
touchdown zone lighting
transmissometer detector
transmissometer projector
transponder
true course
true heading
true north
turn and bank indicator
turn coordinator
ultra-high frequency (UHF)
variation
very high frequency (VHF)
visual approach
visual approach slope indicator (VASI)
visual aural range (VAR)
visual navigation
VORTAC
waypoint
wide area augmentation system (WAAS)
wind correction angle
world aeronautical charts (WACs)

REVIEW QUESTIONS

1. What are the differences between sectional and world aeronautical charts?
2. What is the difference between "dead reckoning" and pilotage?
3. What are the operating limitations of NDB, VOR, RNAV, LORAN, and GNSS?
4. What is the purpose of each segment of an instrument approach procedure?
5. What are the basic principles of the instrument landing system?
6. What are the basic principles of operations of both WAAS and LAAS?
7. What are the limitations of the GNSS?
8. What are the functions of runway edge, embedded, and approach lighting systems?

3

Air Traffic Control System Structure

Checkpoints

After studying this chapter, you should be able to:

1. Identify which aircraft are separated by the ATC system.
2. State the differences among classes of airspace.
3. Describe the purpose of a departure procedure (DP).
4. Define the function of positive controlled, controlled, and uncontrolled airspace.
5. Describe the services offered to pilots in each airspace class.
6. Describe operations under special VFR.

Airspace Classification

The airspace above the United States has been categorized by the FAA into different classes, with specific requirements and different operating rules for each class of airspace. The intent of the classification scheme is to provide maximum pilot flexibility with acceptable levels of risk appropriate to the type of operation and traffic density within that class of airspace. In general, the classification scheme is designed to provide maximum separation and active control in areas of dense or high-speed flight operations. In areas of light traffic, if acceptable weather conditions exist, the pilots themselves can provide much of the needed traffic separation.

General Categories of Airspace

The airspace classification scheme essentially provides for three general categories of airspace: one in which ATC separates all aircraft, whether IFR or VFR, known as **positive controlled airspace (PCA)**; airspace in which ATC separates IFR aircraft but, weather permitting, VFR pilots provide their own separation, generally known as **controlled airspace**; and airspace within which the pilots provide all separation, known as **uncontrolled airspace.**

Positive Controlled Airspace Within positive controlled airspace, the FAA either absolutely prohibits VFR flight operations, or if permitted, separates both VFR and IFR aircraft. Positive controlled airspace is reserved for either very high-altitude flights — above 18,000 feet mean sea level (MSL) — or around high-density areas.

Controlled Airspace Within airspace generally designated as controlled airspace (which is most but not all of the airspace overlying the continental United States), ATC separation services are provided by the FAA to IFR aircraft. IFR aircraft are authorized to fly into clouds or areas of reduced visibility and are provided ATC assistance to remain separated from other IFR aircraft. IFR aircraft (when operating in areas where weather conditions and traffic density permit other aircraft to be safely observed and avoided) are still responsible for separating themselves from VFR aircraft. VFR aircraft operating in controlled airspace are still responsible for separating themselves from both IFR and other VFR aircraft.

Uncontrolled Airspace In uncontrolled airspace, ATC separation services are not provided by the FAA; all aircraft, whether IFR or VFR, are obligated to provide their own separation, regardless of weather condition.

Controlled Versus Uncontrolled Airspace

One of the primary differences between airspace types is that air traffic control separation services can be offered to pilots operating only in controlled airspace. Additional services, such as traffic advisories and safety alerts, can be offered to aircraft flying in uncontrolled airspace, but only on a **workload-permitting** basis.

Early in this century, most of the airspace above the United States was designated as uncontrolled. Only the **federal airways** and the airspace around very busy airports was designated as controlled. But as air traffic increased, and the technical air traffic control capabilities of the federal government improved, additional segments of the nation's airspace have been designated as controlled airspace. About the only uncontrolled airspace left above the continental United States is below 1,200 feet **above ground level** (**AGL**) away from busy airports.

Various names for these airspace categories and the rules for operation within each area have evolved as they were created. In time, due to the unique development of ATC within the United States, the names and rules of operation within each area have become inconsistent with ICAO standards in use throughout the rest of the world. By 1992, the FAA had identified and created operational rules for close to 20 different categories of airspace.

Airspace Review

In 1982, in an effort to standardize and simplify U.S. airspace, the FAA and representative industry groups formed the National Airspace Review (NAR) committee to begin a comprehensive review of the nation's airspace system. In conjunction with an ICAO review commission, the NAR committee recommended that airspace over the United States be reclassified into one of six classes. These recommendations were accepted by the FAA and implementation began in 1993. These new categories of airspace are known as Class A, B, C, D, E, and G airspace. Class F airspace exists as an ICAO classification with no equivalent existing in the United States.

As currently defined, Class A, B, C, D, and E airspace is generally some form of controlled airspace. Class G airspace is designated as uncontrolled. In general, Class A airspace is the most restrictive, and ATC provides maximum services and separation. Class G airspace, on the other hand, is the least restrictive, and few ATC services are provided. Class B, C, D, and E airspace spans the range of services.

IFR Flight in Controlled Airspace

Within controlled airspace, air traffic controllers are required to separate IFR aircraft and participating VFR aircraft using the procedures specified in the ***Air Traffic Control Handbook***. (These procedures are discussed in detail in Chapters 7 and 9.) Since nonparticipating VFR aircraft may also be permitted to operate in some areas of controlled airspace, it remains the responsibility of participating pilots to see and avoid these aircraft, regardless of the services being provided by the air traffic controller.

Before beginning an IFR flight in controlled airspace, the pilot is required to file a flight plan (see Figure 3-1) with the FAA and receive a clearance from an ATC facility. A general aviation or corporate pilot usually files the IFR flight plan with a flight service station specialist, using a home computer or over the telephone, and the information is then forwarded to the air route traffic control center (ARTCC) with jurisdiction over the departure airport. Airline flight plans are normally filed directly with the ARTCC using stored flight plan information. A recent development in computer networking permits private weather-briefing corporations to directly file general aviation or corporate air-

Form Approved: OMB No. 04-R0072

DEPARTMENT OF TRANSPORTATION
FEDERAL AVIATION ADMINISTRATION

FLIGHT PLAN

CIVIL AIRCRAFT PILOTS. FAR Part 91 requires you file an IFR flight plan to operate under instrument flight rules in controlled airspace. Failure to file could result in a civil penalty not to exceed $1,000 for each violation (Section 901 of the Federal Aviation Act of 1958, as amended). Filing of a VFR flight plan is recommended as a good operating practice. See also Part 99 for requirements concerning DVFR flight plans.

1 TYPE	2 AIRCRAFT IDENTIFICATION	3 AIRCRAFT TYPE/ SPECIAL EQUIPMENT	4 TRUE AIRSPEED	5 DEPARTURE POINT	6 DEPARTURE TIME		7 CRUISING ALTITUDE
VFR					PROPOSED (Z)	ACTUAL (Z)	
IFR							
DVFR			KTS				

8 ROUTE OF FLIGHT

9 DESTINATION (Name of airport and city)	10 EST. TIME ENROUTE		11 REMARKS
	HOURS	MINUTES	

12 FUEL ON BOARD		13 ALTERNATE AIRPORT(S)	14 PILOT'S NAME, ADDRESS & TELEPHONE NUMBER & AIRCRAFT HOME BASE	15 NUMBER ABOARD
HOURS	MINUTES			

16 COLOR OF AIRCRAFT

CLOSE VFR FLIGHT PLAN WITH__________FSS ON ARRIVAL

FAA Form 7233-1 (5-77)

Figure 3-1. Sample flight plan form.

craft flight plans into the ARTCC computer using the **Direct User Access Terminal (DUAT)** system.

If a pilot needs to file a flight plan while airborne, the ATC facility in contact with the pilot transmits the flight plan information to the proper ATC facility. The minimum required information that must be received from the pilot when filing a flight plan is specified in the *Air Traffic Control Handbook*. This information includes the following:

1. The aircraft identification number. This is either the aircraft's assigned serial number, if it is a general aviation or corporate flight, or the airline name and flight number.

2. The aircraft type and navigation equipment installed on the aircraft. The aircraft type is abbreviated, utilizing the codes found in the *Air Traffic Control Handbook*. An expanded list is included in the appendix to this book. Examples of these abbreviations include:

Piper Cherokee	PA28
Boeing 747	B747
Lear	LR35
Cessna Skyhawk	C172

The pilot must also identify the navigational capabilities of the aircraft by appending a unique suffix code to the aircraft type. The equipment codes are found in the handbook. The aircraft type is separated from the equipment code with a slash. The equipment codes are as follows:

	NO DME
/X	No transponder
/T	Transponder with no Mode C
/U	Transponder with Mode C
	DME
/D	No transponder
/B	Transponder with no Mode C
/A	Transponder with Mode C
	TACAN ONLY
/M	No transponder
/N	Transponder with no Mode C
/P	Transponder with Mode C
	AREA NAVIGATION (RNAV)
/Y	LORAN, VOR/DME, or INS with no transponder
/C	LORAN, VOR/DME, or INS, transponder with no Mode C
/I	LORAN, VOR/DME, or INS, transponder with Mode C
	ADVANCED RNAV WITH TRANSPONDER AND MODE C (If an aircraft is unable to operate with a transponder and/or Mode C, it will revert to the appropriate code listed above under Area Navigation.)
/E	Flight Management System (FMS) with en route, terminal, and approach capability
/F	A single FMS with en route, terminal, and approach capability (U.S. and U.S. territories only unless otherwise authorized.)
/G	Global Positioning System (GPS)/Global Navigation Satellite System (GNSS) equipped aircraft with en route and terminal capability
/R	Required Navigational Performance (Denotes capability to operate in RNP-designated airspace and routes.)
/W	Reduced Vertical Separation Minima (RVSM)
/Q	Required Navigation Performance (RNP) and Reduced Vertical Separation Minima (RVSM) (Indicate approval for application of RNP and RVSM separation standards.) It should be noted that /Q is for automation purposes only and will not be filed by system users. FAA processors will convert the combination of /R + /W to = /Q.

3. The requested route of flight. This must include the airway and navigation aid identifiers. The entire route of flight must be specified. When changing from one airway to another, the intersection fix must be specified. If no airway is being used, only the navaids need to be specified.

4. Requested cruising altitude.
5. Destination airport.

Pilots operating IFR in the air traffic control system must then be issued a clearance by ATC prior to beginning their flight. This clearance must include the following information and should be communicated to the pilot in this sequence:

1. Aircraft identification number. (Sample phraseology: "Cherokee five one four papa uniform," "United seven thirty-one," "Comair fifteen forty-three.")

2. Clearance limit. This is the farthest location the aircraft is cleared to fly to. Although the clearance limit is normally the destination airport, it may be an intermediate navigation aid or intersection located along the route of flight. If the clearance limit is not the destination airport and the pilot does not receive an additional clearance before reaching the clearance limit, the aircraft will enter a holding pattern at that point. ("Cleared to the Lafayette Purdue University Airport," "Cleared to the Boiler VOR," "Cleared to the Staks intersection.")

3. Departure procedure. If the assigned route of flight does not begin at the departure airport, it is necessary for the controller to assign a **departure procedure (DP)** so that the aircraft can intercept the route of flight. Departure procedures may also be used to ensure that the aircraft avoids areas of obstructions or high-density traffic. Departure procedures direct the pilot to turn or fly a particular heading or route. If a particular departure instruction is routinely issued to most of the departing aircraft, it may be incorporated into and published as a departure procedure (DP) (see Figures 3-2 and 3-3). DPs are constructed by the FAA and are published and sold by the same agencies that publish instrument approach charts. Routine use of DPs relieves controllers from repeating the same departure clearance to every aircraft. If a DP is to be used in a departure clearance, the controller assigns the DP procedure by simply including its name in the clearance. ("After departure, turn left heading three five zero," "After departure, fly runway heading," "O'Hare one departure.")

4. Route of flight. The route of flight issued includes any airways or VOR radials that the pilot will utilize when navigating to the clearance limit ("Via victor two fifty-one," "Via direct Danville," "Via the Boiler one eight five radial and the Danville zero niner radial.")

5. Altitude assignment. The controller should attempt to issue the pilot an altitude that conforms with the procedures contained in the handbook. The proper use of such altitudes will organize the flow of traffic and reduce the hazard of midair collisions, since each aircraft operating at the same altitude will be traveling in roughly the same direction. If circumstances require that a different altitude be issued to an aircraft, the controller is permitted to assign a nonstandard altitude, but advance coordination with adjacent ATC facilities must be accomplished, advising the next controller that the aircraft is not at the proper altitude. Table 3-1 gives handbook guidelines for altitude assignment.

When issuing clearances, a controller should never assign an altitude lower than the **minimum en route altitude (MEA).** The controller should also attempt to assign an altitude as close as possible to that filed in the original flight plan. To meet these two requirements, the controller may assign a sequence of crossing altitudes that will

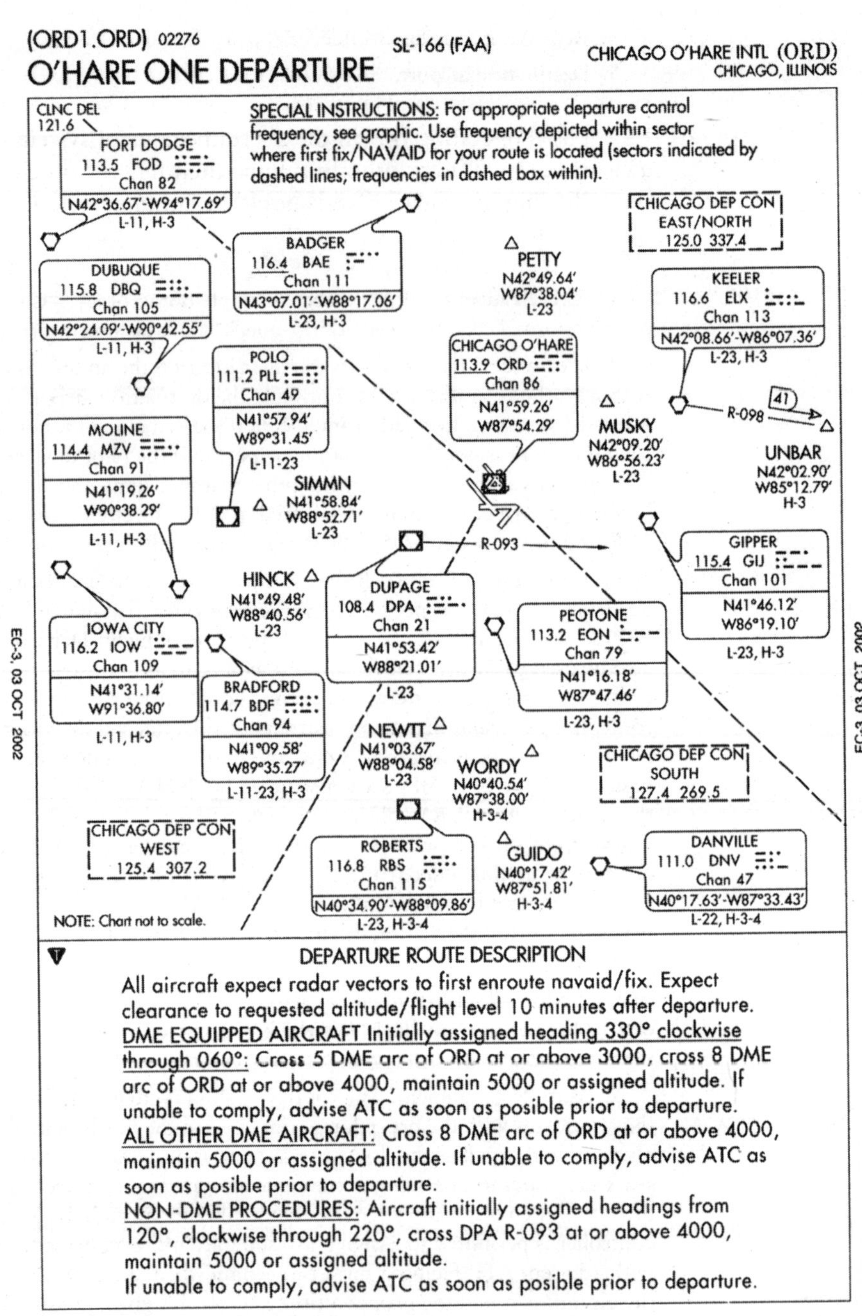

Figure 3-2. Standard departure procedure (vector) for Chicago O'Hare International Airport.

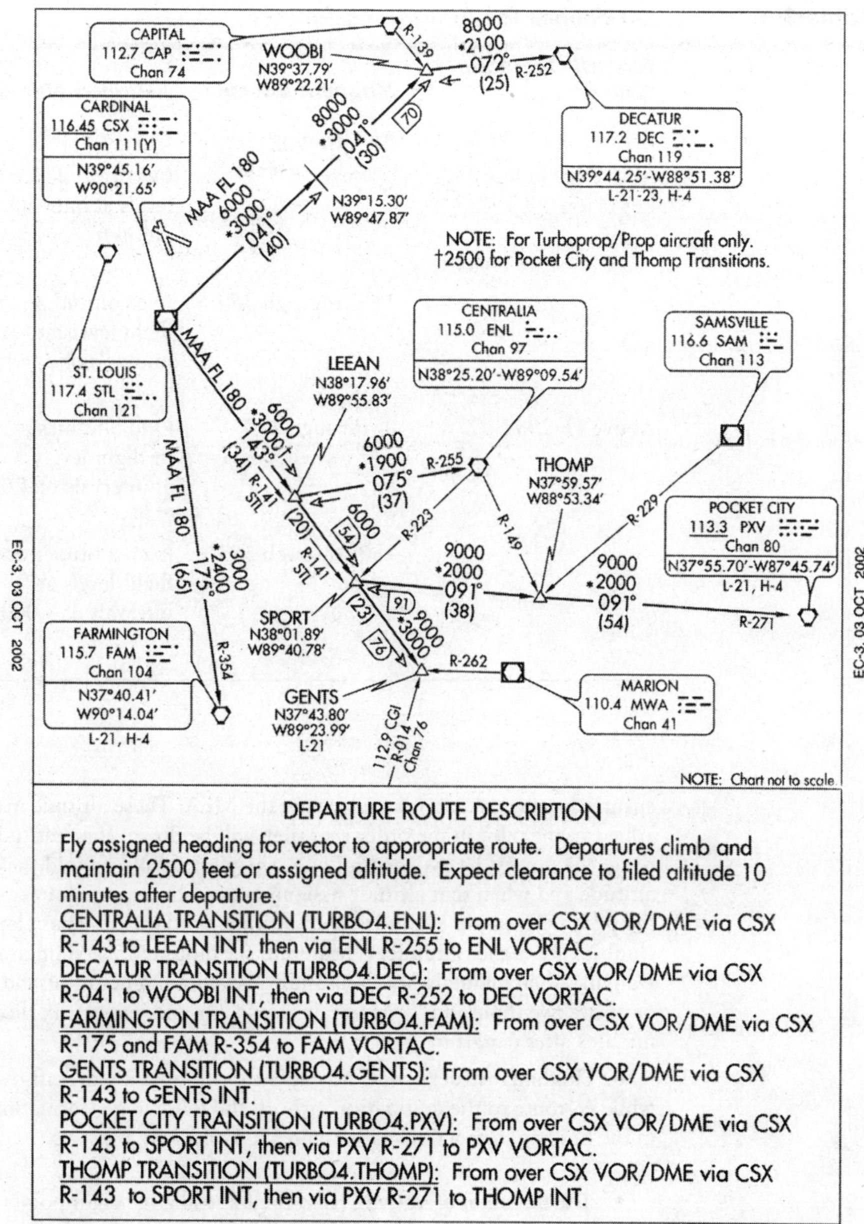

Figure 3-3. Departure procedure (pilot navigation) for Alton, Illinois

Table 3-1. Guidelines for Altitude Assignment

Aircraft Operating Altitude	Magnetic Course	Assigned Altitude	Example
Below 3,000 feet AGL	Any heading	Any altitude	
Above 3,000 feet AGL but below FL 290	0° through 179°	Odd altitudes or flight levels at intervals of 2,000 ft.	3,000 ft. 5,000 ft. FL 250 FL 270
	180° through 359°	Even altitudes or flight levels at intervals of 2,000 ft.	4,000 ft. 6,000 ft. FL 240 FL 260
Above FL 290	0° through 179°	Odd altitudes or flight levels at intervals of 4,000 ft.	FL 290 FL 330 FL 370 FL 410
	180° through 359°	Even altitudes or flight levels at intervals of 4,000 ft.	FL 310 FL 350 FL 390 FL 430

ensure that the aircraft is never below the MEA. These altitude instructions should be issued to the pilot in the order that they will be flown. If an altitude lower than an en route MEA must be assigned initially, the pilot should be told the expected final altitude and when that altitude assignment can be expected. In case of radio failure, the pilot will remain at the assigned altitude until the time has elapsed and will then climb to the higher altitude. ("Maintain six thousand," "Maintain four thousand until Danville, then maintain six thousand," "Maintain niner thousand. Cross Danville at or above five thousand," "Maintain four thousand; expect six thousand one zero minutes after departure.")

6. Holding instructions. If it is necessary to hold an aircraft over a particular fix while en route to the destination airport, the following information must be included in the holding instructions (see Figures 3-4 and 3-5):

- The direction of holding from the fix, using the eight points of the compass: north, northeast, east, southeast, etc.
- The name of the holding fix.
- The radial, course, bearing, azimuth, airway, or route on which the aircraft is to hold.
- The direction of the turns in the holding pattern if a nonstandard holding pattern will be used. A standard holding pattern requires right-hand turns; a nonstandard pattern utilizes left turns.

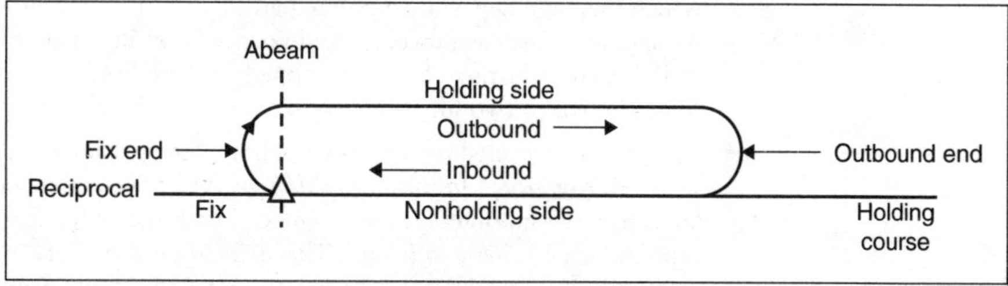

Figure 3-4. Holding-pattern description.

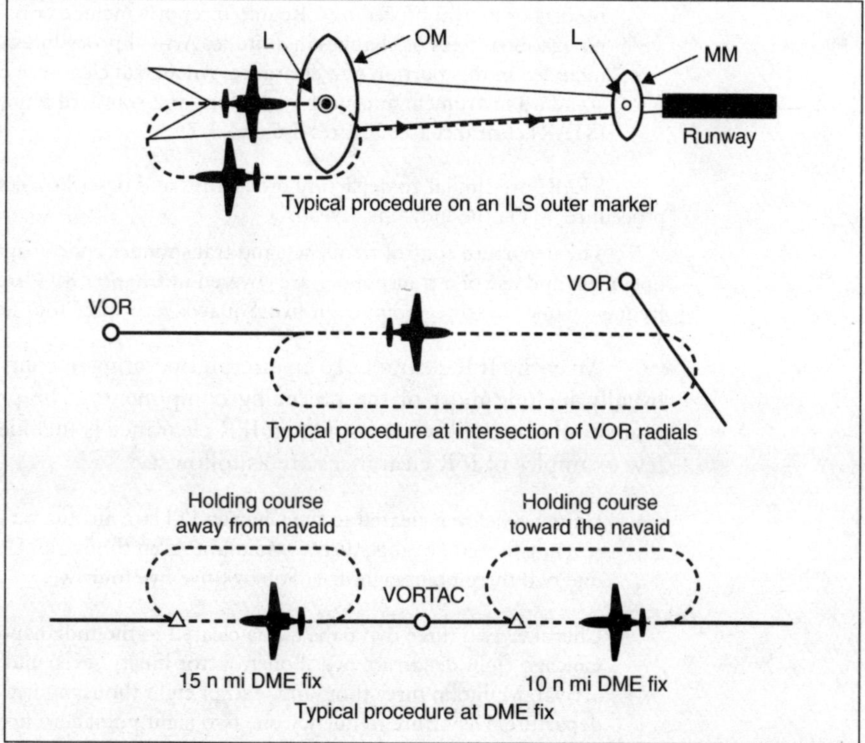

Figure 3-5. Examples of holding.

- The holding-pattern length if a nonstandard holding pattern is being used. A standard holding pattern has a 1-minute inbound leg length (1½ minutes inbound leg length if the aircraft is holding above 14,000 feet).
- The **expect further clearance** (EFC) time. If pilots lose radio contact with ATC, they are expected to remain in the holding pattern until the EFC time, after

which they will depart the holding pattern and continue along the route of flight issued in the last clearance. ("Hold northwest of Boiler on the three two three radial. Expect further clearance at one five three five," "Hold southwest of Vages on victor two fifty-one. Expect further clearance at two three four one.")

- The pilot is expected to enter the holding pattern using the procedures described in the *Aeronautical Information Manual.* The pilot will maneuver the aircraft so as to track inbound on the assigned course and will attempt to make the inbound leg 1 minute in length. This is the only way that a pilot can hold and accurately time the inbound leg length. Air traffic controllers should never issue holding instructions that require a pilot to hold outbound from the holding fix. Since the inbound leg would not be located along any defined course, it would be impossible for the pilot to hold properly.
- Any additional clearance information. This information might include position reports or arrival procedures. Required reports include crossing certain navigational fixes or changes in altitude. Arrival procedures may also be included in this portion of a clearance. An arrival clearance could be either a standard instrument approach procedure or a standard terminal arrival route (STAR) clearance (see Figures 3-6 and 3-7).

STARs are similar to departure procedures and describe a common arrival procedure. ("Via the Indy one arrival.")

7. The departure control frequency and transponder code assignment. (The operation and use of a transponder are covered in Chapter 8.) ("Departure control frequency one two three point eight five. Squawk zero three four five.")

An entire IFR clearance to an aircraft operating in controlled airspace will usually include most of the preceding components. The proper phraseology that should be used when issuing an IFR clearance is included in Chapter 4. A few examples of IFR clearances are as follows:

> United six eleven cleared to the Chicago O'Hare airport via direct Boiler, victor seven, Chicago Heights, direct. Maintain seven thousand. Departure frequency one two three point eight five. Squawk five five four five.

> Cherokee two three two papa alpha cleared to the Indianapolis airport via the Chicago eight departure over Boiler, victor ninety-seven and the Indy seven arrival. Maintain three thousand, except eight thousand five minutes after departure. Departure frequency one two eight point zero five, squawk five five four three.

Clearance Amendments As the IFR flight progresses toward the destination airport, the clearance may need to be amended by ATC. The entire clearance need not be repeated, only those items that have been changed by the controller. For example:

> Five four papa uniform, climb and maintain seven thousand.

> United six eleven cleared to the Indianapolis airport via victor ninety-seven west.

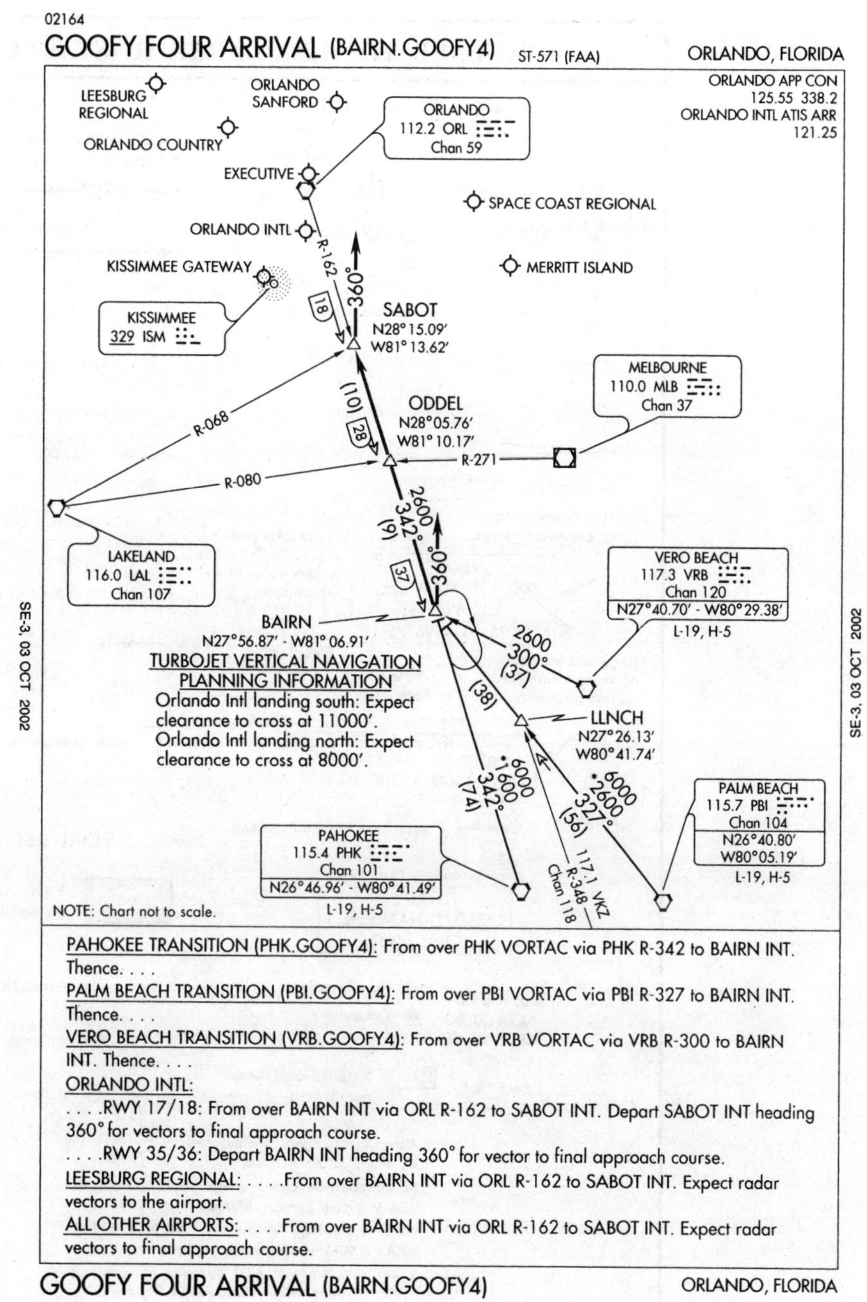

Figure 3-6. Standard terminal arrival route chart for Orlando, Florida.

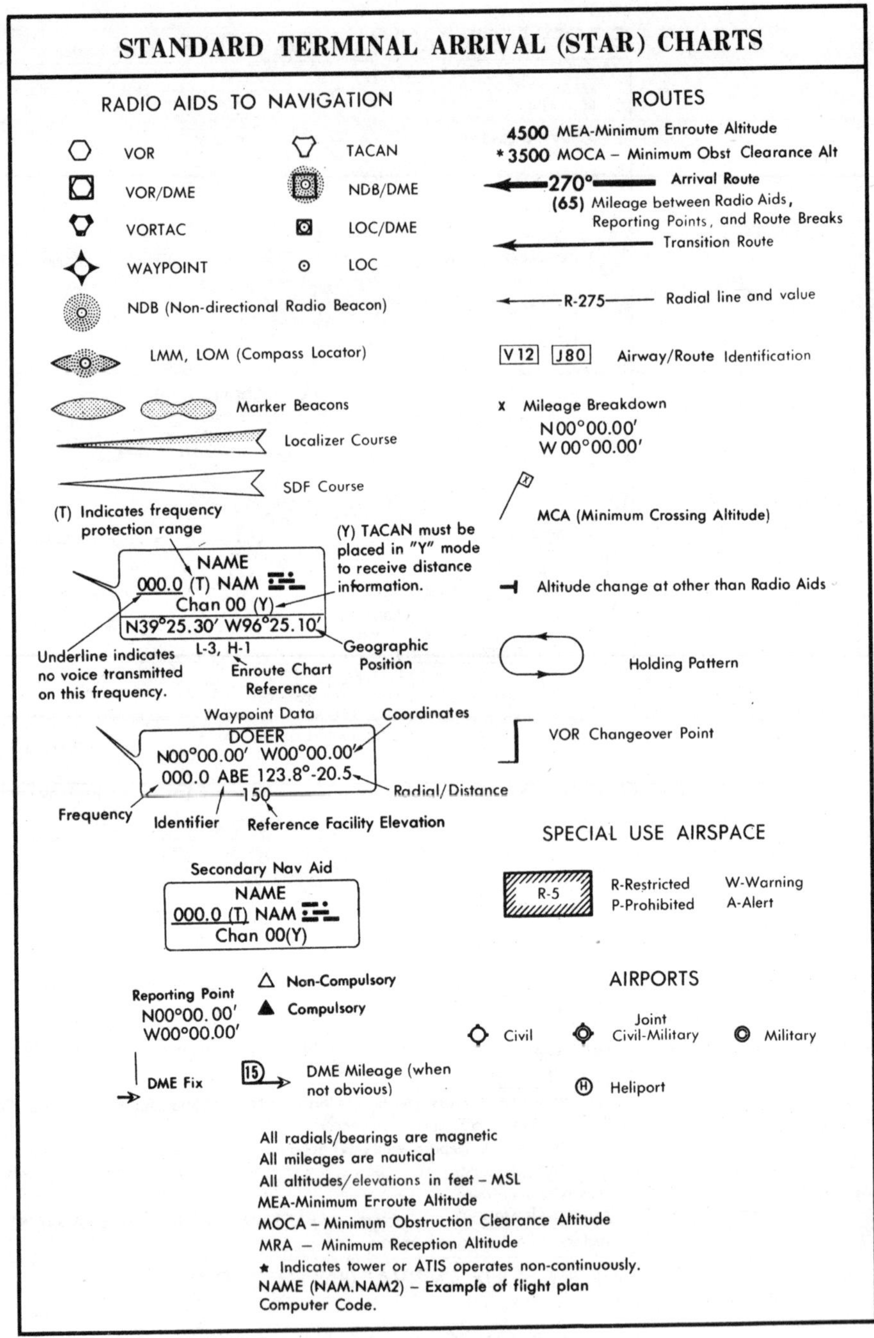

Figure 3-7. Standard terminal arrival route chart legend.

American two thirty-one, descend and maintain four thousand, cross two zero DME southeast of Boiler at or below niner thousand.

VFR Flight in Controlled Airspace

In controlled airspace the FAA offers both separation and additional ATC services to pilots. But depending on the type of flight and the category of airspace involved, the pilot may not be required to use these services or even to contact air traffic control facilities. Within controlled airspace, IFR flights are required to receive these services, but VFR flights may not be. In general, as long as VFR pilots can meet the weather minima outlined by Federal Aviation Regulation (FAR) 91 and are not entering any restricted class airspace, no contact with ATC is required.

VFR pilots may fly in controlled airspace as long as they comply with the regulations included in FAR 91. According to these regulations:

1. VFR pilots must generally provide their own separation from other VFR and IFR aircraft and the terrain.

2. VFR pilots are not required to file a flight plan or contact ATC unless they are planning to enter an area of restricted class airspace where contact is mandatory. VFR flight plans are voluntary and are used by the FAA only to assist in locating lost or overdue aircraft.

3. The weather conditions during flight must meet the criteria specified in FAR 91.155. VFR pilots must also maintain the minimum cloud distance stipulated in the FARs. The minimum visibility and distance from the clouds vary with the aircraft's cruising altitude and the class of airspace within which the flight is operating.

Aircraft Cruising Altitude	***Flight Visibility***	***Distance from Clouds***
1,200 ft. or less above the surface, regardless of MSL altitude	1 statute mile	500 ft. below 1,000 ft. above 2,000 ft. horizontal
More than 1,200 ft. above the surface but less than 10,000 ft. MSL	3 statute miles	500 ft. below 1,000 ft. above 2,000 ft. horizontal
More than 1,200 ft. above the surface and at or above 10,000 ft. MSL	5 statute miles	1,000 ft. below 1,000 ft. above 1 statute mile horizontal

These minima are designed to maximize the chances of a VFR pilot seeing and avoiding other VFR and IFR aircraft. If the pilot is unable to comply with these minima, VFR flight cannot legally be conducted. The pilot must then either land or receive an IFR or a special VFR clearance to continue the flight. Special VFR clearances permit VFR pilots to fly in certain weather conditions that do not meet minimum VFR criteria. VFR aircraft operating under special VFR clearances are afforded IFR separation, however.

4. VFR pilots operating in controlled airspace are required to fly at the proper altitude for the direction of flight unless otherwise requested by ATC. The following table lists these altitudes.

Magnetic Course	Below 18,000 ft. MSL	At or Above 18,000 ft. but Below FL 290	At or Above FL 290
0° to 179°	Odd thousands plus 500 ft. (3,500, 5,500, etc.)	Odd flight levels plus 500 ft. (FL 195, 215, etc.)	4,000 ft. intervals (FL 300, 340, etc.)
180° to 359°	Even thousands plus 500 ft. (2,500, 4,500, etc.)	Even flight levels plus 500 ft. (FL 185, 205, etc.)	4,000 ft. intervals (FL 320, 360, etc.)

These altitudes were chosen to minimize the potential for midair collisions between two aircraft flying in opposite directions. Whenever assigning altitudes to VFR aircraft, controllers should attempt to comply with this regulation. But if traffic conditions dictate, controllers are permitted to assign a nonstandard cruising altitude to VFR aircraft receiving ATC services. When these ATC services are terminated, however, the VFR pilot should be advised to return the aircraft to the proper altitude as soon as is feasible.

Airspace Classes

All the airspace above the United States has been designated by the FARs into one of six classes (see Table 3-2). In each of these areas both VFR and IFR pilots must comply with the regulations that have been previously mentioned in this chapter. In addition to those regulations, supplemental rules may apply to both VFR and IFR pilots who wish to operate in specific areas.

Class A Airspace

Class A airspace is generally defined as that airspace extending from 18,000 feet MSL up to and including FL 600, including the airspace overlying the waters within 12 nautical miles of the coast of the 48 contiguous states and Alaska and the designated international airspace beyond 12 nautical miles of the coast of the 48 contiguous states and Alaska within areas of domestic radio navigational signal or ATC radar coverage and within which domestic procedures are applied. Class A airspace is not specifically charted. Class A airspace evolved from the jet advisory areas that were created to provide advisory services to civilian and military turbojet aircraft operating at high altitudes. When the jet advisory areas were first created, they extended from FL 240 to FL 410 and projected 14 nautical miles laterally on either side of every jet route. It was believed that pilots would be unable to see and avoid any other VFR or IFR aircraft operating at the same altitude at the high airspeeds at which these aircraft routinely operated.

Within jet advisory areas, air traffic controllers were required to use radar to constantly monitor every IFR aircraft operating on a jet route and issue any

Table 3-2. Airspace Classification

Airspace Features	Class A	Class B	Class C	Class D	Class E	Class G
Former Airspace Equivalent	Positive Control Area	Terminal Control Area (TCA)	Airport Radar Service Area (ARSA)	Airport Traffic Area (ATA) and Control Zone (CZ)	General Controlled Airspace	Uncontrolled Airspace
Operations Permitted	IFR	IFR and VFR	IFR and VFR	IFR and VFR	IFR and VFR	IFR and VFR
Entry Requirements	ATC clearance	ATC clearance	ATC clearance for IFR. All require radio contact.	ATC clearance for IFR. All require radio contact.	ATC clearance for IFR. All IFR require radio contact.	None
Minimum Pilot Qualifications	Instrument Rating	Private or student certificate	Student certificate	Student certificate	Student certificate	Student certificate
Two-way Radio Communications	Yes	Yes	Yes	Yes	Yes for IFR	No
VFR Minimum Visibility	N/A	3 statute miles	3 statute miles	3 statute miles	3 statute miles	1 statute mile
VFR Minimum Distance from Clouds	N/A	Clear of clouds	500' below, 1,000' above, and 2,000' horizontal	500' below, 1,000' above, and 2,000' horizontal	500' below, 1,000' above, and 2,000' horizontal	Clear of clouds
Aircraft Separation	All	All	IFR, SVFR, and runway operations	IFR, SVFR, and runway operations	IFR and SVFR	None
Traffic Advisories	N/A	N/A	Yes	Workload permitting	Workload permitting	Workload permitting
Safety Alerts	Yes	Yes	Yes	Yes	Yes	Yes

heading changes (known as **vectors**) necessary to ensure that the IFR aircraft remained separated from any other aircraft observed on the controller's radar display.

The controllers were not usually in radio contact with the VFR aircraft, so it was impossible to determine their altitude, route of flight, or intentions. Because the actions of these aircraft could not be predicted, the controllers were forced to issue numerous unnecessary vectors to IFR aircraft to ensure that they would remain safely separated. Although this procedure might seem to decrease the probability of a midair collision, in many cases it actually made the situation more dangerous. Since the intentions of the VFR pilots were unknown, it was possible that heading changes could be issued to the IFR pilot at precisely the same moment that the VFR pilot began to maneuver to avoid the collision. This might create a situation even more dangerous than if no heading change had been issued at all.

It was soon obvious that unless the controller could be in direct radio contact with every aircraft operating in the vicinity of the jet routes, it would be impossible to positively separate IFR from VFR aircraft. In an attempt to rectify this problem, the FAA established the positive control area (PCA) in 1966. In 1993, the PCA was reclassified as Class A airspace.

Class A airspace is all the airspace located above the United States extending from 18,000 feet MSL to FL 600 (see Figure 3-8).

FAR 91.135 requires that every aircraft operating within Class A airspace operate under instrument flight rules and receive a clearance from ATC. This ATC separation of all aircraft is known as positive separation. To operate within Class A airspace, pilots must comply with the following regulations from FAR 91.

The pilot must be rated for instrument flight.

The aircraft must be operated under instrument flight rules at a route and at an altitude assigned by ATC.

All aircraft must be equipped as specified in FAR 91.215.

The creation of this airspace ensured that every aircraft operating above 18,000 feet MSL was provided separation services by air traffic controllers. Since the creation of Class A airspace, high-altitude midair collisions have become extremely rare in this country. Figure 3-9 summarizes all the airspace classifications over the United States.

Class B Airspace

Even though the establishment of Class A airspace virtually eliminated high-altitude midair collisions, as traffic increased around airports, low-altitude collisions began to occur with increasing frequency. The FAA responded by creating a low-altitude version of Class A airspace called a **terminal control area (TCA)**, which has since been reclassified as Class B airspace.

Class B airspace is defined as the airspace that extends from the surface of the Earth up to 10,000 feet MSL surrounding the nation's busiest airports in terms of IFR operations or passenger enplanements. The configuration of each

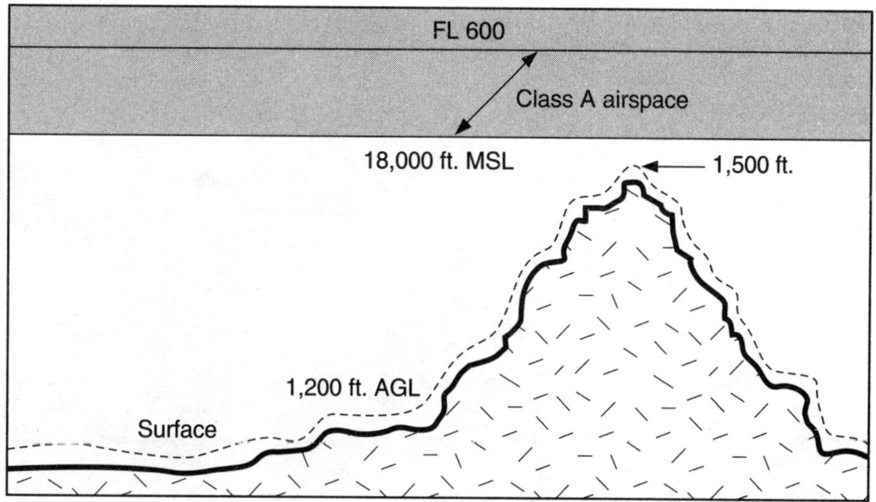

Figure 3-8. Class A airspace.

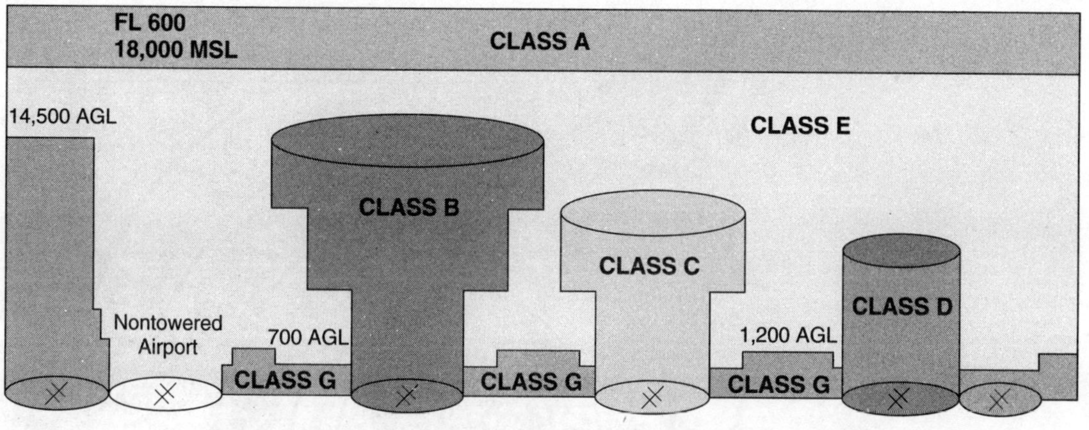

Figure 3-9. U.S. airspace classifications.

Class B airspace area is individually tailored and consists of a surface area and two or more layers (some Class B airspace areas resemble upside-down wedding cakes) and is designed to contain all published instrument procedures once an aircraft enters the airspace (see Figure 3-10). An ATC clearance is required for all aircraft to operate in the area, and all aircraft so cleared receive separation services within the airspace.

Each successive layer of Class B airspace extends out from the central airport, with the floor of each layer raised to a slightly higher altitude. This design

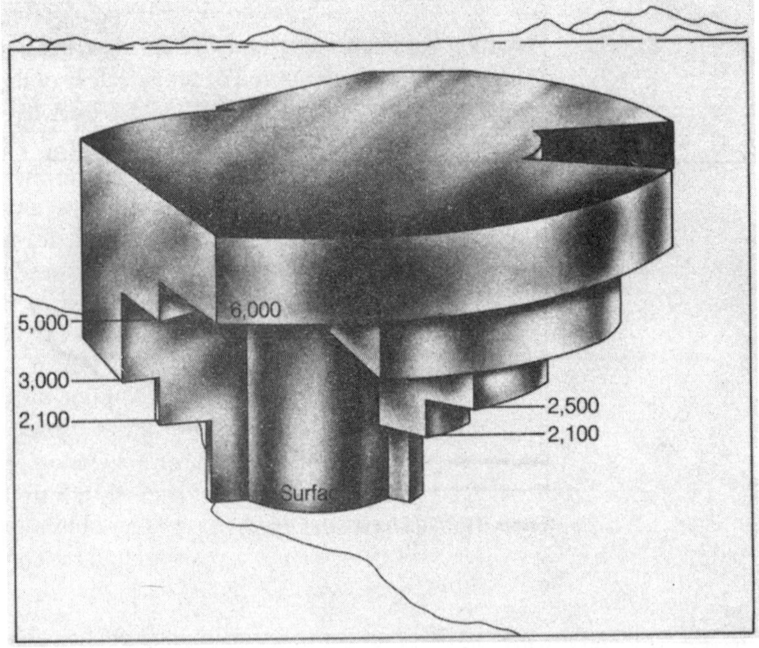

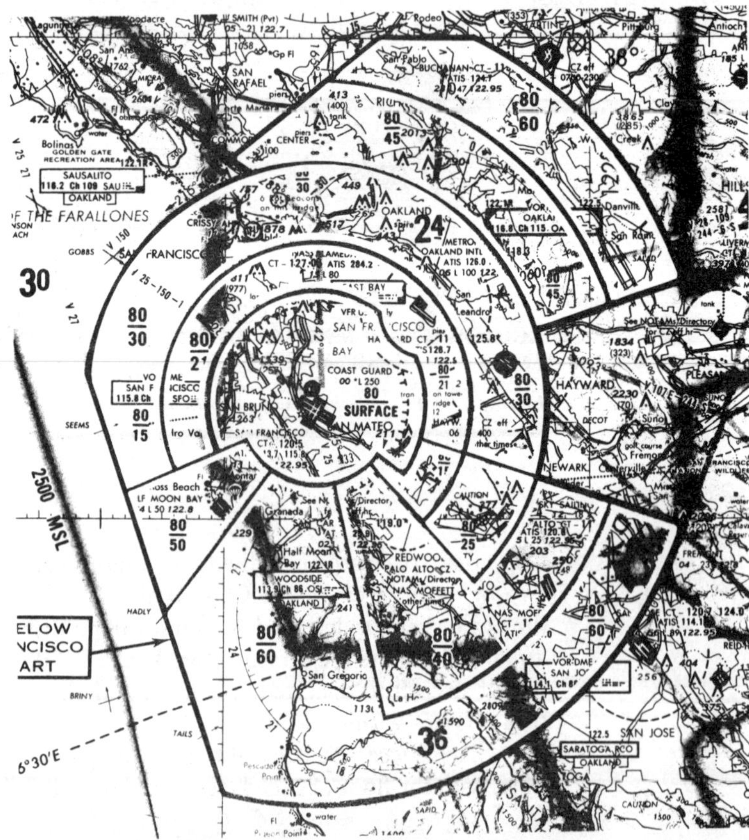

Figure 3-10. Graphic view of Class B airspace and the same airspace as depicted on a sectional chart.

provides the controller with sufficient airspace to vector aircraft to an instrument approach at the primary airport.

The separation procedures applied to aircraft operating within Class B airspace are similar to those applied to aircraft operating in Class A airspace. Prior to entering this airspace, both IFR and VFR pilots are required by FAR 91.131 to receive a clearance from the controlling ATC facility. While operating within the confines of Class B airspace, every pilot is required, if at all possible, to comply with the instructions issued by the controller. Air traffic controllers are responsible for the positive separation of every aircraft within Class B airspace, whether operating under instrument or visual flight rules. This generally means that aircraft operating at the same altitude must be kept at least 3 nautical miles apart. This separation need not be applied if there is at least 1,000 feet of altitude between the aircraft. If either of the aircraft is VFR, the separation may usually be reduced to 1 1/2 mile lateral or 500 feet vertical separation. If both aircraft are VFR or if one is VFR and the other is a small IFR, either 500 feet of vertical separation must be used, or the controller must ensure that the radar targets do not touch. This is known as **target resolution.**

While operating within or, in some cases, near Class B airspace, every pilot must comply with the following FAR 91 regulations:

> Every aircraft must be equipped with appropriate communication and navigation radio equipment. This includes a two-way radio transceiver, VOR or TACAN navigation capability, and a transponder. [A transponder permits the controller to positively identify any particular aircraft when using radar for ATC separation. Transponders are discussed in detail in Chapter 8.]
>
> Aircraft may not operate within the airspace underlying Class B airspace at an indicated airspeed greater than 200 knots.
>
> Unless specifically authorized by the controller, every turbine-powered aircraft operating to or from the primary airport must operate above the floor while within the lateral confines of the Class B airspace.
>
> Every aircraft entering Class B airspace or operating within 30 nautical miles of the primary airport must be equipped with a mode C altitude encoder. This device permits the aircraft's altitude to be shown directly on the controller's radar display.

Pilots operating on IFR flight plans do not need to specifically request permission to enter Class B airspace. VFR pilots, however, must request permission from the ATC facility prior to entering the airspace. Until permission is received from the controller, the VFR pilot is required to remain clear of Class B airspace.

IFR aircraft operating within Class B airspace have priority over VFR aircraft. Air traffic controllers are permitted to deny VFR aircraft clearances if conditions are such that, in the opinion of the controller, the entry of the VFR aircraft might compromise safety. These conditions include, but are not limited to, weather, traffic conditions, controller workload, and equipment limitations. But if the controller concludes that VFR operations can be safely approved, a

VFR clearance to enter may be issued to the pilot. Upon receiving the clearance, and after entering, the VFR pilot is required to comply with any instruction issued by the controller but must also observe the basic VFR flight rules. At no time may the VFR pilot disregard VFR flight rules while attempting to comply with a controller's request. If the pilot believes that the controller's instructions might cause a violation of any VFR flight rule, the pilot is authorized by FARs 91.3 and 91.123 to disregard that instruction but must inform the controller as soon as feasible.

The following terminal areas around the country are currently designated by FAR 71 as Class B airspace:

- Atlanta, Georgia
- Baltimore, Maryland
- Boston, Massachusetts
- Charlotte, North Carolina
- Chicago, Illinois
- Cincinnati, Ohio
- Cleveland, Ohio
- Dallas, Texas
- Denver, Colorado
- Detroit, Michigan
- Honolulu, Hawaii
- Houston, Texas
- Kansas City, Missouri
- Las Vegas, Nevada
- Los Angeles, California
- Memphis, Tennessee
- Miami, Florida
- Minneapolis, Minnesota
- New Orleans, Louisiana
- New York, New York
- Orlando, Florida
- Philadelphia, Pennsylvania
- Phoenix, Arizona
- Pittsburgh, Pennsylvania
- Saint Louis, Missouri
- Salt Lake City, Utah
- San Diego, California
- San Francisco, California
- Seattle, Washington
- Tampa, Florida
- Washington, D.C.

Class C Airspace

Airspace now known as Class C airspace was implemented in 1984 as **airport radar service areas (ARSAs)** to provide separation to aircraft flying within the vicinity of medium-sized airports that did not qualify for a TCA. The ARSA concept was developed by a task group formed by the FAA and composed of representatives from various industry groups and the Federal Aviation Administration.

This task force was commissioned to develop procedures that could improve safety around medium-activity airports. Some members of the group preferred to increase the number of TCAs, whereas other members pointed out that TCAs were in general too cumbersome, restrictive, discriminating, and confusing to general aviation pilots.

The committee examined the **terminal radar service area (TRSA)** program, which the FAA has been using around radar-equipped airports since the early 1970s. TRSAs were developed by the FAA to decrease the number of midair collisions around some of the busier airports. Within TRSA airspace, air traffic controllers were required to separate every IFR and participating VFR aircraft. But any VFR pilot who declined to participate in the TRSA program was not required to contact the appropriate ATC facility and henceforth was not provided any separation service. TRSAs are in many ways similar to the jet advisory areas that were replaced by Class A airspace.

The task force concluded that TRSAs were confusing to pilots since each one is individually shaped to take into consideration local terrain and traffic conditions. They also discovered that pilots were confused about the services offered to them while in the TRSA. The task force determined that TRSAs did not adequately separate aircraft, since pilot participation is not mandatory. Nonparticipating aircraft flying through a TRSA without contacting controllers negate many of the safety benefits of the program.

The task force agreed that the existing TRSA program should be replaced. They also concurred that a broad expansion of the Class B airspace concept was neither necessary nor advisable. After careful consideration, the task force eventually recommended that a new type of radar service area be created to eliminate many of the problems inherent in the TRSA program, without discriminating against VFR pilots. The task force recommended that this new radar service area be a standardized shape and require mandatory pilot participation. The only equipment required on board an aircraft entering this new area would be a two-way radio. (A requirement for a transponder with altitude encoding capability was added at a later point.)

Pilots would be permitted into this area on a first come, first served basis. To improve the efficiency and flexibility of the new radar service area, aircraft separation standards would be developed that would take into account different pilot capabilities. Positive separation would be provided between two IFR aircraft, whereas VFR pilots would be expected to operate using the see-and-be-seen concept of aircraft avoidance.

The FAA utilized these and other recommendations and implemented the ARSA program, now known as Class C airspace. As developed by the FAA, Class C airspace is a standard-shaped area that extends from the Earth's surface, or from an intermediate altitude, up to a higher altitude approximately 4,000 feet above ground level. Within Class C airspace, every aircraft, both IFR and VFR, is subject to the operating rules and pilot and equipment requirements specified in FAR 91. These requirements are similar to, but less restrictive than, the requirements to enter Class A airspace. Student pilot entry into Class A airspace is restricted, whereas student pilots are permitted to operate within Class C airspace under the same rules of operation as any VFR pilot.

Class C airspace is defined as the airspace that extends from the surface to 4,000 feet above the airport elevation (charted using MSL) surrounding those airports that have an operational control tower, are serviced by a radar approach control, and have a certain number of IFR operations or passenger enplanements (see Figures 3-11 and 3-12). Although the configuration of each Class C airspace area is individually tailored, the airspace usually consists of a

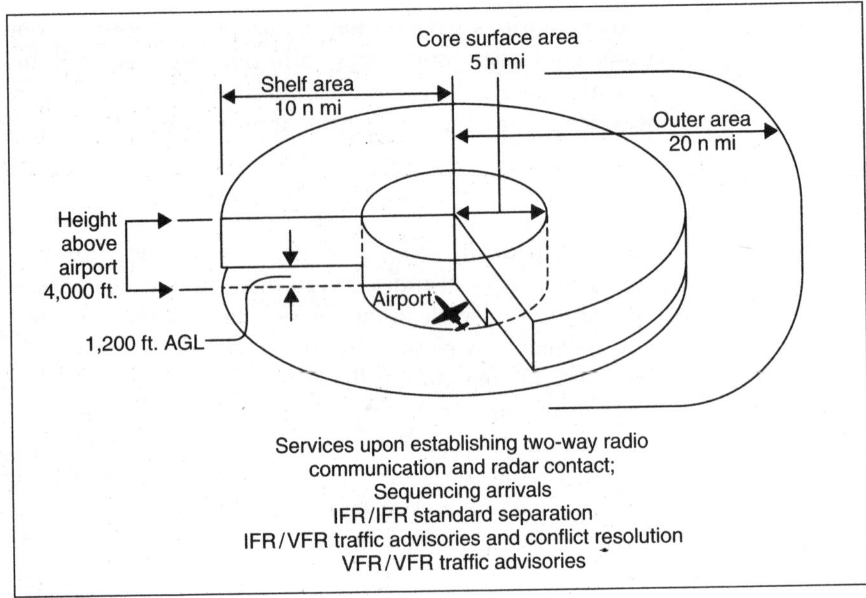

Figure 3-11. Description of Class C airspace.

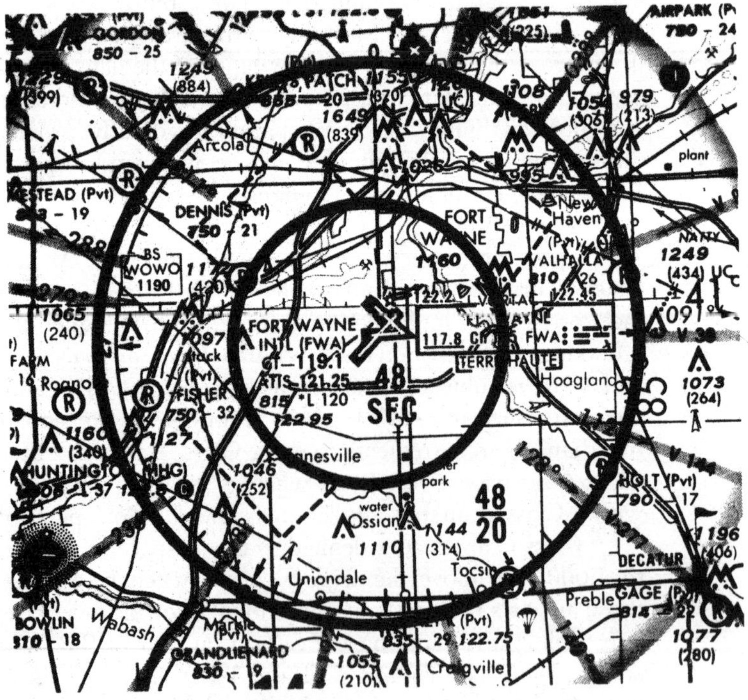

Figure 3-12. Class C airspace as depicted on a sectional chart.

5 nautical mile radius core surface area that extends from the surface up to 4,000 feet above the airport elevation, and a 10 nautical mile radius shelf area that extends from 1,200 feet to 4,000 feet above the airport. An outer area extends 20 nautical miles outward from the center of the primary airport and extends from the lower limits of radar/radio coverage up to the ceiling of the approach control's delegated airspace.

Once the aircraft enters Class C airspace, the pilot is required to comply with any instruction issued by the controller but must still comply with the visibility and cloud avoidance requirements of FAR 91. At no time may a VFR pilot disregard the basic VFR rules when trying to comply with the controller's clearance or subsequent instructions. If the pilot perceives that a controller's request might force a violation of any of the visual flight rules, the pilot is authorized by FAR 91 to disregard that instruction but must inform the controller as soon as possible.

Any VFR or IFR pilot who wishes to enter Class C airspace must comply with the following requirements:

> The pilot must establish communications with the appropriate air traffic control facility prior to entering. Unless the pilot is instructed to remain clear, the establishment of communication with the controller authorizes pilot entry into Class C airspace.
>
> While within Class C airspace, the pilot is required to comply with any of the instructions issued by the controller, unless these instructions will cause the pilot to violate a federal regulation, in which case the pilot is authorized to disregard the offending instruction.
>
> The aircraft must be equipped with an operable mode C transponder.

Class D Airspace

Class D airspace is defined as the airspace extending from the surface to 2,500 feet above the airport elevation (charted using MSL) surrounding those airports that have an operational control tower (see Figure 3-13). The configuration of each Class D airspace area is individually tailored, and when instrument procedures are published, the airspace will normally be designed to contain the procedures.

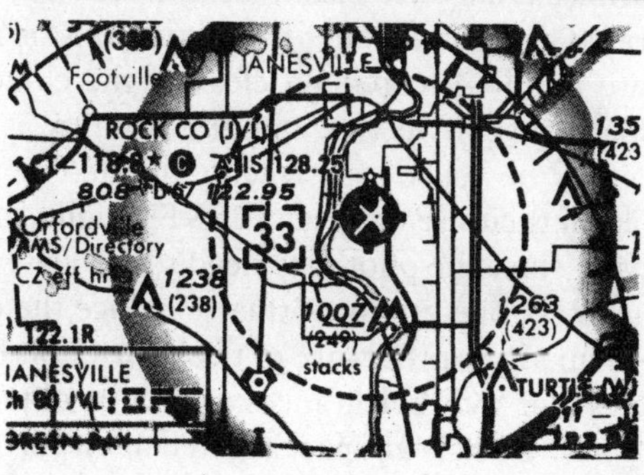

Figure 3-13. Class D airspace surrounding an airport with an operating control tower.

Pilots are required to establish two-way radio communication with the air traffic control tower providing ATC services prior to entry and thereafter maintain those communications while in the Class D airspace. At those airports where the control tower does not operate 24 hours a day, the operating hours of the tower are listed on the appropriate charts and publications. During the hours the tower is not in operation, Class E surface area rules or a combination of Class E rules to 700 feet above ground level and Class G rules to the surface are applicable as appropriate. Class D airspace areas are depicted on sectional and terminal charts with blue segmented lines and on IFR en route low-altitude charts with a boxed D.

Arrival extensions for instrument approach procedures may be Class D or Class E airspace. As a general rule, if the extensions are all 2 miles or less in length, they remain part of the Class D surface area. However, if any one extension is greater than 2 miles, then all extensions become Class E. Due to the speed differential among aircraft operating near the airport, unless authorized or required by ATC, aircraft operating within Class D airspace at or below 2,500 feet above the surface are not permitted to operate at indicated airspeeds of more than 200 knots (230 mph).

IFR aircraft are authorized to operate in Class D airspace if their ATC clearance routes them through one. VFR pilots are permitted to fly through Class D airspace as long as the basic VFR weather minima described in FAR 91 exist, the required cloud separation distances can be maintained, and permission has been received from the control tower. An additional requirement for VFR flight is that the cloud ceiling must be at least 1,000 feet above the ground if the pilot wishes to operate below the ceiling. This requirement ensures that VFR pilots will be able to maintain a distance of at least 500 feet below the clouds and 500 feet above the surface of the Earth, which is a FAR 91 requirement for VFR flight.

VFR pilots may operate above the 1,000 foot ceiling as long as they are able to climb above the ceiling while maintaining VFR conditions and can maintain the basic FAR 91 weather minima while flying above the ceiling. If the ceiling in the Class D airspace is less than 1,000 feet, or if the visibility is less than 3 miles, a VFR pilot is not permitted to operate within the air space. In these conditions, the pilot may request a **special VFR (SVFR)** clearance to operate.

An SVFR clearance is a hybrid clearance in which VFR pilots navigate visually but are separated by the controller from other IFR or SVFR aircraft. Special VFR aircraft are required to remain clear of the clouds while operating within Class D airspace but can operate with the visibility as low as 1 mile. Special VFR clearances may be issued only when requested by the pilot and when traffic conditions permit their use. In general, SVFR flights are allocated a fairly large block of airspace, since the pilot may need to navigate around clouds and obstructions. Special VFR operations normally reduce the number of IFR aircraft that can land at an airport. Because of this impact on IFR flights, FAR 91 appendix D mandates that SVFR clearances cannot be obtained at some of the nation's busiest airports. Class D airspace located at airports with control tow-

Figure 3-14. Airport where no SVFR is allowed.

ers is indicated on VFR navigational charts using blue dashed lines. Class D airspace located at airports without control towers is indicated on VFR navigational charts using magenta dashes. Airspace where SVFR clearances cannot be issued are depicted on VFR navigation charts with the words NO SVFR in the airport data block (see Figure 3-14).

Class E Airspace

Generally, if the airspace is not Class A, B, C, or D, and it is controlled airspace, it is designated as Class E airspace. Class E airspace below 14,500 feet MSL is charted on sectional, terminal, world, and IFR en route low-altitude charts. Class E airspace generally has no defined vertical limit, but rather it extends upward to the overlying or adjacent controlled airspace.

There are seven general forms of Class E airspace, all of which are defined to ensure that aircraft operating on IFR flight plans can remain in controlled airspace during their entire flight. Any normal IFR flight that were to leave the confines of controlled airspace would no longer be offered ATC services, thereby negating the entire concept of air traffic control and separation. The seven forms of Class E airspace are:

1. Surface area designated for an airport. When designated as a surface area for an airport, sufficient Class E airspace will be designated around the airport to contain all instrument procedures.
2. Surface area extensions. Class E airspace areas can serve as extensions to Class B, C, and D surface areas designated for an airport (see Figure 3-15). Such airspace provides controlled airspace to contain standard instrument approach procedures without imposing a communications requirement on pilots operating under VFR.
3. Airspace used for transitions. Class E airspace areas beginning at either 700 or 1,200 feet AGL are used to transition to/from the terminal or en route environment.
4. En route domestic areas. Class E airspace areas that extend upward from a specified altitude and are en route domestic airspace areas provide controlled airspace in those areas where there is a requirement to provide IFR en route ATC services but the federal airway system is inadequate. Most of the United States airspace east of the Rocky Mountains and above 1,200 feet AGL has been designated as Class E airspace.
5. Offshore airspace areas. Class E airspace areas that extend upward from a specified altitude to, but not including, 18,000 feet MSL are designated as offshore

Figure 3-15. Surface area extensions.

airspace areas. These areas provide controlled airspace beyond 12 miles from the coast of the United States in those areas where there is a requirement to provide IFR en route ATC services and within which the United States is applying domestic procedures.

6. Unless designated at a lower altitude, Class E airspace begins at 14,500 MSL to, but not including, 18,000 feet MSL overlying the 48 contiguous states including the waters within 12 miles from the coast of the 48 contiguous states; the District of Columbia; Alaska, including the waters within 12 miles from the coast of Alaska, and that airspace above FL 600; excluding the Alaska peninsula west of long. 160°00′00″W; and the airspace below 1,500 feet above the surface of the Earth unless specifically so designated. This used to be known as the continental control area.

7. Federal airways. Federal airways are Class E airspace areas and, unless otherwise specified, extend upward from 1,200 feet to, but not including, 18,000 feet MSL. This includes the colored airway system that uses NDBs for navigation and the VOR airways.

Federal Airways

FARs 71 and 93 define the structure of the federal airway system. The federal airways are divided into two general types: colored airways and the VOR airway system. The colored airways utilize NDBs and four-course radio ranges for navigation. With the exception of coastal areas, colored airways no longer exist within the continental United States but are still used in Alaska and Canada. They are sometimes utilized to provide temporary airways when a VOR malfunctions or is being relocated. VOR-based airways have been the standard for aviation navigation in the continental United States since the late 1950s.

Every federal airway is designated by the FARs as either a low-altitude airway or a jet route. Low-altitude airways are defined in FAR 71 and utilize both low- and high-altitude VORs for navigation. All low-altitude airways are assigned distinctive route numbers that are prefixed with the letter V and are known as **victor airways** (since victor is the phonetic pronunciation of the letter V). For example, V-251 is known as "Victor two fifty-one."

Low-altitude airways extend from 1,200 feet above the surface of the Earth up to, but not including, 18,000 feet above MSL. Jet routes begin at 18,000 feet MSL and extend up to and including 45,000 feet MSL. High-altitude jet routes utilize high-altitude VORs exclusively, are assigned a route number, and are prefixed with the letter J. These airways are referred to as jet routes or simply jay routes. For example, J-155 would be pronounced as "Jay one fifty-five."

The National Ocean Service publishes both low- and high-altitude charts that depict federal airways. Figure 3-16 provides an example of a low-altitude chart; legends for reading the chart are provided in Appendix A. There are no airways or jet routes above 45,000 feet MSL. High-performance aircraft operating at these altitudes either use RNAV or fly directly from one VOR to the next.

Flight Levels Since aircraft using high-altitude airways are usually traveling at high airspeeds, it is difficult to ensure that every aircraft operating within a given area is utilizing the same altimeter setting. It is imperative that every altimeter measure altitude above the same reference plane (mean sea level). If two aircraft using different altimeter settings were flying in close proximity, they could conceivably be at the same altitude even though their altimeters indicated different altitudes. Improperly set altimeters increase the potential for near misses and actual midair collisions.

This particular problem is solved for low-altitude aircraft by requiring that the pilot set the altimeter to the current station pressure at the controlling ATC facility. This procedure ensures that every aircraft operating within the same area is using the same altimeter setting. This method is not so useful for aircraft operating at high altitudes, since they are usually flying at a fairly high airspeed, requiring pilots to constantly adjust their altimeter setting every few minutes as they pass from one area to another. The possibility that pilots could inadvertently use an incorrect altimeter setting increases every time they readjust the altimeter. The potential collision probability also increases any time a pilot fails to readjust the altimeter or when a controller fails to inform the pilot of the new altimeter setting.

Since pilots operating high-altitude aircraft are not as concerned about their actual altitude above the ground as low-altitude pilots are, this potential collision problem can be solved by requiring pilots to reset their altimeters to 29.92 inches of mercury when operating their aircraft at or above 18,000 feet MSL. The setting of 29.92 inches is known as **standard atmospheric pressure,** and 18,000 feet MSL is known as the **transition level.**

Setting the altimeter to standard pressure when operating at or above the transition level ensures that every aircraft is using the same altimeter setting and measuring altitude from a common datum. The only problem with this procedure is that the altimeter is no longer indicating the true altitude above mean sea level, which makes it difficult to determine the aircraft's true altitude above an obstruction. Fortunately, few obstructions occur at these altitudes. Pilots flying near very high obstructions are routinely assigned altitudes sufficiently high to guarantee obstacle clearance.

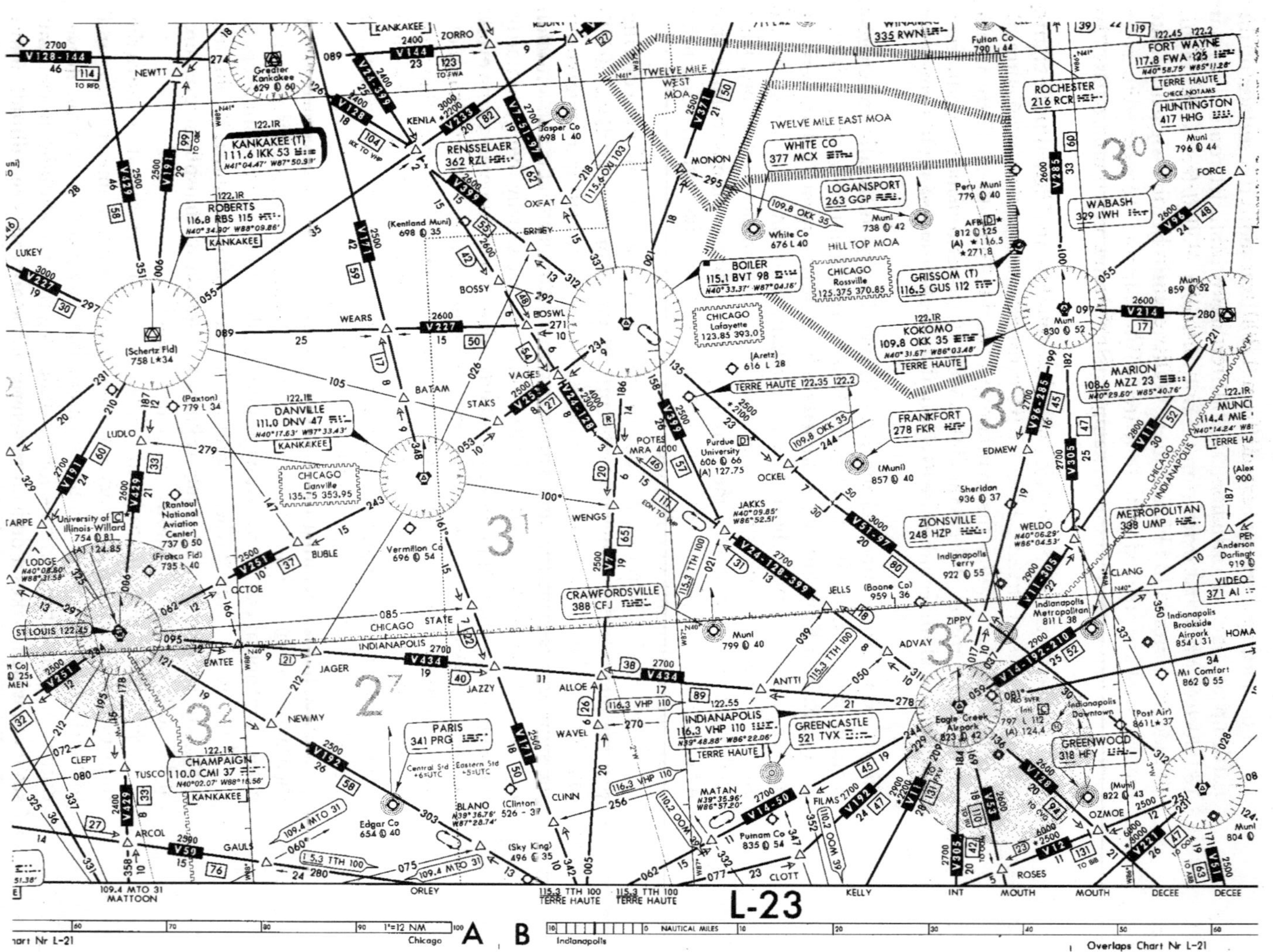

Figure 3-16. Sample low-altitude en route chart.

To reduce the possibility of a pilot mistakenly using the local altimeter setting when flying on a jet route, any cruising altitude at or above 18,000 feet MSL is known as a **flight level (FL)**. A flight level is defined as a level of constant atmospheric pressure related to a reference datum of 29.92 inches of mercury. Each flight level is stated using three digits that represent hundreds of feet; for example, FL 250 represents a barometric altimeter indication of 25,000 feet.

Because every aircraft operating at or above 18,000 feet is using a common altimeter setting, it can be safely assumed that an aircraft operating at FL 250 will always be 1,000 feet below an aircraft operating at FL 260. These two aircraft may not actually be at 25,000 feet and 26,000 feet MSL, respectively, but that is unimportant at these altitudes. The ATC system is primarily concerned that the aircraft are separated by at least 1,000 feet. As aircraft descend through the transition level (FL 180), pilots reset their altimeter to the local barometric pressure to again accurately indicate the aircraft's altitude above mean sea level.

This becomes increasingly important as the aircraft gets closer to the ground. The procedure of resetting the altimeter to 29.92 when passing through the transition level is utilized worldwide, but the transition level altitude varies among ICAO member nations. It is at 18,000 feet MSL in North America but may be as low as 3,000 feet MSL in some European countries. This may cause a problem when controllers are separating aircraft whose pilots are certified in another country and are accustomed to resetting their altimeter to standard pressure at a different transition altitude. Problems can also occur at airspace boundaries between countries with different transition levels.

Flight levels are necessary to ensure that proper separation is being applied to aircraft operating at high altitudes, but whenever the local altimeter setting is less than 29.92 inches, FL 180 may actually be less than 1,000 feet above 17,000 feet MSL. Whenever the local altimeter setting is less than 29.92, FL 180 must be considered as unusable. If the local barometric pressure drops below 28.92 inches, additional flight levels may become unusable. The following table from the *Air Traffic Control Handbook* demonstrates the lowest usable flight level that may be assigned to aircraft based on the local altimeter setting.

Altimeter Setting	***Lowest Usable Flight Level***
29.92 in. or higher	FL 180
29.91 in. to 28.92 in.	FL 190
28.91 in. to 27.92 in.	FL 200

Airway Dimensions

The area reserved for aircraft operating along a federal airway includes the airspace extending laterally 4 nautical miles on either side of the airway's centerline. If the airway is more than 102 nautical miles from VOR to VOR, it is widened to take into consideration the spreading of the radials as they emanate from the VOR. At a point 51 nautical miles from the VOR, the boundaries of the airway begin to include the airspace between two lines that diverge from the

VOR at an angle of 4.5° on either side of the airway centerline. If the airway changes direction, it also includes that airspace enclosed by extending the boundary lines of each segment of the airway. The midway point of the airway is known as the **changeover point (COP)**. This point is defined as the fix between the two navigational aids that define that particular segment of the airway. The changeover point is where the pilot ceases to track from the first VOR and begins to track to the next VOR. Changeover points are not depicted on navigational charts unless they are located somewhere other than the exact midpoint of the airway.

Class G Airspace

Airspace defined as Class G airspace is uncontrolled airspace within which ATC separation services will not be provided to any aircraft, whether IFR or VFR. The regulations for flight in uncontrolled airspace are quite specific and place the burden of separation on the pilot. Most of the uncontrolled airspace in this country is located away from major airports below 1,200 feet AGL. The following procedures must be followed by any pilot flying in uncontrolled airspace.

Uncontrolled Airspace IFR Flight IFR flight may be legally conducted in uncontrolled airspace, although no ATC separation services can be provided by the FAA. A pilot flying in IFR conditions in uncontrolled airspace assumes the entire responsibility for air traffic separation and terrain avoidance. Properly qualified pilots may legally operate under instrument flight rules in uncontrolled airspace as long as they adhere to the applicable FARs. Most of these regulations are found in FAR 91.

Pilots operating in uncontrolled airspace under instrument flight rules are not required to file a flight plan, nor will they receive a clearance or separation services from ATC. In fact, air traffic controllers are prohibited from issuing clearances or providing air traffic separation to IFR aircraft operating in uncontrolled airspace. Since controllers are not informed of every aircraft operating in uncontrolled airspace, it is impossible for them to provide separation to these aircraft. In general, pilots wishing to conduct IFR flight in uncontrolled airspace must comply with the following regulations.

1. The pilot of the aircraft must be properly rated, and the aircraft must be properly equipped for IFR flight as specified in the FARs.
2. The pilot is solely responsible for navigating and avoiding other IFR or VFR aircraft.
3. The pilot is responsible for operating the aircraft a safe distance above the ground.

FAR 91 requires that pilots operating IFR in uncontrolled airspace maintain an altitude of at least 1,000 feet above any obstruction located within five statute miles of the course to be flown. This rule is not applicable to aircraft landing or taking off, during which it is the pilot's responsibility to operate the aircraft a safe distance above obstacles. In theory, during IFR flight in uncontrolled airspace, the pilot is required to fly at an altitude appropriate for the direction of flight. The altitudes are specified in FAR 91 but are seldom used since pilots rarely fly IFR in uncontrolled airspace for any length of time. IFR flight in uncontrolled airspace is usually limited to arrivals and departures from small airports with limited air traffic control services.

Uncontrolled Airspace VFR Flight VFR pilots operating in uncontrolled airspace must adhere to the applicable regulations contained in FAR 91.155. This regulation specifies the weather conditions that must exist for the pilot to legally operate VFR. The required weather conditions vary depending on the aircraft's cruising altitude and its actual altitude above the ground. To legally fly VFR in uncontrolled airspace, pilots must comply with the following minima from FAR 91.

Cruising Altitude	***Flight Visibility***	***Distance from Clouds***
1,200 ft. or less above the surface, regardless of MSL altitude	1 statute mile	Clear of clouds
More than 1,200 ft. above the surface, but less than 10,000 ft. MSL	1 statute mile	500 ft. below 1,000 ft. above 2,000 ft. horizontal
At or above 10,000 ft. MSL	5 statute miles	1,000 ft. below 1,000 ft. above 1 statute mile horizontals

VFR pilots operating in uncontrolled airspace are not required to file any type of flight plan or to contact any air traffic control facility (unless they are entering a designated area where contact is mandatory). It is the responsibility of VFR pilots to see and avoid any other aircraft that might be within their immediate vicinity, regardless of whether that aircraft is operating under IFR or visual flight rules.

ATC Services in Uncontrolled Airspace Air traffic control separation services are not offered to any aircraft operating in uncontrolled airspace unless an emergency exists. Additional ATC services can be provided, however, on a workload-permitting basis. If a controller finds it necessary to issue a clearance to an aircraft while it is still within uncontrolled airspace, the *Air Traffic Control Handbook* suggests that the following phraseology be used to ensure that the pilot is aware that ATC services will not begin until the aircraft enters controlled airspace:

> [aircraft call sign], upon entering controlled airspace, [the clearance].

For example:

> N512PU, upon entering controlled airspace, fly heading two seven zero and join Victor two fifty-one.

Special Use Airspace

In numerous areas scattered around the United States, it is in the national interest to either restrict or completely prohibit the flight of civilian aircraft. The U.S. government, through the FARs, has designated these areas as **special use airspace.** Special use airspace is designed to either confine unique aircraft operations or to entirely prohibit flight within the specified area. Unless otherwise noted, all of the following examples of special use airspace are published on VFR and IFR navigation charts and are designated in appropriate aeronautical publications.

Prohibited Areas A **prohibited area** is airspace where aircraft operations are absolutely prohibited by law. These areas are directly concerned with either national security or public safety. Among the prohibited areas are the White House, the Capitol Building, and Camp David. FAR 91.133 expressly prohibits either IFR or VFR aircraft from entering such areas without specific (and very rarely granted) authorization. Air traffic controllers are not permitted to authorize civilian aircraft operations within these areas unless an emergency exists (see Figure 3-17).

Every prohibited area is designated using a unique identifying number prefixed with the letter P. Prohibited areas are prominently marked on both IFR and VFR navigation charts to assist pilots in avoiding them. Federal airways are routed around prohibited areas, but VFR pilots must be familiar with their locations and plan their flight path accordingly.

Restricted Areas Locations where aircraft operations are not absolutely prohibited but are subject to various restrictions are labeled restricted areas. They are located where both airborne and ground-based activities are routinely conducted that may be hazardous to either the aircraft or its occupants. These activities include artillery firing, aerial gunnery, and high-energy laser and missile testing. Some restricted areas are in effect 24 hours a day, whereas others operate part-time. The part-time restricted areas, also known as joint use areas, are available for civilian flight whenever they are not active.

The FAA facility that has been given responsibility for the airspace containing a joint use restricted area will be notified by the appropriate agency when the restricted area becomes active. At these times, it becomes the air traffic controller's responsibility to issue clearances to keep IFR aircraft out of the restricted area. VFR aircraft are expected to contact appropriate ATC facilities

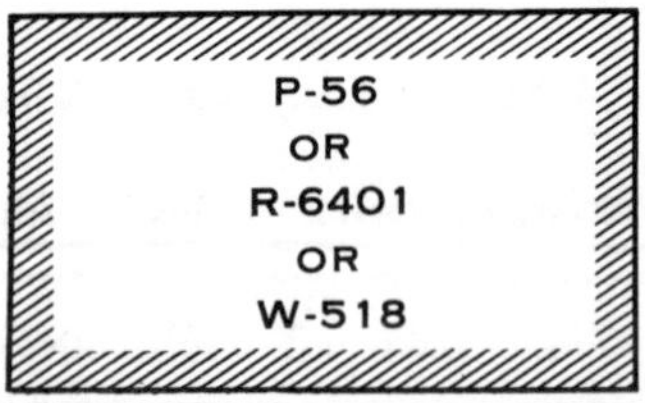

Figure 3-17. Prohibited, restricted, or warning area as depicted on a sectional chart.

when approaching restricted areas to determine their status. VFR pilots are required to provide their own separation from restricted areas, although they may request navigational assistance from ATC facilities. When the restricted area is not active, it will be released by the controlling agency to the appropriate ATC facility, and controllers may permit both IFR and VFR aircraft to use the restricted space. Restricted areas are prominently marked on both VFR and IFR charts and are identified by a unique number prefixed with the letter R (see Figure 3-17).

Warning Areas A **warning area** is airspace located over international waters where operations that may be hazardous to nonparticipating aircraft are routinely conducted. The activities conducted in a warning area are usually similar to those performed in a restricted area (see Figure 3-17). Since warning areas are located in international airspace, neither the United States nor any other government has the right to restrict the flight of aircraft through these areas. Both IFR and VFR aircraft may operate in warning areas but do so at their own risk. International civil aviation organization rules require that signatory nations advise each other when military activities are being conducted within warning areas. Fortunately, most of the developed nations of the world are members of ICAO and abide by this regulation. The military authority conducting the exercise will usually advise the responsible ATC facility of the type of activity and its expected duration.

Temporary Flight Restrictions The FAA may impose **temporary flight restrictions (TFRs)** around any incident or accident that has the potential for attracting a sufficient number of aircraft to create a hazard to either other aircraft in the air or people on the ground. Temporary flight restrictions may be imposed around earthquake, flood, fire, or aircraft crash sites. When a temporary flight restriction is imposed, the FAA notifies pilots by issuing a **notice to airmen (NOTAM).**

These notices are distributed nationwide to FAA air traffic control towers, air route traffic control centers, and flight service stations, who then relay the information to pilots. In addition, NOTAMs are transmitted to the airlines, military services, and many independent pilot-briefing companies who make the information available to their subscribers. When issued, a NOTAM defines the physical location; dimension, and duration of the restriction to flight. The NOTAM usually explains which aircraft are permitted to operate within the TFR. These aircraft include:

> aircraft that are participating in disaster relief and have been approved by the FAA. IFR aircraft properly cleared through the restricted area by ATC.

VFR pilots are required by FAR 91 to avoid these areas unless it is absolutely impossible to do so. IFR aircraft are rerouted by ATC around temporary flight restrictions.

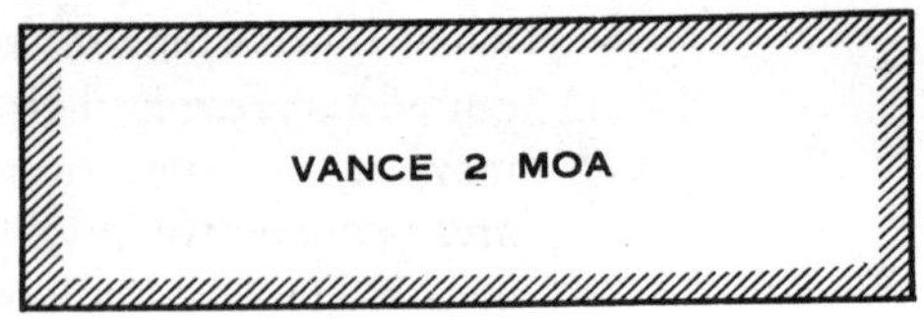

Figure 3-18. Military operations area as depicted on a sectional chart.

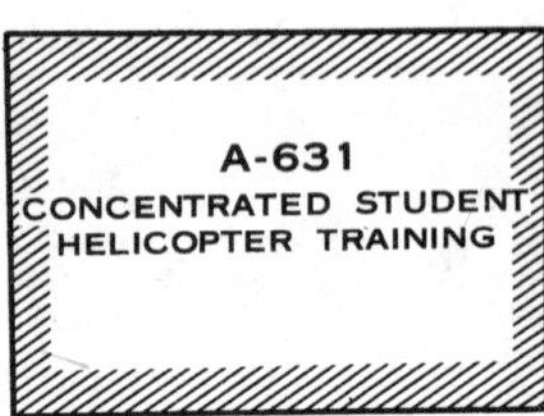

Figure 3-19. Alert area as depicted on a sectional chart.

Military Operations Area A **military operations area (MOA)** is designated airspace where military flight training activities routinely take place that might prove hazardous to civilian aircraft. Some of the flight training being conducted by military aircraft requires acrobatic maneuvers to be practiced on or near a federal airway. Although acrobatic flight along a federal airway is forbidden by FAR 91, the Department of Defense has been exempted from this regulation if the maneuvers are conducted within an MOA. Although military training flights are usually conducted in VFR flight conditions, the rapid changes in aircraft attitude required during these training maneuvers make it extremely difficult for the military pilot to see and avoid civilian aircraft. It is for this reason that military operations areas were created.

When the appropriate military authority advises the FAA that an MOA is active, air traffic controllers are required to reroute IFR aircraft around the MOA. VFR pilots are permitted to enter an MOA at any time but do so at their own risk. MOAs are depicted on both VFR and IFR navigation charts and are given identifying names followed by the letters MOA (see Figure 3-18).

Alert Areas Areas that may contain a large number of high-performance military training aircraft conducting routine training exercises are labeled **alert areas** (see Figure 3-19). Although there are no legal restrictions to civilian aircraft flying through an alert area, both IFR and VFR pilots transiting the area should be aware of the large numbers of VFR military aircraft that may be practicing nonacrobatic high-speed maneuvers there.

Controlled Firing Areas **Controlled firing areas** contain activities that, if not conducted in a controlled environment, could be hazardous to aircraft. These areas are not identified on VFR and IFR charts since the controlling agency suspends its activities whenever nonparticipating aircraft approach the area. Such aircraft are usually detected by the use of spotter aircraft, radar, or ground-based observers. Whenever intrusion of a nonparticipating aircraft into a controlled firing area is detected, the test firings are halted until the aircraft in question has departed the area. Controlled firing areas predominantly affect low-flying aircraft since most test firing is conducted at these altitudes.

National Security Areas **National security areas (NSAs)** consist of airspace established at locations where increased security and safety of ground facilities are required. Pilots are requested to voluntarily avoid flying through NSAs whenever possible. When it is necessary to provide a greater level of security and safety, flight in NSAs may be temporarily prohibited by regulation under the provisions of FAR 99.

Airport Advisory Areas

Airport advisory areas exist at airports where a flight service station is located but where there is no operating air traffic control tower. An airport advisory area is 10 statute miles in radius, centered around the airport. Flight service station personnel will offer weather information and traffic reports to arriving and departing aircraft but will not offer any separation services to aircraft. It is not mandatory that pilots utilize airport advisory services, but it is highly recommended by the FAA that they do so.

Military Training Routes

To remain sufficiently proficient to perform their duties, many military pilots are required to practice low-level, high-speed combat-training flights. The maneuvers performed during these training flights make the see-and-avoid concept of traffic separation difficult without increased vigilance on the part of both military and civilian pilots. To assist civilian pilots to avoid these military aircraft, the FAA and the Department of Defense (DOD) have mutually agreed to participate in the **military training route (MTR)** program. Through this program, designated MTR routes have been agreed to by both the FAA and the DOD and are depicted on VFR navigation charts (see Figure 3-20).

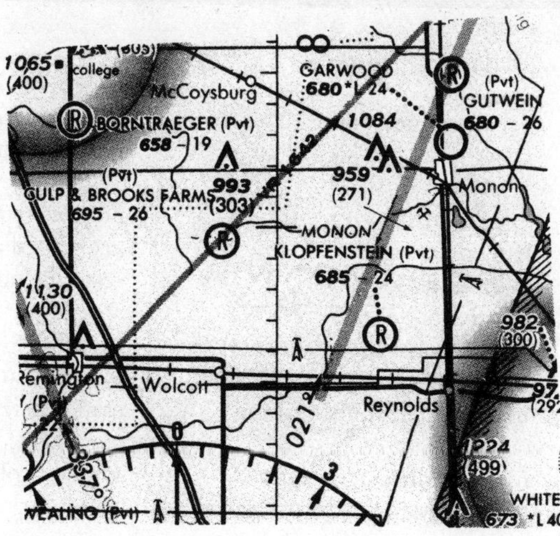

Figure 3-20. Military training route as depicted on a sectional chart.

Every military training route has been assigned a unique identifying designator composed of two letters and either three or four numbers. The first two letters are either IR (instrument rules) or VR (visual rules) for the type of military operation that will be conducted. Military pilots flying on IR-designated routes are provided IFR separation and must remain in contact with FAA controllers during the entire flight. An IR MTR route is flown under instrument flight rules and requires the pilot to file a flight plan and receive an ATC clearance. Military aircraft operating on VR-designated routes use VFR see-and-avoid flight rules. These routes are used only when weather conditions permit the entire flight to be conducted in VFR conditions.

An MTR designator containing three numbers signifies that the pilot will fly the MTR at an altitude that may be both above and below 1,500 feet AGL. Four numbers in the designator means that the entire MTR will be flown at an altitude at or below 1,500 feet AGL. For example:

IR 101	An MTR that will be flown in IFR conditions, with altitude segments that might be both above and below 1,500 feet AGL.
VR 4002	An MTR that will be flown in VFR conditions at or below 1,500 feet AGL.

Civilian aircraft are not prohibited from flying in the vicinity of an MTR, but pilot contact with a nearby ATC facility is recommended. Any flight service station within 100 miles of the MTR route will be advised by the controlling authority when the MTR is active. It is the VFR pilot's responsibility to determine whether the MTR is in use. Civilian IFR aircraft will always be separated from military aircraft operating on IR-designated MTRs but will not be separated from aircraft flying on a VR MTR. It is the civilian IFR pilot's responsibility to remain vigilant and avoid any aircraft using a VR military training route.

KEY TERMS

above ground level (AGL)
air traffic control
Air Traffic Control Handbook
airport advisory areas
airport radar service area (ARSA)
alert areas
area high routes
area low routes
area routes
changeover point (COP)
control areas
controlled airspace
controlled firing areas
departure procedure (DP)
Direct User Access Terminal (DUAT)
expect further clearance (EFC)
federal airways
flight level (FL)
jet advisory area
military operations area (MOA)
military training route (MTR)
minimum en route altitude (MEA)
mode C altitude encoder
national security areas (NSAs)
notice to airmen (NOTAM)
positive controlled airspace (PCA)
positive separation
prohibited area
restricted area
special use airspace
special VFR (SVFR)
standard atmospheric pressure
standard terminal arrival route (STAR)
target resolution
temporary flight restriction (TFR)

terminal control area (TCA)
terminal radar service area (TRSA)
transition level
uncontrolled airspace
vectors
victor airways
warning areas
waypoint
workload permitting

REVIEW QUESTIONS

1. What is the purpose of controlled airspace?
2. What is the difference between a jet route and an airway?
3. What must the pilot do when climbing through the transition level?
4. What is the primary difference between Class A and Class C airspace?
5. How are aircraft separated differently in Class B versus Class C airspace?
6. In what airspace areas are transponders mandatory?

4

Airport Air Traffic Control Communications: Procedures and Phraseology

Checkpoints

After studying this chapter, you should be able to:

1. State the required components of a clearance.
2. Describe what "cleared as filed" means.
3. State which frequency bands are used for aviation communications.
4. State the purpose of coordinated universal time and explain how it is measured.
5. Describe how parallel runways are numbered.
6. Describe the standard measurement for speed in aviation.
7. Identify the function of a pilot's discretion clearance.
8. Describe a holding pattern and identify how it is utilized.
9. Be able to distinguish between proper and improper uses of phraseology.

The safe operation of the nation's air traffic control system ultimately depends on reliable and accurate communication between pilots and air traffic controllers. Virtually every instruction, procedure, or clearance used to separate or assist aircraft relies on written or verbal communication. Any miscommunication between participants in the air traffic control system might contribute to or even be the direct cause of an aircraft accident with a subsequent loss of life. For this reason, proper and correct communications procedures must be observed by both pilots and controllers.

Many of the accidents and incidents that have occurred over the last 50 years can be attributed to improper or misunderstood communications. Although many improvements to the air traffic control communications system have made it less reliant on verbal or written communication, pilots and controllers will continue to rely on human communication until well into the twenty-first century. Thus, controllers must possess a proper understanding of communications procedures and phraseology.

American pilots and controllers are fortunate that the International Civil Aviation Organization (ICAO) has designated English as the international language for ATC communications worldwide. This standard reduces the number of words and communications procedures that American controllers need to learn. But air traffic controllers should realize that although foreign pilots are able to communicate using English, they probably do not have full command of the language. Thus, phraseology and slang not approved by ICAO or the FAA should never be used when communicating with foreign pilots. It is also recommended that standard phraseology be used when communicating with American pilots or controllers. Using standard procedures will help reduce the risk of miscommunication.

Radio Communication

Ever since radio communications equipment was installed in the Cleveland (Ohio) control tower in 1936, radio has become the primary means of pilot-controller communication in the U.S. air traffic control system. Although the type of radio equipment has changed since those days, the basic principles of radio communication remain the same today.

Simplex Versus Duplex

The earliest type of radio communication used in the air traffic control system was one-way. Controllers could communicate with pilots but not vice versa. Since the required radio equipment in those early years was quite bulky and heavy, the airlines were reluctant to install both a navigation receiver and a communications transmitter on each aircraft. Thus, most aircraft were equipped only with a navigation receiver.

The ground-based navaids were eventually modified to permit controllers to transmit instructions using the navigation aid frequencies. At first this communication rendered the navaid useless while the controller was transmitting, but later advances permitted the controller to transmit using the navaid while still allowing the pilot to use the ground station for navigation.

As the benefits of radio communication became increasingly evident, aircraft operators chose to add transmitting equipment to their planes. The equipment operated on a different set of frequencies to eliminate any possible interference with the ground-based navaids. This development created its own set of problems, however. The addition of a separate transmitter and receiver markedly increased the weight of the aircraft, and adding separate transmitters and receivers in each control tower required an additional expenditure. Furthermore, during the transition from the navaid-based communication system, aircraft not equipped with transceivers would be unable to communicate with the control towers.

An interim solution was to install receiving equipment in the control towers and transmitting equipment in the aircraft. This system still used the ground-based navaids for tower-to-aircraft communication but used the newly installed radios for aircraft-to-tower communication. To eliminate navaid interference, the aircraft transmitters used a different frequency than the ground-based navaids. This two-frequency system is known as **duplex communication** (see Figure 4-1).

Duplex was used in the air traffic control system for many years and is still used in some parts of the United States. In particular, FAA flight service stations are usually equipped to receive on one frequency while transmitting to the aircraft over a local VORTAC. The duplex system has disadvantages, however,

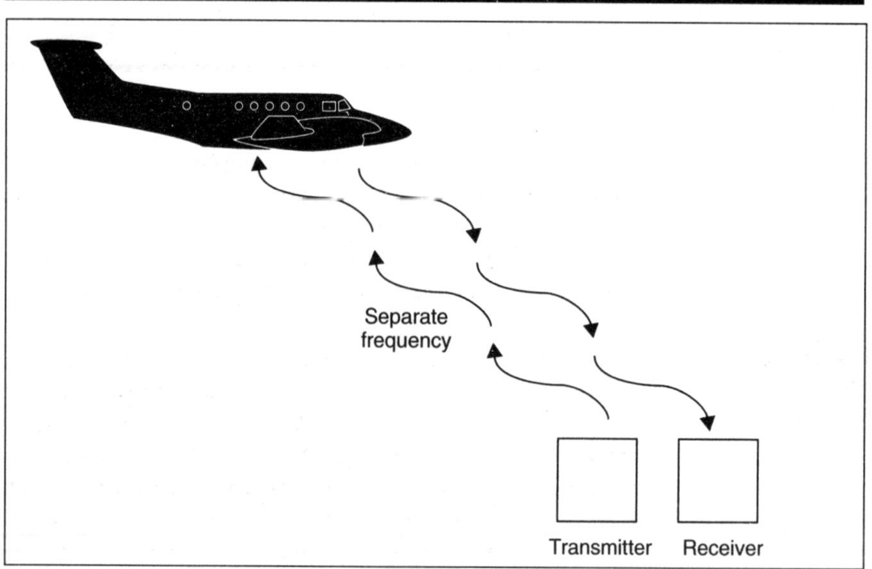

Figure 4-1. Duplex transmission principles.

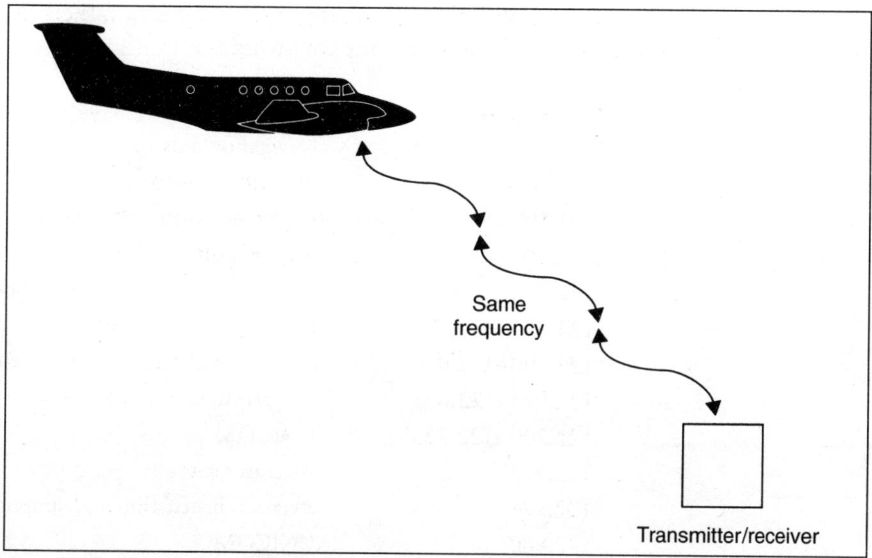

Figure 4.2. Simplex transmission principles.

that spurred the development of a radio system that would permit pilots to communicate with controllers using one discrete frequency. This system was finally implemented within the ATC system and is known as **simplex communications** (see Figure 4-2). For the most part, every ATC facility in the United States relies primarily on simplex communications.

Frequency Assignments

Various international agreements allocate certain radio frequency bands for use in aeronautical communications. These frequency bands exist primarily in the high (HF), very high (VHF), and ultra-high (UHF) spectrums. High frequencies are primarily used for long-range communication, since these frequencies are not line of sight and can follow the curvature of the Earth. Only a few ATC facilities, such as ARTCCs with oceanic responsibility, find a need to use these frequencies.

Most U.S. ATC facilities use both VHF and UHF for routine air-to-ground communication. UHF radio equipment is primarily used by military aircraft, whereas VHF is used by both military and civilian aircraft. The frequencies to be used in ATC communications are assigned by the **Federal Communications Commission (FCC)** in cooperation with the FAA. Since there is not a sufficient number of available frequencies in either the VHF or UHF spectrum to permit every ATC facility to operate using a separate frequency, the FCC often assigns the same frequency to two or more ATC facilities. Because the radio transmissions from high-altitude aircraft travel farther than those from low-flying aircraft, the FCC must carefully determine any potential interference problems before assigning these frequencies.

To simplify the task of assigning frequencies, the FCC has assigned these blocks of VHF bands for the following uses:

Frequencies	Use
108.000–117.950	Navigation aids
118.000–121.400	Air traffic control
121.500	Emergency search and rescue
121.600–121.925	Airport utility and ELT test
121.950	Aviation instructional and support
121.975	FSS private aircraft advisory
122.000–122.050	En route flight advisory service (EFAS)
122.075–122.675	FSS private aircraft advisory
122.700–122.725	UNICOM
122.750	Aircraft air-to-air
122.775	Aviation instruction and support
122.800	UNICOM
122.825	Domestic VHF
122.850	Multicom
122.875	Domestic VHF
122.900	Multicom
122.925	Multicom
122.950	Unicom
122.975–123.000	Unicom
123.050–123.075	Unicom
123.100	Aeronautical search and rescue
123.125–123.275	Flight test stations
123.300	Aviation support
123.325–123.475	Flight test stations
123.500	Aviation support
123.525–123.575	Flight test stations
123.600–123.650	FSS air carrier advisory
123.675–128.800	Air traffic control
126.200	Air traffic control (military common)
128.825–132.000	Domestic VHF (operational control)
132.025–136.975	Air traffic control

Radio Operation

Most air traffic controllers use radio equipment to perform their ATC duties. This equipment may be either fairly simple or very complex, depending on the capabilities of the facility. In general, each controller is assigned one or more radio frequencies for communications with pilots and has access to telephone equipment that permits communication with other controllers in the same facility or in adjacent facilities. The design of the **voice switching system** installed in most ATC facilities is sophisticated enough to permit such communication effortlessly.

Most controllers are outfitted with a boom mike and headset assembly that permits them to move freely around the facility while still remaining in contact with the pilots. Other controllers may use standard microphones and speakers or telephone handsets provided by the local telephone company (which is known throughout the FAA by the generic term **TELCO**). Each controller has a switching panel to choose whether to communicate with other controllers or to the pilot over the radio. The system is designed so that when the controller is communicating on one particular channel, any message sent to him or her on either the radio or another landline is routed through an overhead speaker. Most facilities are equipped such that every frequency assigned to that facility can be used by any controller there.

Standard Phraseology for Verbal Communications

To ensure that miscommunication is kept to a minimum, it is imperative that controllers use the standard phraseology and procedures that have been recommended by ICAO and the FAA. When communicating with pilots or other controllers, a controller should always use the following message format:

1. Identification of the aircraft or controller being contacted. This serves to alert the intended receiver of the upcoming transmission.
2. Identification of the calling controller. This serves to identify who is initiating the communication.
3. The contents of the message. The message format should conform to standards approved by the FAA.
4. Termination. In communications with another ATC facility, the message should be terminated with the controller's assigned operating initials. This procedure simplifies identification of the controller if a subsequent investigation is necessary.

Certain letters and numbers may sound similar to each other when spoken over low-fidelity radio or telephone equipment. In addition, accents and dialects may make it difficult to discern and identify the exact content of a message. To alleviate this problem, a standard for pronunciation of letters and numbers has been approved by ICAO and adopted by the FAA. This standard is presented in Table 4-1. The standardized pronunciations should be used by controllers whenever communicating with pilots or other controllers. Air traffic controllers should also use the following standardized phraseology when passing along control instructions or various information to pilots or to other controllers.

Numbers Each number should be enunciated separately unless **group form** pronunciation is stipulated. For example:

Number	***Group Form Pronunciation***	***Separate Pronunciation***
1	One	One
10	Ten	One zero

Table 4-1. Standard Phraseology for Numbers and Letters

Character	Word	Pronunciation
0	Zero	Zee-ro
1	One	Wun
2	Two	Too
3	Three	Tree
4	Four	Fow-er
5	Five	Fife
6	Six	Six
7	Seven	Sev-en
8	Eight	Ait
9	Nine	Nin-er
A	Alpha	Alfah
B	Bravo	Brahvoh
C	Charlie	Charlee
D	Delta	Delta
E	Echo	Eckoh
F	Foxtrot	Fokstrot
G	Golf	Golf
H	Hotel	Hohtell
I	India	Indee-ah
J	Juliett	Jewlee-ett
K	Kilo	Keyloh
L	Lima	Leemah
M	Mike	Mike
N	November	November
O	Oscar	Osscah
P	Papa	Pahpah
Q	Quebec	Kehbeck
R	Romeo	Rowme-oh
S	Sierra	Seeairah
T	Tango	Tanggo
U	Uniform	Younee-form
V	Victor	Viktah
W	Whiskey	Wisskey
X	X-ray	Ecksray
Y	Yankee	Yangkey
Z	Zulu	Zooloo

Number	Group Form Pronunciation	Separate Pronunciation
15	Fifteen	One five
132	One thirty-two	One three two
569	Five sixty-nine	Five six niner

Unless otherwise specified, when serial numbers are pronounced, each digit should be enunciated individually.

Altitudes Unless otherwise specified, every altitude used in the ATC system is measured above mean sea level (MSL). The only routine exception is cloud ceilings, which are measured above ground level (AGL). A controller who must issue an AGL altitude to a pilot should advise the pilot that the altitude is above ground level. Altitudes should be separated into thousands and hundreds, and the thousands should be pronounced separate from the hundreds. Each digit of the thousands number should be enunciated individually, whereas the hundreds should be pronounced in group form:

Altitude	Pronunciation
3,900	Three thousand niner hundred
12,500	One two thousand five hundred

Flight Levels Flight levels should be preceded by the words "flight level," and each number should be enunciated individually:

Flight Level	Pronunciation
180	Flight level one eight zero
390	Flight level three niner zero

Minimum Descent or Decision Height Altitudes Minimum descent or decision height altitudes published on instrument approach procedure charts should be prefixed with the type of altitude, and each number in the altitude should be enunciated individually:

Altitude	Pronunciation
MDA 1,950	minimum descent altitude one niner five zero
DH 620	decision height six two zero

Time Since numerous ATC procedures require the use of time, a common system of time measurement is essential to the safe operation of the ATC system. The FAA and ICAO have agreed that local time is not to be used within the ATC system. Instead, every ATC facility around the world must use the same time standard, known as **coordinated universal time (UTC)**. UTC is the same as local time in Greenwich, England, which is located on the 0° line of longi-

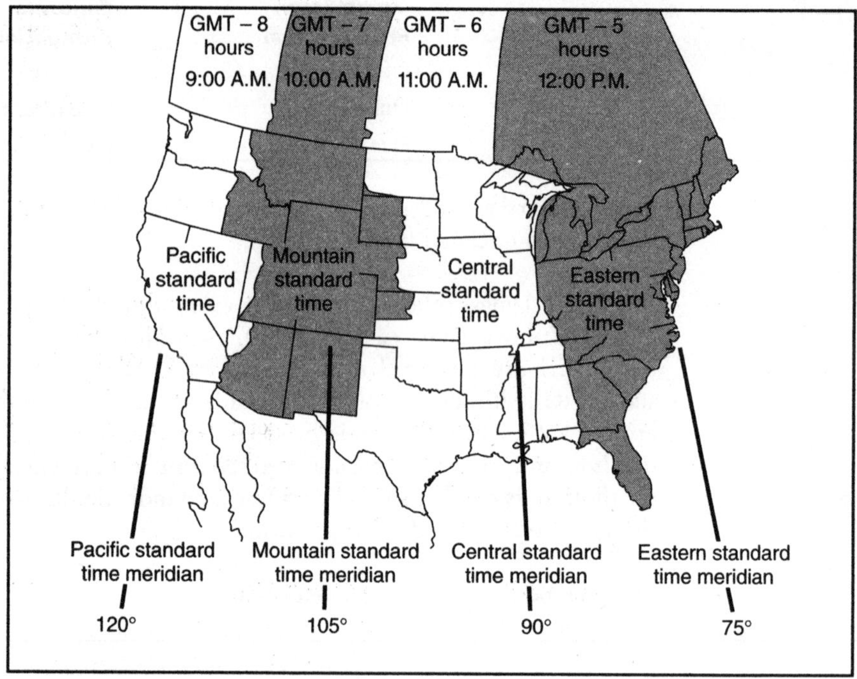

Figure 4-3. Time zones across the United States.

tude, also known as the **prime meridian.** UTC was previously known as **Greenwich mean time (GMT).**

The use of UTC around the world eliminates the question of which time zone a facility or aircraft is located in (see Figure 4-3). In addition, UTC eliminates the need for "A.M." and "P.M." by using a 24-hour clock system. UTC is always issued as a four-digit number, and the word "o'clock" is never pronounced. The conversion from a 12-hour clock to a 24-hour clock is fairly simple:

> Any time that has fewer than four digits should be prefixed with a zero.
>
> Any time between midnight and noon (A.M.) is not converted to a 24-hour clock.
>
> Any time between noon and midnight (P.M.) always has twelve hours added to it to differentiate it from A.M. time.

For example, 6:20 A.M. becomes 0620, and 6:20 P.M. becomes 1820. Local time is converted to UTC by either adding or subtracting the number of hours indicated in the following chart:

Time Zone	***Difference***
Eastern standard time (EST)	5 hours

Time Zone	Difference
Eastern daylight time (EDT)	4 hours
Central standard time (CST)	6 hours
Central daylight time (CDT)	5 hours
Mountain standard time (MST)	7 hours
Mountain daylight time (MDT)	6 hours
Pacific standard time (PST)	8 hours
Pacific daylight time (PDT)	7 hours
Alaskan standard time (AST)	9 hours
Alaskan daylight time (ADT)	8 hours

To convert from local time to UTC, convert the local time to a 24-hour clock, then add the required time difference. To convert from UTC to local time, subtract the difference and convert from a 24-hour to a 12-hour format. For example:

4:35 A.M. (EST) is 0435 (EST), which is 0935 (UTC)
9:13 P.M. (PDT) is 2113 (PDT), which is 0413 (UTC)
1125 (UTC) is 0425 (MST), which is 4:25 A.M. (MST)

To prevent any confusion when issuing time to the pilot, the controller should suffix any UTC time with the word "zulu" and any local time with the word "local." Any issuance of time should also be preceded by the word "time." When issuing time, the controller should enunciate each digit individually:

Time (12-hour clock)	Time (24-hour clock)	Pronunciation
6:20 A.M.	0620	Time zero six two zero zulu
1:35 P.M.	1335	Time one three three five zulu

Altimeter Settings The pilot must be issued the proper barometric pressure so that the aircraft's altimeter can be properly adjusted to indicate altitude above mean sea level. The controller should issue these altimeter settings by individually enunciating every digit without pronouncing the decimal point; the altimeter setting should be preceded by the word "altimeter":

Altimeter Setting	Pronunciation
29.92	Altimeter two niner niner two
20.16	Altimeter two zero one six

Care should be taken when issuing altimeter settings to foreign pilots. Pilots from countries that have converted to the metric system no longer measure barometric pressure in inches of mercury but in **millibars.** It is the foreign pilot's responsibility to convert the issued altimeter setting to millibars or to request a metric altimeter setting from the controller.

Qualimetrics, Inc.

Figure 4-4. *A digital wind direction and velocity indicator.*

Figure 4-5. *An analog wind direction and velocity indicator.*

Wind Direction and Velocity Wind direction at airports is always determined in reference to magnetic north and indicates the direction that the wind is blowing *from*. The direction is always rounded off to the nearest 10°. Thus, a wind blowing from north to south is a 360° wind; a wind from the east is a 90° wind. The international standard for measuring wind velocity requires that wind speeds be measured in **knots;** 1 knot equals approximately 1.15 miles per hour. Wind direction and velocity information is always preceded by the word "wind," with each digit of the wind direction enunciated individually. The wind direction is then followed by the word "at" and the wind velocity in knots, with each digit enunciated individually. If the wind measurement devices (see Figures 4-4 and 4-5) are inoperative, the wind speed and direction are preceded by the word "estimated." If the wind direction is constantly changing, the word "variable" is suffixed to the average wind direction. If the wind velocity is constantly changing, the word "gusts" and the peak speed are suffixed to the wind speed. Here are some examples:

Wind Direction	***Wind Speed***	***Pronunciation***
From the north	15 knots	Wind three six zero at one five
From the east	10 knots with occasional gusts to 25 knots	Wind zero niner zero at one zero gusts to two five
Variable from the southeast	12 knots with occasional gusts to 35 knots	Wind one five zero variable at one two gusts to three five

Wind Direction	**Wind Speed**	**Pronunciation**
Estimated from the southwest	estimated at 15 knots	Estimated wind two three zero at one five

Headings Aircraft headings are also measured in reference to magnetic north. If the heading contains fewer than three digits, it should be preceded by a sufficient number of zeros to make a three-digit number. Aircraft headings should always be preceded by the word "heading," with each of the three digits enunciated individually. Here are some examples:

Heading	**Pronunciation**
005°	Heading zero zero five
090°	Heading zero niner zero
255°	Heading two five five

Runway Numbers Runways are also numbered in reference to their magnetic heading. The runway's number is its magnetic heading rounded to the nearest 10° with leading and trailing zeros removed. For example, a runway heading north would have a magnetic heading of 360°. Dropping the trailing zero makes this runway number 36. Since the other end of the runway heads the opposite direction (south, which is a heading of 180°), it is runway 18. Each digit of a runway number is enunciated individually. Runway designations are always prefixed with the word "runway," followed by the runway number and a suffix, if necessary. For example:

Runway Heading	**Runway Number**	**Pronunciation**
090°	9	Runway niner
261°	26	Runway two six
138°	14R	Runway one four right
	14C	Runway one four center
	14L	Runway one four left

If two or three runways are constructed parallel to each other, the suffixes L for "left," R for "right," and C for "center" are used to differentiate the runways from each other (see Figure 4-6). If there are four or more parallel runways, some may be given a new number fairly close to their magnetic heading (an example is Los Angeles International Airport, which has four parallel runways numbered 25L, 25R, 24L, and 24R).

Radio Frequencies When issuing radio frequencies, the controller should enunciate each digit individually. Current VHF communications radios use 25 kHz spacing between assigned frequencies. For instance, the next usable frequency above 119.600 is 119.625, followed by 119.650, 119.675, and 119.700. The first number after the decimal is always pronounced, whether or not it is a zero. But if the second number after the decimal is a zero, it is not

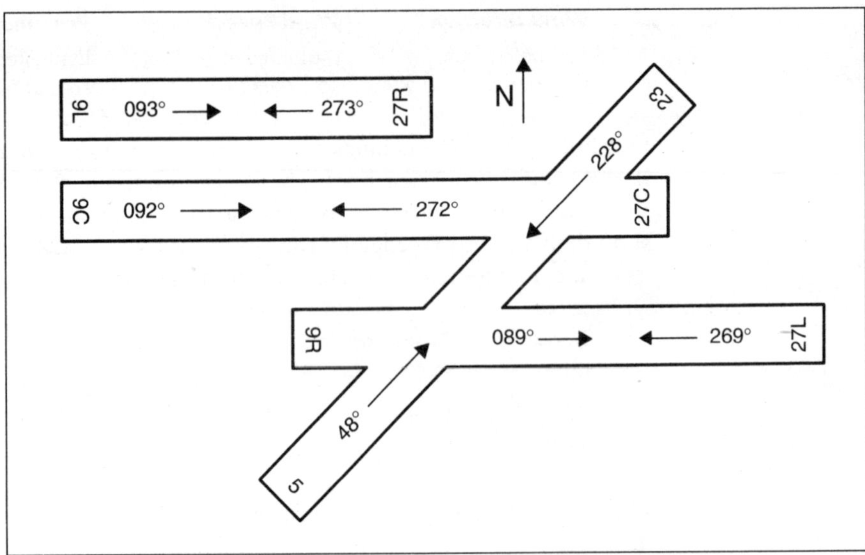

Figure 4-6. Runway numbering.

pronounced. The third number after the decimal is never pronounced, since it is always either a zero or a five and can be assumed. L/MF frequencies used by nondirectional beacons are always pronounced as whole numbers. VHF and UHF communication and navigation frequencies always use the decimal point. The decimal should be pronounced as "point." For L/MF frequencies, the number should be suffixed with the word "kilohertz." Here are some examples:

Frequency	***Pronunciation***
119.600 mHz	One one niner point six
343.000 mHz	Three four three point zero
123.050 mHz	One two three point zero five
131.725 mHz	One three one point seven two
401 kHz	Four zero one kilohertz

The FAA communications standard differs somewhat from that recommended by ICAO. Most ICAO member nations use the word "decimal" instead of "point." For example, using ICAO procedures, 123.050 would be pronounced as "One two three decimal zero five."

MLS or TACAN Channels Microwave landing system and TACAN station frequencies are not issued explicitly. Channel numbers are used instead. MLS and TACAN channels are issued as two- or three-digit numbers, with each digit being enunciated individually. For example:

Channel	***Pronunciation***
MLS channel 530	M-L-S channel five three zero
TACAN channel 90	TACAN channel niner zero

Speeds Aircraft speeds, like wind speeds, are always measured in knots. This occasionally causes some confusion with older general aviation aircraft equipped with airspeed indicators that indicate in miles per hour. Care should be taken when issuing speeds to small aircraft to ensure that the pilots realize that the requested airspeed is measured in knots. A rule of thumb is that an airspeed in miles per hour is about 15 percent higher than the equivalent airspeed in knots. Thus, 100 knots is about 115 miles per hour. Airspeeds are always expressed with each digit being enunciated individually and suffixed with the word "knots," as in the following examples:

Speed	***Pronunciation***
250	Two five zero knots
95	Niner five knots

Air Traffic Control Facilities ATC facilities are identified by name, using the name of the city where the facility is located followed by the type of facility or the operating position being communicated with:

Facility Type	***Pronunciation***
Local control	Tower
Ground control	Ground
Clearance delivery	Clearance
Air route traffic control center	Center
Flight service station	Radio
Approach control	Approach
Departure control	Departure
Flight watch	Flight watch

If a particular city has two or more airports, the airport name is used instead of the city name. Approach controls and centers are always named after the largest nearby city. Navy airports are always prefixed with "navy" to differentiate them from civilian facilities. Here are some examples:

Lafayette Tower
Chicago approach
Indianapolis center
Navy Glenview tower
Terre Haute radio

Route and Navigation Aid Descriptions Airways are always described with the route identification pronounced in group form. The route number is prefixed with "victor" if it is a low-altitude airway or "jay" if it is a jet route. RNAV-based airways are suffixed with "romeo." For example:

Route	**Pronunciation**
V12	Victor twelve
J97	Jay ninety-seven

Radials that emanate from a VOR should be pronounced as a three-digit number with each digit being enunciated individually (similar to the way aircraft headings are pronounced). The radial number is prefixed with the VOR name and is always suffixed with the word "radial" (the word "degree" is never used when describing radials):

Boiler one four three radial
Indianapolis three six zero radial
Champaign zero zero six radial

Bearings from nondirectional beacons (NDBs) are expressed as magnetic bearings from the station and are suffixed with the station's identifying name and the words "radio beacon" or "outer compass locator" as appropriate:

Three five five bearing from the Pully radio beacon
Two seven eight bearing from the Earle outer compass locator

Intersections located along an airway are described using either (1) the five-letter approved intersection name (found in FAA order 7350.5, "Location Identifiers"), or (2) the VOR radial and DME distance from the VOR. Here are some examples:

Staks intersection
Flite waypoint
Boiler zero niner zero radial one two mile fix

ATC Communications Procedures

The communications procedures that should be used by air traffic controllers are detailed in the *Air Traffic Control Handbook*. Although individual circumstances may require modification of these procedures, adhering to them will help eliminate confusion and potential problems.

The remainder of this chapter describes the most common phrases used by air traffic controllers, including how and when to use each phrase and some examples of proper phraseology. The terms may be used when communicating in writing as well as orally. To increase efficiency and conserve space when writing these phrases, standard operating procedure requires that controllers abbreviate them. The approved abbreviation appears in parentheses after each phrase.

Clearance

Any IFR or participating VFR aircraft operating within controlled airspace must be **cleared** (C) prior to participating in the ATC system. A clearance authorizes a pilot to proceed to a certain point or to perform a specific maneuver. When issuing a clearance or a control instruction, the controller must identify the aircraft, identify the ATC facility, and then issue the clearance or instruction. This instruction could be a clearance to take off or land, to perform an instrument approach procedure, or to proceed to an airport or navigational fix, as in the following examples:

Phraseology	**Explanation**
United seven twelve runway two four cleared for takeoff.	This authorizes the pilot to take off using runway 24.
Beech eight delta mike, after departure, turn left and proceed direct to the Boiler VOR, runway one zero cleared for takeoff.	This clearance directs the pilot to turn left after takeoff from runway 10 and proceed to the Boiler VOR.
Delta one ninety-one, after departure turn right heading one two zero, runway three five cleared for takeoff.	After departing runway 35, the pilot will turn right to a heading of 120°.
American nine twenty-one cleared to land runway niner.	This authorizes the pilot to make a full-stop landing on runway 9.
Aztec seven eight one cleared for touch and go runway two three.	A touch and go clearance permits the aircraft to land on the runway but take off again before actually coming to a stop. This maneuver is usually used by students practicing takeoffs and landings.
Mooney three six charlie cleared for stop and go runway five.	A stop and go clearance is similar to a touch and go except that the aircraft comes to a full stop on the runway prior to beginning its takeoff run.
Sport zero two romeo cleared for low approach runway three two.	In a low approach the pilot approaches to land on the runway but does not actually make contact with the surface. Upon reaching the desired altitude, the pilot begins a climb and departs.

Phraseology	***Explanation***
Bellanca two bravo zulu cleared for the option runway two eight left.	An option clearance permits the pilot to perform a landing, touch and go, stop and go, or low approach. The pilot does not normally inform the controller which option has been chosen. This maneuver is used in flight training to permit flight instructors to evaluate a student's performance under changing conditions.
Cherokee two three one papa uniform cleared through the airport traffic area.	This authorizes the pilot to transition through the airport traffic area without landing at an airport.
King Air four papa uniform cleared for ILS runway one zero approach.	This authorizes the pilot to conduct the published ILS approach for runway 10. This does not authorize landing on the runway. An additional clearance is necessary for landing.
Queen Air seven tango yankee cleared for approach.	This clearance authorizes the pilot to conduct any instrument approach procedure at the designated airport.

The word "cleared" is also used when issuing IFR clearances to aircraft prior to departure. An IFR clearance must include the following items (those marked with an asterisk are not required in every clearance and are used only when necessary):

1. Aircraft identification
2. The word "cleared"
3. The clearance limit
*4. Departure instructions
5. The route of flight
6. Altitude assignments
*7. Holding instructions
*8. Any additional information
9. Frequency and transponder code information

Each of these items is discussed in detail in the following sections, with examples of the proper phraseology provided.

Aircraft Identification

Aircraft are identified using standard procedures that help eliminate confusion and misdirected instructions. It is vitally important that control information directed to one aircraft be received by the pilots of that aircraft. It is also exceedingly important that the controller be certain which aircraft he or she is communicating with. If the pilot of one aircraft were to follow the instructions issued to another or if the controller were unsure of which aircraft had just made a position report, the air traffic control system would be unable to function properly.

The assigned aircraft identification call signs used by pilots and controllers vary depending on the type of operation the aircraft is involved in. If the aircraft is a scheduled airline flight operating under FAR 121 or 125, the FAA has authorized the use of a distinctive airline name that should be used when communicating with that aircraft. In addition to this name, every airline flight has been issued a flight number by the airline itself. The approved aircraft identification consists of the airline name, followed by the flight number, pronounced in group form (such as "Comair twenty-six eleven").

Most authorized airline names are easily recognizable, although a few are somewhat unusual. These approved airline names have been selected to ensure that no two sound similar. Every airline has also been issued a three-letter designator to be used in written communications concerning the aircraft. A list of air carrier names and their three-letter identifiers can be found in the *Contractions Handbook* published by the FAA. Here are some examples from the handbook.

Airline Name	**Three-Letter Identifier**	**Call Sign**
Air Canada	ACA	Air Canada
Air France	AFR	Airfrans
Alaska	ASA	Alaska
America West	AWE	Cactus
American	AAL	American
American Trans Air	AMT	Amtran
British Airways	BAW	Speedbird
Continental	COA	Continental
Delta	DAL	Delta
Federal Express	FDX	Fedex
Japan Air Lines	JAL	Japanair
KLM	KLM	KLM
Northwest	NWA	Northwest
Southwest	SWA	Southwest
United	UAL	United
United Parcel	UPS	UPS
US Airways	USA	USA

General aviation aircraft call signs consist of the type of aircraft plus a unique serial number assigned by the FAA. The call sign may contain up to five

Table 4-2. General Aviation Aircraft Call Signals

Aircraft Serial Number	Aircraft Type	Pronunciation
N231PA	Piper Cherokee	Cherokee two three one papa alpha (Cherokee one papa alpha)
N98556	Cessna Citation	Citation niner eight five five six (Citation five five six)
N5102R	Beech Sport	Sport five one zero two romeo (Sport zero two romeo)
CF-AMG	Dassault Falcon	Falcon C-F-A-M-G (Falcon A-M-G)

numbers or letters. The approved aircraft type can be found in Appendix B of FAAH7110.65. When the call sign is pronounced, each character is enunciated individually. Every U.S. aircraft's serial number is preceded by the letter N, signifying that it is registered in the United States. During routine communications, this letter is usually not pronounced but can be used if the pilot wishes to. Aircraft registered in other countries have aircraft identification numbers or letters preceded with a different letter or series of letters.

After initial communication has been established with aircraft, they may be identified using the last three characters of their assigned serial number if no confusion will result. In Table 4-2 these abbreviated call signs are enclosed in parentheses.

If two aircraft have similar last three characters, the full call sign should be used to help eliminate any confusion.

General aviation aircraft being used for special purposes are permitted to use special call sign prefixes that identify their mission. These approved prefixes are found in the FAA handbook. Here are some examples:

Type of Operation	Prefix	Phraseology
Air ambulance	Lifeguard	Lifeguard Cessna two five one lima november
Air taxi	Tango	Tango Aztec niner niner three five eight

Military aircraft are assigned a variety of call signs that may include five numbers, one word followed by numbers, or two letters followed by numbers. Each word is pronounced in full with the letters and numbers enunciated individually. The aircraft's call sign is always prefixed with the name of the military service, as in the following examples:

Call Sign	Military Service	Pronunciation
R23956	Army	Army two three niner five six
VV1963	Navy	Navy one niner six three
A14932	Air Force	Air Force one four niner three two
CAF95	Canadian	Canadian niner five

The approved identification prefixes (found in FAAH 7110.65) are as follows:

Prefix	Military Service
A	U.S. Air Force
C	U.S. Coast Guard
CAF	Canadian Armed Force
CAM	Canadian Armed Force (Transport Command)
CTG	Canadian Coast Guard
E	Medical Air Evacuation
F	Flight Check
G	National Guard
L	LOGAIR (USAF civilian contract flight)
M	MAC (Military Airlift Command)
R	U.S. Army
S	Special Air Mission
VM	U.S. Marine Corps
VV	U.S. Navy

To assist air traffic controllers in identifying military training flights that may require special handling, flights being piloted by students can be suffixed with the letter Z ("zulu").

Presidential aircraft have been assigned call signs that alert controllers that special handling of the aircraft may be required. Anytime the president of the United States is aboard a military aircraft, the call sign becomes a combination of the military service name and the word "one" (such as "Air Force one," "Marine one," "Navy one"). If the president is aboard a civilian aircraft, the aircraft's call sign becomes "Executive one." If a member of the president's family is on board an aircraft but the president is not, the call sign is suffixed with the letter F ("foxtrot"). An aircraft carrying the vice president is identified using a similar procedure but with the word "two" instead of "one." Aircraft with the vice president's family are identified using the "foxtrot" suffix.

Destination Airport or Intermediate Fix

It is preferable for the aircraft to be cleared to the pilot's filed destination airport. This procedure enables the pilot to plan the entire flight and provides a route to the destination in case of radio failure. If the controller is unable to issue a clearance to the destination airport, the pilot should be cleared to an intermediate fix and then informed of the expected route. If a delay is likely at the intermediate fix, the pilot should be informed of the approximate time that may be spent holding at the fix.

Departure Instructions

Every departing IFR aircraft must be issued an initial route that will lead from the airport to the route contained in the clearance. This may be either a published SID route or a heading. The heading should be preceded by one of the following phrases: "turn right heading" (TR), "turn left heading" (TL), or "fly heading" (FH). When issued a "fly heading," the pilot is expected to turn to the assigned heading in whatever direction results in the shortest turn. This phraseology is normally used when the aircraft's current heading is unknown. If the controller assigns a particular direction to turn (left or right), the pilot is required to turn in that direction, regardless of whether it will result in the shortest turn. Here are some examples:

Pronunciation	***Written Version***
Cessna niner papa uniform, turn right heading three five zero	N9PU TR 350
Jetstream five six three, fly heading one one zero	N563, FH 110

A departing aircraft must be assigned a heading to fly until the pilot intercepts the assigned airway or route of flight. Normally, the controller will assign the pilot a heading to fly until the aircraft joins an airway, **intercepts** [$\overline{>}$] a course or radial, or can navigate **direct** [-D→] to the navaid. For example:

Pronunciation	***Written Version***
United six eleven, turn right heading one five zero join victor ninety-seven	UAL611 TR 150 $\overline{>}$ V97
Comair twenty-five forty-one, fly heading two niner zero, join victor two fifty-one	COM2541 FH290 $\overline{>}$ V251
Kingair three papa uniform, fly runway heading until able direct the Kokomo VOR	N3PU FRH -D→ OKK

Route of Flight

The route of flight must consist of an airway, a series of airways, or a series of navaids that lead to the clearance limit. If the route issued to the pilot is exactly the same as the route filed in the IFR flight plan, the controller can substitute the phrase **cleared as filed** (CAF) instead. However, if the ATC facility at the departure airport is not equipped with radar, the first airway that will be used by the pilot should be appended to the "cleared as filed" clearance. This procedure ensures that even if a mistake has been made and the pilot flies a different route than the controller expects, at least the initial route of flight will be correct. If there is a problem later on, it will occur in an area of radar coverage, where the error can be observed and easily corrected.

If just a minor change is made to the pilot's filed route of flight, the changed portion of the route should be issued, followed by the words "then as filed." But if any major changes have been made to the pilot's filed route of flight, the route portion of the IFR clearance should be prefixed with the phrase "unable routing requested." This alerts the pilot that major changes have been made.

Once the aircraft is in flight, if any part of the clearance needs to be amended, only the amended portion of the clearance should be issued to the pilot. Here are some examples.

Pronunciation	***Written Version***
Comair seventeen fourteen, unable routing requested, cleared to the Chicago O'Hare Airport via direct Boiler, victor seven Chicago Heights, direct.	COM1714 -D→ BVT V7 CGT
Northwest two twenty cleared to the Los Angeles Airport as filed.	NWA 220 CAF LAX
Beech eight delta mike cleared to the Chicago Midway Airport via direct Knox, then as filed.	N8DM -D→ OXI CAF MDW

Altitude Assignment

Altitude assignments may be issued to pilots in a number of ways. The following phrases are used to clarify whether the pilot is to remain at a specific altitude or is permitted to climb and descend without the controller's permission.

Maintain Both IFR and participating VFR pilots are assigned an altitude at which they are required to fly. IFR pilots are required to **maintain** (-M→) this altitude, whereas VFR pilots must make every attempt to do so but are permitted to change altitude to remain in VFR conditions. When IFR pilots are assigned a new altitude to maintain, they are required by FAR 91 to advise the controller when they depart their previously assigned altitude. Unless specifically requested, they are *not* required to report when they reach their newly assigned altitude.

A clearance to maintain an altitude may be modified to include the prefixes "climb and" [↑] or "descend and" [↓]. These prefixes should be used when requesting that an aircraft change from one altitude to another. Here are some examples of "maintain" phraseology:

Pronunciation	***Written Version***
Sport zero two romeo, maintain three thousand	N02R -M→ 30
Eastern six fifty-six, climb and maintain niner thousand	EAL656 ↑ 90
Clipper six ninety, descend and maintain flight level three five zero	PAL690 ↓ 350

The word "maintain" may also be used when requesting that a pilot remain in certain weather conditions. If necessary, VFR pilots may be issued a clearance to **maintain VFR.** Since VFR pilots are not permitted by FAR 91 to fly IFR in controlled airspace without a clearance, this clearance is essentially advisory in nature. In essence, it reminds the pilot that an IFR clearance has not been issued or is no longer effective and that the aircraft must remain in VFR conditions. Controllers are not authorized to issue a "maintain VFR" clearance to aircraft operating under an IFR flight plan unless the pilot specifically requests it. A VFR clearance to an IFR aircraft is usually used whenever an IFR-rated pilot wishes to depart on an IFR clearance but upon reaching VFR conditions plans to cancel the IFR clearance and proceed VFR.

In other circumstances, the pilot may want to remain on an IFR clearance but be authorized to maintain flight in VFR conditions and to deviate from the assigned altitude. The pilot does not wish to cancel the IFR clearance since it may be needed later in the flight. This type of flight is known as **VFR on top.** With this type of clearance, the pilot is authorized to change altitudes as long as VFR conditions can be maintained. A pilot desiring this type of clearance would be advised to "maintain VFR on top." Such VFR clearance relinquishes the controller's responsibility for separating this aircraft from other IFR aircraft. The pilot assumes the responsibility for remaining in VFR conditions and for seeing and avoiding other aircraft, both VFR and IFR. If a pilot requests that an IFR clearance be reissued at some time in the future, the controller must comply with the request as soon as possible and then assume IFR separation responsibility for that aircraft.

Cruise A **cruise clearance** is used by air traffic controllers to authorize an IFR aircraft to operate at any altitude between the assigned altitude and the minimum IFR altitude. This clearance permits the pilot to level off and operate at any intermediate altitude within this assigned block of airspace. However, once the pilot begins to descend and verbally reports this descent to the controller, he or she may not return to any vacated altitude without additional ATC clearance. A "cruise" (→) clearance also authorizes the pilot to conduct any instrument approach procedure published for the destination airport. Cruise clearances are rarely used but may be assigned to aircraft approaching smaller, less busy airports that do not have operating air traffic control towers. Here is an example of the phraseology: "Cessna niner three uniform, cleared to the Champaign Airport, cruise six thousand" (N93U CMI → 60).

Cross At There may be situations in which it is operationally advantageous to require an aircraft to cross a particular navigational fix at a predetermined altitude. When this is required, the controller requests that the pilot "cross" (X) the fix "at" (@), "at or above" (⇞), or "at or below" (⇟) a specified altitude. This procedure is used whenever it is critically important, either for separation or to comply with ATC procedures, that the aircraft meet the altitude restriction. Whenever a crossing restriction has been issued, the pilot may change altitude at any desired rate but must ensure that the crossing restriction is met. If the controller requires the pilot to change altitude at the aircraft's optimal rate of climb or descent, the controller should precede the clearance with the phrase "descend now."

Pilot's Discretion Whenever a new altitude is assigned, the pilot is expected to climb or descend at an optimal rate consistent with the aircraft's performance. When the aircraft is within 1,000 feet of the assigned altitude, the pilot should attempt to decrease the climb or descent rate to approximately 500 feet per minute. The only exceptions to this procedure are when a crossing restriction has been issued and when the pilot is permitted to climb or descend at **pilot's discretion.**

If the phrase "at pilot's discretion" (PD) is used by the controller in conjunction with an altitude assignment, the pilot is given the option of when to begin the climb or descent. When authorized to change altitude at pilot's discretion, the pilot is permitted to level off at any intermediate altitude before reaching the assigned altitude but is not permitted to return to any altitude previously vacated. An altitude change in conjunction with pilot's discretion gives the pilot the opportunity to fly the aircraft in the most efficient manner, saving both fuel and time. Here are some examples of phraseology:

Pronunciation	***Explanation***
Air Force one five seven, descend at pilot's discretion, maintain flight level two zero zero	Air Force 157 may begin the descent at any point and at whatever rate the pilot wishes. The aircraft may level off at any intermediate altitude but must eventually descend to FL 200 and cannot return to any previously vacated altitude.
Comanche five niner papa, descend and maintain three thousand, cross Vages at or below five thousand	Comanche 59P may begin the descent at any point and at whatever rate the pilot wishes. The aircraft may level off at any intermediate altitude but must cross Vages at or below 5,000 feet. The aircraft must eventually maintain 3,000 feet and cannot return to any altitude previously vacated.
Gulfstream eight november mike, climb and maintain flight level two five zero, cross Potes at one three thousand	Gulfstream 8NM may climb at any rate up to FL 250 and may temporarily level off at any altitude but must cross the Potes intersection at 13,000 feet.
Mooney eight mike november, descend now to four thousand, cross the Boiler VOR at or below six thousand	Mooney 8MN must initiate a descent upon receipt of the clearance and must descend at an optimal rate for that aircraft. The aircraft must cross the Boiler VOR at or below 6,000 feet and must maintain 4,000 feet. The pilot may *not* temporarily level off at any intermediate altitude but may reduce the aircraft's rate of descent to 500 feet per minute upon reaching 5,000 feet.

Required Reports

The controller may request reports besides position and altitude from the pilot. A clearance may include requests to report crossing, reaching, or leaving.

Report Crossing Following a **report crossing (RX)** request, the pilot will advise the controller when the aircraft crosses the requested fix or intersection. Examples of phraseology include:

> Falcon four two quebec, report crossing Staks (N42Q RX STAKS)
>
> King air four papa uniform, report crossing the Danville one two seven radial, three six mile fix (N4PU RX DNV 127/36)

Report Reaching Following a **report reaching (RR)** request, the pilot will advise the controller when the aircraft has leveled off at the newly assigned altitude. For example:

> Dehavilland one six echo, climb and maintain seven thousand, report reaching (N16E ↑ 70 RR)
>
> Fairchild, eight sierra victor, report reaching flight level one niner zero (M8SV RR 190)

Report Leaving A **report leaving (RL)** clearance is used by the controller to require a pilot to report passing through any intermediate altitude. FAR 91 requires that every pilot advise the controller when *leaving* a previously assigned altitude but not when reaching an assigned altitude. "Report leaving" may be phrased as follows:

> Lear seven golf juliett, descend and maintain six thousand, report leaving flight level one niner zero, report leaving one one thousand (N7GJ ↓ 60 RL190 RL 110)

Holding Instructions

If traffic conditions warrant, pilots may be cleared by air traffic controllers to enter a **holding pattern.** Holding patterns may be necessary when aircraft must remain clear of a specific controller's area because of traffic saturation at the destination airport. Holding patterns require that the pilot fly a modified racetrack pattern in reference to a fix or a navaid. Holding patterns vary in size depending on the aircraft type and the holding altitude. Holding patterns are used primarily in areas without radar coverage. The proper application of holding patterns when separating aircraft is discussed in Chapter 7. The phraseology that air traffic controllers should use when issuing a holding instruction is as follows:

1. State the direction of holding from the fix. This is the location of the inbound course in relation to the holding fix or navigation aid. The direction of holding is issued using one of the eight points of the compass. ("Cherokee two papa uniform, hold west.")

2. State the name of the holding fix to be used. This is the fix or the navigation aid that the aircraft will actually hold at. It can be a VOR, an NDB, an intersection of two VOR radials, an intersection of two NDB bearings, an intersection defined using a VOR radial and an NDB bearing, a DME fix, or any intersection that lies along the final approach course of an instrument approach ("of Boiler," "of Staks," "of the Boiler two seven zero radial, one two mile fix").

3. State the radial, course, bearing, azimuth, or route on which the aircraft will hold ("on victor nine," "on the two seven zero radial," "on the one two three bearing to the Earle outer compass locator," "on the localizer course").

4. State the holding-pattern leg length in miles if DME or RNAV is to be used or in minutes if a nonstandard holding pattern is required. If this section is omitted in the clearance, the pilot will use a standard holding pattern, which is defined as a 1-minute inbound leg if holding is accomplished at or below 14,000 feet MSL or a 1 1/2-minute inbound leg if holding is accomplished above 14,000 feet MSL ("two-minute legs," "seven-mile legs").

5. State the direction of the holding pattern turns if a nonstandard (left turn) holding pattern is necessary. If this phrase is omitted by the controller, the pilot is expected to use right turns while in the holding pattern. ("Left turns.")

6. State the projected time (UTC) when the controller estimates that the pilot will be permitted to exit the holding pattern and continue on course. This is known as the **expect further clearance** (EFC) time. If radio communication between the pilot and the controller is lost, the pilot will depart the holding pattern and continue on course when the EFC time has passed. When the holding instructions are originally issued, the controller should also inform the pilot of the current UTC time. ("Expect further clearance at one two five five zulu, time now one two zero five zulu.")

Here are two examples of full holding messages (see Figures 4-7 and 4-8):

> Sport zero two romeo, hold west of the Earle outer compass locator on the localizer, two-minute legs, left turns, expect further clearance at zero niner zero zero, time now zero eight four five.

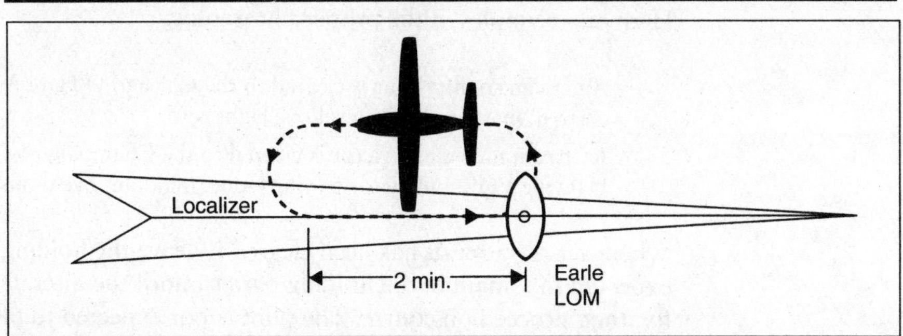

Figure 4-7. Example of an aircraft holding west of the Earle LOM on the localizer course, using two-minute legs and left turns.

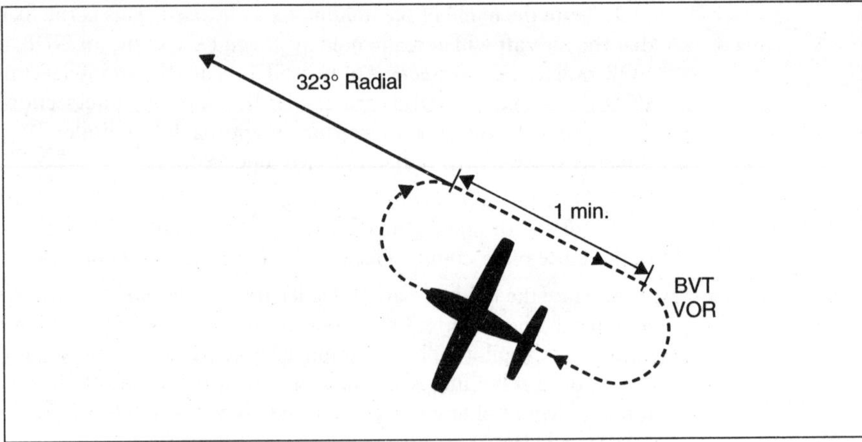

Figure 4-8. Example of an aircraft holding northwest of the BVT VOR on the 323° radial.

> United six eleven, hold northwest of the Boiler VOR on the three two three radial, expect further clearance at one one two five zulu, time now one one zero five.

Whenever the controller determines that the aircraft can be permitted to leave the holding fix and continue on course, the following procedure should be used:

1. Issue the new clearance limit.
2. Issue the route of flight to the clearance limit. If there has been no change in the route since the aircraft entered the holding pattern, the phrase "via last routing cleared" may be used.
3. Restate the assigned altitude.

Here are examples of the proper phraseology:

> American six fifty-four is cleared to the Chicago O'Hare Airport via last routing cleared, maintain flight level one eight zero.
>
> Jetstream nine alpha victor is cleared to the Champaign VOR via direct the Danville VOR and victor two fifty-one, maintain five thousand.

Whenever the aircraft has been cleared to leave the holding pattern, the pilot is expected to remain in the holding pattern until the aircraft crosses the holding fix, then proceed on course. The pilot is not expected to take any shortcuts.

Additional Communications Phraseology

When appended to a controller's transmission, the word "acknowledge" requests that the pilot inform the controller that the message in question has been received:

CONTROLLER: Cessna two mike november, cleared to land. Acknowledge.
PILOT: Cessna two mike november understands cleared to land.

The word "affirmative" means the same as "yes" but is more understandable when spoken over the radio.

The word "negative" means the same as "no" but is more understandable when spoken over the radio.

The term "say intentions" is a request for the pilot to advise the controller of his or her intentions after a maneuver is performed:

CONTROLLER: Sport zero two romeo, say intentions after this touch and go.
PILOT: Sport zero two romeo would like to depart to the east.

When only one pilot is flying an aircraft, it is particularly helpful to the pilot to be given advance notice concerning instructions that might be received in a later clearance. Such instructions are preceded by the word "expect." This information is used by the pilot for planning purposes in case of radio communication failure. Here are some examples:

> Jetstream seven bravo charlie cleared to the Danville Airport via victor two fifty-one. Climb and maintain six thousand. Expect the ILS runway one seven approach at Danville.
>
> Westwind six bravo victor, descend and maintain one zero thousand, expect lower altitude in five miles.

A variety of other standardized phrases and abbreviations are used by air traffic controllers while performing their duties. Some of the more common abbreviations are included in Table 4-3. Other phrases and abbreviations used by controllers can be found either in FAAH 7110.65 or in the facility directives.

If all the communications procedures described in this chapter are used by both air traffic controllers and pilots, the risk of miscommunication and the resulting potential for an accident or incident can be significantly reduced. In the light of this fact, air traffic controllers should routinely use standard communications techniques when conversing with pilots and other controllers, resisting the urge to use slang or CB radio language.

Table 4-3. Some Standard ATC Abbreviations

Abbreviation	Meaning
A	Cleared to airport of intended landing
B	ARTCC clearance delivered
BC	ILS back course approach
CAF	Cleared as filed
CT	Contact approach
D	Cleared to depart from the fix
F	Cleared to the fix
FA	Final approach
I	Initial approach
ILS	ILS approach
L	Cleared to land
MA	Missed approach
MLS	MLS approach
N	Clearance not delivered
NDB	NDB approach
O	Cleared to the outer marker
OTP	VFR on top conditions
PA	Precision approach
PD	Pilot's discretion
PT	Procedure turn
Q	Cleared to fly specified sectors of a navaid
RH	Runway heading
RP	Report passing
RX	Report crossing
SA	Surveillance approach
SI	Straight-in approach
T	Cleared through an intermediate point
TA	TACAN approach
TL	Turn left
TR	Turn right
V	Cleared over the fix
VA	Visual approach
VR	VOR approach
X	Cleared to cross
Z	Tower jurisdiction

KEY TERMS

cleared (C)
cleared as filed (CAF)
coordinated universal time (UTC)
cruise clearance
direct
duplex communications
expect further clearance (EFC)
Federal Communications Commission (FCC)
Greenwich mean time (GMT)
group form
holding pattern
intercept
knots
maintain
maintain VFR
millibars
pilot's discretion
prime meridian
report crossing (RX)
report leaving (RL)
report reaching (RR)
simplex communications
TELCO
VFR on top
voice switching system

REVIEW QUESTIONS

1. How is an air traffic control clearance issued?
2. How is each letter of the alphabet and each number phonetically pronounced in aviation?
3. How are runways, airports, and airways identified?
4. How is time referenced in aviation?
5. What is a holding pattern and how is it used?

5

Air Traffic Control Procedures and Organization

Checkpoints

After studying this chapter, you should be able to:

1. State the general responsibilities of an air route traffic control center, an air traffic control tower, and a terminal radar approach control facility.
2. Describe the function of a letter of agreement.
3. Describe the function of a facility directive.
4. Explain what is meant by "transfer of communication."
5. Explain what is meant by "transfer of control."
6. Explain a handoff.
7. Identify the responsibilities of a controller in an ARTCC.
8. Identify the responsibilities of a controller in a control tower.

Separation Responsibilities in Controlled Airspace

The Federal Aviation Administration is designated, by act of Congress, as the federal agency with authority for the separation of both civilian and military aircraft within the controlled airspace overlying the United States. To carry out this function, the FAA has divided the nation's assigned airspace into 24 areas and assigned aircraft separation responsibility within these areas to 24 air route traffic control centers (ARTCCs) (see Figure 5-1). Personnel at each ARTCC have the ultimate responsibility for separating every IFR and participating VFR aircraft operating within controlled airspace. Nonparticipating aircraft or aircraft operating in uncontrolled airspace are not offered separation services.

The basic function of the ARTCC is to separate aircraft traveling *between* airports. When a particular airport is congested and an FAA or military ATC facility is located at or near the airport, it is usually more efficient for the local ATC facility to be given responsibility for separating the aircraft operating in the immediate vicinity of the airport. If, after performing a study of the local airspace and traffic structure, the FAA determines that both safety and efficiency would be increased if the smaller facility were assigned responsibility for

Michael Nolan

Figure 5-1. The Indianapolis ARTCC, typical of most ARTCC facilities.

Michael Nolan

Figure 5-2. Indianapolis air traffic control tower.

the airspace, the ARTCC usually delegates aircraft separation responsibility to that facility. If it is an FAA facility, it is usually either an **air traffic control tower (ATCT)** (see Figure 5-2) or a **terminal radar approach control (TRACON).** If it is a military facility, it is usually a control tower or a **radar approach control (RAPCON).**

This transfer of separation responsibility from the ARTCC to the smaller facility is formally authorized through a **letter of agreement (LOA).** Such a letter between air traffic control facilities specifically declares:

The physical dimensions of the airspace involved.

The approved altitudes and airways used by aircraft that will cross the boundary between the two facilities.

The procedures used by air traffic controllers when an aircraft progresses from one facility's area of responsibility into the next.

Letters of agreement are also established between adjacent ARTCCs and control towers that describe the boundaries of each facility's area of responsibility and the procedures that should be used when aircraft cross this boundary (see Figure 5-3).

FORT WORTH ARTCC AND HOUSTON ARTCC

LETTER OF AGREEMENT

Effective as of 7/6/02

PURPOSE: This agreement between Fort Worth Air Route Traffic Control Center (ZFW) and Houston Air Route Traffic Control Center (ZHU) covers coordination procedures and is supplementary to the procedures in the FAA Order 7110.65.

CANCELLATION: Any and all previous LOAs between ZFW and ZHU are canceled.

PROCEDURES:

Houston shall ensure that:

- Aircraft landing in the Dallas/Fort Worth terminal area and departing from the Houston terminal area and east portion of Houston ARTCC are cleared to Cedar Creek VORTAC (CQY) and cross the ARTCC boundary at or below FL 270.

- Aircraft landing in the Dallas/Fort Worth terminal area and departing from the Austin terminal area and west portion of Houston ARTCC are cleared to Glen Rose VORTAC (CQY) and cross the ARTCC boundary at or below FL 230.

Fort Worth shall ensure that:

- Aircraft landing in the Houston terminal area and departing from the Dallas/Fort Worth terminal area are cleared to Navasota VORTAC (TNV) and cross the ARTCC boundary at or below FL 270.

- Aircraft landing in the Austin terminal area and departing from the Dallas/Fort Worth terminal area are cleared to Centex VORTAC (CWK) and cross the ARTCC boundary at or below FL 190, descending to 11,000 ft.

EXCEPTIONS: Deviations from the procedures established above may be made only after prior coordination and agreement between the parties involved.

Signed,
Houston ARTCC Chief
Fort Worth ARTCC Chief

Figure 5-3. Letter of agreement.

Air Traffic Control Procedures

When separating aircraft, or when offering any additional ATC services, controllers must use the procedures found in the *Air Traffic Control Handbook.* This FAA handbook was based on guidelines published by the International Civil Aviation Organization (also known as ICAO annexes) but differs from them in some minor areas. FAA-certified air traffic controllers, whether working for the FAA or for another employer, are obligated by law to use the handbook procedures whenever they are performing air traffic control duties.

Department of Defense (DOD) air traffic controllers use their own procedures, which differ somewhat from those used by FAA-certified controllers. In general, military air traffic control procedures are modeled after those contained in the FAA handbook but in some cases permit either the pilot or the controller less flexibility. Since some FAA controllers are assigned to military facilities and some DOD controllers separate civilian aircraft, the handbook contains both the FAA and military ATC procedures. The specific military procedures are described only if they differ from FAA-approved procedures.

To eliminate confusion about which set of procedures to apply when separating aircraft, the FAA and the DOD have mutually agreed that:

> If an FAA facility has the responsibility for providing aircraft separation at a civilian airport, FAA separation procedures shall be applied to both civilian and military aircraft operating within the FAA facility's assigned airspace.
>
> When a military ATC facility has been delegated the responsibility for providing aircraft separation at a military airport, military separation standards shall be applied to both military and civilian aircraft operating within the military ATC facility's assigned airspace.
>
> When an FAA air traffic control facility is located at and is supporting a military base exclusively, the FAA controllers will apply military separation rules to all the aircraft within the FAA facility's assigned airspace.
>
> When an FAA facility is serving both a military base and a civilian airport, military air traffic control procedures will be applied to Department of Defense aircraft whereas FAA procedures will be applied to all civilian aircraft operating within the FAA facility's assigned airspace.

Military Use of Civilian Airspace

The armed forces of the United States periodically conduct training exercises that cannot be accomplished within the confines of restricted and military operation areas. These training exercises, including air intercept and midair refueling training, may require that reduced aircraft separation be applied, which could pose a hazard to civilian aircraft. Procedures have been developed by the Department of Defense and the FAA to permit these exercises to be conducted while still maintaining safe separation between military and nonparticipating civilian aircraft. When such exercises need to be performed, the Department of Defense forwards a request to the FAA to designate and reserve a specific block of airspace where the **military** authority **assumes responsibility for separation of aircraft**, or **MARSA** for short.

Wherever MARSA airspace has been approved by the FAA, the appropriate military authority assumes total responsibility for the separation of every military aircraft operating within its boundaries. FAA controllers are notified of the MARSA reservation and are responsible for rerouting civilian IFR aircraft around the reserved airspace. VFR aircraft are permitted to operate within MARSA airspace as long as the basic VFR weather conditions exist and can be maintained. While operating within MARSA airspace, the VFR pilot is responsible for seeing and avoiding any participating military aircraft. VFR pilots will be advised of the military operations if they are in contact with an ATC facility. Whenever VFR conditions exist, it is also the military pilot's responsibility to see and to avoid any civilian VFR aircraft operating in MARSA airspace.

Air Defense Identification Zones

To protect national security, FAR 99 describes procedures to be used whenever aircraft enter the airspace of the United States from a foreign country. FAR 99 defines six zones of airspace surrounding the United States known as **air defense identification zones (ADIZs)** (see Figure 5-4). These zones are designed to facilitate the early identification and possible interception of any unidentified aircraft inbound to the United States. The six ADIZs are:

Atlantic Coastal ADIZ
Gulf of Mexico Coastal ADIZ
Southern Border Domestic ADIZ
Alaskan Distant Early Warning Identification Zone (DEWIZ)
Hawaiian Coastal ADIZ
Pacific Coastal ADIZ

Pilots penetrating an ADIZ are required to comply with the following regulations, or their aircraft may be considered as unidentified and they may find themselves being intercepted by U.S. government aircraft:

A flight plan must have been filed with the FAA prior to departing from the foreign country. This can be either an IFR or a **defense visual flight rules (DVFR)** flight plan. A DVFR flight plan is a modified VFR flight plan designed for air defense use exclusively. DVFR flight plans require pilots to specifically describe the exact location and time that their aircraft will penetrate the ADIZ.

Any aircraft penetrating an ADIZ must be equipped with a two-way communications radio operating on approved frequencies. This radio may operate in the HF, VHF, or UHF band.

All IFR aircraft must follow normal position-reporting procedures. VFR aircraft must report to the FAA prior to penetrating and ADIZ. (This report must be made 15 to 60 minutes prior to entry, depending on the type of aircraft and its location.)

VFR pilots must penetrate the ADIZ at the exact location and time specified in the DVFR flight plan. Any error exceeding about 10 minutes or 20 miles from

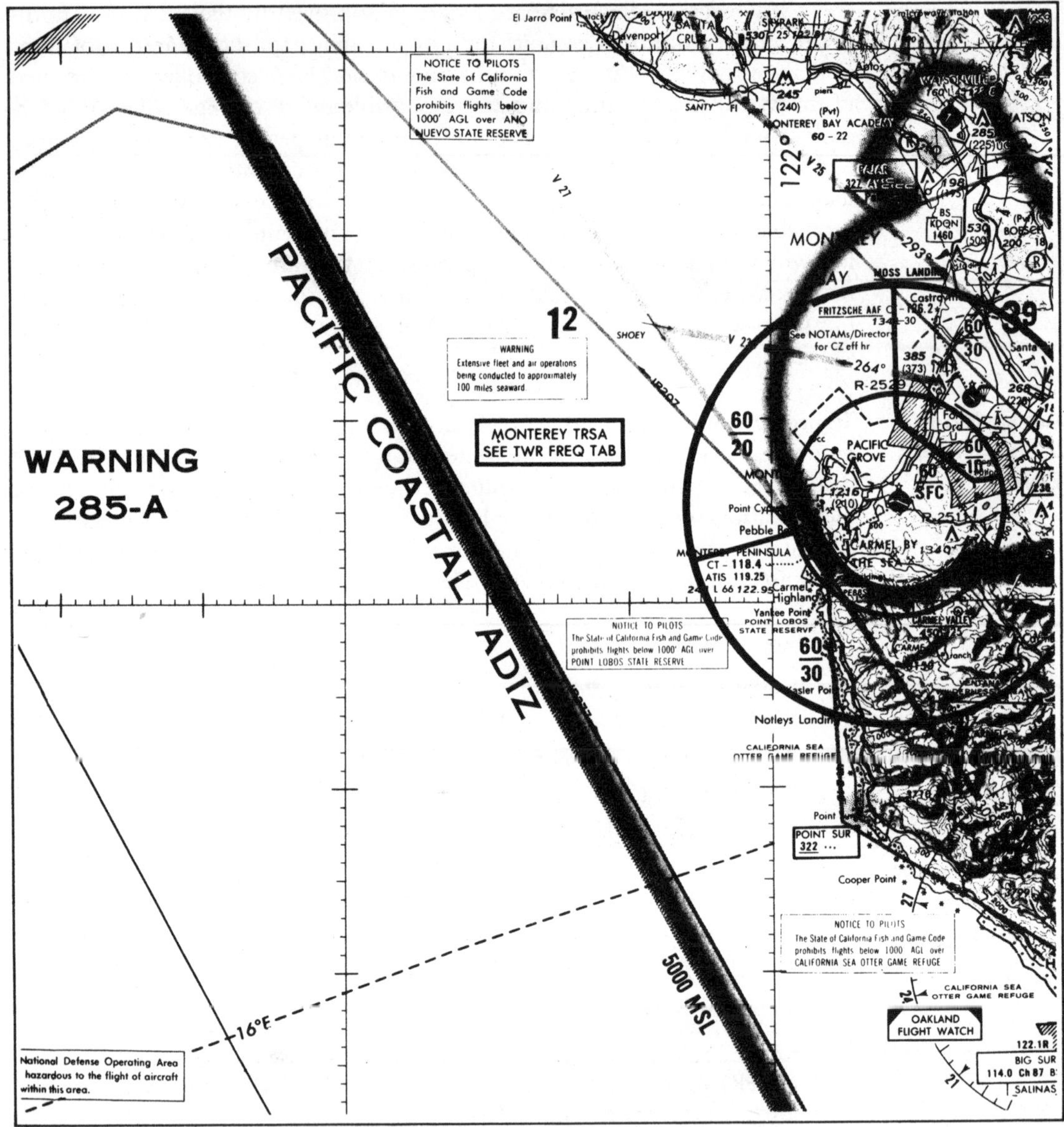

Figure 5-4. An air defense identification zone as depicted on a sectional chart.

what is stated on the flight plan will make the aircraft subject to interception by U.S. government aircraft. These aircraft may be affiliated with the Department of Defense, U.S. Coast Guard, or U.S. Customs Service.

Any pilot who does not observe these procedures will likely be intercepted and ordered to follow the intercepting aircraft to an airfield where a thorough investigation of the pilot, passengers, and aircraft can be conducted. The pilots could be charged with any number of legal violations, including violating the provisions of FAR 99. The air intercept procedures used by these aircraft are described in the *Aeronautical Information Manual.*

There is no ADIZ along the border of the United States and Canada. Because the air defense of the North American continent is maintained jointly by Canadian and American military forces, it is assumed that any unknown aircraft that may have penetrated Canadian national airspace will be intercepted and identified by Canadian military authorities before it reaches the U.S. border. U.S. Customs Service regulations still apply, however, to aircraft flying from Canada to the United States. These regulations include the filing of a flight plan, landing at an international airport, and inspection of the aircraft and passengers by customs agents.

Foreign Air Traffic Control Services

A number of FAA facilities border air traffic control facilities operated by the governments of neighboring countries. These include many of the ATC facilities near the Canadian and Mexican borders. In addition, Alaskan, Hawaiian, Puerto Rican, and Canal Zone facilities may also interact with ATC facilities from other countries. Air route traffic control centers whose jurisdiction includes oceanic flight also interact with foreign ATC facilities. In general, unless otherwise agreed to, U.S. air traffic control responsibility ends at the boundary between the two countries.

In some areas, particularly along the American Canadian border, operational requirements make it advantageous for the ATC service of one country to control traffic within the sovereign airspace of the other country. In some cases, FAA air traffic control facilities have been given responsibility for the separation of aircraft operating within the other nation's airspace, whereas in other areas the foreign country may be authorized to control air traffic within U.S. airspace. When control authority has been granted to the United States, basic FAA air traffic control procedures are applied as long as they do not unduly conflict with the procedures used by the other country. In particular, in 1985 the United States and Canada signed an agreement recognizing the essential safety of each country's air traffic control procedures. The agreement stipulates that each country may use its own ATC procedures even when separating aircraft that are within the other country's airspace.

Since much of the world's airspace lies over international waters, where no nation has the legal right to control or restrict air traffic, ICAO member nations have agreed to assign aircraft separation responsibility within international airspace to specific countries. These chosen countries are responsible for providing air traffic control services using ICAO-approved procedures. ICAO has assigned most of the Gulf of Mexico and about half of both the Atlantic and the Pacific oceanic airspace to the FAA. Because the FAA does not legally have the right to control flights within these areas, they are known as **flight information regions (FIRs)**. All ICAO member nations have agreed to comply with the procedures used by the FAA when it provides ATC services within these FIRs.

Privately Operated ATC Facilities

Because the FAA has limited resources for discharging its mission, it is unable to construct and staff an ATC facility at every airport that wants one. The FAA uses a standard formula, based on a number of factors, to determine whether an ATC facility should be constructed or whether an existing facility should remain in operation. These factors include:

The number of airline flights at the airport.

The number of airline passengers who use the airport.

The total number of flights into the airport.

The total number of IFR flights into the airport.

Any other factor that may warrant the construction of a facility, such as intensive student training, proximity to a larger airport, and so on.

Many municipalities whose airports do not qualify for FAA facilities have decided to construct and operate their own air traffic control towers. These towers are usually constructed at airports that are fairly busy but do not qualify for an FAA-operated tower. Nonfederal towers may also be located at airports with an FAA-operated tower that has been decommissioned. The local airport operating authority may choose to hire and train its own air traffic controllers or may contract this responsibility to a private air traffic control company. In either case, the control tower personnel must be certified by the FAA.

The responsibility of nonfederal control towers is to safely separate aircraft operating within the airport traffic area. Pilots who are landing or departing from one of these towers are still obligated by FAR 91.87 to comply with the instructions issued by nonfederal towers. Since the agencies operating these towers are primarily concerned with separating traffic within the immediate vicinity of the airport, the FAA seldom delegates authority for IFR separation to nonfederal control towers. This responsibility is usually assigned to a nearby FAA or military ATC facility.

Delegation of Responsibility

As stated previously, the FAA has been given the responsibility of separating every aircraft participating in the nation's air traffic control system. The definition of participating aircraft is:

Any aircraft operating under an FAA clearance in controlled airspace, utilizing IFR flight rules.

VFR aircraft operating within areas of designated airspace where air traffic control participation is mandatory (such as Class A, B, C, or D airspace).

The FAA has chosen to distribute this separation responsibility domestically to 22 air route traffic control centers in the United States (see Figure 5-5). These ARTCCs are located in the following cities:

Albuquerque ARTCC	Albuquerque, New Mexico
Anchorage ARTCC	Anchorage, Alaska
Atlanta ARTCC	Hampton, Georgia
Boston ARTCC	Nashua, New Hampshire
Chicago ARTCC	Aurora, Illinois
Cleveland ARTCC	Oberlin, Ohio
Denver ARTCC	Longmont, Colorado
Fort Worth ARTCC	Euless, Texas
Honolulu ARTCC	Honolulu, Hawaii
Houston ARTCC	Houston, Texas
Indianapolis ARTCC	Indianapolis, Indiana
Jacksonville ARTCC	Hilliard, Florida
Kansas City ARTCC	Olathe, Kansas
Los Angeles ARTCC	Palmdale, California
Memphis ARTCC	Memphis, Tennessee
Miami ARTCC	Miami, Florida
Minneapolis ARTCC	Farmington, Minnesota
New York ARTCC	Ronkonkoma, New York
Oakland ARTCC	Fremont, California
Salt Lake City ARTCC	Salt Lake City, Utah
Seattle ARTCC	Auburn, Washington
Washington ARTCC	Leesburg, Virginia

Because an individual controller cannot possibly separate all the aircraft within a particular ARTCC's boundaries, every center is divided into numerous smaller areas called **sectors.** Each of these sectors is fashioned in a logical manner, taking into consideration the airway structure and traffic flows. The process of sectorization is designed to make it easier for the controller to separate all aircraft within the sector. Every ARTCC's airspace is partitioned both vertically and horizontally into from 20 to 80 sectors (see Figure 5-6). The sectors are usually stratified vertically into two or three different levels. The vertical levels are then further partitioned into additional horizontal sectors.

The airspace at most centers is usually stratified into at least two levels: a low-altitude group of sectors extending from the Earth's surface up to 18,000 feet MSL, and a high-altitude group of sectors extending from 18,000 feet MSL (FL 180) to 60,000 feet MSL (FL 600). Busier centers may stratify into three levels, in which the low-altitude sectors extend from the ground to 18,000 feet MSL, the high-altitude sectors from FL 180 to FL 350, and the super-high sectors from FL 360 to FL 600. This vertical stratification coincides with the VOR airway structure. Aircraft operating on low-altitude victor airways are always separated by low-altitude controllers, whereas aircraft operating on high-altitude jet routes are separated by high-altitude controllers.

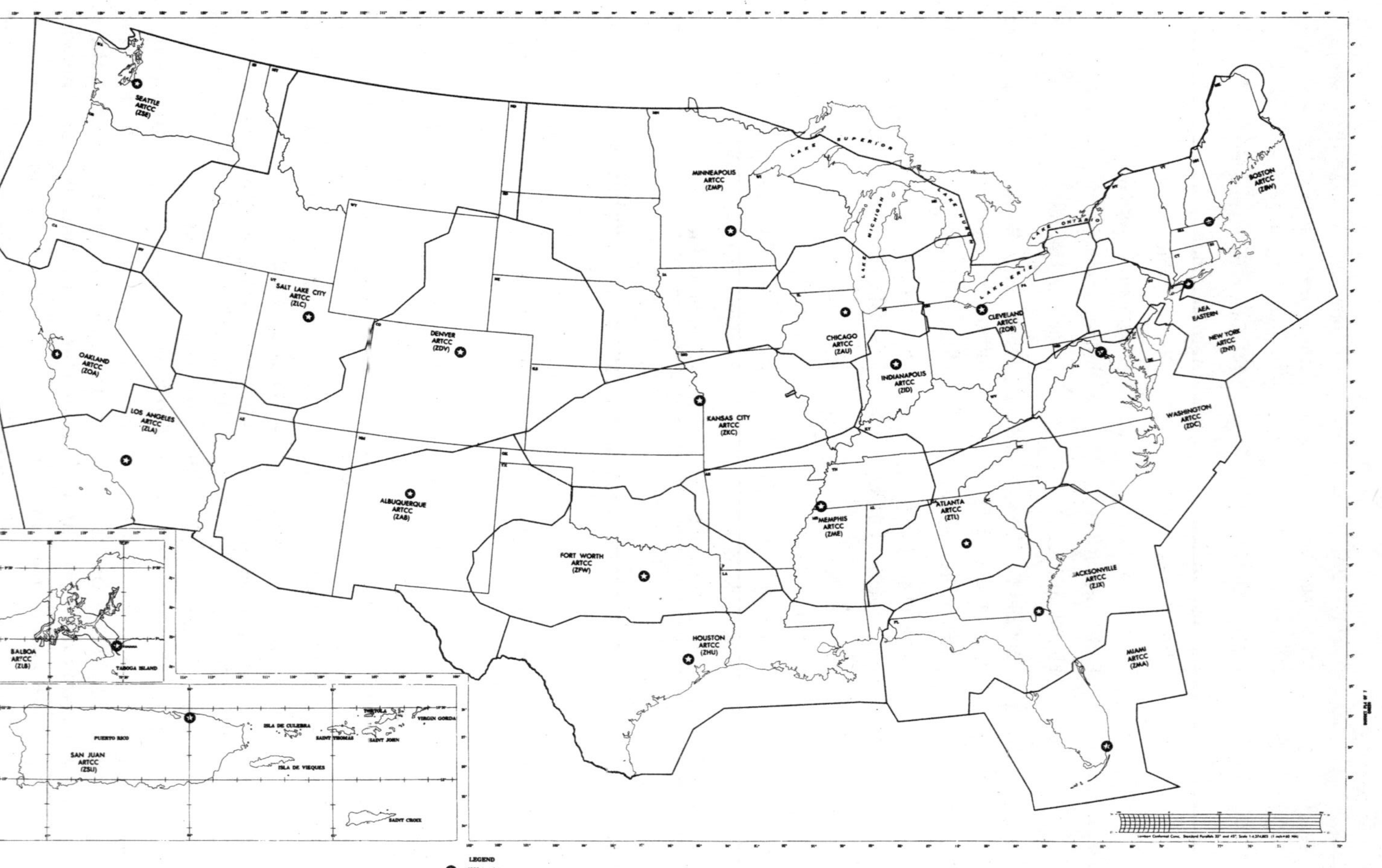

Figure 5-5. Air route traffic control center locations and boundaries.

Figure 5-6. Low-altitude ARTCC sectors.

The physical dimensions of each sector within an ARTCC are specified in the **facility directives.** Facility directives are similar to letters of agreement but apply only to controllers working within a particular facility. Facility directives specify the horizontal and vertical boundaries of each sector and describe the procedures to be used when aircraft cross the boundary between sectors.

Handoff Procedures

When an aircraft crosses a sector boundary, the responsibility for separating that aircraft passes on to the controller in the new sector. The original controller is known as the **transferring controller,** whereas the next controller is called the **receiving controller.** This transfer of separation responsibility is known as the **transfer of control.** Normally, the pilot is directed to contact the receiving controller on a different radio frequency prior to crossing the sector boundary. This is known as the **transfer of communication.** The process of transferring control and communication of an aircraft from one controller to the next is known as a handoff (see Figure 5-7).

Handoffs are necessary when aircraft cross sector boundaries and when an aircraft crosses the boundary between two separate ATC facilities, such as between two centers or between a tower and a center. The FAA handbook specifies that the transfer of communication must occur before the aircraft crosses the sector boundary. This ensures that the receiving controller will be in radio contact with the pilot before the aircraft enters his or her sector. This permits the receiving controller to issue any new control instructions to the pilot before the aircraft crosses the sector boundary.

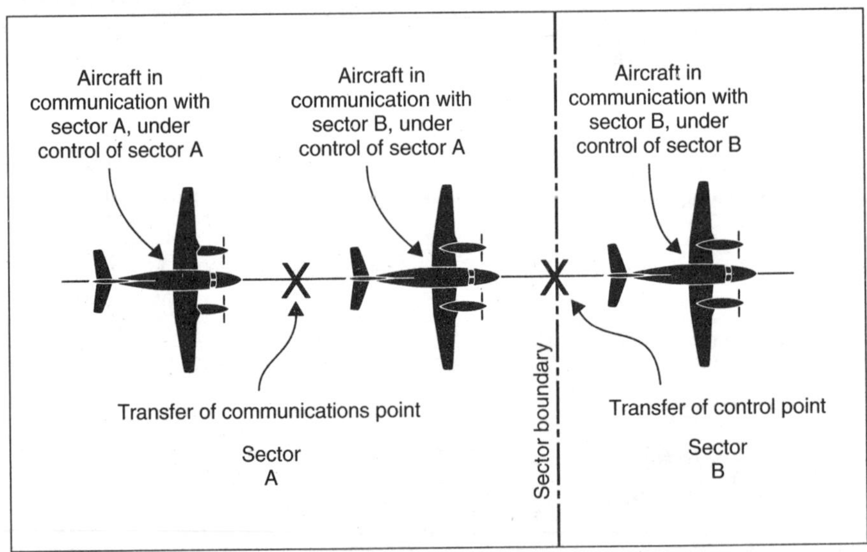

Figure 5-7. Example of transfer of communications and transfer of control.

Transfer of control does not occur until the aircraft actually crosses the boundary; thus, the receiving controller does not have separation responsibility or authority to change either the aircraft's route of flight or altitude until the aircraft crosses the sector boundary. The transferring controller must authorize any changes to the aircraft's route or altitude while it is still in his or her sector. Any clearance issued by the receiving controller cannot instruct the pilot to alter the aircraft's flight path or altitude until the aircraft crosses the boundary or unless the transferring controller approves.

Preferential Routes

The FAA has developed a system of **preferential routes** and altitudes for flight between sectors. Some of these routes are published in the *Airport Facility Directory,* whereas others are described in facility directives. The consistent use of preferential routes and altitudes enhances traffic flows, thereby reducing the controller's workload. When more than one airway extends from one busy airport to another, it is common practice to designate each as a one-way airway. This procedure reduces the chance of a head-on collision at or near a sector boundary.

If there are insufficient airways to designate one-way airways between facilities, specific altitudes will usually be reserved for inbound aircraft, and other available altitudes will be used by outbound aircraft. In most cases, odd-numbered altitudes such as 3,000, 5,000, 7,000, and so on are assigned to aircraft heading generally east, and even-numbered altitudes are assigned to aircraft heading west. The letter of agreement between the two facilities is specific about the procedures, altitudes, and airways to be used as aircraft cross the facility boundaries. Facility directives are just as specific, defining the routes and altitudes that should be used by aircraft crossing sector boundaries within the facility (see Figure 5-8).

Approval Requests and Coordination

In some circumstances the controller may need to hand off an aircraft at a different altitude or on a different airway than specified in the letter of agreement. The circumstances may be bad weather, local traffic conditions, or the pilot's request for a different route or altitude. In these cases, when the procedures specified in the letter of agreement cannot be complied with, the two controllers involved must effect **coordination** before the aircraft crosses the boundary. In the coordination process, one controller asks for and receives permission from the other controller to deviate from the terms of the letter of agreement. When effecting coordination, the transferring controller contacts the receiving controller and requests approval for a route or altitude not specified in the letter of agreement or facility directive. This type of request is known as an **approval request, or APPREQ.**

If the receiving controller determines that the approval request can be accommodated without denigrating safety or delaying other traffic, approval will normally be granted. Approval of an APPREQ is always left to the discretion of the receiving controller, since he or she will ultimately be responsible for the separation of the aircraft once it enters his or her sector.

PREFERRED IFR ROUTES

Terminals	Route	Effective Times (UTC)
MOLINE		
Fort Wayne	V156 BDF V144	0000–2359
South Bend	V156 BDF V144 MAPPS V156	0000–2359
ROCKFORD		
Fort Wayne	V128 SMARS V144	0000–2359
South Bend	V128 SMARS V144 MAPPS V156	0000–2359
SOUTH BEND		
Moline	V156 MAPPS V144 BDF V156 MZV	0000–2359
Rockford	V156 MAPPS V144 SMARS V128	0000–2359
SPRINGFIELD, IL		
Chicago Meigs	PNT MOTIF–STAR	0000–2359
Chicago Midway	PNT MOTIF–STAR	0000–2359
Chicago O'Hare	PNT V227 PLANO	0000–2359
Springfield, MO	V50 UIN V63 SGF V14	0000–2359
Tulsa	V50 UIN V63 SGF V14	0000–2359
TERRE HAUTE		
Kansas City	V50 CAP BQS–STAR	0000–2359
Springfield, MO	V7 PXV V190	0000–2359
Tulsa	V7 PXV V190 SGF V14	0000–2359

HIGH ALTITUDE

Terminals	Route	Effective Times (UTC)
CHICAGO (O'Hare—Midway)		
Atlanta	EON DNV TTH BWG RMG STAR	1100–0300
Baltimore	GIJ J146 J34 DJB J162 MGW EMI–STAR	
Boston	ELX CRL J554 JHW J82 ALB GDM GDM–STAR	
Bradley	ELX CRL J554 JHW J82 WILET RKA292 RKA SWEDA–STAR	
Charleston, WV	EON DNV VHP FLM	1200–0400
Cleveland	GIJ J146 PLAIN VWV WAKEM–STAR	
Columbus, OH	GIJ GIJ092036 FWA J178 APE	
Dallas/Fort Worth	RBS STL RZC FSM BYP	
Dayton	GIJ GIJ092036 FWA ROD DQN	
Denver	IOW DSM J10 LBF SAYGE–STAR or PLL PLL275065 FOD J94 ONL J114 SNY LANDR–STAR	
Detroit Metro–Wayne Co.	PETTY MKG POLAR–STAR	
Fort Lauderdale	GUIDO J73 SZW J43 PIE FORTL–STAR	
Fort Myers	GUIDO J73 SZW J43 PIE SRQ–STAR	
Greensboro	EON DNV VHP FLM HVQ PSK	1200–0400
Harrisburg	GIJ J146 J34 DJB J518 J152 HAR	
Kennedy	ELX CRL J554 JHW J70 LVZ LENDY–STAR	
La Guardia	GIJ J146 MIP MIP–STAR	
Memphis	RBS FAM GQE–STAR	
Miami	GUIDO J73 SZW J43 PIE CYY–STAR	
Nashville	GUIDO J73 PXV CCT HEHAW–STAR	
Newark	ELX CRL J584 SLT FQM–STAR	
Oakland	PLL PLL275065 FOD J94 ONL J148 OAL MOD or PLL PLL275065 FOD J94 LCU J158 MVA	1500–0400
Orlando Executive	EON DNV TTH BWG GQO ATL J89 OTK LEESE–STAR	
Orlando Intl	EON DNV TTH BWG GQO ATL J89 OTK LEESE–STAR	
Philadelphia	GIJ J146 J34 DJB J518 J152 HAR BUNTS–STAR	
Phoenix	MZV STJ J18 LVS J19 ZUN FOSSL–STAR or IOW J192 PWE J64 PUB ALS J102	1200–0400
Pittsburgh	GIJ J146 J34 DJB ACO V337 CUTTA	
Raleigh–Durham	EON DNV VHP299 VHP J24 HVQ BKW ROA SBV–STAR	
St. Petersburg/Clearwater	GUIDO J73 SZW DARBS–STAR	

Figure 5-8. Preferred routes.

Approval requests are used whenever a controller wants to utilize a procedure that conflicts with those contained in letters of agreement or facility directives. A controller can never be granted approval to deviate from the procedures contained in the FAA handbook, however. Application of the procedures included in the handbook is mandatory for controllers.

A typical approval request would be accomplished as follows:

TRANSFERRING CONTROLLER: APPREQ Rummy zero niner at 7,000 over Pines.
RECEIVING CONTROLLER: Rummy zero niner at Pines at 7,000 approved.

In this example, the transferring controller has requested that Air Force Rummy five niner be permitted to enter the receiving controller's airspace at the Pines intersection at an altitude of 7,000 feet. This is apparently either the wrong altitude or a route of flight different from that specified in the letter of agreement between the two facilities. The receiving controller has determined that safety will not be compromised if Air Force Rummy five niner enters the sector at this route and altitude and has granted approval. The transferring controller must then advise the pilot to contact the receiving controller before the aircraft crosses the sector boundary.

The basic rule of air traffic control separation is that *every controller is responsible for the separation of participating aircraft for the duration of time the aircraft is within the controller's sector of responsibility.* Controllers are never permitted to change the route or altitude of an aircraft while it is in another controller's area without the express permission of that controller. Conversely, a controller must always transfer both control and communication before an aircraft crosses the boundary into the receiving controller's airspace, unless approval has been granted by the receiving controller.

Controller Duties in an Air Route Traffic Control Center

Flight Data Controllers

Every sector within an ARTCC usually has one to three controllers assigned to separate the aircraft within that sector. The first position that most controllers in an ARTCC are assigned to is the role of **flight data controller.** The flight data controller is responsible for assisting the other controllers, who actually separate the aircraft. The flight data controller effects coordination with other controllers and passes along pertinent flight information to controllers working in other sectors. The flight data controller's responsibilities are similar to and directly descended from those of the C controller in the old air traffic control centers described in Chapter 1.

Radar Controllers

Every ARTCC sector equipped with radar is staffed by a controller whose responsibility is to separate participating aircraft utilizing a radar-derived display. **Radar controllers** issue altitude, heading, or airspeed changes to keep the aircraft separated and in compliance with the various letters of agreement and facility directives that may apply to that sector.

Radar Associates/ Nonradar Controllers

Every sector within the center is also staffed by a **radar associate/nonradar controller** whose duties are to assist the radar controller when separating aircraft that do not appear on the radar display. The nonradar controller's duties include updating the flight progress strips to accurately reflect every aircraft's position, altitude, and route of flight. The nonradar controller uses this information to separate aircraft that are either too low or too far away to be displayed on the radar. The nonradar controller must be prepared to assume aircraft separation responsibility if the radar display should malfunction. The nonradar controller's duties are similar to those performed by the B controller in the old air traffic control centers.

Air Traffic Control Tower Responsibilities

When it is operationally advantageous for an ARTCC to delegate separation responsibility to an air traffic control tower (ATCT), an appropriate letter of agreement is drafted by representatives of both the tower and the center. This letter of agreement delineates the control tower's area of responsibility and formally transfers the responsibility for aircraft separation to the tower. In most cases, the control tower is delegated the responsibility for separation of participating aircraft operating within about a 40-mile radius of the airport. This airspace usually extends from the Earth's surface up to an altitude of 6,000 to 10,000 feet MSL.

The letter of agreement between the tower and the center also specifies how and where the transfer of control and communication will occur. If the tower's delegated airspace is adjacent to that of another tower or a different center, a letter of agreement is also drafted by representatives from each of these facilities, describing the procedures to be used when handing off aircraft as they cross the facility boundaries.

Since the control tower's designated airspace is usually too large or complex for one controller to safely handle, it is usually divided into smaller sectors, with individual controllers responsible for aircraft separation within each sector. The facility manager, after consulting with the controllers, drafts and distributes a facility directive that defines the operating rules and procedures controllers should use when separating aircraft within the control tower's delegated airspace.

Most control towers have at least three and as many as ten operating positions where controllers might work. Every position has standardized duties and functions, which are described in the remainder of this chapter. Keep in mind, however, that each air traffic facility has its own unique requirements that might modify the generic job responsibilities described here.

Ground Control

The ground controller works in the glass-enclosed portion of the tower known as the **tower cab** and is responsible for the separation of aircraft and vehicles operating on the ramp, taxiways, and any inactive runways. This responsibility includes aircraft taxiing out for takeoff, aircraft taxiing into the terminal building after landing, and any ground vehicles operating on **airport movement areas.** Airport movement areas do not include those areas solely reserved for vehicular traffic such as service roads or boarding areas.

The ground controller is assigned a unique radio frequency to communicate with pilots and vehicle operators. The most common ground control frequency is 121.90 mHz. In congested areas where two or more control towers are located near each other, ground controller transmissions from each airport might overlap, causing pilot misinterpretation. Thus, in such cases each control tower is assigned a different frequency for its ground controllers. These additional frequencies are usually 121.80 or 121.70 mHz.

The duties of the ground controller include:

Providing instructions to taxiing aircraft and ground support vehicles.

Controlling taxiway lighting systems.

Issuing clearances to IFR and participating VFR aircraft.

Coordinating with the local controller when taxiing aircraft need to operate on active runways.

Issuing weather and NOTAM information to taxiing aircraft.

Receiving and relaying IFR departure clearances.

Relaying runway and taxiway condition information to airport management.

At less busy air traffic control towers, the ground controller may also be responsible for coordinating with other facilities and issuing ATC clearances to aircraft prior to departure. At busier control towers, these tasks are assigned to a **clearance delivery controller,** who is assigned a frequency separate from that used by the ground controller. At very busy locations, a flight data controller may also be on duty to assist the ground controller when coordinating with other controllers.

Local Control

The **local controller** is primarily responsible for the separation of aircraft operating within the airport traffic area and those landing on any of the active runways. The local controller is assigned a unique radio frequency that permits communication with these aircraft. The primary responsibility of the local controller is arranging inbound aircraft into a smooth and orderly flow of traffic and sequencing departing aircraft into this flow. The local controller's responsibilities are complicated by the fact that most of the airports in this country do not have sufficient nonintersecting runways to handle the number of aircraft wanting to land or take off. Thus, the local controller may be forced to use two or three runways that intersect with each other.

At very busy facilities, the local controller's workload may be too much for one person to handle. In such cases, the local control position is split into two, with each controller responsible for different runways and assigned separate radio frequencies. Among the duties performed by the local controller are:

- Determining the active runway.
- Issuing landing and takeoff clearances.
- Issuing landing information.
- Sequencing landing aircraft.
- Coordinating with other controllers.
- Issuing weather and NOTAM information to pilots.
- Operating the runway and approach light systems.

Approach and Departure Control

At busy facilities that have been delegated a large amount of airspace from the ARTCC, an **approach and departure control** position is usually designated. This position is commonly referred to simply as the *approach control* position. At smaller, less busy towers, approach control may be the responsibility of one controller stationed in the tower cab itself, but at larger and busier airports equipped with radar, the approach control may be housed in a separate building located near the tower. This facility is known throughout the FAA as a terminal radar approach control (TRACON). The TRACON may be equipped with up to 20 radar displays and may be staffed by up to 40 controllers at a time. At most facilities, TRACON controllers may also occasionally work in the tower cab, but at some of the larger TRACONs they are assigned strictly to the approach control facility. The airspace controlled by a TRACON is usually too large to be administered by one controller and is divided into smaller, more manageable sectors. The physical dimensions of each sector and the procedures controllers use as aircraft pass from one sector to another are delineated in the appropriate facility directives.

KEY TERMS

air defense identification zone (ADIZ)
air traffic control tower (ATCT)
airport movement areas
approach and departure control
approval request (APPREQ)
clearance delivery controller
coordination
defense visual flight rules (DVFR)
facility directives
flight data controller
flight information regions (FIRs)
ground controller
handoff
letter of agreement (LOA)
local controller
military assumes responsibility for separation of aircraft (MARSA)
preferential routes
radar approach control (RAPCON)
radar associate/nonradar controller
radar controllers
receiving controller
sectors
terminal radar approach control (TRACON)
tower cab
transfer of communication
transfer of control
transferring controller

REVIEW QUESTIONS

1. What procedure is used to distribute air traffic control separation responsibility to different ATC facilities around the United States?
2. How do military and civilian ATC facilities coordinate amongst themselves?
3. How do air traffic control facilities coordinate separation?
4. What are the operational positions within an air route traffic control center?
5. What are the operational positions within an air traffic control tower?

6

Control Tower Procedures

Checkpoints

After studying this chapter, you should be able to:

1. Explain how a controller obtains and amends information from the FDP system.
2. Explain the purpose and operation of the Automatic Terminal Information Service (ATIS).
3. State the duties of a controller in a control tower.
4. Define runway incursions and explain why they should be prevented.
5. Explain how the local controller separates aircraft in the traffic pattern.
6. State the runway separation minima for landing and departing aircraft.
7. Explain wake turbulence and the rules concerning its avoidance.
8. Explain the usage, requirements, and limitations of land and hold short operations (LAHSO).

Control Towers

Air traffic control towers are operated by both the FAA and non-federal agencies to provide separation to aircraft utilizing an airport. The primary responsibility of the control tower is to ensure that sufficient runway separation exists between aircraft landing and departing. Other responsibilities of the control tower include relaying IFR clearances, providing taxi instructions, and assisting airborne aircraft within the immediate vicinity of the airport. These tasks are accomplished using two-way radio equipment to instruct the pilot to land or take off or to adjust the aircraft's flight pattern.

There are three general categories of control towers: **VFR towers, nonradar-approach control towers,** and **radar-approach control towers.** Both radar- and nonradar-approach control towers have been delegated IFR separation responsibility by a letter of agreement between the control tower and the ARTCC. Nonradar-approach controllers are usually located in the tower cab itself and separate IFR aircraft using the nonradar procedures described in detail in Chapter 7. Radar-approach controllers are usually housed in a separate room near the base of the tower. These controllers separate IFR aircraft using radar and the procedures described in Chapter 8. VFR towers are not delegated any significant separation responsibility by the ARTCC. Primary responsibility for IFR separation around VFR towers is retained by the ARTCC or has been delegated to another control tower. Controllers in a VFR tower may be delegated limited responsibility for initially separating IFR departures or separating IFR arrivals from IFR departures. These procedures are covered in Chapter 7.

All three types of control towers are responsible for the separation of aircraft taking off or landing at the airport. Only the procedures and techniques used by controllers to separate aircraft operating within the airport traffic area or on the airport surface are discussed in this chapter.

The duties of personnel assigned to a control tower have been subdivided into four categories: flight data, clearance delivery, ground control, and local control. In a busy tower, these responsibilities may be assigned to four or more individual controllers, whereas at less busy facilities these responsibilities may be combined into fewer positions.

Flight Data Controller Duties

The **flight data controller** assists the other controllers in the tower and performs the clerical duties inherent in the operation of any facility. As noted in Chapter 5, this position is normally the first one assigned to a new controller at the facility.

The basic responsibilities and duties performed by a flight data controller include:

> Receiving and relaying IFR departure clearances to the clearance delivery controller.

Operating the flight data processing equipment.

Relaying weather and NOTAM information to other positions of operation.

Aiding other tower controllers by relaying any directed information.

Collecting, tabulating, and storing daily records.

Preparing the Automatic Terminal Information Service (ATIS) recordings.

Processing field condition reports.

Receiving and Relaying IFR Departure Clearances

The flight data controller is responsible for obtaining IFR clearances from the ARTCC and relaying them to the clearance delivery controller. These clearances are received over the telephone or through automated procedures. IFR clearances obtained by telephone are handwritten, whereas those obtained automatically are printed mechanically on a **flight data input/output (FDIO)** device.

Clearances are printed in a standard format on forms known as **flight progress strips** (or *flight strips*), as shown in Figure 6-1. After obtaining the IFR clearance, the flight data controller passes the strip to the clearance delivery controller. To facilitate accurate interpretation, flight strips are printed using standard markings and abbreviations, ensuring that specific information will always be found in the same place. These locations of flight strips are known as **fields.** The approved field contents and format can be found in the *Air Traffic Control Handbook*.

Flight progress strips used in control towers are formatted differently from those used in the ARTCCs but contain essentially the same information. A sample flight progress strip used in a control tower is shown in Figure 6-2 with the appropriate field numbers notated.

The format for flight strips differs somewhat depending on whether the aircraft involved is a departure, an arrival, or an over flight (an aircraft that

N186MC	3465	OKK	+OKK FWA MOTER DTW+			
BE2Ø/R	P2Ø4Ø		OKK SVM DTW			
979	17Ø					

Figure 6-1. Sample terminal flight progress strip.

1 2 2A 3 4	5	8	9 9B	10	11	12
	6	8A		13	14	15
	7	8B	9A 9C	16	17	18

Figure 6-2. Fields on a terminal flight progress strip.

passes through the airspace delegated to the tower but is not planning to land). Since the flight data controller in the tower will primarily be concerned with departing aircraft, that type of flight strip is discussed here.

A flight progress strip for a departing aircraft includes the following information, by field number:

1. Aircraft identification. The aircraft identification consists of the approved identification as discussed in Chapter 4.

2. Revision number (FDIO strip only). When the first flight progress strip has been printed for this aircraft, a number 1 appears in this location. If the pilot's flight plan is changed, or if the ARTCC amends the pilot's clearance, a new flight strip is printed with a number 2 in the field. The old strip should be destroyed. If by some chance it is not destroyed, the revision number will help establish which strip contains the most current flight plan information.

2A. Strip request originator. At FDIO-equipped locations this indicates the sector or position that requested the strip.

3. Type of aircraft. The type of aircraft is indicated using the conventions covered in Chapter 4. If more than one aircraft is included in the clearance, the number of aircraft involved precedes the aircraft type, separated by a slash (multiple aircraft flying under the same IFR clearance are known as a *flight*). If the aircraft's gross weight is over 255,000 pounds, it is considered a *heavy* aircraft and usually creates a phenomenon known as wake turbulence. This turbulence, which can be dangerous to following aircraft, is discussed later in this chapter. A heavy aircraft is identified with an H preceding the aircraft type on the flight strip. Examples of aircraft types include:

2/F16	Two F16 fighters
H/B747	A heavy Boeing 747

To assist subsequent controllers, an equipment suffix is added to the aircraft type. The type of equipment onboard the aircraft is determined using the information provided by the pilot upon filing the IFR flight plan. The equipment suffix printed on the flight strip will usually be one of the following:

	No transponder	**Transponder without altitude encoding**	**Transponder with altitude encoding**
No DME	/X	/T	/U
DME	/D	/B	/A
TACAN Only	/M	/N	/P
Flight Management Systems (FMS)			/E
GPS/GNSS			/G
Required Navigation Performance (RNP)			/R

	No transponder	Transponder without altitude encoding	Transponder with altitude encoding
Reduced Vertical Separation Minima (RVSM)			/W
RNP and RVSM aircraft			/Q

4. Computer identification number (FDIO only). If the flight progress strip has been computer generated and printed, a unique computer identification number will be printed in this field. This number is unique to the aircraft and can be used in place of the aircraft identification number when using FDIO equipment to obtain additional information about the aircraft.

5. Assigned transponder code. The computer located in the ARTCC will assign a transponder code to this flight. The transponder code is allocated automatically according to the **National Beacon Code Allocation Plan (NBCAP).** Since two aircraft cannot be assigned the same transponder code while within the boundaries of the same ARTCC, the NBCAP computer program attempts to assign each aircraft a transponder code that will not be the same as that assigned to another aircraft. The NBCAP plan reserves some codes that cannot be assigned to IFR flights. These transponder codes include:

1200	Reserved for VFR aircraft not in contact with an ATC facility.
7500	Reserved for aircraft being hijacked.
7600	Reserved for aircraft experiencing radio communications failure.
7700	Reserved for aircraft experiencing some type of emergency.

6. Proposed departure time. This is the proposed UTC departure time that the pilot filed in the original flight plan.

7. Requested altitude. This is the altitude requested in the pilot's original flight plan. To conserve space on the flight progress strip, the last two zeros in the altitude are dropped. For example:

Printed Altitude	***Actual altitude***
50	5,000 feet
100	10,000 feet
240	Flight level 240 (24,000 feet)

8. Departure airport. This is the airport that the aircraft will be departing from. It is printed as a three-character identifier. Every airport that has a published instrument approach has been issued an identifier. Some of the more common identifiers include:

ORD	O'Hare International, Chicago, Illinois
JFK	John F. Kennedy International, New York
ATL	Hartsfield International, Atlanta, Georgia

A more complete list is included in Appendix C.

8A and 8B. Optional use.

9. Route of flight and destination airport. The clearance limit is either the destination airport or an intermediate en route fix. The route to be flown includes any airways or VORs that the pilot will be using. If the route is to be flown using area navigation (RNAV), either the waypoint names or their latitude-longitude coordinates will be included. If no airway is designated between two VORs, it is assumed that the pilot will fly directly from one VOR to the next.

This field may also include any preferential routes that have been assigned by the ARTCC computer. A preferential route may be a departure, an en route, or an arrival route. Whenever the computer places a preferential route on the flight strip, it should replace the route of flight filed by the pilot. Preferential routes can be identified on the flight progress strip since they are bracketed with + symbols. This field may also contain the abbreviation FRC, which stands for "full route clearance." This abbreviation is added to the flight plan whenever the pilot's requested route of flight was changed by a controller, *without the knowledge of the pilot.* This information will be used by the clearance delivery controller.

9A, 9B, and 9C. Optional use.

10–18. These fields include any items that may be specified in the facility directives, including actual departure time, departure runway, or any other pertinent information. Standard symbols have been developed for use in these situations. These symbols may be found in the FAA handbook. A sampling is provided in Table 6-1.

The flight data controller should check each flight progress strip to ensure that all the appropriate information has been obtained. It is the flight data controller's responsibility to obtain a corrected flight progress strip, if necessary.

Operating the Flight Data Processing Equipment

In 1961 when President John F. Kennedy created the Project Beacon task force, controllers were still hand printing flight progress strips and passing along flight information to other controllers using teletypes and party-line telephone equipment. A significant portion of a controller's time was spent communicating with other controllers, requesting and passing along this essential flight information. The Project Beacon task force recommended that the FAA develop a computerized flight information system to automatically update and print out flight progress strips. Such a system was developed and finally installed by IBM in the early 1970s. By the mid-1980s this system had become outdated, and the FAA replaced it with a new computer system called the **flight data processing (FDP)** system.

The flight data processing system uses computers located at each of the ARTCCs to store and update aircraft flight plan information. Whenever a pilot files an IFR flight plan with any air traffic control facility, the information contained in the flight plan is transmitted to and stored in the computer. A half hour prior to the pilot's proposed departure time, the computer assigns the aircraft a transponder code and causes a flight progress strip to be printed on an FDIO printer at the departure airport. At facilities not equipped with FDIO, the flight progress strip is printed at the appropriate ARTCC sector, and the flight

Table 6-1. Symbols Used in Flight Strip Fields

Symbol	Meaning
T→	Depart (direction, if specified)
↑	Climb and maintain
↓	Descend and maintain
→	Cruise
@	At
×	Cross
—M→	Maintain
$\overline{>}$	Join or intercept airway
=	While in controlled airspace
△	While in control area
△↘	Enter control area
△↗	Out of control area
NW ⦸ ⦸ NE ○→ E	Cleared to enter, depart, or pass through a control zone (direction of flight is indicated by an arrow and the appropriate compass letter)
250K	Assigned airspeed
>	Before
<	After or past
/	Until
⤈	At or below
⤉	At or above
V<	Clearance void time
℄	Pilot cancelled flight plan
C	Contact (facility) on the appropriate frequency
RV	Radar vector

data controller in the tower must telephone the ARTCC and request the appropriate flight information. This information must then be handwritten by the flight data controller onto a flight progress strip.

Departure Message When the aircraft departs, the FDIO is used to send a departure message to the computer at the center. A departure message may be sent either manually or automatically.

To manually transmit a departure message, the flight data controller types the departure aircraft's identification and time of departure into the FDIO. This information is then sent to the computer. The controller may use the aircraft's call sign, transponder code, or computer identification number to identify any particular aircraft. The departure time is always entered as UTC time. If no time is entered in the departure message, the current time is assumed by the computer. A departure message is preceded by the characters "DM" when being entered into the FDIO. For example:

DM UA611 0313	United airlines flight 611 departed at 0313 UTC.
DM 561	The aircraft assigned computer identification number 561 departing at current UTC time.

If the control tower is equipped with the **automated radar terminal system (ARTS)** or STARS the departure message will be automatically sent to the ARTCC computer whenever the secondary radar receiver detects the transmission from the aircraft's transponder (ARTS is discussed in detail in Chapter 8). Upon receipt of the departure message, the ARTCC computer begins to automatically calculate the aircraft's future position and prints a flight progress strip for every controller who will eventually be responsible for separating the aircraft. The computer transmits the flight progress strip to each sector approximately 20 to 30 minutes before the aircraft is scheduled to enter that sector.

Amending Flight Progress Strips Using FDIO The flight progress strip is normally printed in the control tower 30 minutes before the pilot's proposed departure time. If for any reason the flight strip has not been printed when the pilot is ready to depart, the flight data controller may be asked to obtain a flight strip using the FDIO. This is accomplished through the use of a **strip request (SR)** message. To request a flight strip through the FDIO, the controller must type the letters SR followed by the aircraft's call sign (for example, SR UA611).

If one of the fields on the flight strip contains incorrect information or if the pilot requests a change to the flight plan, the flight data controller may be asked to amend the strip to incorporate the new information. The controller does so by using the FDIO to send an **amendment** (AM) message. The proper procedure is to type the letters AM followed by the aircraft's identification, the number of the field that needs to be changed, and the new information for that field. For example, AM UA611 7 120 changes the pilot's requested altitude (field 7) to 12,000 feet.

If the aircraft's route of flight or altitude is amended, a new flight progress strip is automatically sent to every subsequent sector. If the aircraft's route of flight will cause it to cross into another ARTCC's area of responsibility, the appropriate flight information is automatically transmitted to the computer within that ARTCC. When the aircraft leaves the ARTCCs area or lands at the arrival airport, the flight information is erased from the computer's memory, permitting that aircraft's transponder code to be allocated to another aircraft.

Relaying Weather and NOTAM Information

The flight data controller is required to acquire and disseminate appropriate weather information to other controllers or to the **National Weather Service** (NWS). If the NWS office is located at the airport, its personnel are usually responsible for taking routine weather observations. The controllers in the tower are required to make only tower visibility observations. If no NWS office is located at the airport, the tower controllers are likely to be responsible for performing all of the necessary weather observations, which they forward to the nearest NWS facility.

The controllers in the tower are also responsible for soliciting **pilot reports (PIREPs)** from pilots operating within the vicinity of the control tower. PIREPs are an essential means of passing along actual flight conditions to other pilots and the NWS. The flight data controller is also responsible for disseminating this weather information to pilots through the use of **Automatic Terminal Information Service (ATIS)** equipment.

ATIS is a continuous-loop audiotape usually recorded by the flight data controller and transmitted on a VHF frequency for pilot reception ATIS recordings inform both arriving and departing pilots of weather conditions and other pertinent information at the airport. Pilot reception of ATIS information relieves the ground or approach controller of repeating weather conditions and noncontrol information to every aircraft. Recordings are made at least once every hour but may be made more often if weather conditions change rapidly. The following information should be included in an ATIS recording.

1. The name of the airport.
2. The ATIS phonetic alphabet code. Every ATIS recording is assigned a code letter that identifies it. The code begins with the letter A and is incremented as new ATIS recordings are made. When pilots make initial contact with a controller, they will advise that they have received "Information (code letter)." Whenever a new ATIS recording is made, it is flight data controller's responsibility to inform the other controllers in the facility of the new ATIS code letter. Because pilots may listen to the ATIS recording 10 to 20 minutes prior to contacting a controller, this procedure will identify whether the pilot has received the latest ATIS information.
3. The UTC time that the weather observation was taken. This may not be the actual time that the ATIS recording is taped, as there is usually a delay between the weather observation and the taping.
4. The cloud ceiling. The ceiling is measured in feet above the ground and is either measured or estimated. Measured ceilings are determined using a **ceilometer.***
5. The visibility in miles and/or fractions of a mile.*
6. The current temperature in degrees Fahrenheit.
7. The current dewpoint temperature in degrees Fahrenheit.
8. Wind direction and speed.
9. The altimeter setting.
10. The instrument approach procedure(s) currently in use.
11. The runways(s) used for arrivals.

---

*Items 4 and 5 may be replaced by the phrase "better than five thousand and five," if the ceiling is higher than 5,000 feet and the visibility is greater than 5 miles.

12. The runway(s) used for departures.
13. Pertinent NOTAMS or weather advisories. These include any severe weather advisories, navigation aid disruptions, unlit obstacles in the vicinity of the airport, or any other problems that could affect the safety of flight. A list of NOTAM abbreviations is included in the "Common Abbreviations" section at the end of this book.
14. Braking action reports (if appropriate).
15. Low-level wind-shear advisories (if appropriate).
16. Remarks or other information. This may include VFR arrival frequencies, radio frequencies that have been temporarily changed, runway friction measurement values, bird activity advisories, and part-time tower operation.
17. Instructions for the pilot to advise the controller that the ATIS recording has been received. A typical ATIS recording would be taped in the following sequence:

    Lansing Airport information charlie, one five five zero zulu weather, measured ceiling six hundred overcast, visibility five, light snow, temperature three one, dew point two seven, wind one six zero at one zero, altimeter two niner five five. ILS runway two eight left approach in use, landing and departing runways two eight left and two eight right. Notice to airmen, taxiway bravo is closed. VFR arrivals contact Lansing approach control on one two five point niner. Advise the controller on initial contact that you have information charlie.

Clearance Delivery Controller Duties

The clearance delivery controller is responsible for obtaining and relaying departure clearances to pilots. The clearance should include:

Aircraft identification

Clearance limit

Departure procedure

Route of flight

Altitude

Departure frequency

Transponder code

The clearance delivery controller is also responsible for amending clearances as necessary. Aircraft clearances may need to be amended to conform to any of the procedures spelled out in letters of agreement or facility directives. Typical amendments might include temporary altitude restrictions or temporary changes in the aircraft's route of flight.

Temporary altitude restrictions may be placed on a pilot to ease the coordination required between the local and the departure controller. At most radar-equipped facilities, the local controller has been delegated the responsibility for initially separating departing aircraft. To ensure that these departures

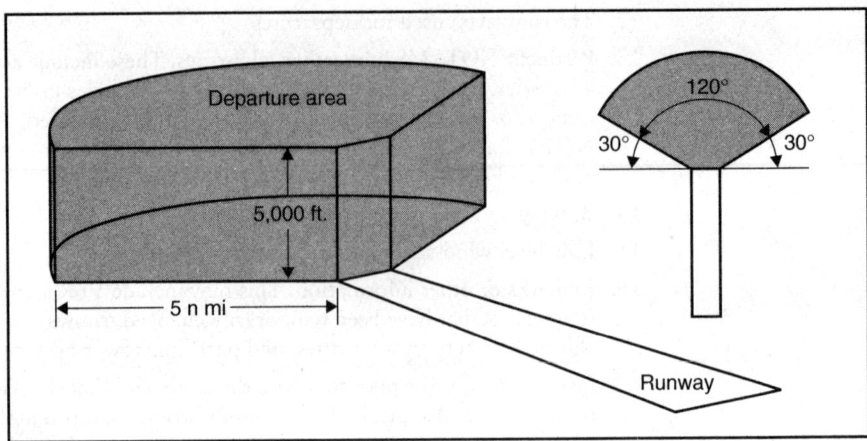

Figure 6-3. Example of a local controller's delegated departure area.

are properly separated from other aircraft within the approach controller's airspace, facility directives usually describe a specific area that may be used only for departures. Arriving aircraft may not enter this airspace without approval from the local controller. Every facility has its own unique requirements that affect the shape of this departure area, but it is usually an area 40° to 180° wide extending from the Earth's surface up to about 5,000 feet AGL (see Figure 6-3).

The facility directives usually state that the local controller may depart aircraft into this area *without* prior coordination with the departure controller. The departure controller must keep arriving aircraft out of this area unless the local controller grants approval. This procedure automatically provides for initial separation of aircraft. Once aircraft have departed, the local controller advises the pilot to contact the departure controller, who has the authority to amend the aircraft's clearance as necessary.

It is the clearance delivery controller's responsibility to temporarily amend the pilot's clearance to comply with these departure restrictions, which usually consist of restricting the pilot's altitude to the upper limit of the departure area. To reassure the pilot that this restriction is temporary and that the requested altitude will probably be granted at a later time, and to conform with the FARs, the clearance delivery controller must advise the pilot to expect his or her final altitude at some later time. This interval is specified in the facility directives and is usually 5 or 10 minutes. For example, if a pilot requests a cruising altitude of 15,000 feet MSL but the upper limit of the departure area is 5,000 feet MSL, the clearance delivery controller would advise the pilot to "maintain five thousand, expect one five thousand one zero minutes after departure."

The clearance delivery controller must also ensure that the pilot's route of flight is accurate and conforms to any preferential routes that may have been established. If the route must be changed, the controller must issue the new route to the pilot and amend the route of flight using the FDIO equipment.

Ground Controller Duties

The **ground controller** is responsible for the safety of aircraft that are taxiing on **taxiways** or **inactive runways.** The ground controller issues instructions to aircraft taxiing to or from runways, or to vehicles operating around the airport. The ground controller is permitted to exercise this control authority only in areas where traffic can be observed and controlled. The controller is not responsible for aircraft taxiing where they cannot be observed from the control tower, such as aircraft parking areas, hangars, and terminal boarding. Aircraft operating within these nonmovement areas cannot be offered any ground control services. Aircraft and vehicles operating within these areas may proceed without contacting the ground controller.

To ensure that the ground controller is always communicating with the correct pilot, the aircraft's position must be positively determined before issuing any instructions. This position determination can be made through the use of visual observation, a pilot report, or airport surface radar.

After determining the aircraft's location, the ground controller should issue positive instructions to the pilot. These instructions should include the aircraft identification, the name of the ground controller's facility, the route to be used while taxiing, and any restrictions applicable to the pilot. Here are some examples of phraseology:

United six eleven, Lafayette ground, taxi to runway one zero.

Cherokee two one four papa alpha, taxi to runway three five via taxiway bravo and charlie.

American niner twenty-one, taxi to the terminal via the new scenic taxiway.

To avoid confusion when issuing taxi instructions to pilots, the ground controller should never use the word "cleared." The only person who should use this word is the clearance delivery controller when issuing a clearance or the local controller when clearing aircraft for takeoff or landing. Because the fidelity of aircraft communications equipment is low and the noise level in the tower and cockpit is fairly high, it is possible for the pilot to misinterpret "Cleared to taxi to runway three one" to mean "Cleared for takeoff runway three one." Obviously these are two very different clearances. But if the weather conditions are such that neither the ground nor the local controller can see the aircraft, this misinterpretation might prove to be very dangerous.

Preventing Runway Incursions

One of the primary responsibilities of the ground controller is to ensure that vehicles and taxiing aircraft remain clear of the active runways. If an aircraft or vehicle must cross an active runway, the ground controller must receive permission for that operation from the local controller. If an aircraft should inadvertently taxi onto an active runway without the local controller's knowledge, an accident could result. Such accidental entry, known as a **runway incursion,** should be avoided at all costs.

One of the best ways to prevent a runway incursion is to use and understand the appropriate phraseology for communicating with taxiing aircraft. When the clearance to the aircraft begins with "Taxi to (runway number)," the pilot is authorized to cross any and every taxiway and runway along the route. The pilot does not know which of these runways are active and assumes that any required coordination has been accomplished. If the aircraft is required to taxi across an active runway en route to the departure runway, the ground controller must coordinate with the local controller to receive permission to cross the active runway. If that permission is not received, the ground controller must advise the pilot to stop prior to the runway. This is known as holding short of the runway.

To differentiate a "hold short" type of clearance from the others, the phraseology of this clearance has been somewhat modified. It includes the aircraft identification, the facility name, the departure runway number, the taxi route, the words "hold short," the position to hold, and the reason for holding short. For example:

> Comair twenty-three eleven, Lafayette ground, runway five, hold short of runway one zero, traffic landing on one zero.
>
> Clipper four fourteen, runway three two right via the cargo and the old scenic taxiway, hold short of runway two seven left, traffic landing two seven left.
>
> Kingair four papa uniform, runway one zero, hold short of the parallel taxiway, traffic inbound on the taxiway.

If the aircraft must cross an active runway, the ground controller must receive permission from the local controller and advise when the operation is complete. For example (see Figure 6-4):

AMERICAN 810:	O'Hare ground, American eight ten ready to taxi.
GROUND CONTROLLER:	American eight ten, O'Hare ground, runway two-seven right via taxiway hotel, hold short of runway three two right traffic departing runway three two right.
AMERICAN 810:	American eight ten, roger, taxi to runway two-seven right, hold short of runway three two right.
(as the aircraft approaches runway 32R)	
GROUND CONTROLLER:	(to the local controller): Cross three two right at hotel?
LOCAL CONTROLLER:	Cross runway three two right at hotel.
GROUND CONTROLLER:	American eight ten, cross runway three two right.
AMERICAN 810:	American eight ten, roger.

If the aircraft must taxi quickly across the runway, the ground controller should use either "taxi without delay" or "immediately." "Taxi without delay" advises the pilot to cross the runway safely but using a minimum of time. "Immediately" should be used only in an imminent emergency.

After the aircraft has crossed the active runway, the ground controller must advise the local controller that the crossing is complete, either verbally or through any visual means specified in the facility directives.

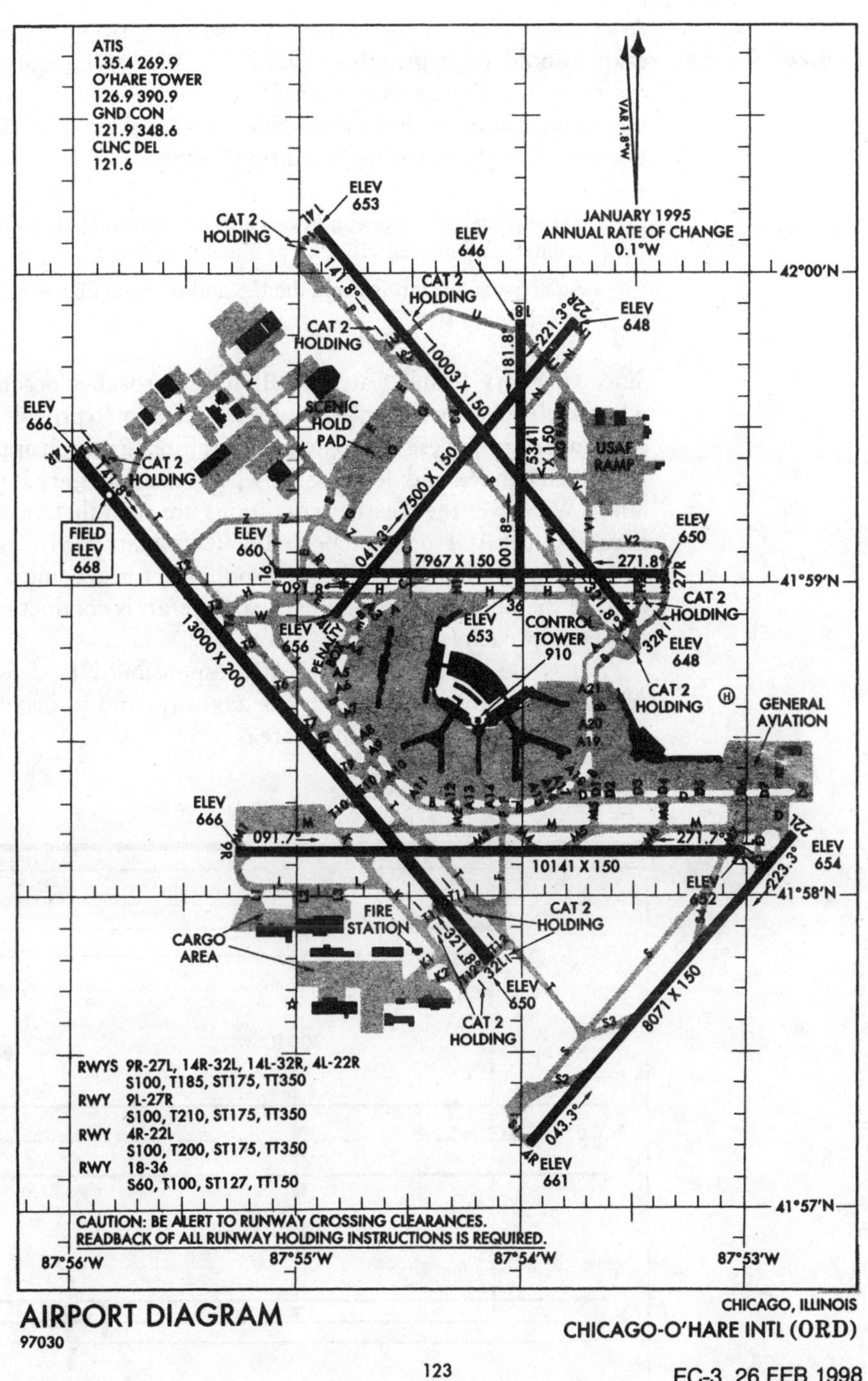

Figure 6-4. Airport taxi chart for O'Hare International Airport.

Protecting Critical Areas

Areas other than the active runway where the ground controller may want aircraft to hold short include the localizer, glide slope, and precision approach critical areas. The ground controller should not authorize any aircraft or vehicular operation within the confines of a localizer or glide slope critical area when both of the following conditions occur:

> The reported weather conditions at the airport include a lower than 800-foot ceiling *or* a reported visibility of less than 2 miles.
>
> An arriving aircraft is using the ILS and is located between the outer marker and the airport.

Since Category II and Category III ILS approaches permit the pilot to land when visibilities are extremely low, it is necessary to provide additional obstacle clearance during these approaches. Therefore, precision approach critical areas have been defined and demarcated wherever a Category II or III ILS is in operation. Whenever the weather conditions are such that either the ceiling is less than 200 feet AGL or the reported RVR visibility for the runway is 2,000 feet or less, the ground controller is responsible for keeping aircraft and vehicles clear of the obstacle critical area as an aircraft is conducting an approach or a missed approach (see Figure 6-5).

It is the airport management's responsibility to determine whether ILS critical areas affect any runways or taxiways and to install appropriate signs and markings to delineate these areas.

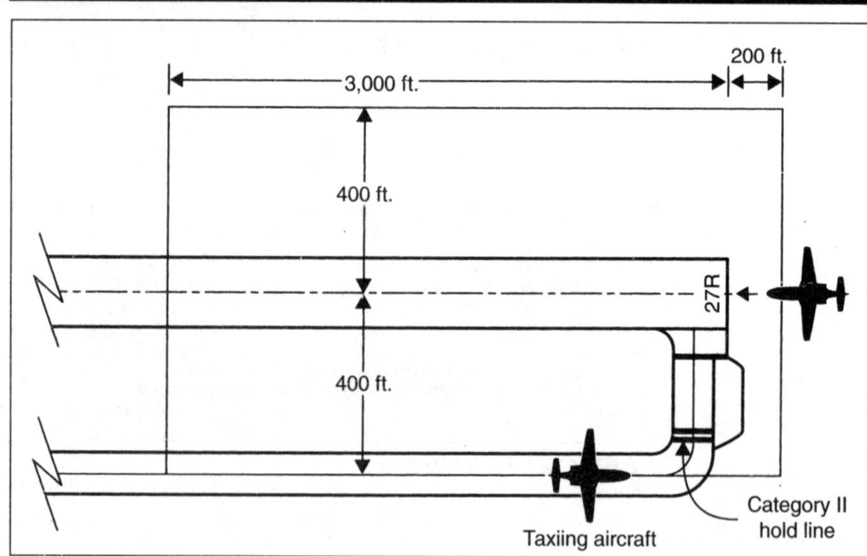

Figure 6-5. Obstacle critical area.

Local Controller Duties

It is the responsibility of the **local controller** to safely sequence arrivals and departures at the airport. The primary responsibility of the local controller is to ensure that proper **runway separation** exists between aircraft. The local controller issues appropriate instructions to arriving and departing aircraft to ensure this runway separation. It is *not* the local controller's responsibility to separate VFR aircraft inbound to the airport, although the controller may offer assistance and issue traffic advisories. It is assumed that the pilots will apply the see-and-be-seen rules of traffic avoidance.

Runway Separation

For the purpose of runway separation, every aircraft is classified by **aircraft category.** Aircraft categories are determined as follows:

CATEGORY I	Lightweight, single-engine, propeller-driven personal aircraft. This category includes the Cessna 152 and 172, Piper Cherokee, and Bellanca Viking. It does not include high-performance single-engine aircraft such as the T-28.
CATEGORY II	Lightweight, twin-engine, propeller-driven aircraft weighing 12,500 pounds or less. This category includes aircraft such as the Twin Commanche, Piper Seneca, and Cessna 320 but does not include larger aircraft such as the Lockheed Lodestar or Douglas DC-3.
CATEGORY III	All other aircraft not included in either Category I or II. This category includes high-performance single-engine, large twin-engine, four-engine propeller-driven, and turbojet aircraft. Category III includes aircraft such as the Douglas DC-3 and DC-6, Cessna Citation, Boeing 727, and McDonnell Douglas MD-80.

Departing Aircraft Separation The local controller is required to separate departing aircraft using the same runway by ensuring that an aircraft does not begin its takeoff roll until at least one of the following conditions exists:

1. The preceding landing aircraft has taxied off of the runway.
2. The preceding departing aircraft is airborne and has crossed the departure end of the runway or has turned to avoid any conflict (see Figure 6-6). If the local controller can determine runway distance using landmarks or runway markings, the first aircraft need only be airborne before the second aircraft begins its takeoff roll if the following minimum distance exists between the aircraft involved (see Figure 6-7):
   a. If both aircraft are Category I, a 3,000-foot separation interval may be used.
   b. If a Category II aircraft precedes the Category I, a 3,000-foot separation interval may be used.

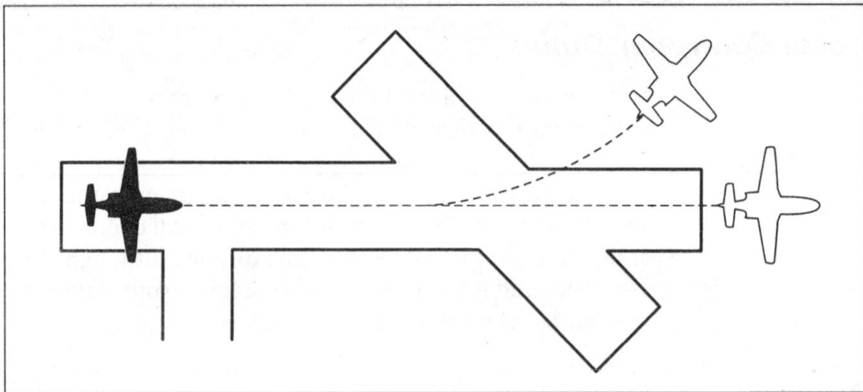

Figure 6-6. Preceding aircraft must have crossed the departure threshold or turned to avoid a conflict before the following aircraft can depart.

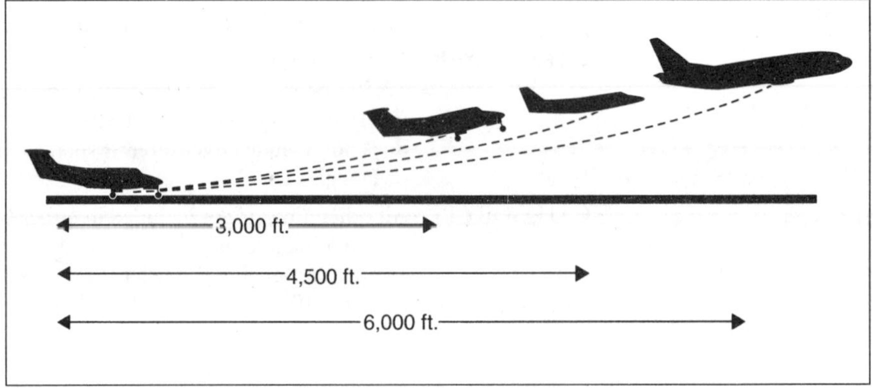

Figure 6-7. The following aircraft must wait until the preceding aircraft is both airborne and a specified distance down the runway before it can begin its takeoff roll.

c. If the succeeding or both of the aircraft are Category II, a 4,500-foot separation interval must be used.

d. If either of the aircraft is a Category III aircraft, a 6,000-foot separation interval must be used.

Thus, if a Piper Cherokee (Category I) departs and is followed by a Cessna 152 (Category I), the local controller must not permit the Cessna to begin its takeoff roll until the Piper has crossed the departure end of the runway, has turned to avoid a conflict, or is airborne and is at least 3,000 feet down the runway. But if the Piper is followed by a Cessna 310 (Category II), the local controller must not permit the Cessna 310 to begin its takeoff roll until the Piper has crossed the departure end of the runway, or has turned to avoid a conflict, or is airborne and is at least 4,500 feet down the runway. If the Cessna 310 were to precede the Piper, however, only 3,000 feet of separation would be needed.

To increase runway utilization, it may be advantageous to have the aircraft on the runway, in position to depart, waiting for the preceding aircraft to complete its departure. When this procedure is used, the pilot can be advised to "taxi into position and hold." The controller should then state the reason that the departure clearance is being withheld:

> Bellanca six eight charlie, runway two three, taxi into position and hold, traffic landing runway one zero.
>
> Clipper one seventeen, runway two one center taxi into position and hold traffic crossing the runway at midfield.

Controllers should be careful when using this clearance to ensure that the pilot does not misinterpret the instruction as a takeoff clearance. It is for this reason that the word "cleared" should never be included in a clearance to hold short or to taxi into position and hold.

The instruction "cleared for takeoff" clears the pilot to perform a normal takeoff on the runway specified. If more than one runway is active, the runway number should precede the clearance. If additional departure instructions are necessary, they should also precede the takeoff clearance:

> United seven twenty-five, cleared for takeoff.
>
> Kingair six papa uniform, runway two three, cleared for takeoff.
>
> Cessna six niner eight, after departure fly heading one eight zero, runway one five, cleared for takeoff.

When issuing a "cleared for immediate takeoff," the controller is expecting the pilot to minimize any delay in departing. The pilot will do his or her best to comply with this clearance but certain procedures may have to be performed by the pilot while the airplane is still on the runway. If there is any doubt about safe separation, one of the following alternative clearances should be used:

"Cleared for immediate takeoff or hold short"	The controller is advising the pilot that an immediate departure is required. A pilot who feels that a safe departure can be accomplished will proceed; otherwise, he or she will hold the aircraft short of the runway.
"Cleared for immediate takeoff or taxi off the runway"	This clearance is most often used when an aircraft has been taxied into position to hold and departure clearance has been delayed.

Here are some examples of phraseology:

> Piedmont two fifty, runway one seven, cleared for immediate takeoff or hold short, traffic one mile on final.
>
> TWA six ninety-one, runway one eight, cleared for immediate takeoff or taxi off the runway, traffic is a DC-9 landing runway two four.

Intersecting Runway Separation If the departing aircraft is taking off on a runway that intersects another active runway, or if the flight path of the aircraft will intersect another runway, the local controller must ensure that the aircraft does not begin the takeoff roll until at least one of the following conditions exist:

1. A preceding, landing aircraft has:
   a. taxied off the landing runway, or
   b. completed the landing roll and has advised the local controller that it will stop prior to the runway intersection, or
   c. passed the intersection (see Figure 6-8).
2. A preceding, departing aircraft is airborne and has passed the intersection or is turning prior to the intersection to avert a conflict (see Figure 6-9).

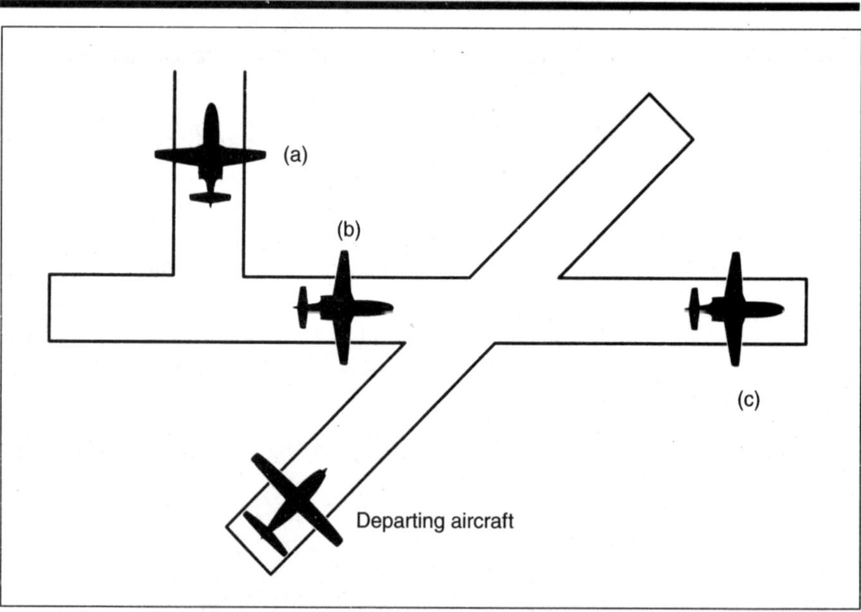

Figure 6-8. Before the departing aircraft on the intersecting runway can be cleared for takeoff, the arriving aircraft must have (a) landed and turned off of the runway, (b) landed and advised that it will hold short of the intersection, or (c) passed through the intersection.

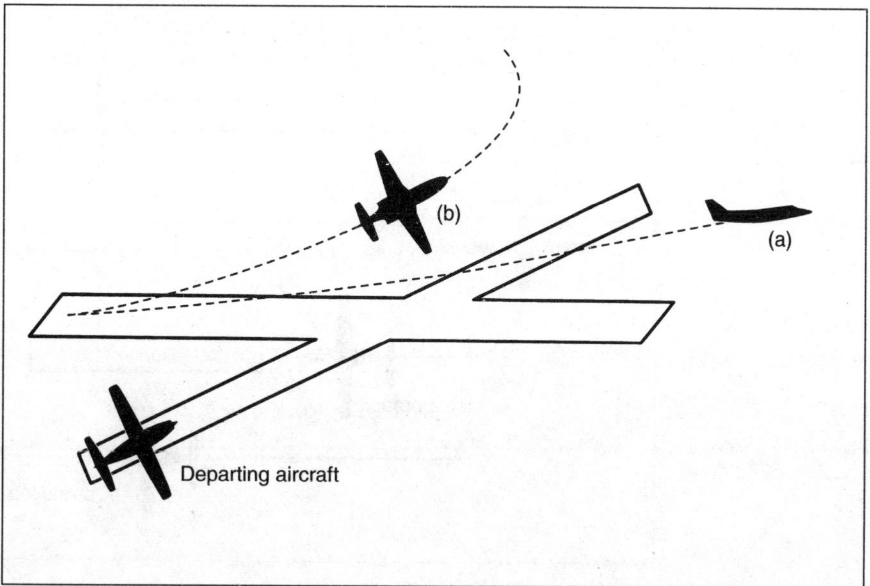

Figure 6-9. An aircraft departing on an intersecting runway must wait till the preceding departure has either (a) passed through the intersection or (b) turned to avoid a conflict.

Anticipated Separation The local controller need not actually wait for the appropriate separation interval to clear an aircraft for takeoff. If there is reasonable assurance that the correct separation will exist before the departing aircraft actually begins its takeoff roll, the clearance may be issued at that time. This is known as **anticipated separation.** In accordance with the FAA handbook, air traffic controllers are permitted to issue both anticipated arrival and departure clearances if proper separation can be expected when needed.

Arriving Aircraft

IFR pilots use a standard instrument approach when arriving at the airport, whereas VFR pilots approach the airport using all or a portion of a standardized **traffic pattern.** It is the local controller's responsibility to properly space these two types of inbound aircraft while also sequencing departures into the traffic flow. A VFR traffic pattern consists of five portions known as **traffic pattern legs:**

UPWIND	A flight path parallel to the landing runway in the direction of landings and departures.
CROSSWIND	A flight path at right angles to the landing runway on the departure end.
DOWNWIND	A flight path parallel to the landing runway in the direction opposite to landing.

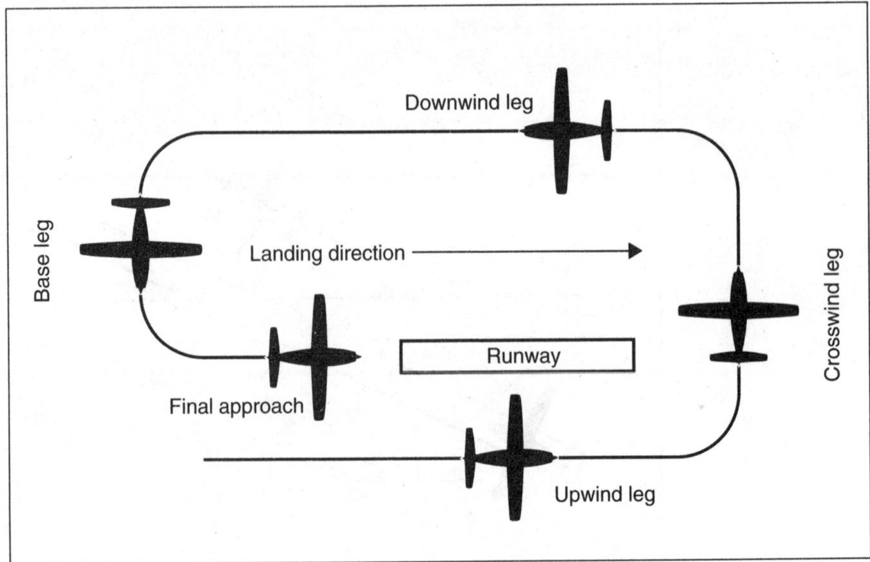

Figure 6-10. Traffic pattern legs. This example depicts an aircraft in left traffic.

BASE	A flight path at right angles to the landing runway off its approach end and extending from the downwind leg to the intersection of the extended runway center line.
FINAL	A flight path in the direction of landing along the runway centerline extending from the base leg to the runway.

A typical traffic pattern is shown in Figure 6-10.

If the turns performed by the aircraft in the pattern are to the left, the traffic pattern is known as **left traffic.** If all turns are made to the right, it is known as **right traffic.** Unless specified in the facility directives, either left or right traffic can be used for any runway at a tower-controlled airport; left traffic is considered standard at uncontrolled airports.

The local controller is required to apply runway separation standards to arriving aircraft just like departures. This requirement is accomplished by requiring the pilots to adjust their flight pattern as necessary to provide the following separation for single runways and intersecting runways.

Single Runway Separation If only one runway is in use, the local controller must separate arriving aircraft from other aircraft by ensuring that the arriving aircraft does not cross the landing threshold until at least one of the following conditions exists:

1. If the preceding aircraft is an arrival, it has landed and taxied off of the runway (see Figure 6-11). Between sunrise and sunset, the preceding aircraft need not

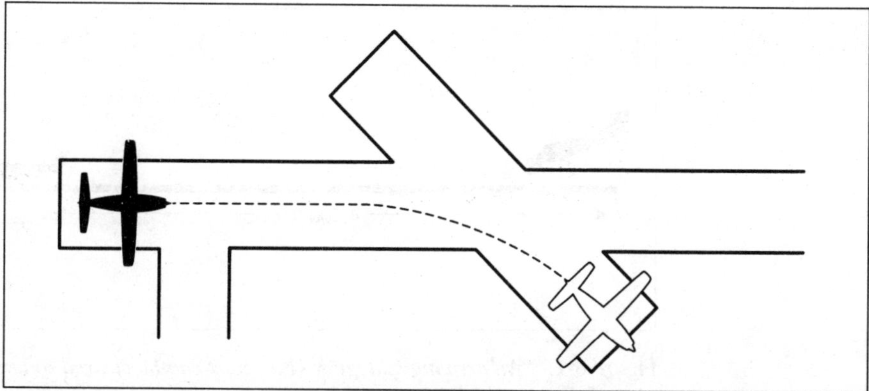

Figure 6-11. An arriving aircraft may not cross the landing threshold until the preceding aircraft has landed and turned off of the runway.

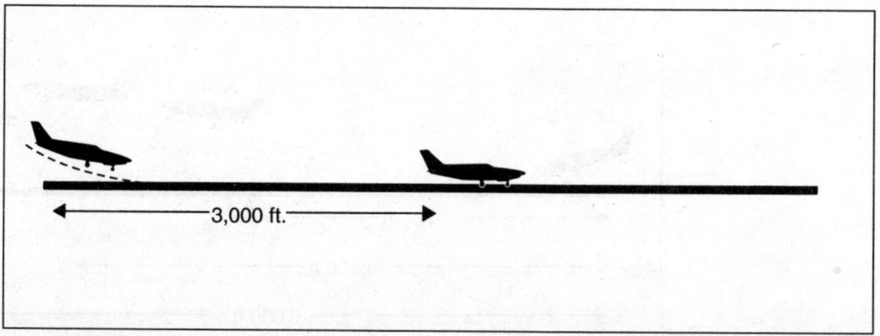

Figure 6-12. An arriving Category I aircraft can be cleared to land if the preceding aircraft has touched down, is a Category I aircraft, and is at least 3,000 feet down the runway.

have taxied off of the runway if the distance between the two aircraft can be determined using landmarks or runway markings, and the following minimums can be maintained:

a. 3,000 feet, if a Category I aircraft is landing behind either a Category I or a Category II aircraft (see Figure 6-12).

b. 4,500 feet if a Category II aircraft is landing behind either a Category I or a Category II aircraft (see Figure 6-13).

2. If the preceding aircraft is a departure, it must have already crossed the departure end of the runway. This minimum can be disregarded if the departing aircraft is airborne and is at least the following distance from the landing threshold:

   a. 3,000 feet if a Category I aircraft is landing behind either a Category I or a Category II aircraft.

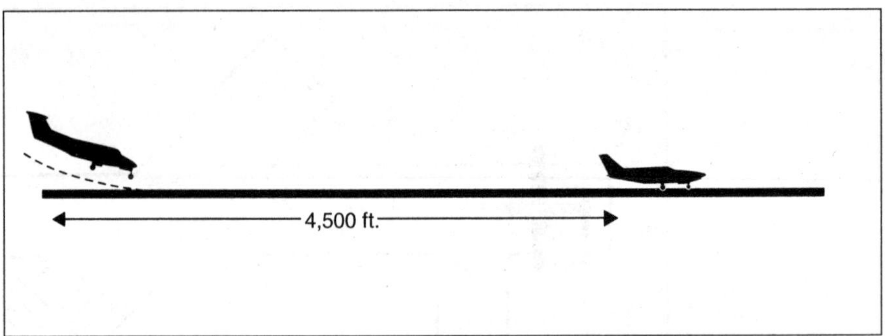

Figure 6-13. An arriving Category II aircraft can be cleared to land if the preceding aircraft has touched down, is a Category I or II aircraft, and is at least 4,500 feet down the runway.

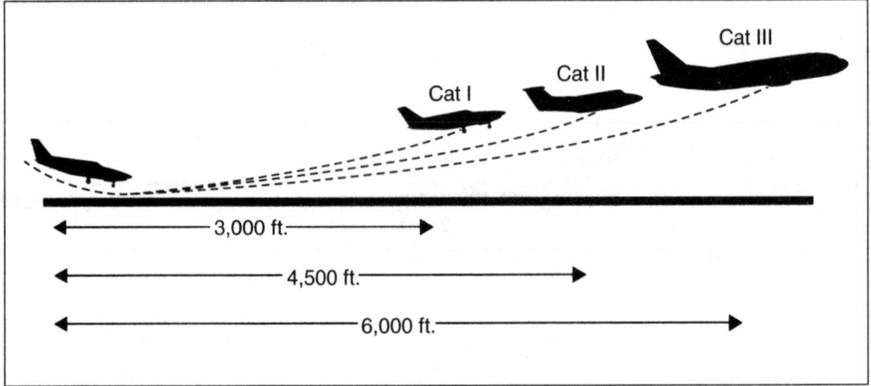

Figure 6-14. The following aircraft must wait until the preceding aircraft is airborne and has traveled a specified distance down the runway before it can cross the landing threshold.

b. 4,500 feet if a Category II aircraft is landing behind either a Category I or a Category II aircraft.

c. 6,000 feet if either of the aircraft is a Category III aircraft (see Figure 6-14).

Intersecting Runway Separation If intersecting runways are in use, a landing aircraft must be sequenced so as not to cross the landing threshold until at least one of the following conditions exists:

1. A departing aircraft from an intersecting runway has either crossed the intersection or has turned to avert any conflict (see Figure 6-15).

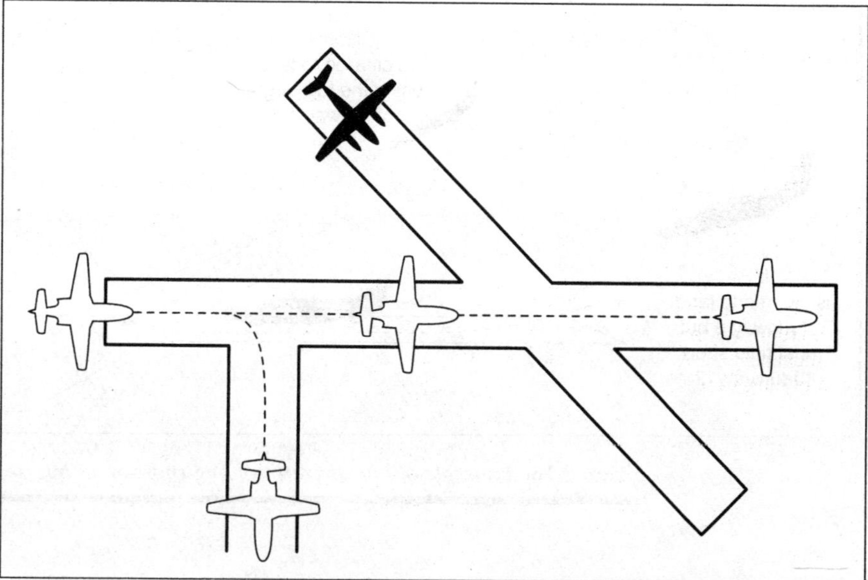

Figure 6-15. Example of an intersecting runway departure separation.

2. An aircraft landing on the intersecting runway has taxied off the landing runway, has crossed the runway intersection, or has completed the landing roll and advised the local controller that the aircraft will hold short of the intersecting runway.
3. When approved in the facility directives, the local controller may authorize an aircraft to land on a runway that intersects the departure runway when *all* of the following conditions can be met:
   a. VFR conditions exist at the airport.
   b. The aircraft has been instructed to hold short of the intersecting runway, has been informed of the traffic departing on the intersecting runway, and has acknowledged the instruction.
   c. The departing aircraft has been advised that the other aircraft will be holding short.
   d. Both runways are clear and dry with no reports that the braking action is less than "good."
   e. The aircraft instructed to hold short has no tailwind.
   f. If requested by the pilot, the distance from the landing threshold to the intersection has been issued by the local controller.

Facility directives specifically state which intersections may be utilized and which **aircraft group** is authorized to hold short. The aircraft group number can be found in Appendix B of the FAA handbook. Here is an example of the phraseology to be used when landing aircraft are holding short (see Figure 6-16):

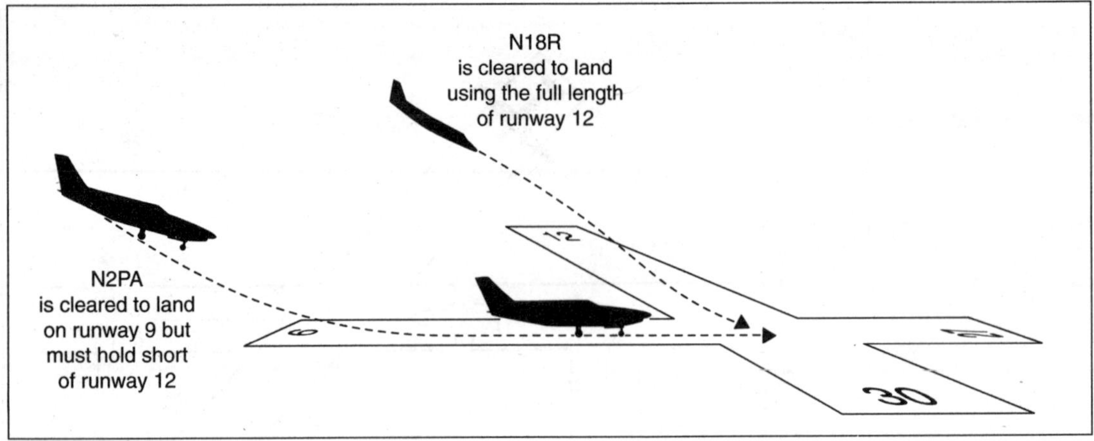

Figure 6-16. Example of one aircraft holding short of an intersecting runway.

LOCAL CONTROLLER:	Cherokee two papa alpha, cleared to land runway niner, hold short of runway one two, traffic landing runway one two.
CHEROKEE 2PA:	Cherokee two papa alpha, roger.
LOCAL CONTROLLER:	Sport one eight romeo, cleared to land runway one two, traffic landing runway niner will hold short of the intersection.
SPORT 18R:	Sport one eight romeo, roger.

Land and Hold Short Operations

At selected controlled airports where appropriate data have been published, air traffic controllers may utilize an expanded procedure whereby they may clear a pilot to land and hold short of an intersecting runway, an intersecting taxiway, or some other designated point on the runway. This operation is known as a **land and hold short operation (LAHSO).** Once procedures have been developed and approved and appropriate runway or taxiway signage has been installed, controllers may routinely issue LAHSO clearances. LAHSO procedures improve the efficiency of certain airports by essentially eliminating crossing runways. Although the runways (or runway and taxiway) in question may in reality cross each other, by requesting that a pilot hold short, the controller can move traffic as if the runways were physically disconnected from each other.

Pilots may accept LAHSO clearances provided that they determine that their aircraft can safely land and stop within the available landing distance (ALD). ALD data are published in the special notices section of the Airport Facility Directory (AFD) (see Table 6-2). Controllers can also provide ALD data to the pilot upon request. Although controllers may routinely issue LAHSO clearances, the pilot in command has the final authority to accept or decline any land and hold short clearance. The safety and operation of the aircraft remain the responsibility of the pilot. Pilots are expected to decline a LAHSO clearance if they determine it will compromise safety.

Table 6-2. LAHSO Runway Distance Information

City/Airport	Landing Runway	Hold Short Point	Measured Distance in Feet
Champaign/CMI	04L	14R/32L	3,600
	36	14R/32RL	3,950
Chicago/ORD	09R	Taxiway Sierra	9,300
	14R	09R/27L	9,800
	27L	14R/32L	6,500

When LAHSO operations are being conducted, that information is included on the taped ATIS broadcasts to pilots. The pilots should, as part of their pre-flight briefing, review any applicable LAHSO procedures and check to see whether their aircraft can meet the LAHSO requirements.

Here is an example of the phraseology to be used:

AIR TRAFFIC CONTROLLER:	Cherokee two three four papa uniform, cleared to land runway one zero, hold short of taxiway bravo for crossing traffic, traffic is a Cessna one seventy-two.
N234PU:	Cherokee two three four papa uniform, wilco, cleared to land runway one zero to hold short of taxiway bravo.
AIR TRAFFIC CONTROLLER:	Cherokee two five two mike november, cross runway one zero at taxiway bravo, landing aircraft will hold short.
N252MN:	Cherokee two five two mike november, wilco, cross runway one zero at bravo, landing traffic to hold.

Spacing Aircraft

The local controller may use any of the following phrases to achieve proper spacing of aircraft in the traffic pattern:

1. "Enter (pattern leg) runway (runway number)." The controller uses this phrase to direct the pilot to enter one of the five identified pattern legs. For example:

   Cessna niner papa uniform, enter left downwind runway two three.

2. "Report (position)." For the purposes of identifying and spacing aircraft, the pilot can be requested to make various position reports. The controller may request distance from the airport, distance from the runway, distance from a prominent landmark, or entry into the pattern. This request is usually combined with the previous instruction:

   Diamond eight delta mike, report three miles north of the airport.

   Cherokee two papa uniform, enter left downwind for runway five, report over the red and white water tower.

   Sport zero two romeo, report two mile final runway one zero.

3. "Number (sequence number, runway)." This phrase advises the pilot of the planned landing sequence for the aircraft. The pilot assumes that the preceding aircraft is landing on the same runway unless stated otherwise. If the local controller is using more than one runway for arrivals, the pilot should be advised of the sequence for the airport and for the arrival runway. This instruction is usually used in conjunction with a "follow" phrase.

4. "Follow (description and location)." Once the preceding aircraft has been located and identified, it is the pilot's responsibility to provide the proper spacing in the traffic pattern. The local controller should advise the pilot of the location and type of the preceding aircraft to make it easier to locate and follow.

5. "Traffic is (description and location) landing (runway)." If the landing aircraft is sequenced behind an aircraft landing on a different runway, the pilot should be advised of the type and location of the preceding aircraft in order to provide proper spacing. The controller may refer to a local landmark or to the pilot's aircraft when pointing out the preceding traffic:

> Cherokee niner alpha uniform, number two runway two three.
>
> Bellanca six alpha victor, follow the twin Cessna ahead and to your right.
>
> Twin Beech one seven three, enter and report left base for runway two three, number two for the airport, traffic is a Cessna on a quarter mile final runway one zero.

Spacing Instructions

The following instructions can be used to either increase or decrease the spacing between aircraft in the traffic pattern.

"Extend Downwind/Upwind" The pilot can be requested to extend either the downwind or the upwind leg a specified distance or until over a prominent landmark. The pilot should never be requested to extend the crosswind leg unless it is absolutely necessary. Extending the crosswind leg will result in the downwind leg being flown far enough from the airport that the pilot may be unable to glide to the runway in case of engine failure. An extension to the base leg is impossible since the distance from the downwind leg to the final leg is fixed. Here are examples of the phraseology:

> United six sixteen, extend downwind one mile to give room for a departure.
>
> Cessna one niner foxtrot, extend upwind to the lake.

"Short Approach" A **short approach** is a request for the pilot to shorten the downwind leg as much as possible, which results in an equivalent reduction in the length of the final approach leg (see Figure 6-17). Because the pilot is still required to fly a pattern within the capabilities of the aircraft, this request may not provide consistent results. Some pilots may be able to fly a very short pattern, whereas others are unable to do so.

"Make Left (or Right)" In a normal traffic pattern, the pilot makes a 90° turn when transitioning from one leg to another. One method of increasing the spac-

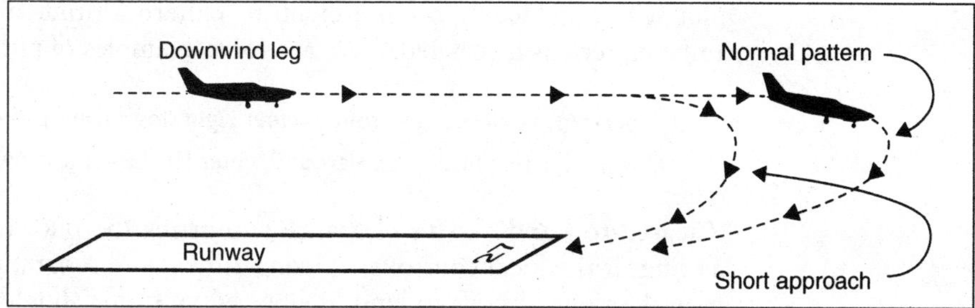

Figure 6-17. Example of an aircraft conducting a short approach.

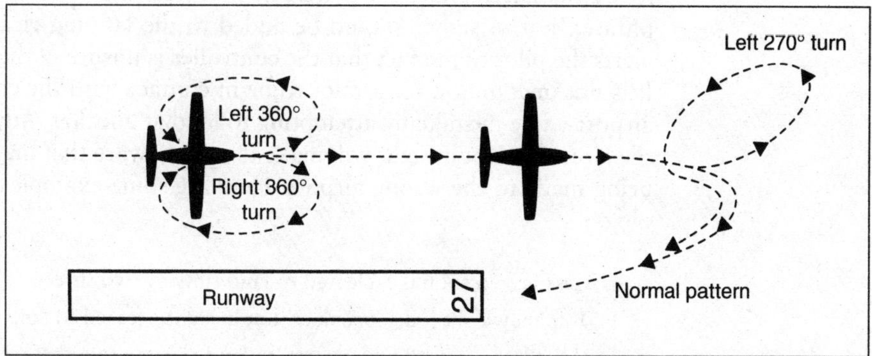

Figure 6-18. Example of a 360° turn and a 270° turn while in the traffic pattern.

ing between two aircraft is to request that the pilot turn 270° in the "wrong" direction when transitioning to the next leg (see Figure 6-18). For instance, if the pilot is on a right downwind, a request to "make a left two seventy to base" will result in a longer turn and increased separation. If the pilot is not transitioning from one leg to another and increased spacing is necessary, a 360° turn in either direction may be requested ("Sport one two romeo, make a right three sixty"). Caution should be used when issuing such instructions, since they can be potentially disorienting to pilots. Controllers should also refrain from using these methods when the aircraft has begun to descend from pattern altitude and is on either the base or final leg. It can be dangerous for a pilot to perform these maneuvers at low airspeeds while close to the ground.

"Go Around" If it is apparent that proper runway separation cannot be achieved and neither aircraft's traffic pattern can be adjusted, it will be necessary to cancel landing clearance for one of the aircraft. In this case, the local controller determines which aircraft's landing clearance should be cancelled and instructs that aircraft to "go around." Upon receipt of this instruction, the pilot will immediately begin a climb to pattern altitude and will reenter the traffic pattern as instructed. Here are some examples of phraseology:

> American six eleven, go around, enter right downwind runway two seven left.
>
> Cessna niner eight delta, go around, enter left base for runway two five.

"Cleared to Land" This clearance authorizes the pilot to make a full-stop landing. If the local controller is using anticipated separation and has cleared more than one aircraft to land, the preceding traffic should be included in the landing clearance. Any restrictions or requests should precede this clearance. These might include instructions to hold short of a runway or to plan to turn off of the runway at a designated taxiway. If the local controller should be able to see the landing aircraft but cannot do so either visually or using radar, the phrase "not in sight" should be added to the landing clearance. This phrase alerts the pilot to the fact that the controller is unsure of the aircraft's position. It is not uncommon for a pilot to be in contact with the control tower at one airport while mistakenly attempting to land at another. Advising the pilot that the aircraft is not in sight will make the pilot aware that the approach might be being made to the wrong airport. Here are some examples of landing phraseology:

> Cessna two six mike, cleared to land runway two three.
>
> Tomahawk six four november, not in sight, cleared to land runway one zero.
>
> United one twenty-five, cleared to land runway two three, traffic landing runway one zero.
>
> Clipper four seventeen, cleared to land runway one four right, hold short of runway niner right, traffic landing runway niner right.

After the aircraft has landed, the local controller should advise the pilot where to exit the runway and what frequency to use for contacting the ground controller.

"Cleared for Touch and Go" A **touch and go clearance** permits an aircraft to land on the runway but to take off again before actually coming to a stop. This maneuver is usually used by students practicing takeoffs and landings. An aircraft performing a touch and go is considered an arriving aircraft until actually touching down and then is considered a departure.

"Cleared for Stop and Go" A **stop and go clearance** is similar to a touch and go except that the aircraft comes to a full stop on the runway before beginning its takeoff run. A stop and go is also considered an arriving aircraft until coming to a complete stop, after which it is considered a departure.

"Cleared for Low Approach" In a **low approach,** the pilot approaches to land on the runway but does not actually make contact with the runway surface. Upon reaching the desired altitude, the pilot begins a climb. Low approaches are usually used by pilots practicing instrument approaches. In many cases, the pilot may wish to execute the published missed approach procedure. When it is desirable to determine the pilot's intentions prior to issuing this clearance, the controller may ask the pilot, "State your intentions." An aircraft conducting a low approach is considered an arriving aircraft until it crosses the landing threshold, after which it is considered a departure.

"Cleared for the Option" An **option clearance** permits the pilot to perform a landing, touch and go, stop and go, or low approach. The pilot will not normally inform the controller which option he or she has chosen. This maneuver is normally used in flight training to permit a flight instructor to evaluate a student's performance under changing conditions. If the controller is unable to approve all the options, the following phraseology should be used to restrict the pilot to the options that can be safely accommodated:

> Sport one three romeo, unable option, make a full stop landing.
>
> Cessna three niner eight, unable stop and go, other options approved.

Runway Selection

Since aircraft landing into the wind touch down at lower ground speeds that shorten the landing roll, most pilots, when given a choice, prefer to land or depart on a runway as nearly aligned with the wind as possible. Unless otherwise specified by facility directives, it is usually the local controller's responsibility to decide which runway becomes the active runway. Local controllers should comply with the following guidelines from the FAA handbook when selecting active runways:

1. Whenever the wind speed is greater than 5 knots, use the runway most nearly aligned with the wind.
2. The **calm wind runway** should be used whenever the wind is less than 5 knots. The calm wind runway will be specified by the airport management and is contained in the facility directives. This runway is chosen to maximize arrivals and departures while minimizing the noise impact on local dwellings.
3. The local controller can use any other runway when it is operationally advantageous to do so.
4. If a runway use program has been designated for the facility, the runways specified in the program should be used as the active runways.

Runway Use Programs

To minimize the noise impact of landing and departing aircraft, the FAA has implemented a nationwide **Aviation Noise Abatement Policy.** This policy places the primary responsibility for planning and implementing a noise abatement program on the operator of each airport. The **runway use program** put into place may be either informal or formal.

Informal runway use programs primarily affect aircraft that weigh more than 12,500 pounds. At airports with informal runway use programs, the controllers will assign these aircraft to the runway chosen by airport management whenever all of the following conditions can be met:

The wind direction is within 90° of the runway heading.

The wind does not exceed 15 knots.

The runway is clear and dry, which means that there is no snow, ice, slush, or water on the runway.

If pilots wish to use a different runway than that specified in the informal runway use program, they are expected to inform the controller of that fact. Air traffic controllers are required to honor these requests but will advise the pilot that the runway is "noise sensitive."

If airport management wishes to have aircraft use specific runways even when the runway conditions exceed those listed earlier, a **formal runway use program** must be initiated. A formal program requires that aircraft operators, airport management, and the FAA consummate a letter of agreement specifying the preferential runways and the weather conditions that must exist to use those runways. The establishment of a letter of agreement ensures that everyone concerned completely understands the conditions of the runway use program. The letter of agreement specifies that although pilots are expected to comply with these procedures, pilot requests for other runways will be honoured but the pilot will be advised that the previously assigned runway is specified in the formal runway use program.

Helicopter Operations

Helicopters can taxi around the airport by ground taxiing, hover taxiing, or air taxiing. **Ground taxiing** of a helicopter is similar to that of a taxiing plane. Only those helicopters equipped with landing gear are able to ground taxi. In **hover taxiing** the helicopter actually lifts off of the ground and remains airborne while maneuvering around the airport. A hover-taxiing helicopter usually remains within about 50 feet of the ground and proceeds at airspeeds less than 20 knots. Helicopters that are **air taxiing** operate below 100 feet and proceed at speeds in excess of 20 knots.

Each type of taxiing has its advantages and disadvantages. Ground taxiing is the most fuel efficient of the three and creates less air turbulence around and behind the helicopter. Hover taxiing is much faster than ground taxiing but creates a high level of air turbulence both below and behind the helicopter. Air taxiing is the fastest method and actually creates less air turbulence since the helicopter is at a greater altitude and most of the air turbulence is directed backward. Whenever a helicopter is taxiing, aircraft in the vicinity should be advised that it could be creating wake turbulence.

Helicopters are unique in that they may descend and climb with little or no forward movement. Nevertheless, helicopter pilots must be careful that they never depart from the safe flight envelope. For a pilot to properly control the aircraft in case of an engine failure, a helicopter must have sufficient speed, altitude, or a combination of the two to safely perform a maneuver known as an **autorotation.** An autorotation is similar to a glide in a fixed-wing aircraft. During an autorotation, the helicopter descends at a rapid rate but is able to reduce that rate of descent just prior to touchdown. Typically, a hovering helicopter needs about 600 feet of altitude to safely perform an autorotation. A helicopter travelling forward at a speed of about 40 knots needs virtually no altitude to autorotate. Keeping these factors in mind, helicopter pilots normally prefer to approach for landing in a manner similar to fixed-wing pilots. The only difference is that the helicopter does not need to use the entire length of the runway to decelerate.

Wake Turbulence

Every aircraft in flight trails an area of unstable air behind it known as **wake turbulence.** This turbulence was originally attributed to "prop wash" but is now known to be caused in part by a pair of counter-rotating vortices trailing from the wing tips (see Figure 6-19). These vortices are a by-product of the lift produced by the wing, which is generated by the creation of a pressure differential between the lower and the upper wing surfaces. High pressure is created below the wing, while low pressure is created above. The resultant upward pressure on the wing, known as *lift,* causes whirling vortices of airflow to be

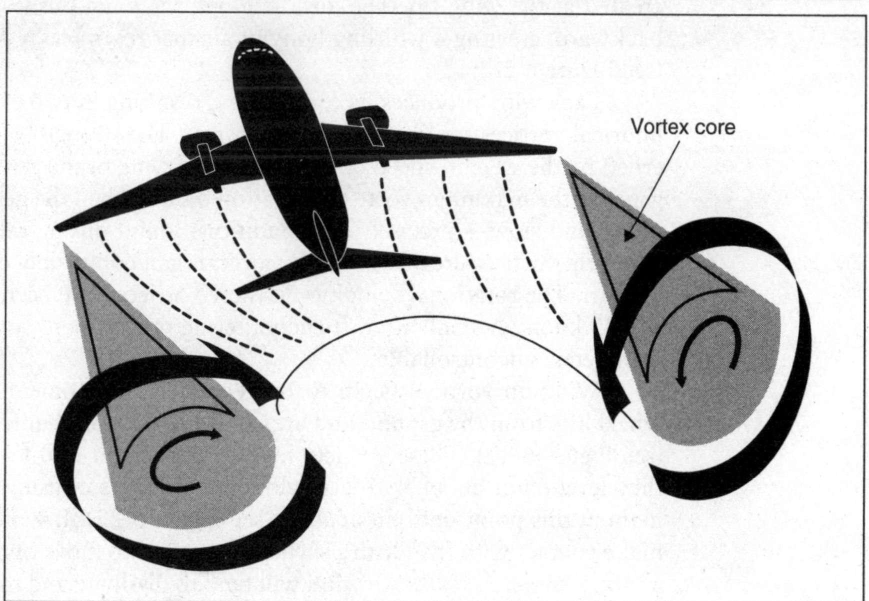

Figure 6-19. Wake turbulence behind an aircraft.

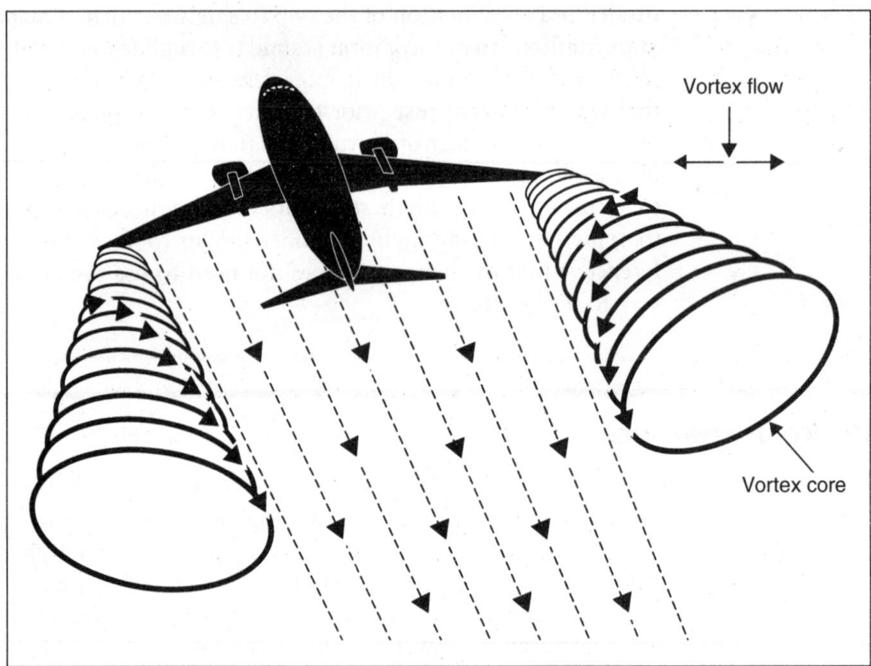

Figure 6-20. Example of vortex rotation and movement behind an aircraft.

created at the wing tip. The airflow along the wing pushes this upward flow backward, creating a whirling body of air that resembles a horizontal tornado (see Figure 6-20).

Each wing produces its own vortex, resulting in two counter-rotating cylindrical vortices trailing from each aircraft. The strength of the vortex is governed by the weight, speed, and shape of the wing of the generating aircraft. In general, the maximum vortex generation occurs when the generating aircraft is heavy and slow—precisely the conditions found during takeoff and landing. Wing-tip vortices created by larger aircraft can completely encompass smaller aircraft. The rotational velocities in these vortices have been measured as high as 133 knots. A small aircraft encountering one of these vortices may become completely uncontrollable.

Wing-tip vortices begin to be generated the moment an aircraft's nose wheel lifts from the ground and are continually created until the aircraft lands (see Figure 6-21). These vortices tend to descend at 500 feet per minute until they level off at about 900 feet below the aircraft's cruising altitude. They remain at this point until dissipating (see Figure 6-22). If while descending they make contact with the Earth's surface, they tend to move outward at a speed of about 5 knots. Any surface wind will tend to dissipate and move these vortices. A crosswind will tend to increase the speed of the downwind vortex while im-

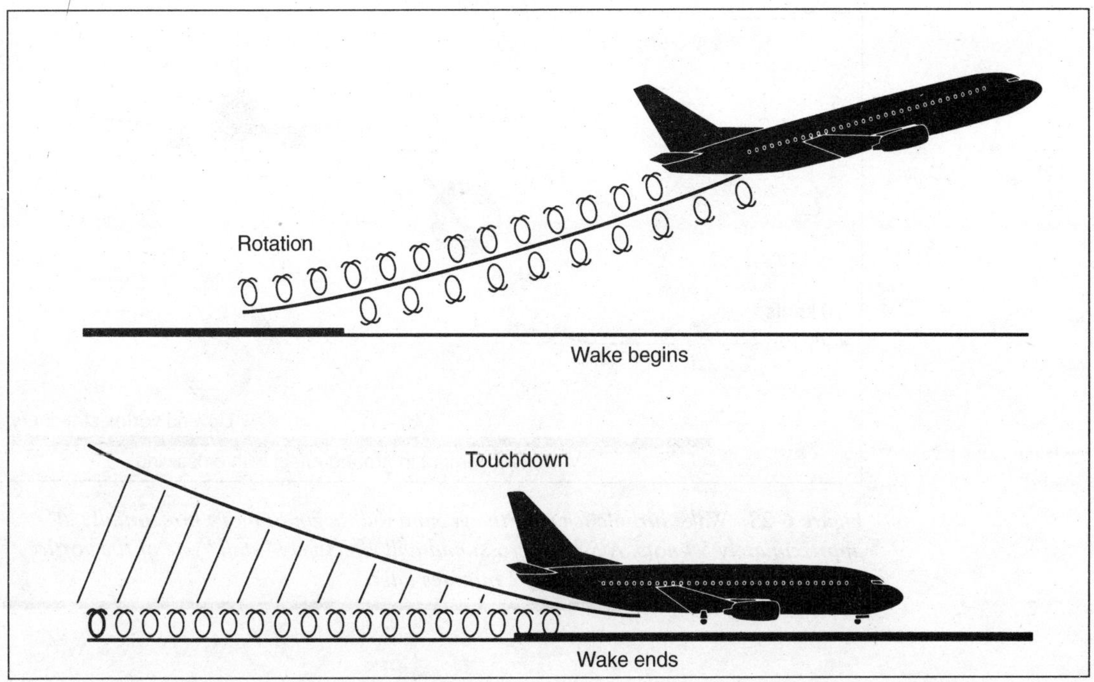

Figure 6-21. Wake turbulence starts when a departing aircraft's nose wheel leaves the ground. It stops when a landing aircraft's nose wheel touches the ground.

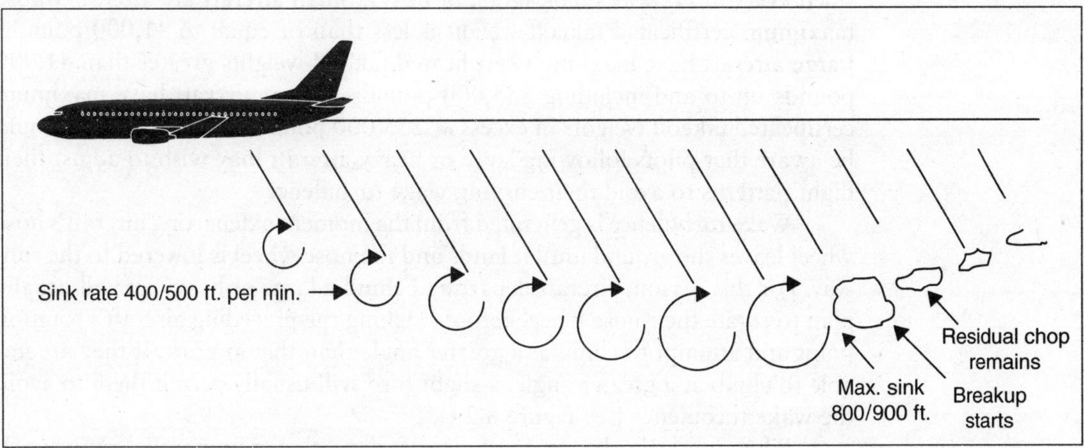

Figure 6-22. Wake turbulence behind an aircraft as it descends and dissipates. The wake turbulence will descend at 500 feet per minute until it begins to break up approximately 900 to 1,000 feet below the cruising altitude of the originating aircraft.

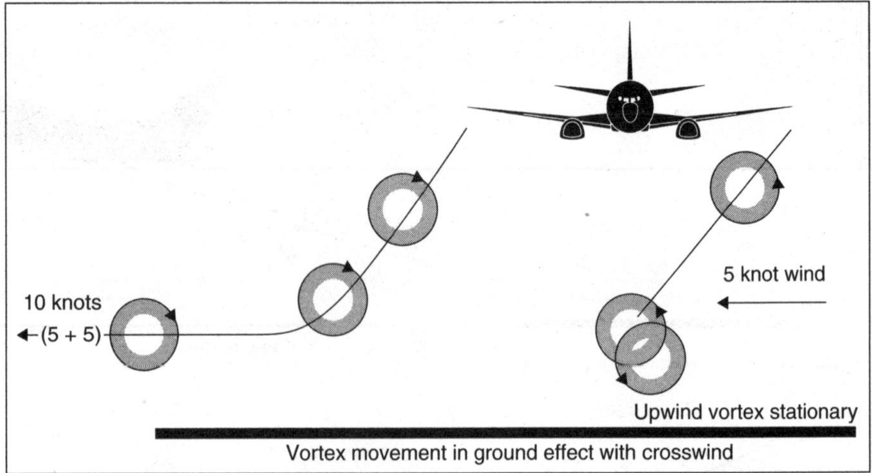

Figure 6-23. Wake turbulence near the ground will begin to move horizontally at approximately 5 knots. A 5-knot crosswind will effectively "stall" one of the vortices over the runway until it dissipates 2 minutes later.

peding the progress of the upwind vortex (see Figure 6-23). A cross wind between 3 and 7 knots may prevent the upwind vortex from moving.

Although it is primarily the pilot's responsibility to avoid wake turbulence, controllers are required to assist the pilots of smaller aircraft whenever they fly behind an aircraft that could be creating potentially dangerous wake turbulence. For the purposes of wake turbulence separation minima, the FAA has classified every aircraft as small, large, or heavy. **Small aircraft** are aircraft whose maximum certificated takeoff weight is less than or equal to 41,000 pounds. **Large aircraft** have maximum certificated takeoff weights greater than 41,000 pounds up to and including 255,000 pounds. **Heavy aircraft** have maximum certificated takeoff weights in excess of 255,000 pounds. The controller should be aware that pilots following large or heavy aircraft may wish to adjust their flight patterns to avoid their ensuing wake turbulence.

Wake turbulence is generated from the moment a departing aircraft's nose wheel leaves the ground until it lands and the nose wheel is lowered to the runway. For this reason, aircraft departing behind a large or heavy jet will usually plan to rotate their nose wheel before reaching the preceding aircraft's rotation point and attempt to climb at a greater angle than that aircraft. If they are unable to climb at a greater angle, a slight turn will usually permit them to avoid the wake turbulence (see Figure 6-24).

Pilots must also be aware of aircraft departing from parallel runways. If the parallel runways are less than 2,500 feet apart, it is quite possible that the wing-tip vortices may drift from one runway to the other (see Figure 6-25). In these cases, the pilot of the smaller aircraft will attempt to rotate prior to the point of rotation of the heavy aircraft.

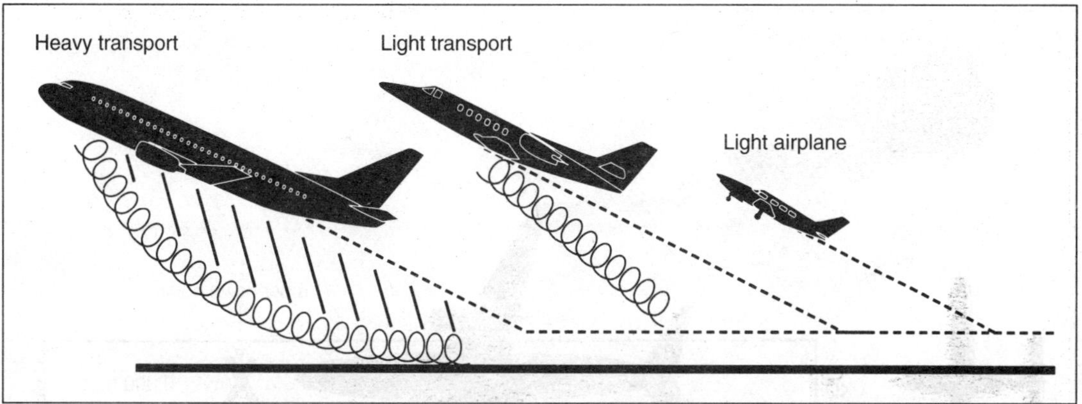

Figure 6-24. To avoid wake turbulence, departing aircraft should rotate prior to the point of rotation of the preceding aircraft and should climb at a steeper angle.

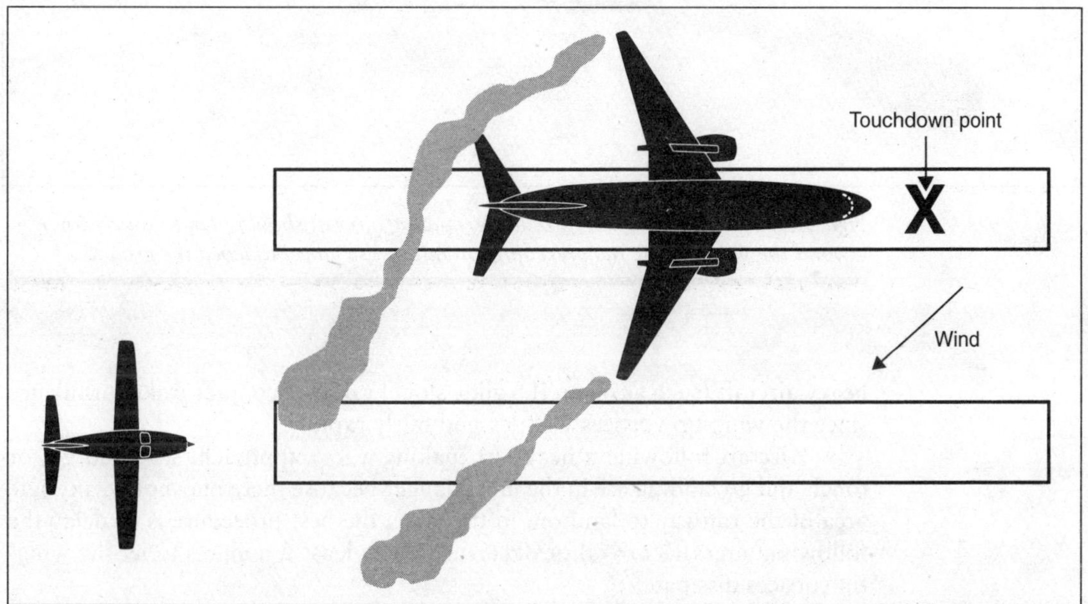

Figure 6-25. Aircraft landing on parallel runways could encounter wake turbulence "blown" over from the parallel runway.

Since the wake turbulence caused by an arriving aircraft ceases when the heavy aircraft's nose wheel settles to the ground upon landing, pilots of smaller aircraft following heavy aircraft will attempt to remain above the flight path of the heavy aircraft and land beyond the point where the heavy aircraft touched down (see Figure 6-26). Small aircraft following the same flight path as the

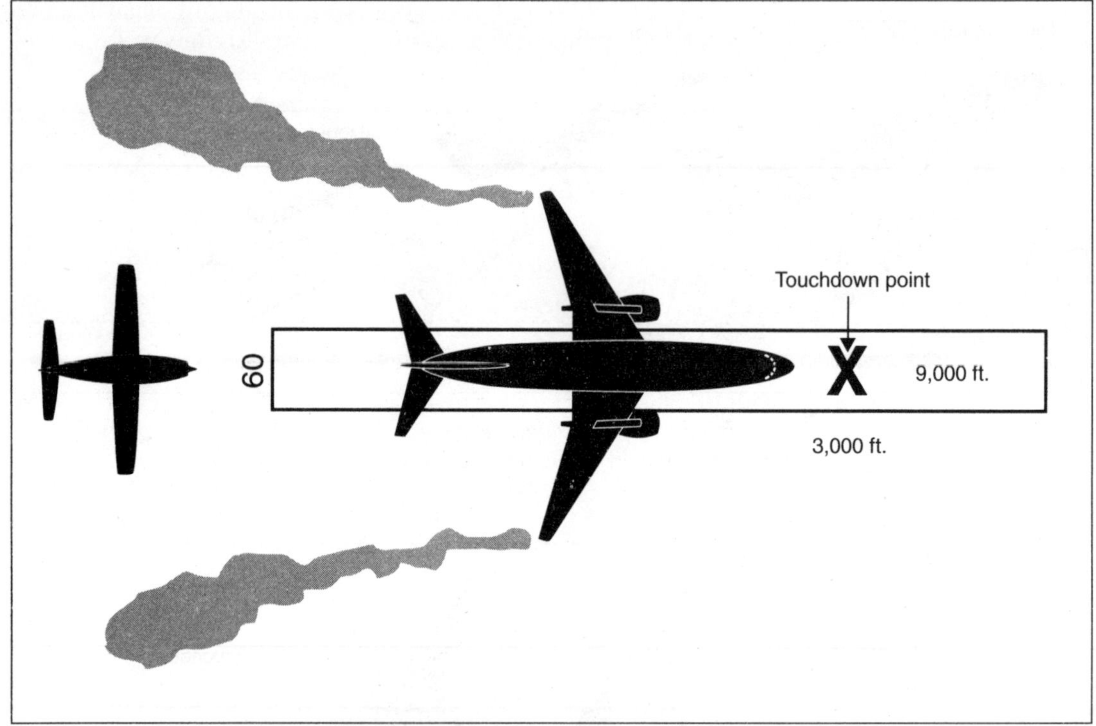

Figure 6-26. To avoid wake turbulence, landing aircraft should plan to touch down beyond the point where the preceding aircraft's nose wheel touched the ground.

heavy aircraft (such as on an ILS glide slope) rarely encounter wake turbulence since the wing-tip vortices will descend fairly rapidly.

Aircraft following a heavy jet making a low approach, stop and go, or touch and go landing are in the most danger because there may not be any safe area of the runway to land on. In this case, the best procedure is to delay the following aircraft's arrival or departure for at least 2 minutes to let the wing-tip vortices dissipate.

Tower controllers must apply the following procedures to small aircraft following larger aircraft creating potentially dangerous wake turbulence. (Controllers in an approach control or in an ARTCC have a different set of procedures they must comply with. Those procedures are explained in detail in Chapter 9.)

Since wake turbulence tends to dissipate in a matter of minutes, time is used as a means of ensuring that a following aircraft does not encounter any severe wake turbulence. In general, the following aircraft will usually be delayed by either a 2- or 3-minute interval wherever dangerous wake turbulence might exist.

Two minutes of separation must be applied to any aircraft departing behind a heavy aircraft utilizing the same runway or a parallel runway if the runways are separated by less than 2,500 feet. Two minutes of separation must also be applied to an aircraft whose flight path will cross that of a heavy jet departing from an intersecting runway. The pilot of the following aircraft may waive this wake turbulence separation by stating "Request waiver of the two-minute interval" or by making a similar statement. This request means that the pilot has accepted responsibility for wake turbulence separation.

Three minutes of separation must be provided to any small aircraft departing behind a large aircraft whenever the small aircraft is departing from an intersection or in the opposite direction on the same runway. This interval may be waived upon pilot request. A 3-minute interval will also be provided to any small aircraft departing behind a heavy aircraft whenever the small aircraft is departing from an intersection or in the opposite direction on the same runway. This 3-minute interval may *not* be waived by the pilot.

A second type of wake turbulence, produced by turbine engines, propellers, and helicopter rotor blades, is fairly localized and not long lasting but can be just as dangerous to an unsuspecting pilot. The wake turbulence found behind a turbine engine can overturn or hurl a small aircraft hundreds of feet. Controllers must always remember that the cockpit of a small aircraft is fairly noisy and the pilot may not be able to hear the engine noise of a nearby jet. The pilots of small aircraft should thus be warned whenever they are taxiing behind jet aircraft.

KEY TERMS

air taxiing
aircraft category
aircraft group
amendment (AM)
anticipated separation
automated radar terminal system (ARTS)
Automatic Terminal Information Service (ATIS)
autorotation
Aviation Noise Abatement Policy
calm wind runway
ceilometer
clearance delivery controller
critical areas
departure message
fields
flight data controller
flight data input/output (FDIO)
flight data processing (FDP)
flight progress strips
formal runway use program
ground controller
ground taxiing
heavy aircraft
holding short
hover taxiing
inactive runways
informal runway use program
land and hold short operation (LAHSO)
large aircraft
left traffic
local controller
low approach
National Beacon Code Allocation Plan (NBCAP)
National Weather Service (NWS)
nonradar-approach control towers
option clearance
pilot reports (PIREPs)
radar-approach control towers
right traffic
runway incursion
runway separation
runway use program
short approach
small aircraft
stop and go clearance
strip request (SR)
taxiways
touch and go clearance
tower visibility
traffic pattern
traffic pattern legs
VFR towers
wake turbulence

REVIEW QUESTIONS

1. What are the four operating positions in a control tower, and what are the duties assigned to each?
2. What are the separation minima for departing aircraft?
3. What are the separation minima for arriving aircraft?
4. What may the pilot be asked to hold short of during LAHSO procedures?

7

Nonradar En Route and Terminal Separation

Checkpoints

After studying this chapter, you should be able to:

1. State the four methods of nonradar separation.
2. Define the dimensions of the area generally assigned to each aircraft.
3. Generally state how nonradar separation is applied to aircraft.
4. Be familiar with the methods of marking an aircraft's reported position on a flight progress strip.
5. Determine the appropriate holding pattern to be used in any given situation.

Design of Separation Procedures

Before the widespread installation and use of radar for air traffic control, controllers could not accurately determine the location of the aircraft they were attempting to separate. In most cases, the required separation was accomplished by instructing the pilot to change course or altitude or to enter a holding pattern. Direct radio communication between the controller and the pilot was not always possible, and pilot-controller communication was passed through intermediaries, such as airline radio operators or Interstate Airway Communication Stations. The resultant time delay, as long as 30 minutes, further complicated the controller's task. Some of this delay was alleviated through the development and use of remote radio transmitters and receivers known as **remote communication air/ground (RCAG)** devices. RCAG devices permitted controllers to communicate with pilots whose aircraft were beyond the range of the radio transmitters at the control facility. RCAG units used telephone circuitry connected to remote radio transmitters and receivers.

Even with the help of RCAG equipment, controllers were still unable to accurately determine an aircraft's position and had to rely on pilot reports and handwritten flight progress strips to separate aircraft. Using procedures developed by the CAA and relying on flight strips to remember each aircraft's approximate position, the controller could crudely effect aircraft separation.

The flight progress strips became an invaluable tool in helping controllers perform their separation duties. Through the use of standardized procedures and markings, trained controllers could use these strips to visualize the relative position of every aircraft and apply the proper separation procedures. Whenever the controller issued an instruction to a pilot or the pilot made a position or altitude report, the controller wrote the information on the flight progress strip. The strip could then be readily interpreted to determine the status of the aircraft. This information aided the controller in visualizing the position of each aircraft and made it much easier for other controllers to evaluate the airspace. The flight progress strip also became a valuable record of the instructions issued by the controller and any reports made by the pilot, in case of an investigation following an incident or an accident. It was the controller's responsibility to constantly update the information contained on the flight progress strip to be able to visualize both the present and, more important, the future position of every aircraft being separated.

When the first ATCUs were conceived in the 1930s, controllers had few hard-and-fast rules for separating aircraft. Commonsense rules and the experience gained as the air traffic control system matured formed the basis of separation procedures. But as traffic increased and the air traffic control system grew in size and complexity, the CAA began to develop a set of rules and procedures to be used by air traffic controllers.

It might appear fairly simple to develop such procedures, but in actuality it is far more difficult than one might expect. Air traffic control procedures specialists are employed by the FAA to develop these separation procedures and must consider many variables to ensure that aircraft that seem to be separated, actually are. They must take every variable that affects the air traffic control system into consideration when developing these procedures. Some of these variables include ground-based navigation equipment error, airborne navigation equipment error, different navigation systems in use by the pilots, winds aloft, and communications delay.

The designers of the air traffic control system must consider the worst-case scenario and must ensure that even with the maximum possible error in each component of the system a sufficient margin of safety will still exist. Consider the following example. Suppose that there are two fictitious intersections, alpha and bravo, defined as the intersection of radials emanating from two VORTACs (uniform and victor). Alpha is the intersection of the 180° radial of the uniform VORTAC and the 270° radial of the victor VORTAC, whereas bravo is the intersection of the 170° radial of the uniform VORTAC and the 260° radial of the victor VORTAC.

Looking at the illustration in Figure 7-1, your first impression might be that each of these intersections is in a different physical location and that an aircraft directly over the alpha intersection would be separated from an aircraft over the bravo intersection. But an evaluation by a trained airspace procedures specialist would reveal that this might not be the case. The expert would immediately realize that because the VOR equipment on board an aircraft is permitted to be accurate within ±6° to be certified for IFR flight, each aircraft might not be located directly over each intersection. If we assume that the maximum VOR receiver error exists in each aircraft's navigation system, it becomes apparent that an aircraft reporting over the alpha intersection may not be located exactly on the 180° degree radial of the uniform VORTAC but may in fact be anywhere between the 174° and the 186° radial. Using the same margin of error, the aircraft might also be located anywhere between the 264° and the 276° radial of the victor VORTAC. When determining where an aircraft that has reported over the alpha intersection is actually located, the air traffic controller must assume that the aircraft might be located, anywhere within the shaded area shown in Figure 7-2. If the position of the second aircraft is then calculated using the same margin of error, it becomes apparent that an aircraft that has reported over the alpha intersection may not actually be separated from an aircraft that has just reported over the bravo intersection—they may in fact be located in the same approximate position and about to collide.

This is a simplistic example of just one of the problems that airspace planners and air traffic controllers face when trying to effect safe separation using nonradar techniques. When every variable is taken into account, aircraft are not always precisely located over the intersections where their pilots report. To take all variables into consideration, the controller must reserve a block of airspace for each aircraft. The size of this block is partially determined by the variables mentioned earlier and by such other factors as the aircraft's performance, altitude, navigation system, and distance from the navaid. Because of

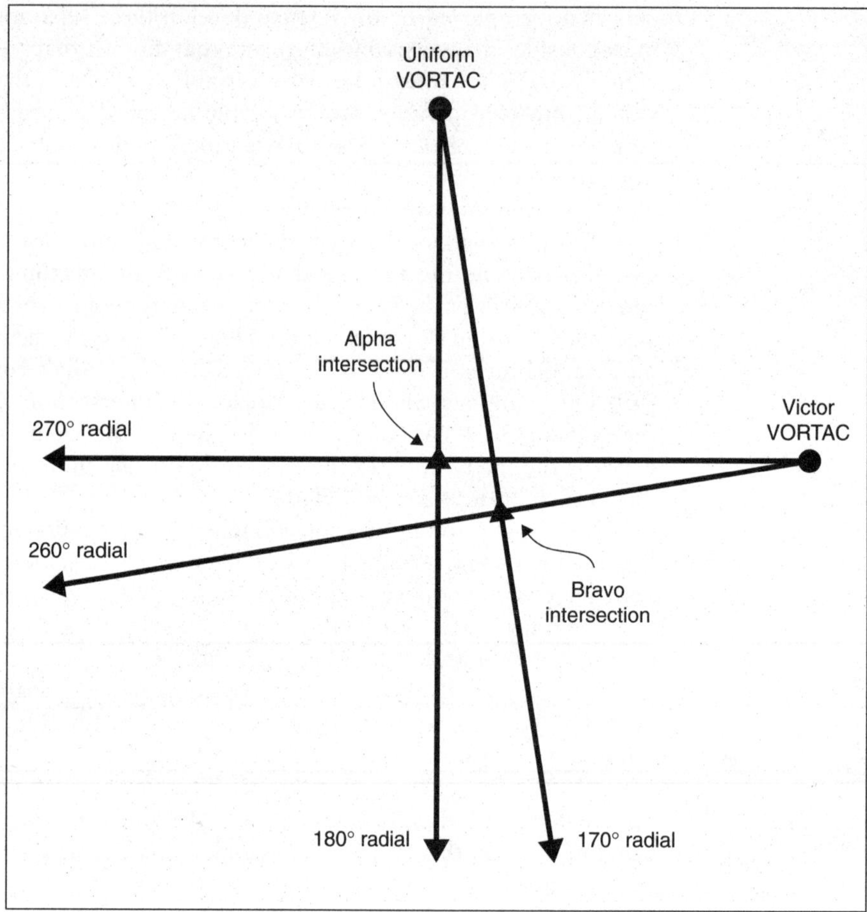

Figure 7-1. At first glance it appears that the alpha and bravo intersections are located in two different places.

the variables inherent in this type of separation, the controller must assume that the aircraft could be located anywhere within this block of airspace and must then separate every block of airspace accordingly.

Because each aircraft might be located anywhere within its reserved area, it is necessary for the controller to separate each aircraft's reserved airspace from the airspace reserved for other aircraft. Areas of reserved airspace may butt up against one another, but they must never be allowed to overlap, since any overlap might permit two aircraft to actually come into contact. The separation of each aircraft's reserved airspace is the only way to ensure that the aircraft within that airspace remain safely separated.

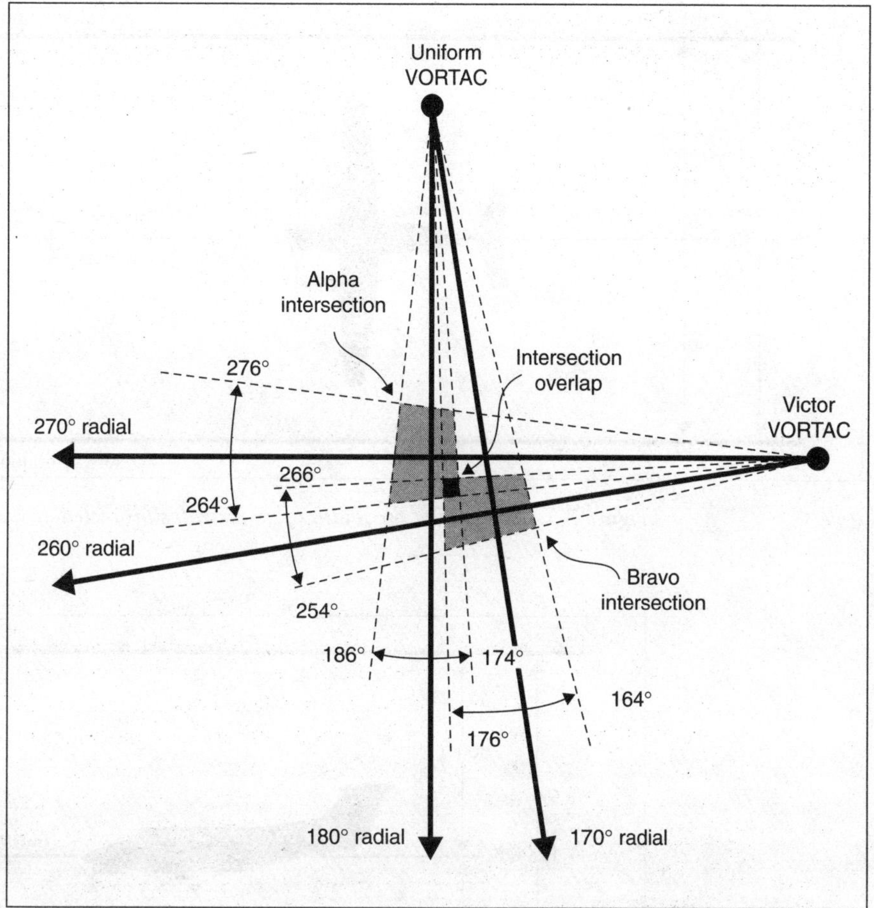

Figure 7-2. In reality, when navigation receiver tolerances are taken into consideration, the alpha and bravo intersections may actually overlap. Thus, an aircraft reporting over alpha may conflict with an aircraft reporting over bravo.

Airspace Dimensions

The *width* of an aircraft's reserved block of airspace is normally the width of the airway that the aircraft is navigating on. As long as the aircraft is within 51 nautical miles of the navigation aid providing guidance for that airway segment, the airway is 8 nautical miles wide—4 nautical miles on either side of the airway centerline (see Figure 7-3). The width of any airway segment greater than 51 miles from the navaid is defined as the area between two lines that diverge at an angle of 4.5°, centered on the airway centerline. Whenever an aircraft is

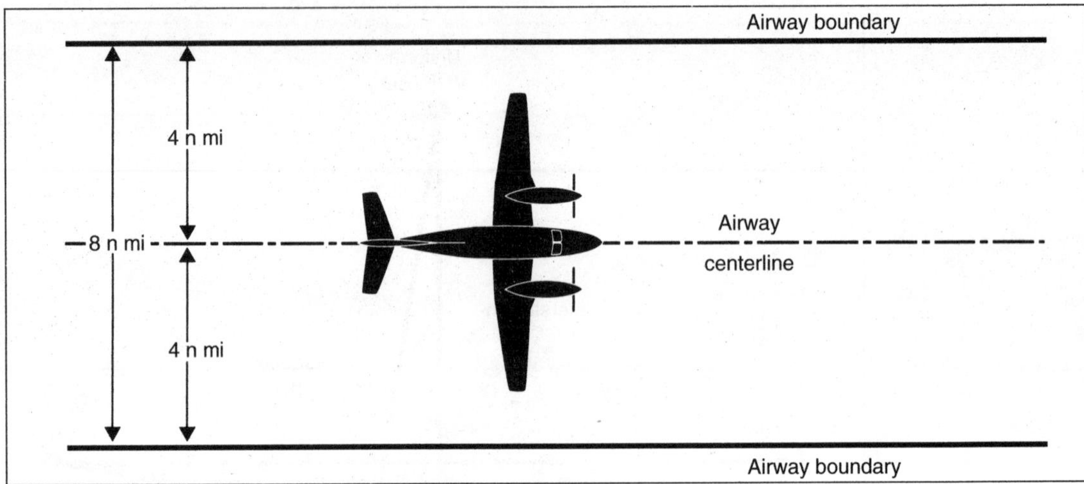

Figure 7-3. Example of the width of an aircraft's protected airspace.

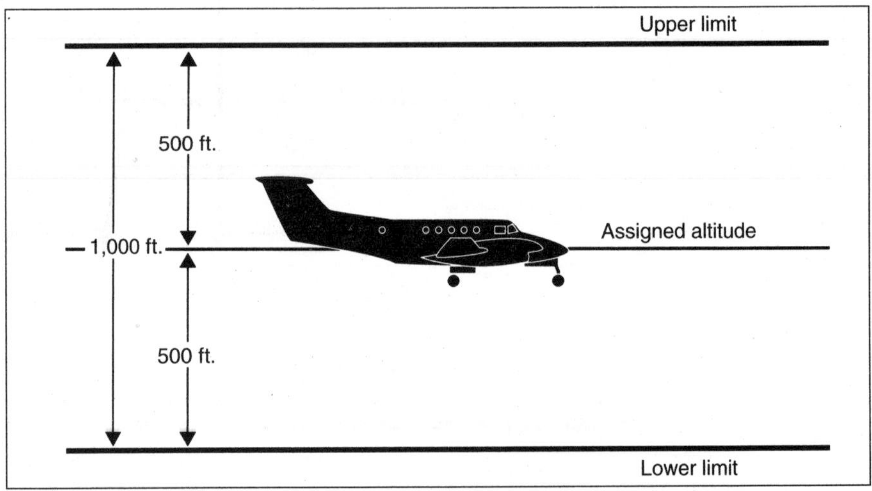

Figure 7-4. Example of the depth of an aircraft's protected airspace.

cleared to operate on an airway, the entire width of that airway is reserved for that aircraft.

The *depth* of the reserved block of airspace for aircraft operating at or below FL 290 is 1,000 feet. This area extends from 500 feet above the aircraft to 500 feet below the aircraft (see Figure 7-4). If the aircraft is operating above FL 290, the depth of the reserved airspace becomes 2,000 feet, extending 1,000 feet both above and below the aircraft. The distance is increased above FL 290 because aneroid altimeters are unable to measure small pressure changes at high altitudes, and the additional distance is needed to provide for accurate separation.

The *length* of the reserved airspace extends some specified time or distance ahead of the aircraft's last position. This reserved area usually extends 10 minutes in front of the aircraft. If the aircraft is either DME or RNAV equipped, the time requirement may be converted into a reserved distance of 20 nautical miles. Under certain circumstances, if the speed of the aircraft can be accurately determined, the time or distance requirement may be reduced.

The exact dimensions of the reserved airspace and the procedures applied by the controller vary depending on the aircraft's speed, navigational capability, altitude, and distance from the navigation aid. Thus, specific conditions that apply to individual situations are covered throughout this chapter.

Separation Procedures

When separating aircraft participating in the air traffic control system, the controller is required to ensure that the airspace reserved for one aircraft does not overlap the airspace reserved for another. If an overlap does occur, even if the two aircraft are miles apart, it is presumed that adequate separation does not exist and that a **separation error** has occurred. Controllers use four methods for separating aircraft: vertical, lateral, longitudinal, and visual separation. To ensure that the aircraft are in fact separated, the controller needs to apply at least one of these methods at any given time.

Since most of these methods are based on pilot report of position or altitude, the successful application of nonradar separation procedures depends on the accuracy of pilot reports. If a controller has any reason to suspect that a report may be in error, he or she must resolve the situation as soon as possible.

Although the widespread installation and use of radar has reduced the need for these procedures, they are still used by controllers whenever radar procedures cannot be applied. In many cases, even radar controllers will use some of these methods, since they may be easier to apply than radar separation procedures. Nonradar separation methods are still the primary means of separating air traffic in areas of limited or nonexistent radar coverage. And radar controllers must remain proficient in these methods in case of a radar malfunction or failure.

The primary principle to be observed when applying nonradar separation procedures is that any two aircraft are presumed *not* to be separated unless separation can be positively proven using one or more of the four methods. The use of any single method is considered proof that separation exists.

Vertical Separation

Basic Vertical Separation Rule **Vertical separation** is one of the easiest ways to separate two aircraft. As long as both aircraft are at altitudes that differ by at least 1,000 feet (2,000 feet when operating above FL 290), they are separated vertically. Since every aircraft's reserved airspace extends from 500 feet above it to 500 feet below, two aircraft separated by at least 1,000 feet are considered to be separated vertically.

The usual method of vertical separation is for the controller to request that the pilot report passing through or leveling off at a particular altitude. Once the pilot has reported an altitude, the controller can assign another altitude to a different aircraft, as long as the two altitudes differ by at least 1,000 feet. The *Aeronautical Information Manual* states that pilots should report *leaving* any previously assigned altitude. It does not state that the pilot must report passing through any intermediate altitude or arriving at the assigned altitude. Controllers must always presume that unless they request otherwise, pilots will report only when their aircraft leaves an assigned altitude. Pilots will not report passing through or leveling off at any other altitude unless the controller makes such a request. The phraseology that should be used by a controller to request an altitude report is as follows:

CONTROLLER:	United seven twenty-one, descend and maintain six thousand, report leaving eight thousand and report reaching six thousand.
UNITED 721:	United seven twenty-one, roger, leaving niner thousand for six thousand. [This pilot report is mandatory, because the previously assigned altitude is now being vacated.] We will report leaving eight thousand and reaching six thousand.
UNITED 721:	United seven twenty-one leaving eight thousand.
CONTROLLER:	United seven twenty-one, roger.
UNITED 721:	United seven twenty-one is level at six thousand.
CONTROLLER:	United seven twenty-one, roger.

Flight Progress Strip Marking Whenever a new altitude is assigned by the controller, it is written on the appropriate flight progress strip. This altitude is written in field 9 on a terminal flight progress strip and field 20 on a flight strip used by center controllers. The last two zeros of the altitude figure are always left off, because of the limited space available on a flight progress strip. Thus, 8,000 feet is written as "80," and 6,000 feet is written as "60." When the pilot reports leaving an altitude, that altitude is lined out on the flight strip with a single line.

Pilots will report passing through intermediate altitudes only if requested to do so by the controller. Such requests must be properly notated by writing the letters **RL** (for **report leaving**) or **RR** (for **report reaching**) next to the appropriate altitude on the flight progress strip. For example, when a pilot reports leaving 8,000 feet, the entire "80 RL" is lined out, signifying that the aircraft has vacated 8,000 feet. But when the pilot reports reaching 6,000 feet, only the "RR" next to the "60" is lined out (see Figure 7-5). This signifies that the aircraft has arrived and is level at 6,000 feet. The "60" is not lined out, as that would signify that the aircraft has vacated 6,000 feet.

Since the flight progress strips are official documents that could be utilized during an investigation, mistakes made when writing on them should never be erased but should always be crossed out with an X. Mistakes should not be

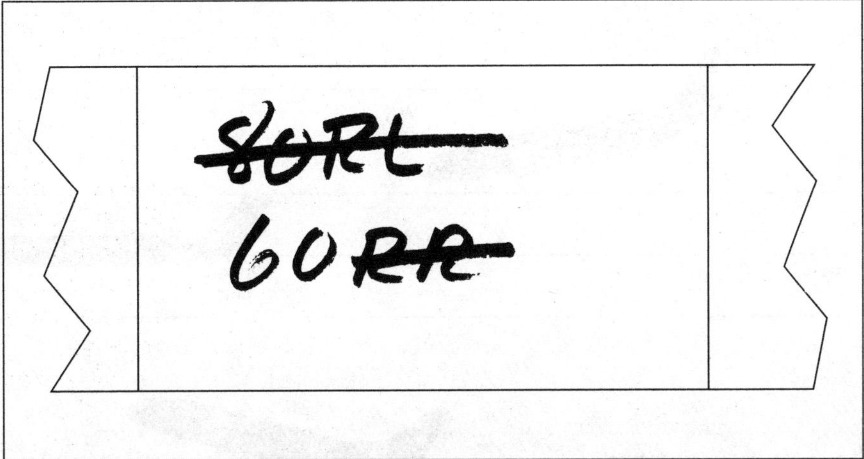

Figure 7-5. Example of proper strip marking to indicate that an aircraft has reported leaving 8,000 feet and has reported reaching 6,000 feet.

lined out because that could indicate that the aircraft in question has actually vacated an altitude.

Rule Application The vertical separation rule can best be shown by the following example. If United 965 is cruising overhead and has reported level at 8,000 feet, American 121 could be cleared for takeoff with a clearance to maintain 7,000 feet. In addition, SWA 877, cruising overhead at 15,000 feet, could be cleared to descend as low as 9,000 feet (see Figure 7-6). In this example, each aircraft is always separated from the others by at least 1,000 feet. If, however, 3 minutes after American 121 departed Continental 342 were to request permission to depart, the Continental flight could not be assigned 6,000 feet. Even though the two aircraft would be assigned different altitudes, the controller would not be able to ensure that at any given moment both aircraft would be separated by at least 1,000 feet. Since the precise altitude of American 121 is unknown, there would be no way to ensure that Continental 342 would be at least 1,000 feet lower. Until American 121's altitude is determined, the Continental flight could not be cleared to depart at all.

In this particular situation, the pilot of American 121 will not make any altitude reports, since the controller has not requested any. The simplest method of separating the two aircraft would be to request that the pilot of American 121 report leaving a series of intermediate altitudes and then to clear the Continental flight to maintain an altitude 1,000 feet below those altitudes.

Exceptions to the Basic Rule According to the FAA handbook, the only time that the basic vertical separation rule may be relaxed is when both aircraft involved are either climbing or descending. The handbook states that if both air-

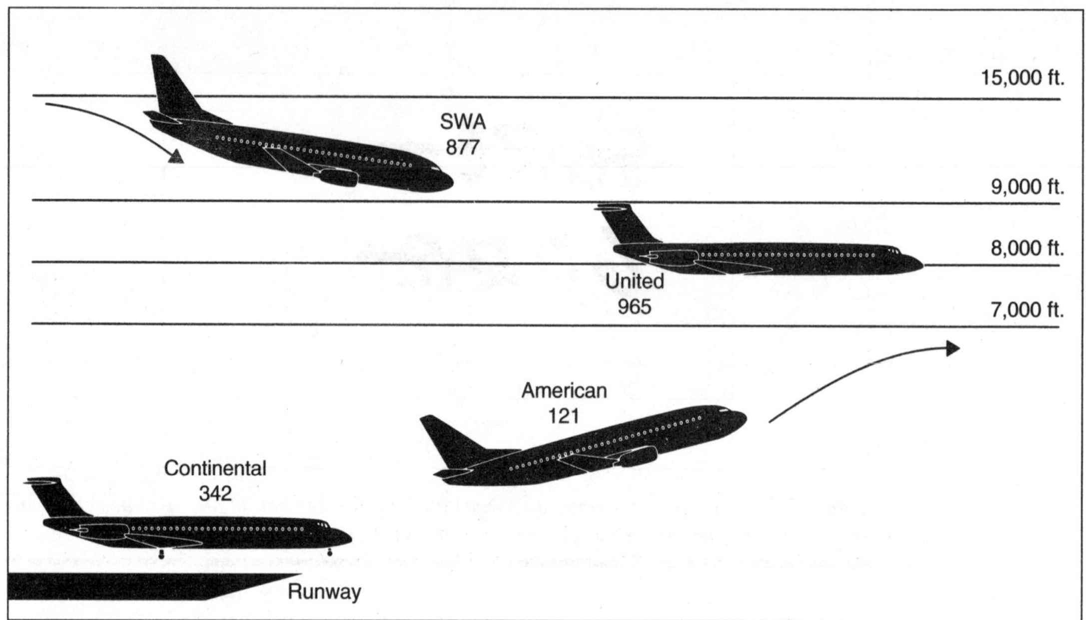

Figure 7-6. United 965 is level at 8,000 feet and is separated from SWA 877, which is descending from 15,000 feet to 9,000 feet. United 965 is also separated from American 121, which is climbing from the airport to 7,000 feet. Since American 121's altitude has not been confirmed, Continental 342 cannot depart.

craft are climbing, once the higher aircraft has reported leaving an assigned altitude and is climbing to another altitude at least 1,000 feet higher, the lower aircraft may be assigned the altitude the first aircraft has reported vacating. The handbook also states that if both aircraft are descending, once the lower aircraft has reported leaving an assigned altitude, the higher aircraft can be assigned the altitude just vacated by the lower aircraft.

Using our example, if the pilot of American 121 were asked to report leaving the altitudes between 2,000 and 6,000 feet, those altitudes could be immediately assigned to Continental 342 as they are vacated. The phraseology for this type of clearance is as follows:

CONTROLLER:	American one twenty-one, climb and maintain seven thousand, report leaving two thousand, three thousand, four thousand, five thousand, and six thousand.
AMERICAN 121:	American one twenty-one, roger. Climb and maintain seven thousand. We will report leaving two thousand, three thousand, four thousand, five thousand, and six thousand.
AMERICAN 121:	American one twenty-one is leaving two thousand.
CONTROLLER:	American one twenty-one, roger. Continental three forty-two, climb and maintain two thousand.

CONTINENTAL 342:	Continental three forty-two, roger.
AMERICAN 121:	American one twenty-one is leaving three thousand.
CONTROLLER:	American one twenty-one, roger. Continental three forty-two, climb and maintain three thousand.
CONTINENTAL 342:	Continental three forty-two is leaving two thousand for three thousand.
CONTROLLER:	Continental three forty-two, roger.
AMERICAN 121:	American one twenty-one is leaving four thousand.
CONTROLLER:	American one twenty-one, roger. Continental three forty-two, climb and maintain four thousand.
CONTINENTAL 342:	Continental three forty-two is leaving three thousand for four thousand.
CONTROLLER:	Continental three forty-two, roger.
AMERICAN 121:	American one twenty-one is leaving five thousand.
CONTROLLER:	American one twenty-one, roger. Continental three forty-two, climb and maintain five thousand.
CONTINENTAL 342:	Continental three forty-two is leaving four thousand for five thousand.
CONTROLLER:	Continental three forty-two, roger.
AMERICAN 121:	American one twenty-one is leaving six thousand.
CONTROLLER:	American one twenty-one, roger. Continental three forty-two, climb and maintain six thousand.
CONTINENTAL 342:	Continental three forty-two is leaving five thousand for six thousand.
CONTROLLER:	Continental three forty-two, roger.

This procedure is common in air traffic control and is known as **stepping up** an aircraft. It can just as easily be applied to aircraft that are descending, in which case it is known as **stepping down** an aircraft.

This exception to the basic vertical separation rule cannot be applied whenever the aircraft are unable to maintain at least a 500-foot-per-minute rate of climb or descent. The *Aeronautical Information Manual* states that the pilot should inform the controller whenever this rate of change cannot be maintained. But the controller should also be cognizant of the performance characteristics of both aircraft. The basic premise for this rule exception is that both aircraft will climb or descend at approximately the same rate, keeping at least 1,000 feet apart. If this separation interval is not likely to be maintained, this exception to the basic rule should not be utilized. It would be potentially dangerous, for example, to try to step up a military fighter jet that is directly below a small civilian training aircraft. If either pilot became careless, or if the altimeter in either aircraft malfunctioned slightly, a midair collision might result.

The handbook also precludes the use of this vertical separation rule exception under the following conditions:

Whenever severe turbulence is being reported in the area, which might make it impossible for either pilot to maintain a consistent climb or descent profile.

Whenever either of the aircraft is participating in military refueling maneuvers.

Whenever the preceding aircraft has been issued a clearance to climb or descend at pilot's discretion. A pilot's discretion clearance does not obligate the pilot to maintain at least a 500-foot-per-minute climb or descent.

Whenever the air traffic controller concludes that 1,000 feet of vertical separation between the aircraft may not be maintained during the procedure. The controller must make this judgment based on pilot reports and knowledge of the aircraft involved. The controller should be wary of using this rule exception whenever separating two aircraft that have widely different climb or descent characteristics.

If any of these conditions exist, the second aircraft cannot be assigned the altitude vacated by the first aircraft until the first reports being established at or passing through an altitude at least 1,000 feet away from the altitude to be assigned to the second aircraft.

Lateral Separation

Because of their ease of use, vertical separation methods are usually applied to aircraft operating along the same route or airway or within the immediate vicinity of an airport. But exclusive use of vertical separation can result in inefficient airspace usage and reduced traffic flows. Thus, controllers should consider using alternative methods of separation whenever possible. One of the methods that can be applied to aircraft operating on different routes is **lateral separation.** Lateral separation presumes that both aircraft are on different routes whose reserved airspaces do not overlap (see Figure 7-7). Two aircraft that are separated laterally may operate at the same altitude.

Basic Lateral Separation Rule Two aircraft are considered to be separated laterally whenever at least one of the following conditions exists:

The two aircraft are operating on different airways or routes whose protected airspaces do not overlap. Since each airway is 8 nautical miles wide, to be separated laterally two aircraft must be operating on different airways whose centerlines are at least 8 nautical miles apart. This assumes that the aircraft are within 51 miles of the navigation aid defining that airway. If the aircraft are greater than 51 miles from the navigation aid, the airways diverge at 4.5°.

The aircraft are holding over different navigation fixes whose defined holding-pattern airspace does not overlap.

The airspace reserved for aircraft operating on airways is described in FAR 71.5. In general, every airway is 8 nautical miles wide unless the changeover point is farther than 51 miles from the navaid. The width of the airway segment greater than 40 miles from the navaid is defined as that area between two lines that diverge at an angle of 4.5°, centered on the airway centerline (see Figure 7-8).

With lateral separation, each aircraft must be established on an airway whose protected airspaces do not overlap. If this cannot be accomplished, one of the other methods of separation (vertical, longitudinal, or visual) must be used until the airways cease to overlap.

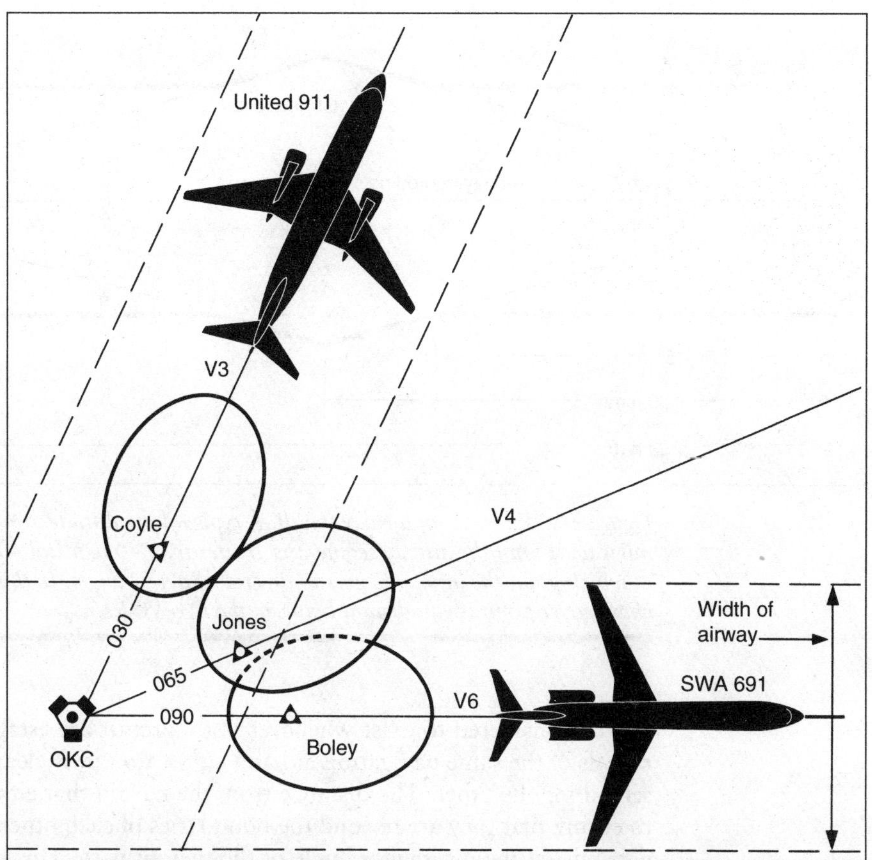

Figure 7-7. Example of lateral separation. United 911 is laterally separated from SWA 691 since their protected airspaces no longer overlap. In addition, aircraft could hold at the Coyle and Jones intersections simultaneously at the same altitude, since their airspaces abut but do not overlap. Aircraft cannot hold simultaneously at the same altitude at Boley and Jones, however, since their airspaces overlap.

It is fairly easy to determine whether lateral separation exists by using navigation charts. If the controller determines that the two airways are at least 8 nautical miles apart, the two airways can be used simultaneously by aircraft operating at the same altitude. If the airways are less than 8 miles apart, they are not separated laterally. In most cases, if the airway boundaries begin to diverge at 4.5°, the airway will have already been plotted and will be drawn on the controller's chart.

Exceptions to the Lateral Separation Rule In a few instances lateral separation can be applied to aircraft operating on airways that are not 8 nautical miles apart. The FAA handbook states that lateral separation between two aircraft

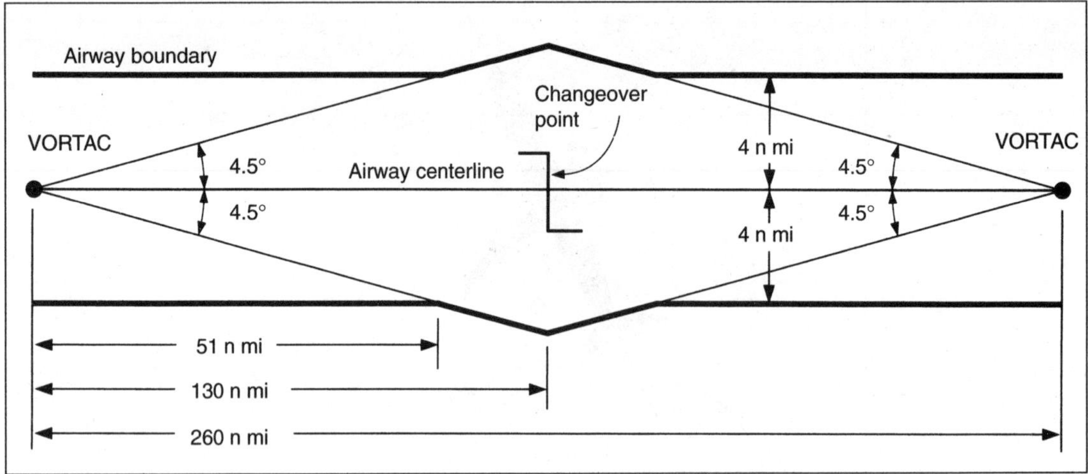

Figure 7-8. Example of airway width. A typical low-altitude airway is 8 nautical miles wide until the airway segment is greater than 40 nautical miles from the VOR, at which point the airway begins to diverge at a 4.5° angle. In this example, the changeover point is equidistant between the two VORs.

can be considered to exist whenever both aircraft are established on different radials of the same navigation aid and *either* aircraft is clear of the airspace reserved for the other. The distance from the navaid that either aircraft must be to ensure that they are beyond the boundaries of each other's reserved airspace depends on the divergence angle of the two airways. This distance determination can be made using Table 7-1 or Table 7-2 (from the handbook). Table 7-1 is used whenever distance is being determined by pilots using non-DME methods. Table 7-2 is used when the pilots are utilizing DME and takes into consideration slant range measurement error.

To properly utilize these tables, the controller must determine the angular difference between the two airways. If this value is not found on the table, the controller must use the next *lowest* angular value. The controller then uses the applicable distance value from the table. Whenever either of the aircraft has flown at least this distance, it is presumed to be clear of the airspace reserved for the other, and lateral separation exists.

This method of lateral separation is best applied to two aircraft crossing the same navigation fix, but that will then diverge, operating on different airways. Prior to crossing the fix, each aircraft must be separated by some method other than lateral separation—most often vertical separation. Once one of the aircraft has crossed the navaid and progressed the prescribed distance along the airway, lateral separation exists and vertical separation may be discontinued. For example, assume that Pan Am 415 is operating westbound on victor 251 and is still east of the VORTAC. Cessna 4152G is operating northbound on victor 171 and is still south of the VORTAC. In this situation, the controller would

Table 7-1. **Degree-Divergence for Non-DME-Equipped Aircraft**

Divergence in Degrees	Distance in Nautical Miles
15	16
20	12
25	10
30	8
35	7
45	6
55	5
90	4

Table 7-2. **Degree-Divergence for DME-Equipped Aircraft**

Divergence in Degrees	Distance in Nautical Miles (aircraft operating below FL 180)	Distance in Nautical Miles (FL 180–FL 450)
15	17	18
20	13	15
25	11	13
30	9	11
35	8	11
45	7	11
55	6	11
90	5	11

be required to apply vertical separation, since the aircrafts' paths will cross. Let us assume that Pan Am 415 has been assigned 6,000 feet and Cessna 4152G has been assigned 7,000 feet. Once the aircraft cross the VORTAC, their respective airways will diverge by 43° (315° minus 272°). Since 43° of divergence is not listed on the tables, the next lower angle, 35°, must be used. If DME can be used, the controller must request that one or both of the pilots report when they are 8 DME from the VORTAC (see Figure 7-9).

Suppose that Pan Am 415 is the aircraft that will reach this point first. Once the pilot makes this position report, the controller can discontinue the use of vertical separation, since lateral separation now exists. If desired by either the controller or the pilot, Pan Am 415 can now be assigned the same altitude as the Cessna, 7,000 feet.

If DME is not being used to determine the aircraft's location, the controller must use the value obtained from Table 7-1 to determine when lateral separation exists. According to the table, 7 miles of distance is needed to ensure

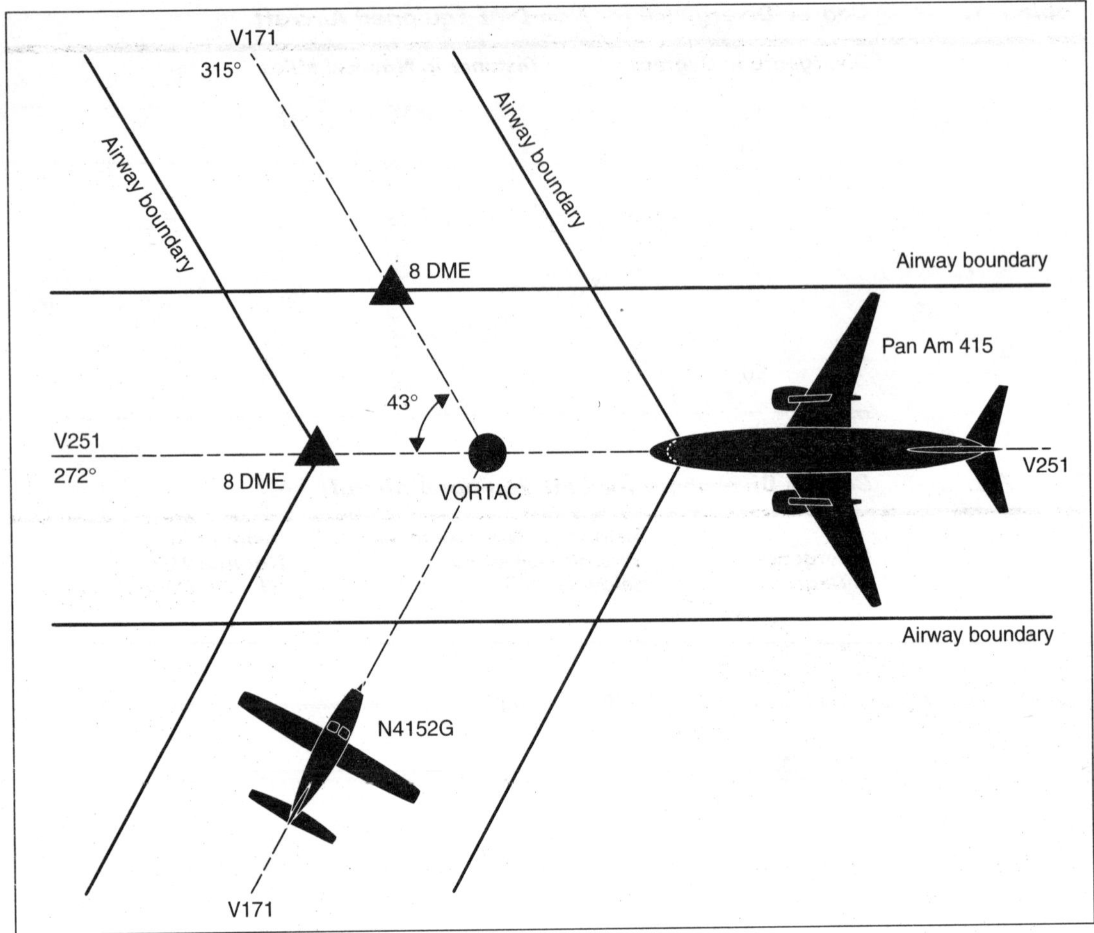

Figure 7-9. The airspace reserved for each aircraft overlaps until at least one of the aircraft is 8 nautical miles from the VORTAC. At that point the airways no longer overlap, and lateral separation can be presumed to exist.

lateral separation. The controller would consult the appropriate navigation chart to locate the closest intersection on the airways that is at least 7 miles from the VOR. Once the pilot reports crossing this intersection, vertical separation could be discontinued, since lateral separation exists.

Holding Patterns

Holding patterns are used whenever insufficient airspace exists for an aircraft to continue toward its destination. While within a holding pattern, an aircraft is restricted to a fairly small area, making it relatively easy for the controller to apply separation. Vertical separation may be applied by clearing aircraft to operate either above or below other holding aircraft. Lateral separation may be applied by ensuring that the airspace reserved for the holding aircraft does not overlap the airspace reserved for the other aircraft.

To properly apply lateral separation to an aircraft within a holding pattern, the controller must determine the airspace that must be reserved for that aircraft. This is accomplished quite easily, taking into account the following factors that may affect the aircraft's performance within the holding pattern:

Indicated airspeed of the holding aircraft.

Navigation aid and aircraft navigational system performance.

Effect of wind on the holding aircraft.

Distance between the navaid and the navigational fix being used for holding.

The altitude being used by the holding aircraft.

The speed of the aircraft is important because faster aircraft cover a greater distance while turning. Because the inbound leg of a standard holding pattern is 1 minute, faster aircraft will cover a greater distance in that time. The FAA uses the **maximum holding airspeeds** in Table 7-3 when determining holding-pattern sizes. These airspeeds are described in detail in the *Aeronautical Information Manual* and in FAA Order 7130.3, "Holding Pattern Criteria."

The accuracy of both airborne and ground-based navigation systems must also be considered when determining the holding-pattern size. As the possibility of navigation error increases, so must the size of the protected airspace reserved for each holding aircraft. The following worst-case errors are assumed when determining holding-pattern sizes:

Ground-based navigation error: ±5°

Airborne navigation error: ±10°

Six-second delay between pilot recognition of holding fix passage and commencing the turn outbound

Holding-pattern sizes are predicated on wind directions that would cause the maximum deviation from the holding pattern. FAA specialists must assume

Table 7-3. Maximum Holding Airspeeds

Altitude Limits	Holding-Pattern Airspeed (in knots of indicated airspeed)*
0–6,000 MSL	200
6,000–14,000 MSL	230
above 14,000 MSL	265

* In certain situations, holding pattern airspeeds as low as 175 knots may be assigned to smaller aircraft.

that a strong crosswind exists at the holding-pattern altitude and that the aircraft may be temporarily blown off-course. Holding-pattern sizes are also predicated on wind velocities of 50 knots at 4,000 feet that increase 3 knots for every 2,000 feet of altitude, with a maximum possible wind velocity of 120 knots. After analyzing wind speeds across the United States, FAA procedures specialists have found that the winds aloft seldom exceed these values.

The effective size of any particular intersection is directly related to its distance from the defining navigation aid. Thus, when deciding a holding-pattern size, the controller must also determine the holding fix distance from each navigation aid and use that which is greater.

The altitude that the aircraft will be holding at also directly affects the size of the holding pattern. An aircraft's **true airspeed** increases as its altitude increases. Because of the decrease in air density as altitude increases, a constant **indicated airspeed** will result in a faster true airspeed. As true airspeed increases, so does the area used by the aircraft when holding.

Holding-Pattern Templates To simplify the controller's task of separating aircraft in holding patterns, the FAA has developed a set of 31 standard holding-pattern sizes, known as **holding-pattern templates.** A number 1 template defines the airspace used for the smallest holding pattern, whereas a number 31 defines the largest holding pattern. The proper holding-pattern template can be selected for any aircraft by using the information contained in FAA Order 7130.3, "Holding Pattern Criteria."

To choose the proper template, the controller must first determine the aircraft's airspeed using Table 7-3. The controller must then select the appropriate template using the Template Selection Charts in Tables 7-4 through 7-7.

To choose the proper template, the controller must ascertain the aircraft airspeed, the maximum holding fix distance from the navaid, and the altitude that the aircraft will be operating at. For example, the controller who wishes to hold a Cessna 210 (which is a single-engine, propeller-driven personal aircraft) at the delta intersection at 6,000 feet should use the following method to determine the proper holding-pattern template:

1. Determine the holding-pattern airspeed that will be used by the aircraft. (In this example, assume a 175 knot maximum airspeed holding pattern is being utilized.)
2. Use the airspeed to determine which pattern selection chart to use. In this case, Template Selection Chart I (Table 7-4) should be utilized, because it was expressly designed for aircraft holding at or below 175 knots IAS.
3. Using a navigation chart, determine the holding fix distance from each navigation aid. Use the largest distance of the two. In this hypothetical case, if delta intersection is 10 DME from alpha VOR and 16 DME from bravo VOR, 16 nautical miles should be used.
4. Refer to the proper template selection chart and locate the aircraft's altitude in the appropriate column. In this example, the 15-29.9 nautical mile distance column should be used.

Table 7-4. **Template Selection Chart I, for Aircraft Holding at or below 175 Knots IAS**

Distance from the Navaid 0–14.9 n mi		Distance from the Navaid 15–29.9 n mi		Distance from the Navaid 30 n mi and over	
Altitude	**Template**	**Altitude**	**Template**	**Altitude**	**Template**
2,000	1	2,000	1	2,000	2
4,000	1	4,000	2	4,000	3
6,000	2	6,000	3	6,000	4
8,000	3	8,000	4	8,000	5
10,000	4	10,000	5	10,000	6
12,000	5	12,000	6	12,000	7
14,000	6	14,000	7	14,000	8
16,000	7	16,000	8	16,000	9
18,000	8	18,000	9	18,000	10
20,000	8	20,000	9	20,000	10
22,000	9	22,000	10	22,000	11
24,000	10	24,000	11	24,000	12
26,000	11	26,000	12	26,000	13
28,000	12	28,000	13	28,000	14
30,000	13	30,000	14	30,000	15

5. Note the template number that should be used. In this example, a number 3 template is specified.

To properly utilize the template, the controller must:

1. Determine whether a right- or a left-turn holding pattern will be utilized. The templates are designed for use with a right-turn holding pattern. If a left turn is desired, the controller must physically turn the template over when tracing the holding pattern.
2. Place the small hole in the template directly over the holding fix. The line extending from the small hole is laid directly over the inbound course to the holding fix.
3. Trace around the holding-pattern template to delineate the airspace reserved for that aircraft. To laterally separate this aircraft, the controller must ensure that this airspace does not overlap that reserved for any other aircraft operating at the same altitude.

A number of factors can be considered to reduce the size of the holding-pattern airspace. One is the aircraft's type of holding-pattern entry and its position

Table 7-5. Template Selection Chart II, for Aircraft Holding Between 175 and 230 Knots IAS

Distance from the Navaid 0–14.9 n mi		Distance from the Navaid 15–29.9 n mi		Distance from the Navaid 30 n mi and over	
Altitude	**Template**	**Altitude**	**Template**	**Altitude**	**Template**
2,000	3	2,000	4	2,000	5
4,000	4	4,000	5	4,000	6
6,000	5	6,000	6	6,000	7
8,000	6	8,000	7	8,000	8
10,000	7	10,000	8	10,000	9
12,000	7	12,000	8	12,000	9
14,000	8	14,000	9	14,000	10
16,000	12	16,000	13	16,000	14
18,000	13	18,000	14	18,000	15
20,000	14	20,000	15	20,000	16
22,000	15	22,000	16	22,000	17
24,000	16	24,000	17	24,000	18
26,000	17	26,000	18	26,000	19
28,000	18	28,000	19	28,000	20
30,000	19	30,000	20	30,000	21
32,000	20	32,000	21	32,000	22
34,000	21	34,000	22	34,000	23
36,000	22	36,000	23	36,000	24
38,000	23	38,000	24	38,000	25
40,000	24	40,000	25	40,000	26
42,000	25	42,000	26	42,000	27
44,000	26	44,000	27	44,000	28
46,000	27	46,000	28	46,000	29
48,000	28	48,000	29	48,000	30
50,000	28	50,000	29	50,000	30

within the holding pattern. Whenever an aircraft enters a holding pattern from a direction other than the inbound course, additional airspace is reserved to permit the aircraft to become established on the inbound course. Normal application of the holding-pattern template takes this additional airspace into consideration. Once the aircraft is established on the inbound course, this extra maneuvering space can be eliminated, thereby reducing the size of the holding-pattern airspace. The dashed line known as the **fix end reduction area** de-

Table 7-6. Template Selection Chart III, for Aircraft Holding at 265 Knots IAS

Distance from the Navaid 0–14.9 n mi		Distance from the Navaid 15–29.9 n mi		Distance from the Navaid 30 n mi and over	
Altitude	**Template**	**Altitude**	**Template**	**Altitude**	**Template**
2,000	7	2,000	8	2,000	9
4,000	8	4,000	9	4,000	10
6,000	9	6,000	10	6,000	11
8,000	10	8,000	11	8,000	12
10,000	11	10,000	12	10,000	13
12,000	12	12,000	13	12,000	14
14,000	13	14,000	14	14,000	15
16,000	15	16,000	16	16,000	17
18,000	16	18,000	17	18,000	18
20,000	17	20,000	18	20,000	19
22,000	18	22,000	19	22,000	20
24,000	19	24,000	20	24,000	21
26,000	20	26,000	21	26,000	22
28,000	21	28,000	22	28,000	23
30,000	22	30,000	23	30,000	24
32,000	23	32,000	24	32,000	25
34,000	24	34,000	25	34,000	26
36,000	25	36,000	26	36,000	27
38,000	26	38,000	27	38,000	28
40,000	27	40,000	28	40,000	29
42,000	28	42,000	29	42,000	30
44,000	28	44,000	29	44,000	30
46,000	29	46,000	30	46,000	31
48,000	31				

lineates this airspace (see Figure 7-10). If the aircraft is initially established on the inbound course, or once the pilot reports being established on the inbound course, this area is no longer reserved for the aircraft. A number of additional holding-pattern reduction areas are located on the outbound end of the holding-pattern template. Use of these reduction areas is infrequent; the proper means of applying these reductions can be obtained from FAA Order 7130.3.

Holding-Pattern Applications This entire procedure seems too cumbersome and too complex for controllers to perform on a routine basis. And it is. In

Table 7-7. Template Selection Chart IV, for Aircraft Holding at or below 310 Knots IAS

Distance from the Navaid 0–14.9 n mi		Distance from the Navaid 15–29.9 n mi		Distance from the Navaid 30 n mi and over	
Altitude	**Template**	**Altitude**	**Template**	**Altitude**	**Template**
2,000	11	2,000	12	2,000	13
4,000	12	4,000	13	4,000	14
6,000	13	6,000	14	6,000	15
8,000	14	8,000	15	8,000	16
10,000	15	10,000	16	10,000	17
12,000	17	12,000	18	12,000	19
14,000	18	14,000	19	14,000	20
16,000	19	16,000	20	16,000	21
18,000	20	18,000	21	18,000	22
20,000	21	20,000	22	20,000	23
22,000	22	22,000	23	22,000	24
24,000	22	24,000	23	24,000	24
26,000	24	26,000	25	26,000	26
28,000	24	28,000	25	28,000	26
30,000	25	30,000	26	30,000	27
32,000	26	32,000	27	32,000	28
34,000	27	34,000	28	34,000	29
36,000	28	36,000	29	36,000	30
38,000	29	38,000	30	38,000	31
40,000	30	40,000	31		

most cases, this process need not be done on a day-to-day basis. Most air traffic control facilities have developed a chart of the local airspace, with every conceivable holding pattern traced on it. Any overlap of holding pattern or airway airspace is noted and prominently displayed (see Figure 7-11). All a controller must do to routinely issue holding instructions is to glance at the chart and determine whether any lateral conflicts exist. If none do, the holding pattern may be safely occupied. In most cases, the only time that the entire holding-pattern procedure must be applied is when an unusual situation occurs or when a particular air traffic control facility's airspace is being modified.

Longitudinal Separation

Whenever two aircraft are flying along the same route, either vertical or **longitudinal separation** methods must be used. Vertical separation is the easier method, but it may also result in an inefficient use of the airspace. This may be

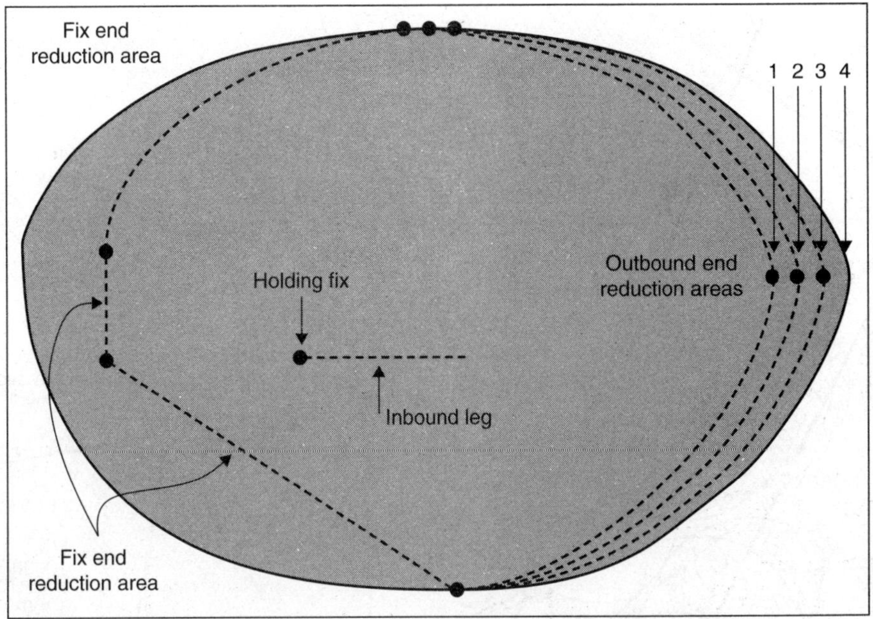

Figure 7-10. A drawing of a holding-pattern template (number 29) showing the locations of the holding fix, the fix end reduction area, and the outbound end reduction areas.

of less concern to controllers in a terminal environment but can greatly reduce the amount of traffic that can operate on highly traveled airways within certain ARTCCs or on transoceanic routes. In these cases, it would be more efficient to clear multiple aircraft to operate along these airways at the same altitude, using longitudinal separation techniques. Longitudinal separation may also be used when two aircraft are operating at different altitudes along the same route but one is changing altitude and must pass through the altitude being used by the other.

Longitudinal separation presumes that both aircraft are operating along the same route or are on routes whose protected airspaces overlap one another (see Figure 7-12). Routes whose protected airspaces overlap (the centerlines of the airways are less than 8 nautical miles apart) are considered for separation purposes to be the same route. For longitudinal separation to be applied to two aircraft, both must be flying at or near the same airspeed or the leading aircraft must be significantly faster than the following aircraft. Situations in which the following aircraft is faster than the leading aircraft usually make it impossible to apply longitudinal separation. If the following aircraft were indeed faster, it would eventually overtake the leading aircraft, thereby incurring a loss of separation.

Longitudinal separation can also be applied to aircraft operating along the same route but in opposite directions. This procedure is fairly complex and

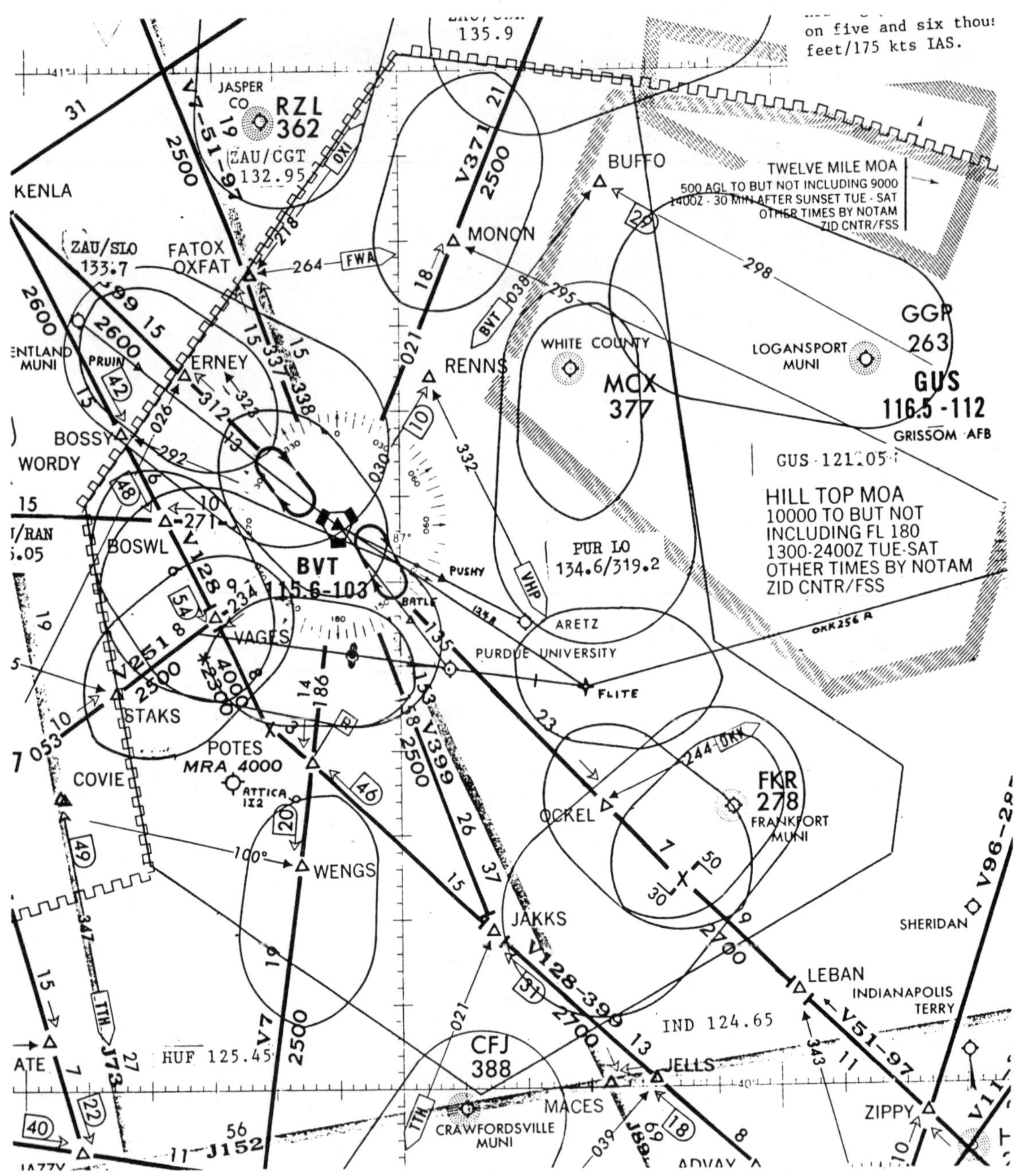

Figure 7-11. An example of various holding patterns plotted on a chart and used by the controllers in the Lafayette (Indiana) airspace.

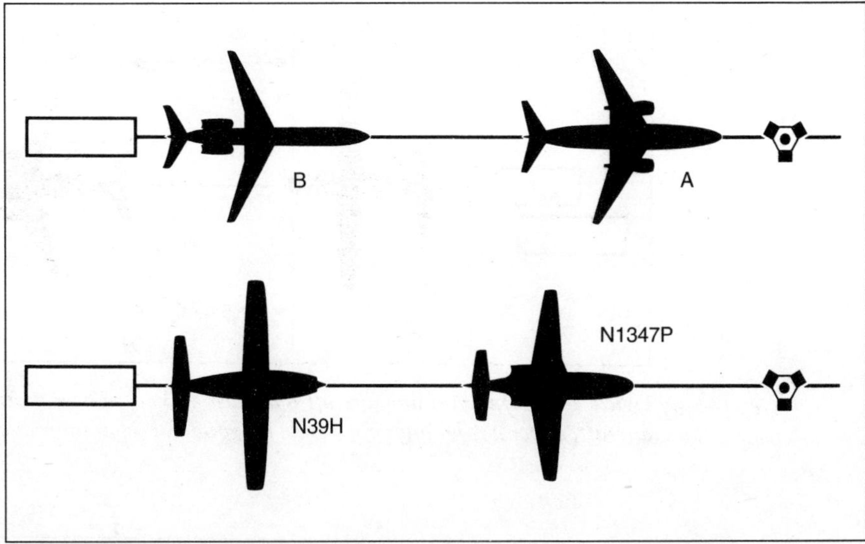

Figure 7-12. Longitudinal separation can be applied between two aircraft on the same route when the leading aircraft is either faster or operating at the same speed as the following aircraft. As the speed differential increases, the separation interval can be decreased.

requires a significant amount of airspace. Thus, terminal controllers will almost never use this method of separation for opposite-direction traffic operating along the same airway.

Longitudinal separation between two aircraft is applied using one of the following methods:

1. Aircraft operating in the same direction.
   a. If the leading aircraft is flying at a true airspeed at least 44 knots faster than that of the following aircraft, at least a 3-minute (or 5-mile, as measured by DME or RNAV) interval of separation must be maintained between the two aircraft (see Figure 7-13).
   b. If the leading aircraft will fly at a true airspeed at least 22 knots faster than that of the following aircraft, at least a 5-minute (or 10-mile, as measured by DME or RNAV) interval of separation must be maintained between the two aircraft (see Figure 7-14).
   c. If either of the aircraft is climbing or descending through the altitude of the other aircraft, at least a 5-minute (or 10-mile) interval of separation must be maintained. This procedure may be applied only when either the leading aircraft is descending or the following aircraft is climbing (see Figure 7-15).
   d. If none of the above conditions can be met, both aircraft must be separated by at least a 10-minute or a 20-mile interval of separation (see Figure 7-16).

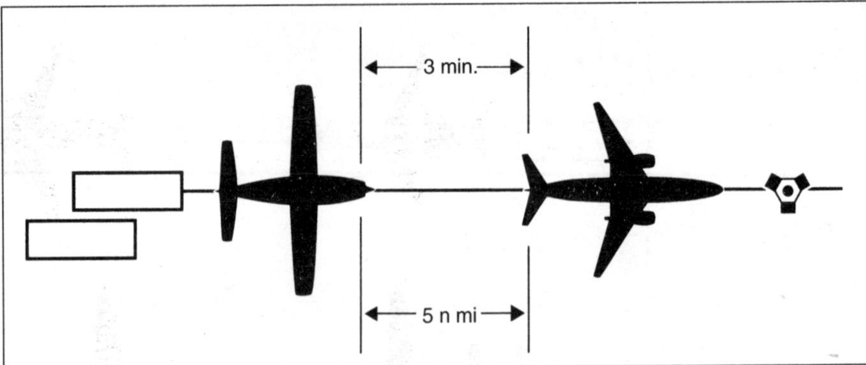

Figure 7-13. If the leading aircraft is at least 44 knots faster than the following aircraft, either a 3-minute or a 5-mile longitudinal separation interval must be utilized.

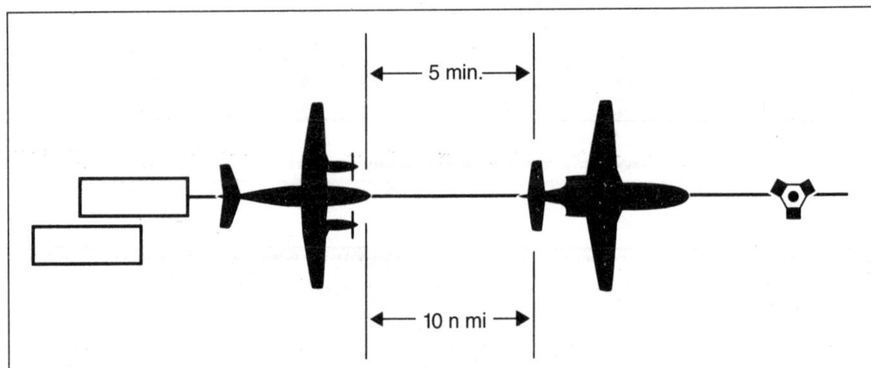

Figure 7-14. If the leading aircraft is at least 22 knots faster than the following aircraft, either a 5-minute or a 10-mile longitudinal separation interval must be utilized.

2. Aircraft operating in opposite directions.

   Aircraft operating along the same route but in opposite directions must be separated vertically from at least 10 minutes prior to and until at least 10 minutes after they are estimated to pass each other. This vertical separation can be discontinued prematurely and longitudinal separation can be applied if:

   a. Both aircraft have passed the same navaid or DME fix, or

   b. Both aircraft have reported passing the same intersection and are now at least 3 minutes apart. This 3-minute interval is used to take into consideration the maximum airborne navigational equipment error.

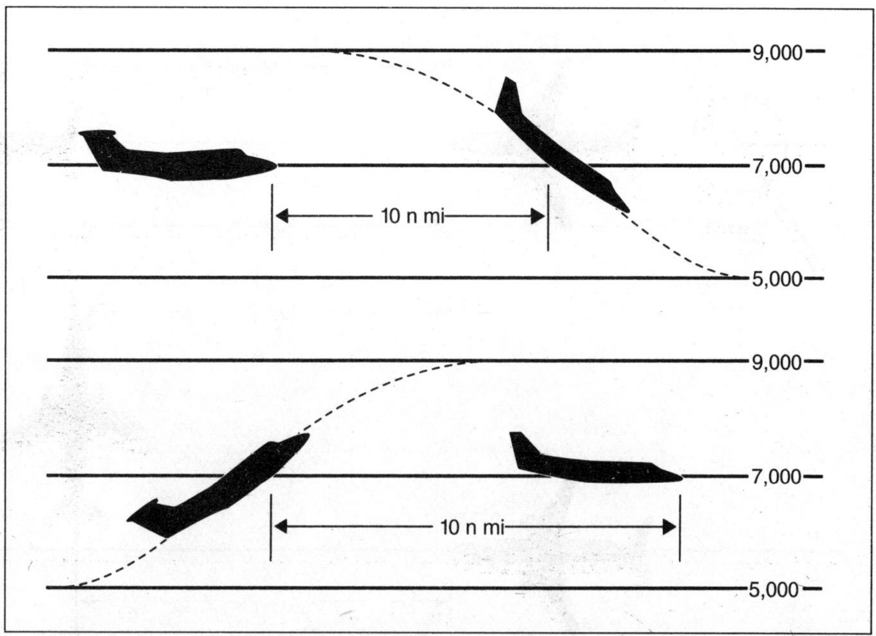

Figure 7-15. If the leading aircraft is descending or the following aircraft is climbing, a 10-mile longitudinal separation interval must be utilized.

If none of these separation criteria can be met, the controller is not permitted to use longitudinal separation and must apply vertical, lateral, or visual separation methods.

Longitudinal separation is usually more efficient than the strict use of vertical separation for busy and congested airways. If, for example, there is only one airway between two airports, the use of vertical separation could severely restrict the number of aircraft operating along that airway. Once a sufficient number of aircraft had departed to utilize every available altitude, no other aircraft could use the airway. And if the controller had inadvertently permitted a relatively slow and low-flying aircraft to be the first departure, it might prove to be impossible to depart any subsequent aircraft, since there would be no available lower altitudes. In such cases, the controller usually finds it advantageous to use longitudinal separation, permitting a number of aircraft to operate along the same route at the same altitude.

Longitudinal separation of aircraft is accomplished by requiring the pilot to do one of the following:

Depart an airport or a navigational fix at a specified time.

Depart an airport or a navigational fix after the preceding aircraft has traveled a specified distance.

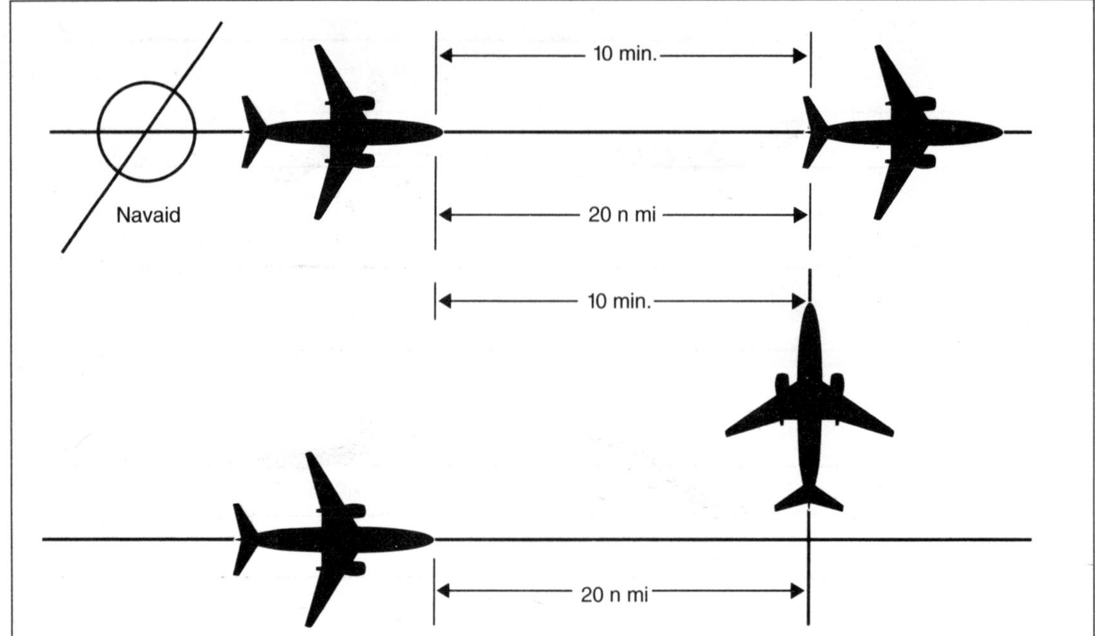

Figure 7-16. If the leading aircraft is operating on the airway at the same speed as the following aircraft or is crossing the airway, a 10-minute or a 20-mile longitudinal separation interval must be utilized.

Arrive at a navigational fix at a specified time.

Cross a navigational fix at a specified time.

Enter a holding pattern at a fix until the preceding aircraft has traveled a specified time or distance.

Change altitude at a navigational fix.

At the time that the particular clearance is issued, the controller will write the appropriate information in designated sections of the flight progress strip. Most of this information is written in field 9 on a terminal flight progress strip and field 25 on a flight strip used by center controllers. In most cases, fields 10 through 18 on a terminal flight strip are also used to record some of this information.

Whenever the pilot makes a required report to the controller, this information is noted on the flight progress strip. In a similar manner to the method that altitude reports are recorded, the controller will usually line out the written request on the flight strip when the position report is made. In most cases, the time that the report was made will also be recorded. Since many of the longitudinal separation procedures rely on time interval separation, time reports must be accurately recorded.

Whenever the controller requests that the pilot **report crossing** a particular fix, the letters **RX,** followed by the name of the fix, are placed on the flight progress strip. If the controller clears the pilot to *cross* a navigational fix *at* a specified time, the controller writes "X @ (time)" on the flight strip. The time is always specified in UTC. The pilot may also be requested to cross the fix *before* a specified time, or *after* a specified time. The > symbol is used to indicate before, whereas the < symbol indicates after. The pilot may also be requested to maintain a certain altitude *until* a specified time or position is reached. This is indicated using the / symbol.

If it is necessary to hold an aircraft, the controller issues the holding instructions to the pilot and places the letter H on the flight strip, followed by the specific holding instructions. The controller may also request the pilot to depart a fix or an airport at a specified time, which is indicated using the letter T.

One of the simplest applications of longitudinal separation is in separating two aircraft departing from the same airport when both pilots wish to fly the same route at the same altitude. If both aircraft will fly at the same true airspeed, the controller will clear one aircraft to depart and will then wait at least 10 minutes before clearing the following aircraft to depart. As long as the leading aircraft will operate at the same speed or faster than the following aircraft, this procedure ensures that there will always be at least a 10-minute interval of separation between the two aircraft.

If both aircraft can determine distance along the airway using DME, the controller may find it operationally advantageous to use a 20-mile interval of separation. In this case the controller asks the pilot of the leading aircraft to report when the aircraft is 20 miles from the airport. At that point, the following aircraft can be released.

The separation interval can be reduced if the leading aircraft is at least 22 knots faster than the following aircraft, in which case the controller is required to separate the aircraft by either a 5-minute or a 10-mile interval. If the leading aircraft is at least 44 knots faster than the following, the separation interval can be reduced to 3 minutes or 5 miles.

But what if the leading aircraft is slower than the following aircraft? In this case, it is virtually impossible to use longitudinal separation, as there is no way to ensure at least a 10-minute interval of separation between the aircraft at all times. Even if the controller waits 15 to 20 minutes after the first aircraft leaves before departing the following aircraft, there is no assurance that both aircraft will always maintain at least a 10-minute separation interval. Eventually the second aircraft will overtake the first. Once they are less than 10 minutes or 20 miles from each other, a loss of separation will have occurred. So it is obvious that it is in the controller's, and the pilot's, best interest that faster aircraft be permitted to depart prior to slower aircraft. Even if the slow aircraft is ready to depart earlier, insofar as ATC system efficiency is concerned it is more efficient to hold the slower aircraft for a short time than to be unable to release the faster aircraft.

Another situation in which it might be more efficient to use longitudinal separation is when the first aircraft is airborne, flying over the airport, while the second is still on the ground but wishing to depart and use the same airway. If there is a navigation fix directly above the airport, it can be used to effect longitudinal separation. If, for instance, the airborne aircraft is at least 44 knots faster than the aircraft on the ground, the controller need only ensure that the second aircraft does not depart until at least 3 minutes since the first aircraft crossed the navigation aid.

Another application of longitudinal separation is in the use of crossing restrictions. If two aircraft were initially on different routes (separated laterally) but are both eventually going to operate along the same route, the faster aircraft might be allowed to operate unrestricted, with the slower aircraft instructed to join the airway at a later time. For example, let us assume that the first aircraft is at least 22 knots faster than the second aircraft, and the second aircraft is in a holding pattern laterally separated from the airway. If the first aircraft reports crossing the VOR at 30 minutes past the hour, the second aircraft could be authorized to depart the holding pattern and to cross the VOR at or after 35 past the hour. This procedure would ensure that at least a 5-minute separation interval is maintained.

Initial Separation of Aircraft

The three methods of separation mentioned thus far, vertical, lateral, and longitudinal, are difficult to apply to aircraft that have just departed an airport or to separation of an arriving and a departing aircraft, because the aircraft's positions and altitudes are most likely constantly changing. Thus, the FAA has developed procedures to initially separate aircraft. **Initial separation procedures** are used only to separate aircraft that are beginning or ending their flight and are within the immediate vicinity of an airport. Since the location of each aircraft can be accurately determined, separation intervals can be temporarily reduced.

These procedures presume that both aircraft are operating from the same airport. Once initial separation methods have been used and the aircraft are established on their respective courses, any of the previously mentioned separation methods must be applied. The use of initial separation procedures does not permit the controller to utilize procedures that might place either aircraft in an unsafe situation. These rules must therefore be used with discretion, taking into consideration aircraft performance and the local air traffic structure.

Initial Separation of Arriving and Departing Aircraft When separating an arrival aircraft from a departure at the same airport, the controller must first determine whether the course of the departing aircraft will diverge from that of the inbound aircraft. **Divergence** occurs whenever the course of one aircraft differs from the other by at least 45°. If the courses differ by less than this angle, they do not diverge, and the controller must presume that for all practical purposes the aircraft will be operating on the same route.

If the departing aircraft is taking off in a direction headed toward an inbound aircraft, the controller may consider the two aircraft to be separated as long as at least one of the following conditions exists:

The departing aircraft must be established on a course that differs by at least 45° from the reciprocal of the final approach course before the inbound aircraft reaches a fix located at least 4 miles from the end of the runway. If no such fix exists, the aircraft must depart at least 3 minutes before the inbound aircraft's estimated time of arrival at the airport (see Figure 7-17).

If the departing aircraft's initial heading does not differ from the reciprocal of the final approach course by at least 45°, the departing aircraft may depart only if it is established on a course that diverges at least 45° from the reciprocal of the inbound course, at least 5 minutes before the inbound aircraft's estimated time of arrival (see Figure 7-18) or at least 5 minutes before the arrival aircraft begins the procedure turn (see Figure 7-19).

Initial Separation of Successive Departing Aircraft When two aircraft are departing from the same airport and their eventual courses will diverge by at least 45°, the controller may use the same-runway, different-runway, or intersecting-runway separation techniques described in the following sections until one of the standard methods of separation (vertical, lateral, or longitudinal) can be applied. The controller must take into consideration both the aircraft's and the pilot's performance capabilities before utilizing these procedures.

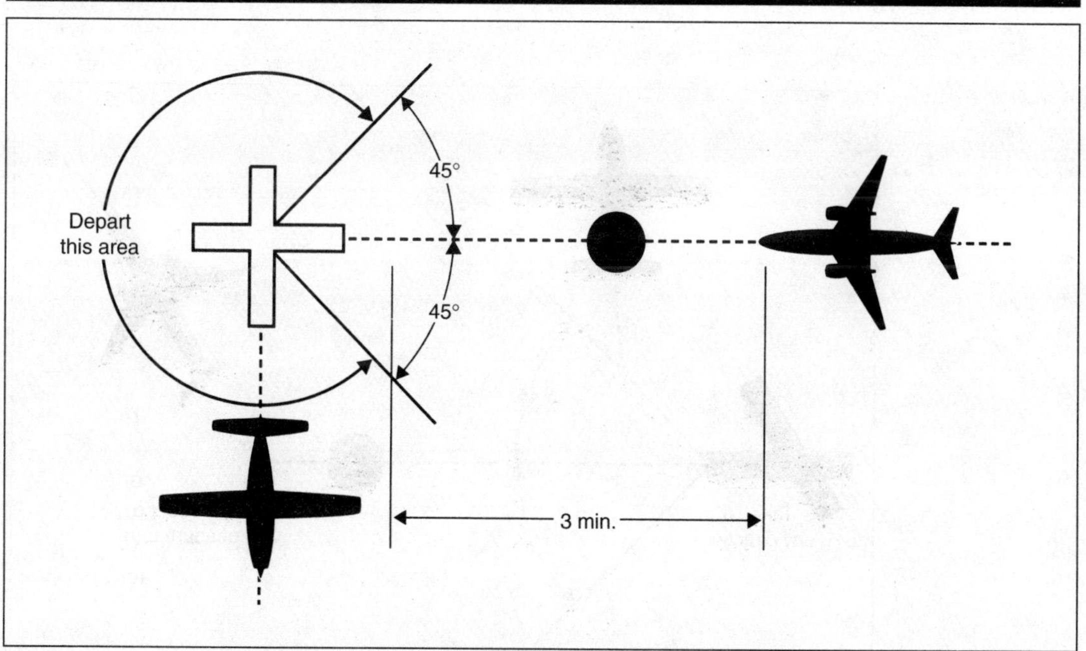

Figure 7-17. A 3-minute separation interval must be applied between an arriving and a departing aircraft whose courses will immediately diverge by at least 45°.

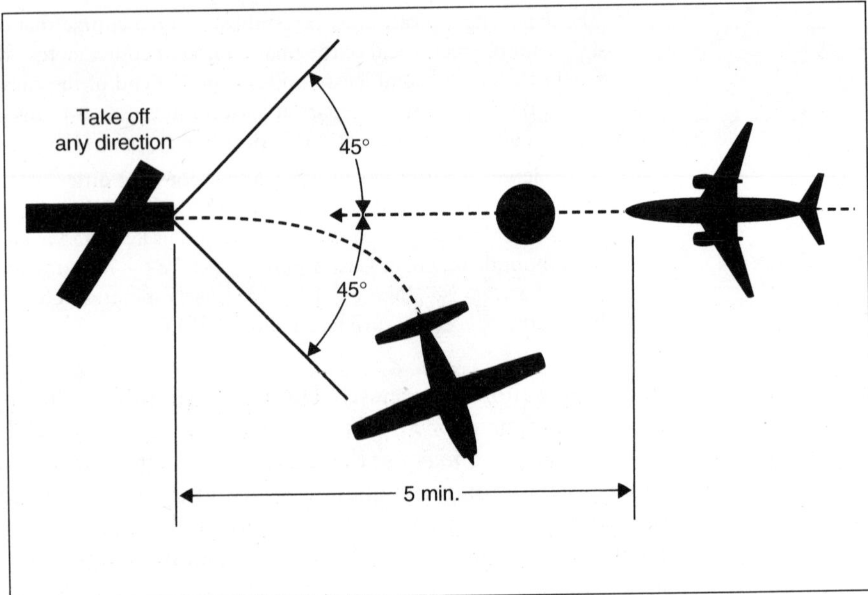

Figure 7-18. A 5-minute separation interval must be applied between an arriving and a departing aircraft whose courses will eventually diverge by at least 45°.

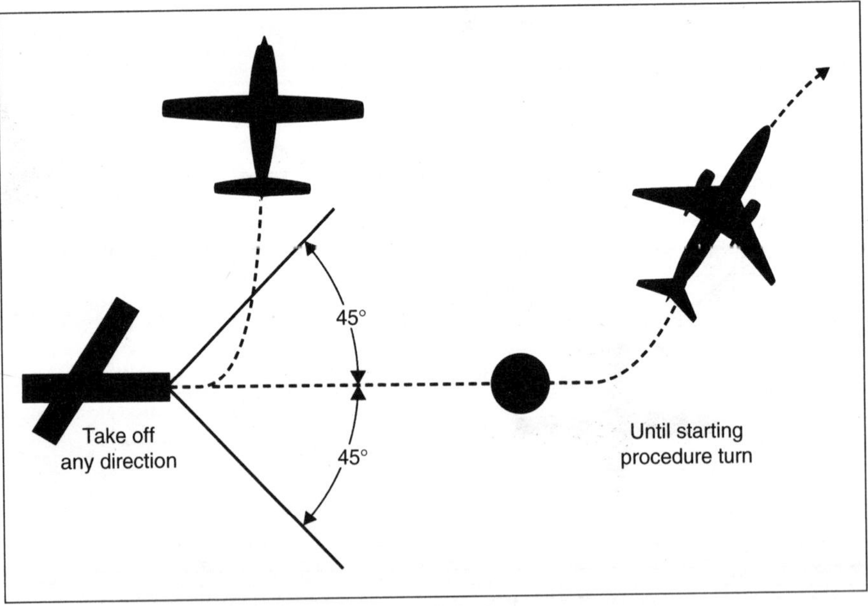

Figure 7-19. A departing aircraft is separated from an arrival if it is established on a diverging course before the arriving aircraft begins the procedure turn.

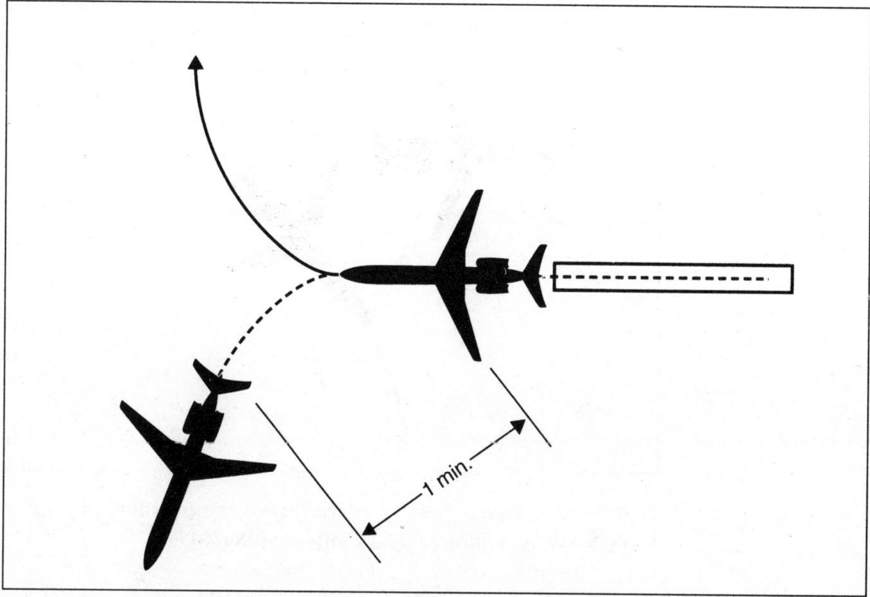

Figure 7-20. Two departing aircraft whose courses will diverge immediately after takeoff can be separated by a 1-minute interval.

Same-Runway Separation If two aircraft are departing from the same runway or from parallel runways separated by less than 3,500 feet, they can be considered to be separated if one of the following criteria can be met (see Figures 7-20 through 7-22):

1. If the two aircraft will fly diverging courses immediately after takeoff, the aircraft must be separated by at least a 1-minute interval.
2. If the two aircraft will not diverge immediately but will diverge within 5 minutes after departure, they must be separated by at least a 2-minute interval.
3. If the two aircraft will diverge within 13 miles of the departure airport, they must be separated by at least a 3-mile interval.

Although seemingly straightforward, these rules are not as easy to apply in practice as it might first appear. For instance, using criterion number one, it would seem that if both aircraft were going to diverge immediately after takeoff, the second aircraft could be cleared for takeoff 1 minute after the first has departed. But note that rule number one specifies that the controller must ensure that at least a 1-minute *continuous* separation interval exist between both aircraft until their courses diverge. For example, assume that two aircraft wish to depart from the same airport utilizing runway 27. The first aircraft is a Piper Cherokee whose heading will be 315° and whose true airspeed is about 100 knots. Assume that the second aircraft is a Lear whose heading will be 270° with a

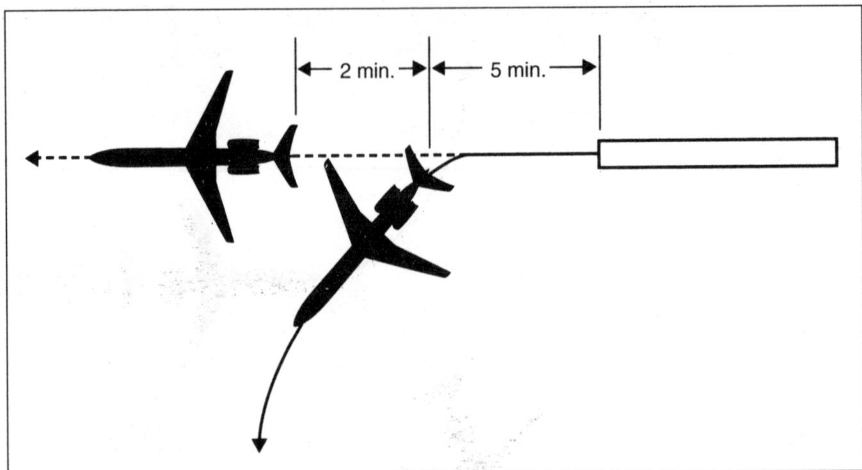

Figure 7-21. Two departing aircraft whose courses will diverge within 5 minutes after takeoff can be separated by a 2-minute interval.

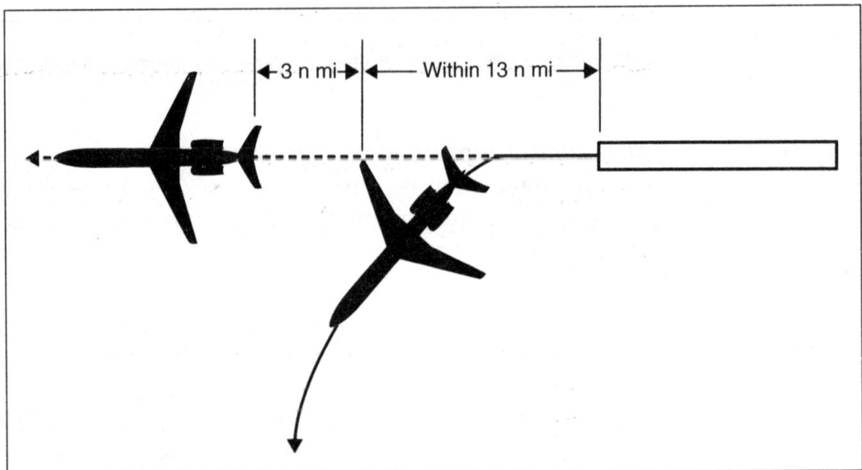

Figure 7-22. Two departing aircraft whose courses will diverge within 13 nautical miles can be separated by a 3-nautical-mile interval.

true airspeed of 250 knots (see Figure 7-23). If the controller waits 1 minute after the Cherokee departs before clearing the Lear for takeoff, a 1-minute separation interval will not be maintained. If the Cherokee pilot delays turning right for a few extra seconds, and considering that the Lear will probably make a fairly wide left turn, it is almost a foregone conclusion that a continuous 1-minute interval of separation will not be maintained between the two aircraft.

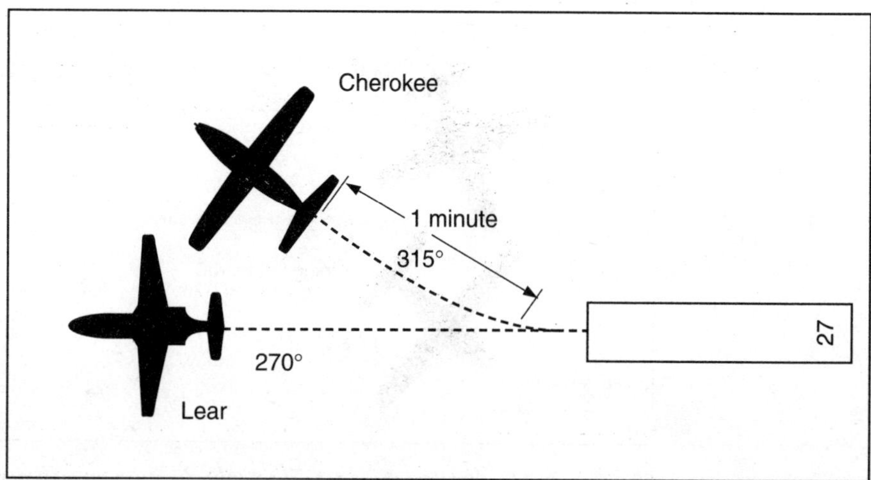

Figure 7-23. A following aircraft may not remain properly separated if it is much faster than the preceding aircraft.

In this case, a very dangerous situation would soon develop. A safer method of separating these two aircraft might be for the controller to request that the Cherokee pilot report when the aircraft is established on the 315° heading, wait for 1 minute, and then clear the Lear for takeoff. This situation could also be more safely handled if the controller cleared the Lear for takeoff first, followed by the Cherokee 1 minute later. In this case, because the Lear is significantly faster than the Cherokee, a continuous 1-minute interval of separation is virtually guaranteed.

Suppose the Cherokee were turning left to a heading of 135° after takeoff, with the Lear turning left to a heading of 180°. In this case, the two aircraft would not diverge immediately after departure but would most likely diverge within 5 minutes. So the controller assumes that rule number two can be safely utilized. Again, if the two aircraft in question are vastly different in size and capability, this may not be the case (see Figure 7-24).

Different-Runway Separation If two aircraft are departing from parallel runways separated by at least 3,500 feet, the controller may authorize simultaneous departures if the aircrafts' courses diverge by at least 45° immediately after takeoff. The controller must ensure separation between these departures and from succeeding departures. If two aircraft departing from parallel runways will not diverge immediately after takeoff, the controller must act as if both aircraft were departing from the same runway and use the separation rules stated above that govern these situations.

If the two aircraft will depart from nonparallel runways, one of the following conditions must exist or the controller must act as if the aircraft were departing from the same runway:

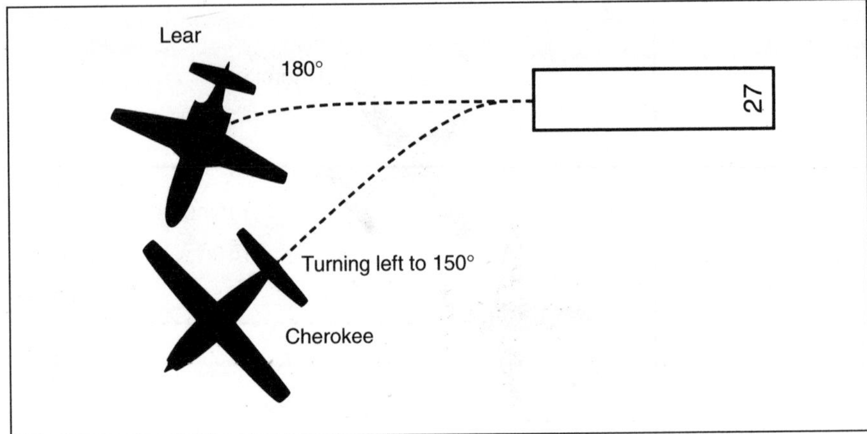

Figure 7-24. Controllers must take into consideration aircraft performance characteristics when separating aircraft.

> The runways diverge by at least 30° and the aircraft's courses immediately diverge by at least 45°.
>
> The runways diverge at less than 30° but by at least 15°, the runways are separated by at least 2,000 feet, and the aircraft will diverge by at least 45° immediately after takeoff.
>
> The runways diverge by less than 15° and are separated by at least 3,000 feet, and both aircraft will diverge immediately after takeoff.

Intersecting-Runway Separation If the two departing aircraft will utilize intersecting runways for departure, the controller may depart these two aircraft if either of the following conditions exists:

> The runways diverge by at least 30°, the preceding aircraft has passed the intersection, and the aircraft will diverge by at least 45° immediately after takeoff.
>
> The runways diverge at less than 30° but by at least 15°, the preceding aircraft has crossed the intersection and has commenced the turn on course, and the two aircraft will diverge by at least 45° immediately after takeoff.

As in the previous examples, if neither of these conditions can be met, the controller must act as if the two aircraft were departing from the same runway.

These rules were designed to be as flexible as possible and to improve efficiency within the immediate vicinity of the airport. If the controller does not find them helpful, he or she can use one of the standard methods of separation. A good, resourceful controller can usually find some way to depart the aircraft with a minimum of delay while still using the airspace efficiently.

Visual Separation

One of the most flexible means of separating aircraft is through **visual separation** techniques. In general, visual separation requires that either of the pilots sees the other aircraft and will provide the required separation or that the controller is able to observe both aircraft and assume the responsibility for providing safe separation. It is obvious from the examples included in this chapter that the application of nonradar separation rules usually results in the inefficient use of airspace and can incur substantial delays to aircraft, both airborne and on the ground. If visual separation can reduce these delays without degrading safety, it is in both the controller's and the pilot's best interest to use this technique.

Visual separation is most often applied by terminal controllers. Because of their large areas of responsibility and the fact that ARTCC controllers usually rely exclusively on radar separation, en route controllers seldom use visual separation. One of the few instances where visual separation is used by center controllers is in conjunction with visual approaches to airports not served by an approach control facility.

The controller may use visual separation as long as radio contact is maintained with at least one of the aircraft involved and at least one of the following conditions can be met:

> The controller can visually identify both aircraft and is willing and able to provide separation.
>
> The pilot of at least one of the aircraft can visually identify the other aircraft and has accepted responsibility for separation, and the pilot of the second aircraft has been informed that visual separation is being applied. If at any time the pilot of the first aircraft advises the controller that visual separation can no longer be maintained, either the second pilot or the controller must accept visual separation responsibility, or the controller must provide another method of separation.

The pilots need not be informed that visual separation is in use if the controller accepts the responsibility for that separation. But if one of the pilots is accepting the responsibility, this fact must be made quite clear. The typical phraseology used by controllers is provided in the following example involving two aircraft inbound for the same runway. United 324 (UAL 324), the first aircraft, is conducting an ILS approach. The second aircraft, Delta 111 (DAL 111), is holding 1,000 feet above the United flight and is waiting to conduct an ILS approach. Since DAL 111 will not be able to begin the approach until UAL 324 has landed, DAL 111 will be required to enter a holding pattern.

To properly utilize visual separation, the controller must ensure that the pilot of DAL 111 has observed UAL 324. Once positive identification is established, the controller must request that the pilot of DAL 111 maintain visual separation, advise the pilot of UAL 324 that visual separation is being used, and receive acknowledgments from both pilots. If any of these criteria cannot be met, visual separation may not be used.

CONTROLLER:	United three twenty-four cleared for the ILS runway one zero approach.
UNITED 324:	United three twenty-four, roger.
CONTROLLER:	Delta one eleven, traffic is a United seven twenty-seven ahead of you at two thousand, conducting the ILS approach, do you have it in sight?
DELTA 111:	Delta one eleven. We have the seven twenty-seven in sight.
CONTROLLER:	Delta one eleven, maintain visual separation from the seven twenty-seven, cleared for the ILS one zero approach.
DELTA 111:	Delta one eleven, roger.
CONTROLLER:	United three twenty-four, traffic is a DC-9 two miles behind you on the ILS maintaining visual separation.
UNITED 324:	United three twenty-four, roger.

When the Delta pilot accepted the clearance, he accepted responsibility for the separation of the two aircraft. The controller still needed to apply standard separation between these aircraft and any other IFR aircraft within the facility's airspace. If the Delta pilot had declined to accept the responsibility, the controller could not use visual separation and would have had to use some other separation technique.

Although visual separation usually provides increased ATC system efficiency, it has some serious shortcomings that both controllers and pilots should be aware of. For instance, both the controller and the pilot must be certain that the other aircraft is identified correctly. In highly congested areas where many similar types of aircraft are inbound to the airport, a misidentification is a distinct possibility. Both the controller and the pilot must also ensure that visual separation can properly be maintained during the entire approach. Even if the pilot is initially able to provide visual separation, distractions can make it difficult for the pilot to maintain that separation. Controllers must realize that pilots are usually very busy during the arrival and departure phases of flight. To ask them to provide separation at this time may be a mistake if they are involved in other, more crucial tasks. In particular, pilots of small civilian and most military fighter aircraft are often just too busy navigating and operating aircraft systems to be able to effectively maintain visual separation.

It is up to the pilot to decline visual separation responsibility whenever, in his or her opinion, it might be unsafe to accept it. The controller should realize that it may be difficult for pilots to maintain visual contact with other aircraft in hazy or foggy weather or when headed directly into the sun. The controller should also be aware that under certain conditions it may be physically impossible for the pilot to remain in visual contact with the other aircraft, such as when one aircraft is directly under another. High-wing aircraft tend to block the pilot's view directly above and to one side while turning, whereas low-wing aircraft block the pilot's view directly below and to the other side when turning. Because of the cockpit window design, most aircraft have limited visibility in the area directly below and behind the cockpit.

Even though it is the pilot's responsibility to maintain visual separation once the clearance has been accepted, it is the controller's moral responsibility to use this method of separation only when there is a reasonably good chance that the pilot will be able to maintain visual contact with the other aircraft.

If a pilot feels that visual separation is not practical, he or she should decline the clearance. A clearance is never effective until the pilot accepts it. The mere issuance of a visual separation clearance does not make it so. The controller must also be prepared to apply nonvisual separation techniques at a moment's notice. If, during a visual separation procedure, the pilot declares that visual separation cannot be maintained, either the controller or the pilot of the other aircraft must be able to provide the required visual separation or a nonvisual separation technique must be immediately employed.

KEY TERMS

divergence
fix end reduction area
holding-pattern templates
indicated airspeed
initial separation procedures
lateral separation
longitudinal separation
maximum holding airspeeds
remote communication air/ground (RCAG)
report crossing (RX)
report leaving (RL)
report reaching (RR)
separation error
stepping down
stepping up
true airspeed
vertical separation
visual separation

REVIEW QUESTIONS

1. What are the four different methods of nonradar separation?
2. What is the purpose of a holding pattern?
3. How does the controller know where each aircraft is?
4. Who can be responsible for visual separation?
5. What variables affect the size of a holding pattern?

8

Theory and Fundamentals of Radar Operation

Checkpoints

After studying this chapter, you should be able to:

1. Describe the operation of a radar system.
2. Explain the need for and the operation of moving target indicator equipment.
3. Describe the advantages of a PRF stagger.
4. Explain the need for and the operation of circular polarization equipment.
5. Describe the differences between primary and secondary radar.
6. Describe the major components and the operation of the air traffic control radar beacon system.
7. Identify and distinguish between the different modes used by the ATCRBS system.
8. Explain the differences among TPX-42, ARTS-II, ARTS-III, NAS-A, DARC, and EARTS.

Air traffic controllers use **radar,** which is similar to broadcast radio, to detect and track aircraft. An acronym for *Ra*dio *D*etection *a*nd *R*anging, radar is not a new development but an improvement of concepts that date back to the late nineteenth century. In 1888 the German physicist Heinrich Hertz demonstrated that radio waves were reflected by objects in the same manner as light waves. In 1904 the German engineer Christian Hulsmeyer was granted a patent on a collision prevention device that used reflected radio waves. In 1917 Nikola Tesla predicted that radio waves would eventually be used to detect solid objects such as ships. In 1922 the Institute of Radio Engineers honored Guglielmo Marconi for proving that these concepts were possible.

In that same year, just three months after Marconi was honored, two research engineers at the Naval Research Laboratory in Anacostia, Maryland, provided proof that these theoretical concepts could be of practical use. The two researchers, A. Hoyt Taylor and his assistant, Leo C. Young, noticed that ships traveling on the Potomac River between an experimental radio transmitter and receiver both blocked and reflected the radio transmissions, as shown in Figure 8-1. Later that year, the two researchers recommended that the U.S. Navy continue this research and concentrate on developing a system to detect hostile naval vessels and attacking aircraft.

The Navy's Bureau of Engineering continued this research, attempting to perfect a system of determining the location of objects using blocked or reflected radio energy. Protecting naval convoys was one of the first demonstrated uses for this new system. Properly equipped escort vessels could blanket the perimeter of a convoy with high-powered radio transmissions that were directed from one ship to the next. Every escort ship was also equipped with a sensitive radio receiver, which was constantly monitored to detect any change or distortion of the transmitted signal. Any disturbance of the radiated energy signified that an unidentified vessel had just passed between two of the escorts.

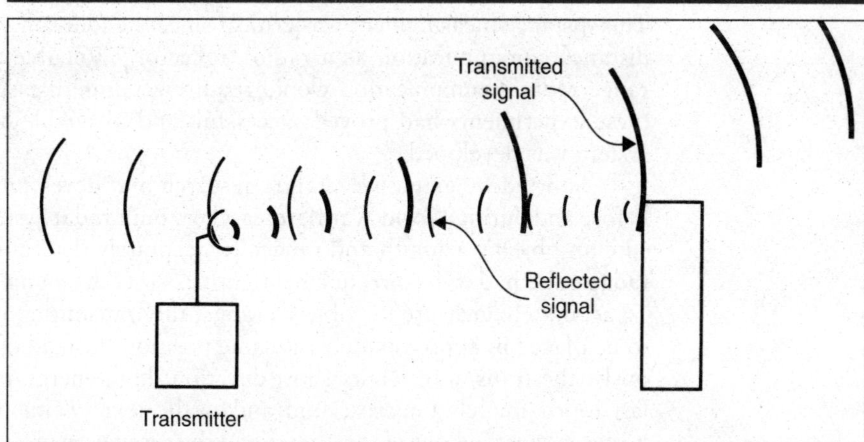

Figure 8-1. Radar operates by transmitting a high-powered radio pulse and "listening" for its reflection. Solid objects both reflect and block radar transmissions.

Another primitive radar system used radio reflections to locate unidentified vessels. The Navy used this system to determine the relative position of enemy vessels that were still a significant distance from the naval convoy. One or more of the escort vessels were equipped with a directional radio transmitter, whose antenna could be manually rotated 360° in azimuth, and a directional receiver, whose antenna rotated synchronously with the transmitter's antenna. The radar operator, equipped with a radar display similar to an oscilloscope, observed the radar indicator as both antennas were rotated. When the transmitter's signal was reflected by a solid object, such as another ship, an electronic pip appeared on the indicator. Then, the radar operator observed the relative position of the antennas to determine the unidentified vessel's bearing from the ship. If more than one escort vessel was able to locate the unidentified vessel using its radar, triangulation could determine the unidentified vessel's exact location.

Since the transmitters operated continuously, this type of radar was known as **continuous wave (CW) radar.** One significant disadvantage of CW radar is that only the reflecting object's bearing, or azimuth, can be determined. A different type of radar system is needed to determine the object's distance, or range.

Development of Pulse Radar

During the early 1920s, Gregory Briet and Merle A. Tuve of the Carnegie Institute of Washington perfected a primitive radar system that used short radar pulses instead of the continuous transmissions used by Taylor and Young. The Carnegie radar system was designed to transmit these short pulses of radio energy straight up into the atmosphere, where they would be reflected by the ionosphere. By measuring the elapsed time between the pulse transmission and reception, Briet and Tuve hoped to determine the actual height of the ionosphere. This distance measurement was critical since the ionosphere was used in long-distance communication as a radio "reflector." Reliable and practical long-range radio communication would require accurate measurements. By 1925, these experiments had proved successful, and a reliable height measurement system was developed.

Spurred by extensive military research and development conducted just before and during World War II, researchers built radar systems that could measure an object's azimuth and range. To accurately determine the azimuth, the radar pulse had to be directed in a tightly focused beam only 1° or 2° in width. To accurately measure the object's range, the transmitter and the receiver had to be placed in approximately the same position. Instead of operating continuously, the transmitter emits short-duration, high-energy pulses. These pulses last approximately 1 microsecond, and for the next 999 microseconds, the transmitter is switched out of the circuit and the receiver is placed into the circuit to listen for any echoes. This procedure is repeated about 1,000 times per second. To determine the range to the reflecting object, the time that elapses between the transmission of the radar pulse and reception of its echo (see Figure 8-2) is measured. This operating principle is much different from CW radar and is known as **pulse-type radar.**

In 1931 Taylor and his team of Navy researchers collaborated with the U.S. Army to develop a primitive radar system that could detect airborne aircraft up to 50 miles away. This system was probably the first use of radar in a

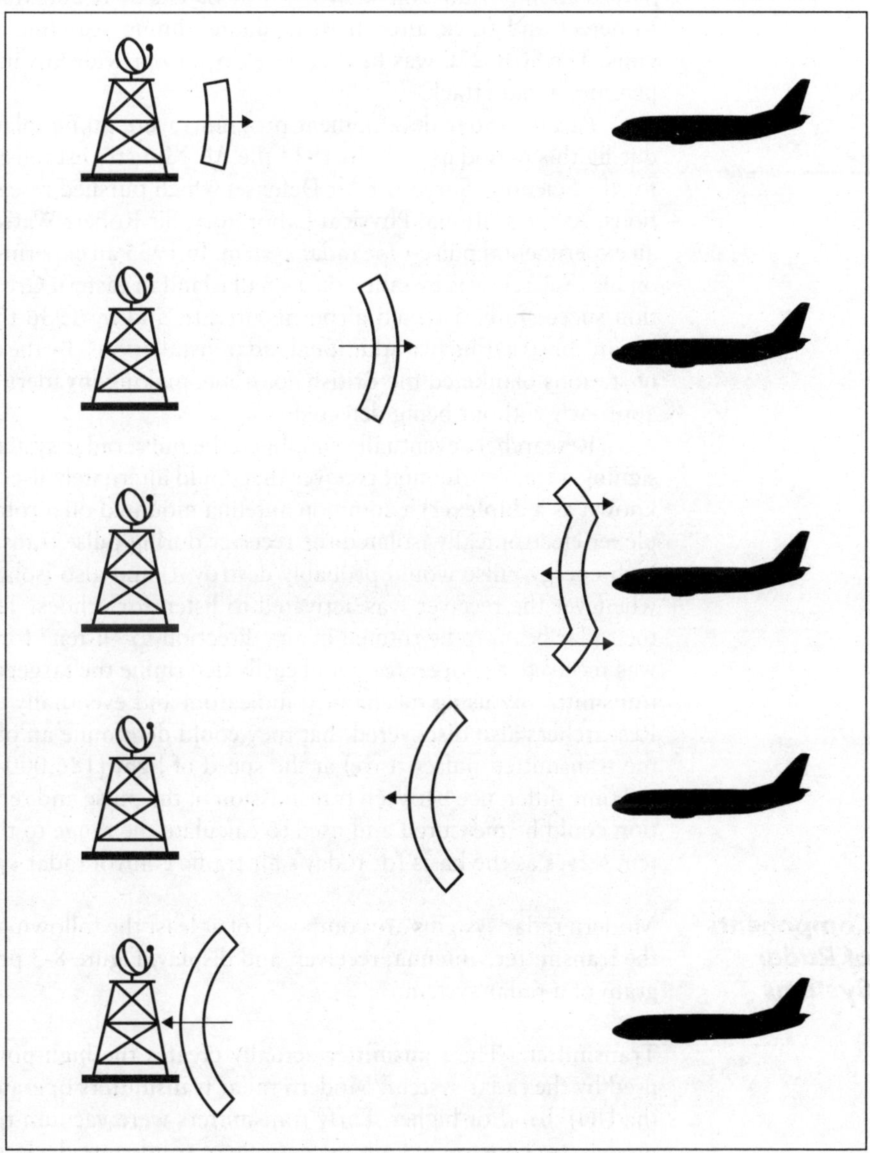

Figure 8-2. An example of a radar pulse being reflected off an aircraft.

functional setting. Following this development, progress in radar development escalated in the United States, Britain, and Germany. In 1936 the U.S. Navy experimentally used radar to control the guns of the battleship *New York*. In 1939 the first commercial contract for naval radar acquisition was let to the Radio Corporation of America (RCA).

In 1937 the U.S. Army Signal Corps experimentally used a low-powered radar to direct searchlights designed to illuminate airborne aircraft. Known as the SCR-270, this system became the core of two different radar systems that played an important role in World War II. The SCR-268 system was developed to detect and track aircraft to facilitate aiming searchlights and antiaircraft guns. The SCR-270 was further developed to provide advance notice of an impending aerial attack.

Parallel radar development programs were taking place in Great Britain during this period as well. In 1934 the Air Ministry established the Committee for the Scientific Survey of Air Defense, which pursued research in many directions. At the National Physical Laboratory, Sir Robert Watson-Watt developed an experimental pulse-type radar system. In 1935 an experimental system based on his research was installed on a small island in eastern Great Britain. This station successfully detected airborne aircraft, and by 1936 the Royal Air Force began constructing five additional radar installations. By the early 1940s a chain of stations blanketed the British coastline, making invaders virtually unable to approach without being detected.

Researchers eventually simplified the pulse radar system operation by designing a transmitter and receiver that could alternately use (by way of a device known as a **duplexer**) a common antenna mounted on a rotating base. The duplexer electronically isolated the receiver during pulse transmission (because a high-energy pulse would probably destroy it) and also isolated the transmitter whenever the receiver was activated to listen for echoes. This system enabled the radar beam to be rotated in any direction to "listen" for echoes. If an echo was received, the operator could easily determine the target's azimuth from the transmitter by using mechanical indicators and eventually electrical readouts. Researchers also discovered that they could determine an object's range. Since the transmitted pulses travel at the speed of light (186,000 miles per second), the time difference between transmission of the pulse and reception of its reflection could be measured and used to calculate the range to the object. This system serves as the basis for today's air traffic control radar systems.

Components of Radar Systems

Modern radar systems are composed of at least the following four components: the transmitter, antenna, receiver, and display. Figure 8-3 presents a block diagram of a radar system.

Transmitter The **transmitter** actually creates the high-powered radio pulses used by the radar system. Modern radar transmitters operate on frequencies in the UHF band or higher. Early transmitters were vacuum-tube designs; newer models are constructed almost entirely of solid-state devices. As noted earlier, the radar pulse lasts about 1 microsecond; the pulses are transmitted rapidly at

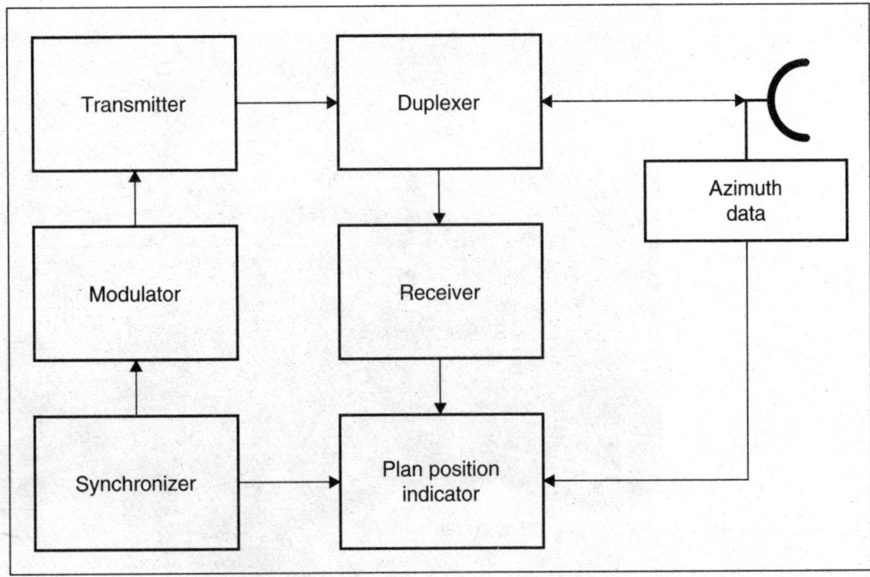

Figure 8-3. Block diagram of a radar system.

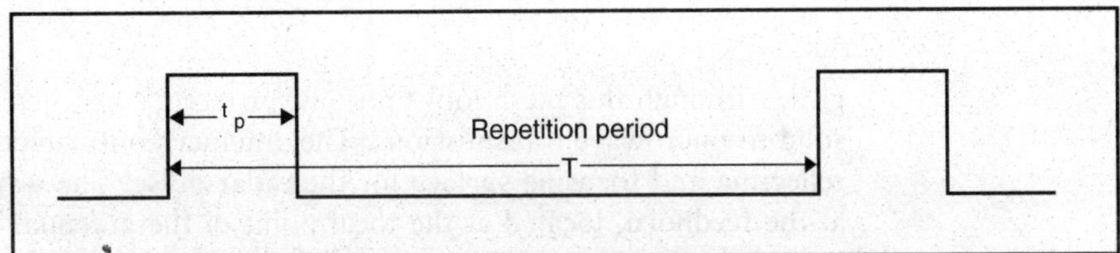

Figure 8-4. Waveform of the pulses transmitted by a radar system. Each pulse is of limited duration, with a substantial quiet period between pulses. The number of pulses transmitted per second determines the system's pulse repetition frequency (PRF). The repetition period is inversely proportional to the PRF.

a rate of about 1,000 pulses per second. This rate of transmission is known as the **pulse repetition rate** or **pulse repetition frequency (PRF)**. An example of the radar system's PRF is shown in Figure 8-4. Some radar units have a variable PRF known as **PRF stagger,** which will be discussed later in this chapter.

Antenna The **antenna** functions as both a transmitting and a receiving device. The radio pulses emitted by the transmitter are routed to the antenna using a **waveguide**—a hollow metal channel that conducts the microwave energy to the antenna. The antenna, which is parabolic in shape, is mounted on a rotor. Most current radar antennas are not solid; they are constructed of a metallic mesh or

Figure 8-5. An airport surveillance radar rotating antenna.

grid. Although this mesh looks porous, properly constructed antennas appear solid to microwave transmissions. The antenna's only function is to provide a reflecting and focusing surface for the radar pulse. The waveguide terminates at the **feedhorn,** located at the focal point of the antenna. The radar pulse is routed through the waveguide from the transmitter and emanates from the feedhorn. An antenna assembly is shown in Figure 8-5. After leaving the feedhorn, the pulse is reflected and focused by the antenna into a narrow, vertical beam approximately 2° wide and 40° high, known as the antenna **boresight** (see Figure 8-6).

If this radar pulse hits an object, it will reflect off the object and part of the reflection will return to the transmitter. Emitting the signal is known as **target illumination,** and the reflection is known as the **echo.**

Receiver Immediately after the transmitter shuts off, the **receiver** is switched into the circuit to listen for any echoes. If the radar pulse is reflected by an object, a small portion of the emitted radio energy will return to the antenna and will be focused back into the feedhorn. This pulse then returns to the receiver through the waveguide. Since radio energy diminishes proportionately to the distance it must travel, by the time the echo has returned it will have lost a considerable amount of power. This pulse must then be amplified, sometimes at least one million fold, so that it can be properly processed and displayed by the receiver.

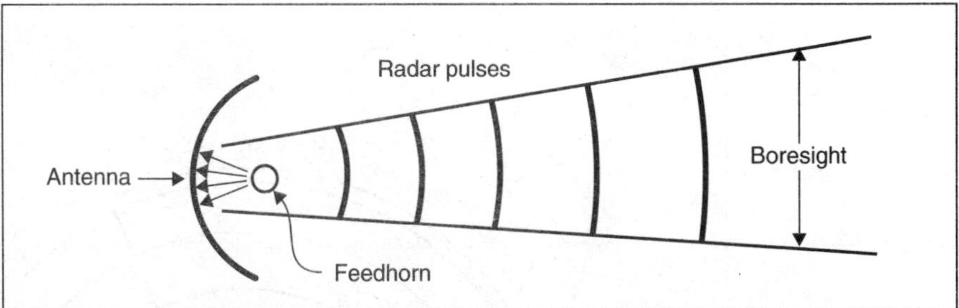

Figure 8-6. Boresight.

Theoretically, the signal can be infinitely amplified, but in actual practice this is not the case. Any electronic amplifier introduces random signals, known as electronic **noise,** into its circuitry during operation. Noise is almost impossible to remove once it is introduced, but it can be managed by incorporating circuitry into the receiver that deletes any signal whose strength falls below a predetermined threshold level. Radar engineers can predict the amount of noise that will be introduced and then set the threshold limit to a value just above the predicted noise level. Unfortunately, this technique also causes the receiver to delete low-energy echoes reflected by distant targets.

Upon receipt of this radar reflection, the radar system measures the time difference between transmission and reception and uses this calculation to determine the object's distance from the antenna. Since radio signals traveling at the speed of light take 6.18 microseconds to travel 1 nautical mile, these pulses will take 12.36 microseconds—6.18 times 2—to travel 1 mile, reflect off an object, and return. This time—12.36 microseconds—is known as a **radar mile.** A radar mile is *not* a distance measurement.

Indicator Once the radar system has received and processed the reflected signal, the object's relative position can be displayed to the controller in many ways. In air traffic control, targets are most often displayed on a cathode ray tube known as a **plan position indicator (PPI), radar scope,** or just scope. The PPI is a circular television-type tube that is about 36 inches in diameter and covered with two types of luminous phosphor. One type emits a low-persistence–high-intensity blue flash; the other type emits a high-persistence–low-intensity orange light. This arrangement causes a fairly bright but transient flash to blossom on the radar screen at the target's location. The flash begins to diminish immediately, and the high-persistence phosphor takes over. This type of phosphor does not cause a very bright light to be emitted, but the light does persist for a long time—in some cases, up to 5 minutes. This phosphor enables the controller to visualize where the aircraft has been, because the flashes still glow faintly on the PPI. Because the flashes are different colors, most PPIs are equipped with an orange filter that is placed in front of the tube to equalize the intensity and the color of the emitted light.

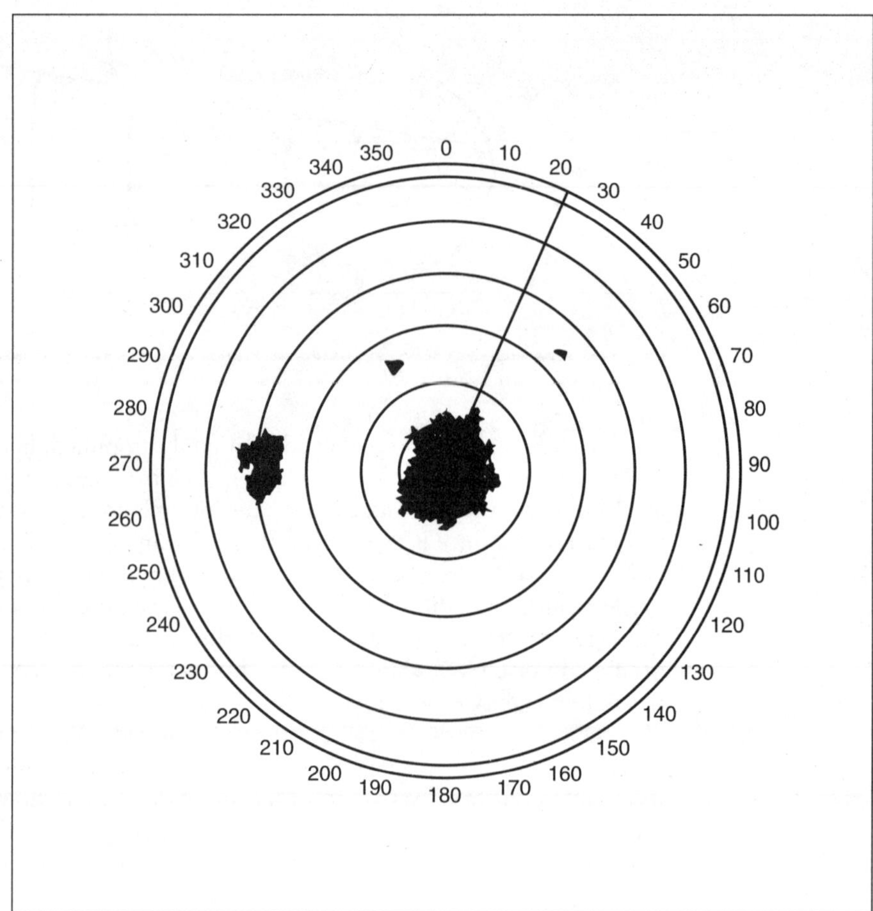

Figure 8-7. Pictorial representation of a plan position indicator.

The center of the PPI, known as the **main bang,** corresponds to the physical location of the radar antenna. The top of the PPI is the area north of the radar antenna; the right side is east, the bottom is south, and the left side is west. As the radar operates, a faint line, known as the **sweep,** emanates from the main bang to the edge of the radar screen (see Figure 8-7). The sweep line corresponds with the antenna's boresight. As the antenna slowly rotates, the sweep is synchronized with the radar antenna and rotates in the same direction and at the same speed. Most radar antennas rotate at a speed of 5 to 15 revolutions per minute.

Since radio waves being reflected from objects farther away from the antenna take longer to return, long-range radars must revolve at slower speeds and at reduced pulse repetition frequencies. If the radars revolve too quickly, by the time the radio signal is returned from a distant object the antenna will have rotated a sufficient distance to prevent reception. When attempting to detect distant targets, sufficient time must be given for echoes to return to the antenna before the next pulse is generated.

Distance from the radar antenna is determined using precise time measurement between the transmission and the reception of any reflection. The reflected object's distance is depicted concentrically as range from the main bang. Its azimuth is determined electronically from the antenna's exact bearing when the echo is received. An object illuminated by the radar signal will reflect radar energy; this radar pulse will cause a small dot—known as an echo, **target,** or **blip**—to appear along the sweep at the object's range and azimuth (see Figure 8-8). The target's exact location can be determined by noting its distance from the main bang (range) and its azimuth from the center of the PPI.

The target's size and intensity varies in relation to its distance from the antenna, the relative conductivity of the atmosphere, and the **radar cross section** of the object. Since radar pulses weaken as they travel, distant objects will reflect

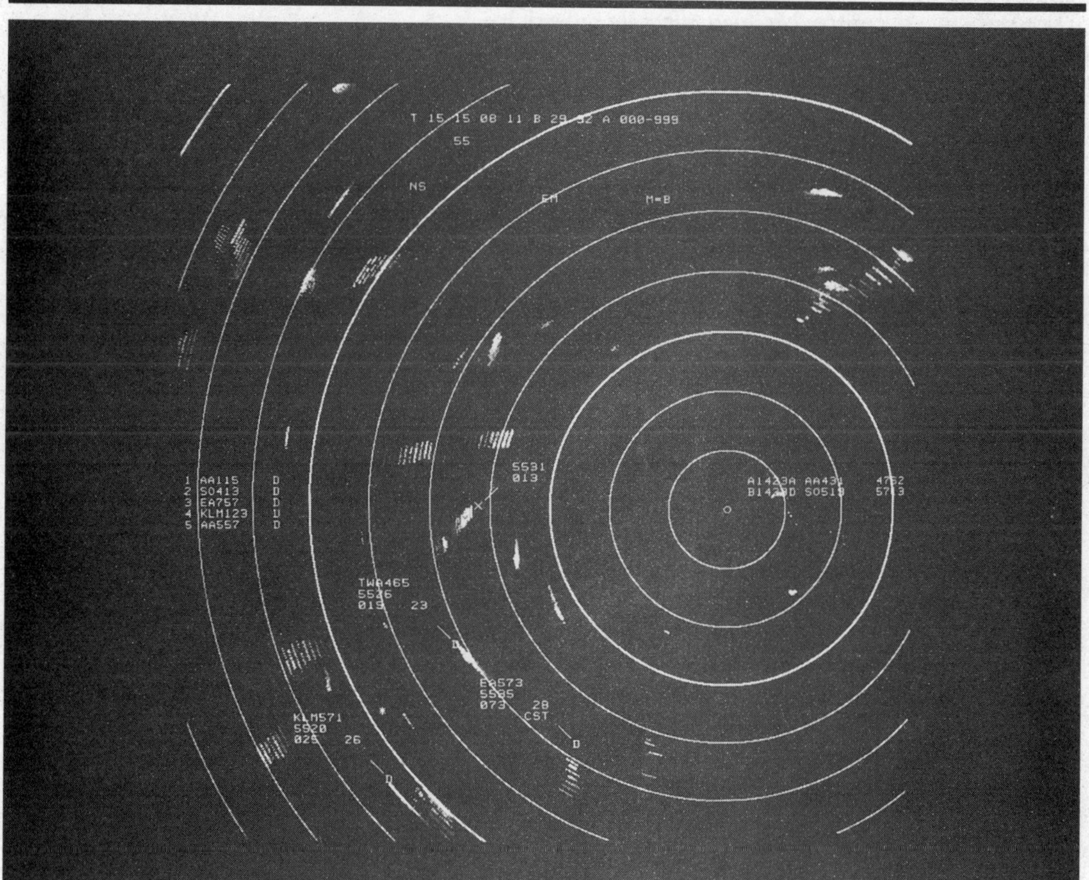

Figure 8-8. Radar display showing range marks, targets, and data blocks.

less energy than nearby objects. The atmosphere's relative conductivity may also interfere with transmission. Airborne obstructions—such as dust, moisture, and precipitation—will cause much of the radar pulse to be blocked or reflected before it can return to the radar antenna. Precipitation, in particular, can easily block or diffuse most of the radar's radiated energy.

The target's radar cross section also helps determine the relative size and intensity of the displayed echo. Radar cross section is a technical measurement of the relative radar reflectivity of an object. Normally, larger objects have greater radar reflectivity, which causes more energy to be reflected back to the radar antenna. Thus, the displayed echo will be brighter and more distinct. But various factors, such as the aircraft's configuration, the type of material used in its construction, and the relative shape of its surface, can significantly affect its reflectivity, thereby changing its radar cross section. Large metal aircraft usually have greater radar cross sections than fabric-covered, wooden, or composite-structure aircraft. Propeller-driven aircraft usually appear larger to a radar transmitter because the rotating propellers appear as large flat disks. Whether the aircraft is heading toward, away from, or tangentially to the radar transmitter can also affect its reflectivity.

The object's distance from the antenna also affects its relative size when displayed on the PPI. Since the radar's boresight is usually only 1° to 2° wide, targets close to the antenna will be illuminated only by a fairly narrow beam of energy. Targets located near the extreme range of the radar system, however, will be illuminated by a fairly wide beam. Since the radar receiver has no means of determining the actual width of the aircraft, its displayed width will be the same as the width of the radar pulse at the object's range.

Ground Clutter

Radar systems used for air traffic control are unable to distinguish between different types of reflecting objects. Objects that are undesirable to display on the PPI also reflect radar energy, known as **ground clutter.** These objects include buildings, terrain, radio and television transmitter antennas, electrical transmission towers, temperature inversions, and precipitation. Cars, buses, boats, and even flocks of birds can reflect radar transmissions. In fact, almost any object, solid or liquid, is capable of reflecting radar energy. Although some of these reflections are useful to the controller, most serve only to clutter up the display and must be filtered out by the receiver. The means of filtering out this clutter will be discussed later in the chapter.

The PPI used to display the radar echoes is designed with a high-persistence screen, which allows the blip to be displayed for a number of antenna revolutions before completely disappearing. The most recent echo will be the brightest, with subsequent echoes being of lower intensity. These lower-intensity echoes are known as **history.** The PPI's persistence permits the controller to determine the object's relative direction of flight and its velocity. Faster aircraft will move farther between each illumination and the PPI will display a greater distance between successive echoes, whereas slower aircraft will leave histories that contain closely spaced echoes.

Table 8-1. Radar Frequency Utilization

Letter Code	Applications	Nominal Wavelength (cm)	Wavelength (cm)	Frequency (mHz)
P	Ground-based early warning Searchlight aiming	100	77–133	225–390
L	DME equipment Transponders TACAN Air route surveillance radar	30	19.35–77.0	390–1,550
S	Airborne search radar Airport surveillance radar	10	5.77–19.35	1,550–5,200
X	Storm detection Precision approach radar Airborne navigation Airborne fire control	3	2.75–5.77	5,200–10,900
K	Cloud detection Airborne navigation Airport surface detection equipment	1	0.834–2.75	10,900–36,000

Transmitter Frequency

As previously mentioned, radar signals are reflected by many unwanted objects. One method of reducing undesirable reflections is to use a transmitting frequency that tends to be reflected by objects such as aircraft but is absorbed by most other objects. Extensive research in this area was conducted during World War II. Researchers classified potential radar transmitting frequencies into five categories. Each band was found to reflect off some objects and to be absorbed by others. The bands were each assigned an identifying letter, selected at random to keep the information secret. Although secrecy is no longer necessary, the identifying letters are still used to distinguish each band (see Table 8-1).

Receiver Controls

Once the receiver processes the reflected signal, it must be adjusted by the controller for proper use. The controller has numerous operating controls that can modify the signal received by the radar system. Some of the controls include receiver gain, moving target indicator, and sensitivity time control (see Figure 8-9).

Receiver Gain

Although the radar system is transmitting a signal with a strength of between 500 and 5,000 kilowatts, the signal reflected by the target may possess only .01 percent of this power. The signal must be amplified for the radar system to

UNISYS Corporation

Figure 8-9. Example of a PPI control panel.

properly process and display it. The radar receiver has a signal amplifier that increases the level of the radar echo. Unfortunately, the amplifier cannot distinguish between wanted and unwanted echoes, and amplifies them all. During amplification, various transient electronic pulses, known as noise, are introduced and amplified as well. Once noise enters the receiver circuitry, it cannot be totally eliminated and might be displayed on the PPI as random targets. To eliminate these spurious signals from the PPI, FAA air traffic control radar receivers are equipped with a fixed **threshold** level that determines which echoes should be displayed. Any echo whose amplified signal strength is below the threshold value will not be plotted, whereas an echo stronger than the threshold value will be displayed. In general, targets with greater radar cross sections will reflect sufficient energy and will be displayed. Although this device tends to eliminate most unwanted signals, it can also eliminate very small aircraft operating at the fringe of the radar system's effective range.

Another method of eliminating unwanted targets is to modify the receiver's level of amplification, or **gain**, thereby effectively selecting which echoes are displayed. If too many nonaircraft targets are being displayed, the controller can decrease the receiver's amplification, thereby reducing the number of targets whose amplified signal exceeds the threshold level. This technique normally eliminates most of the noise, but it can also eliminate some of the small distant targets that should be displayed. If the controller selects a small amount of **receiver gain**, only targets with relatively large cross sections will return sufficient energy to be displayed. In most cases, the controller increases the receiver gain to a value that will display the desired targets but not the noise.

Moving Target Indicator

One of the biggest complications with the use of radar for air traffic control is ground clutter, which occurs whenever the transmitted radar signal is reflected by nearby stationary objects. The PPI displays these reflections whenever their amplified signal strength is above the preset threshold level. Ground clutter tends to completely obscure legitimate targets within about 20 miles of the antenna. In some cases, ground clutter may be so severe that it is impossible to observe any echoes from aircraft close to the radar site. If radar is to be used for air traffic control, ground clutter *must* be suppressed as much as possible without eliminating the echoes from legitimate targets.

One method of reducing ground clutter is to elevate the plane of the antenna a few degrees above the horizon, thus directing the radar pulse above most of the objects creating the ground clutter. But raising the antenna may also eliminate the display of some low-flying aircraft at the extreme limit of the radar's range. Since the frequencies used by radar systems are line of sight, the Earth's natural curvature reduces the radar's ability to detect distant aircraft flying at relatively low altitudes. This factor may not be as critical with short-range radar, but any elevation of long-range radar will severely hamper its ability to detect distant aircraft. At the extreme limit of the radar system's range (approximately 250 nautical miles), the radar may not be able to detect aircraft flying below 5,000 to 10,000 feet. In general, as the angle of the antenna is increased, more ground clutter is eliminated, but the antenna's ability to detect distant aircraft is reduced proportionally. The proper angle of the antenna is usually a compromise between these two extremes.

Another method of reducing ground clutter is through the use of an electronic filtering circuit known as the **moving target indicator (MTI)**, which uses phase-change filtering techniques to eliminate any objects that are not actually moving. This technique assumes that the only objects that an air traffic controller might want to have displayed on the PPI are moving targets. The circuitry measures the change in the object's position between each successive radar pulse by comparing the returned signal's phase with that transmitted by the radar system. A moving object that has **radial velocity** (that is, the object is moving either toward or away from the radar antenna) will cause the phase of the transmitted radar signal to shift as it is reflected. Any object without radial velocity (not moving toward or away from the antenna) will not cause a phase shift to occur. The MTI circuitry in the receiver compares the phase of each returning echo to determine whether the reflecting object has any radial velocity. The MTI system concludes that if an object has no radial velocity, it must not be an aircraft and is not displayed on the PPI. Any object with radial velocity is displayed on the PPI. Through this technique, MTI circuitry can be used to eliminate most of the ground clutter from the radar screen. The MTI cannot, however, eliminate unwanted moving targets such as trains or automobiles. Fortunately, most of these objects are below the plane of the radar's transmitted signal and will not cause any serious problems. Figure 8-10 shows a radar

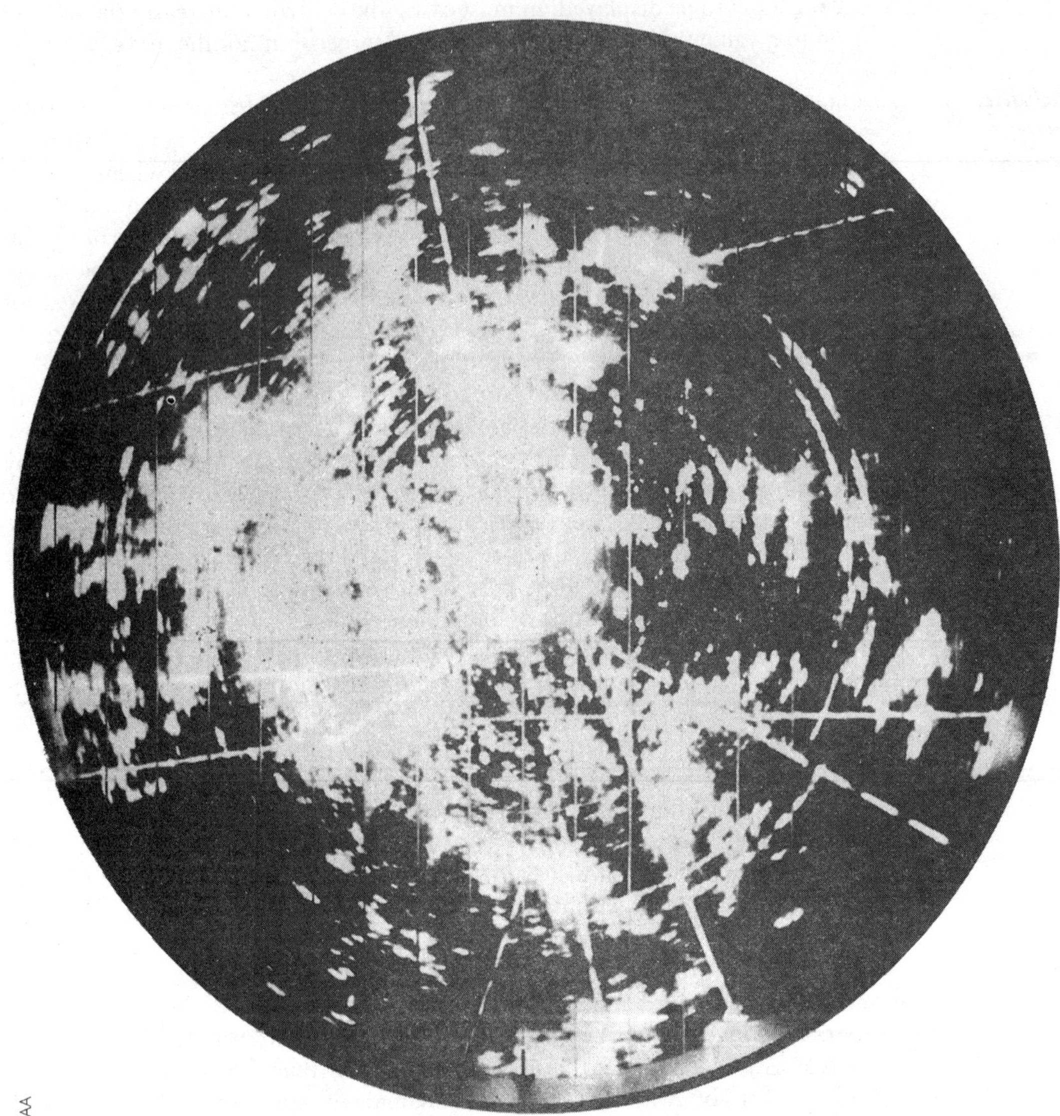

FAA

Figure 8-10. Radar display with the MTI turned off. The resultant ground clutter dominates the display.

display with the MTI turned off; notice the dominance of the ground clutter. Now look at Figure 8-11, a display with the MTI turned on.

Unfortunately, it is a physical principle that some of the reflected energy from a moving target will remain in phase, and the MTI circuitry will eliminate this portion of the echo. Therefore, any object whose echo is being processed

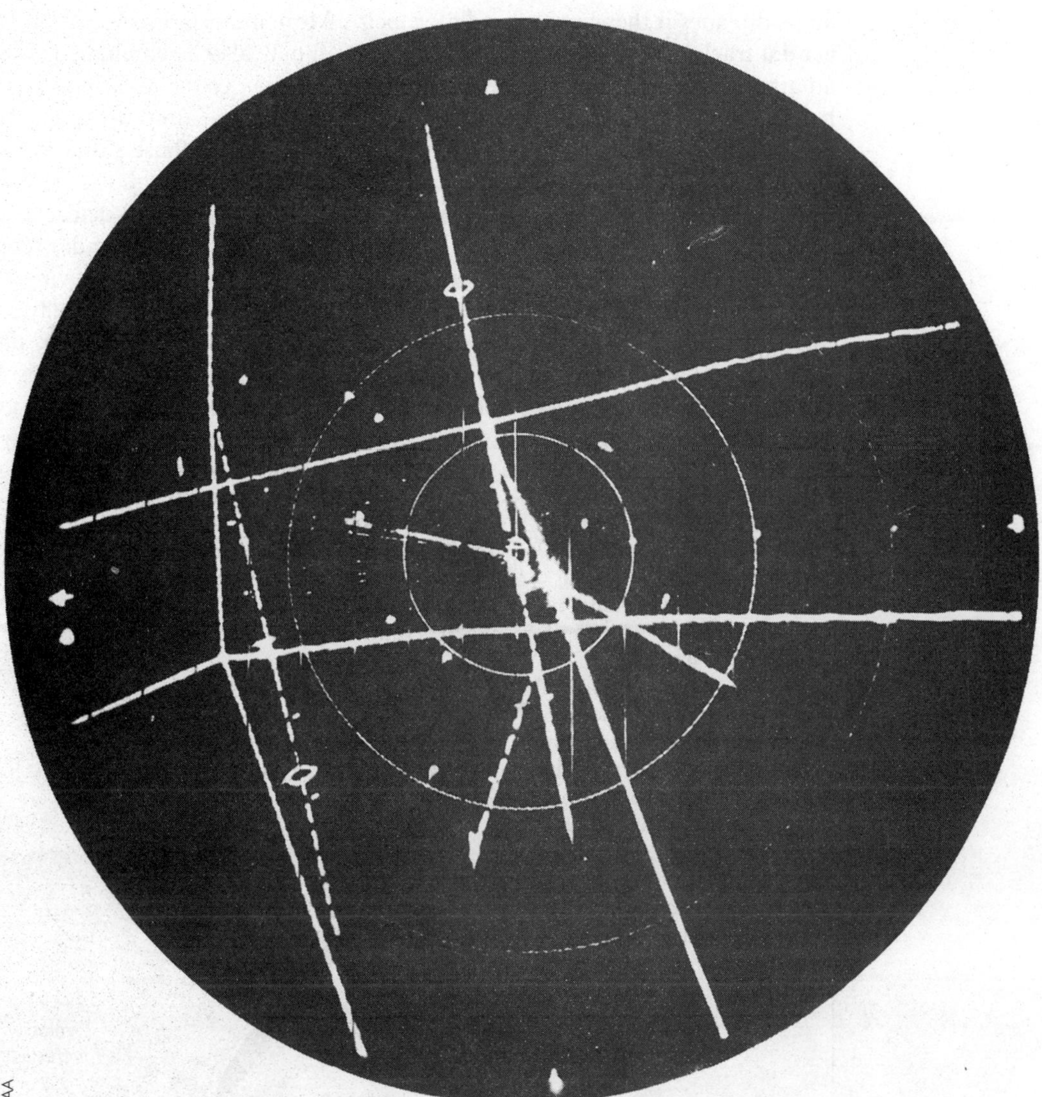

FAA

Figure 8-11. Radar display with the MTI turned on. Ground clutter is virtually eliminated. Only moving targets remain on the display.

through MTI circuitry will appear somewhat dimmer on a PPI. In cases where the reflected energy returning from the object is just barely above the threshold level, the MTI circuitry may inadvertently eliminate it from the display.

The MTI may also remove some desirable targets from the radar display. The MTI circuitry may eliminate hovering helicopters, slow-moving balloons, and sailplanes if they have no radial velocity. More important, aircraft on **tangential tracks** to the antenna's plane of rotation may also be eliminated. When an aircraft's ground track is tangential to the rotation of the radar antenna, it has no effective radial velocity for an instant. Even if the aircraft has forward velocity, it may have no radial velocity, there will be no phase shift, and the MTI circuitry will not display the aircraft on the PPI (see Figure 8-12). When the aircraft's track becomes nontangential, the MTI system will detect its increasing radial velocity and will no longer filter the target from the radar scope.

The use of MTI circuitry will normally remove most slow-moving airborne moisture such as clouds, light rain, and snow from the radar display. Unfortunately, the MTI is unable to remove all of this precipitation return, since some precipitation within a storm has radial velocity. In many cases, it is desirable to observe precipitation returns. Pilots often rely on controllers to give them advice concerning the intensity and the movement of weather displayed on the PPI. If the precipitation begins to obscure aircraft targets, it must be removed from the display. If the MTI circuitry cannot accomplish this, the controller must turn to circular polarization, which will be explained shortly.

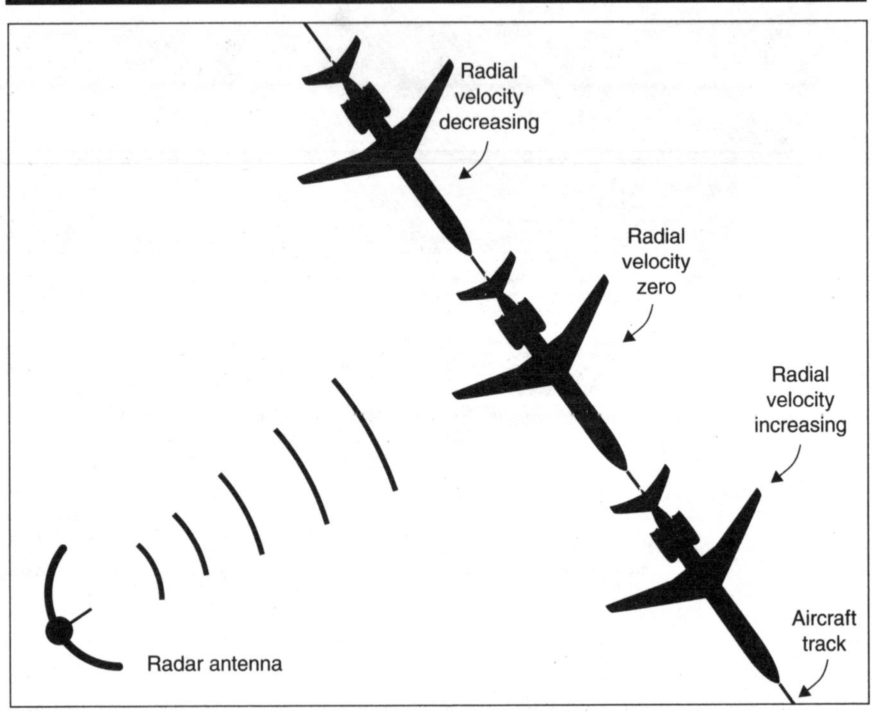

Figure 8-12. As an aircraft flies tangentially to the antenna, its radial velocity begins to decrease. When exactly tangent to the antenna sweep, the radial velocity will become zero and the MTI will remove the aircraft from the display.

The controller can adjust the MTI range on most radars to reduce these unwanted effects. In some cases, the controller may bypass the MTI circuitry when trying to divert an aircraft around weather that is being filtered out or when trying to identify a small aircraft at the extreme range of the radar system. Even though the amount of displayed ground clutter will increase, it may be operationally advantageous. The MTI circuitry can be bypassed using a control known as the MTI gate.

MTI Gate Since the problem of ground clutter is most prevalent at ranges close to the radar antenna, most ATC radar systems are designed with a variable MTI range-setting device known as an **MTI gate.** The MTI gate is a variable control that selects the range limit where radar echoes will be processed through the MTI circuitry. If, for instance, the MTI gate is set at 20 nautical miles, every echo from objects less than 20 miles from the radar antenna will be processed through the MTI circuitry, whereas reflections from objects farther away than 20 nautical miles will not. This procedure reduces ground clutter and the echo intensity of nearby aircraft but retains the full strength of echoes reflected from distant targets. In normal use, the MTI gate is set to as low a value as possible while still attempting to eliminate as much ground clutter as possible.

Blind Speed Since MTI circuitry routinely filters out targets with little or no radial velocity, one could logically conclude that as a target's radial velocity increases, the effect of MTI circuitry on the target intensity decreases. This conclusion is true, up to a point, after which a further increase in radial velocity tends to actually *increase* the effect of MTI. At some point, even though the aircraft still has radial velocity, the MTI may be unable to detect it and may conclude that the target is stationary; this speed is known as the MTI **blind speed.** In technical terms, blind speeds occur whenever a moving target's radial velocity travels exactly one-half, or any multiple thereof, of the wavelength of a radar transmission between pulses. Typical blind speeds for a radar system with a constant pulse repetition frequency might be 250 knots, 500 knots, 750 knots, and so on. At these velocities and PRFs, an aircraft with a radial velocity equal to one of these speeds would travel exactly one-half wavelength between successive radar pulses. A radar system's blind speed can be easily calculated. If the PRF is multiplied by 291 and then divided by the radar transmitting frequency (in mHz), the resultant value will equal the blind speed in knots. For example, a radar system operating at 3,000 mHz with a PRF of 1,200 will have a blind speed of 116 knots:

$$1{,}200 \times 291 \div 3{,}000 \text{ mHz} = 116 \text{ knots}$$

The easiest method for eliminating MTI blind speeds is to vary the rate of transmission of the radar pulses, known as PRF stagger. Radar transmitters equipped with staggered PRFs sequentially use two or more different pulse repetition frequencies. If, for instance, a typical system used a PRF of 1,200 during the first pulse transmission, the blind speed would be 116 knots. But if the PRF were changed to 800 during the next transmission, the blind speed would then be 78 knots. Since it is highly unlikely that any aircraft would be able to instantaneously decelerate from 116 knots to 78 knots in less than one-thousandth of a second, the use of PRF stagger by the FAA has virtually eliminated the problem of blind speeds. Although harmonic blind speeds may still cause the MTI to remove aircraft from the PPI, these speeds are usually in excess of 1,000 knots.

Moving Target Detection

The **moving target detection** (**MTD**) system was developed to address many of the shortcomings of the MTI system. In particular, MTD attempts to mitigate blind speed, tangential course and precipitation display problems inherent in MTI-based radar systems.

MTD is a digital signal processor, which means that the radar returns are converted to digital data, analyzed by various filters, assigned numeric values and then stored in computer memory. During routine operation of a digital radar equipped with MTD, (such as the ASR-9 or ASR-11), reflected radar energy is processed by the MTD system and then stored into computer memory known as memory range cells. The data in the memory range cells is stored in a manner very similar to the random-access memory (RAM) utilized by personal computers.

The primary radar returns stored in the memory range cells are first analyzed by a variety of Doppler filters. These filters operate at different frequencies, with each attempting to determine whether any object located within a particular range cell is moving or not. If no radar energy is returned from a particular azimuth/range, then the number zero is entered into the appropriate range cell. But if any radar return is detected (whether moving or not), the Doppler filters try to resolve whether any or all of the return is coming from a moving object, then assigns a numeric value to that target, which is stored in the appropriate memory range cell. After subsequent processing, the digital value is converted to a visual target, which is plotted on a radar display similar to high-definition video monitors used in the commercial computer industry.

Clutter Map The FAA conducts tests during the initial setup of the radar system to determine the locations of permanent, stationary objects (such as buildings, terrain, and man-made objects) within the radar range. Stationary objects are further identified during normal operation. Using Doppler filters, the MTD system recognizes when a stationary radar return is consistently returned from a fixed location. After a set period of time, if the object never shows any motion, the MTD system determines that the object is stationary and stores a value that relates to the amount of radar energy reflected by the stationary object in the appropriate range cell for that location. When all stationary objects have been identified within the coverage area of the particular radar system, this information is saved in a digital file known as the **clutter map.** A clutter map is a stored series of values that define the nonmoving objects routinely detected by the radar system.

During normal operation of the radar, when the radar return for any particular range cell is analyzed, if a Doppler shift is detected, the radar assumes that a moving target has been detected. Even if some of the radar energy is being reflected from a stationary target located at the same azimuth or distance, the MTD system can utilize the stored values in the clutter map to determine how much radar energy is being reflected from the moving target and how much from the stationary target, and the MTD will display a valid radar return associated with the aircraft.

Geo-Map The clutter map is very effective at removing stationary objects from the radar display but is not able to remove moving non-aircraft radar returns, such as those reflected off of vehicles traveling on the airport's surface or near the airport. These returns are mitigated through the use of the **Geo-Map** function of the MTD system.

The Geo-Map is similar to a clutter map in that it is programmed into the MTD system during installation. The Geo-Map system digitally defines the areas near the airport where vehicular traffic might be operating. During normal radar operation, when a target is identified and assigned a radar return value, the Geo-Map subsystem processes the value to determine whether the radar return frequency pattern resembles that belonging to a vehicle. If the radar return resembles that of a vehicle, and it comes from an area where vehicular traffic is common, the Geo-Map subsystem will inhibit its display on the radar system.

Merge/Tracking

The merge/tracking subsystem correlates the radar returns from the primary radar system to those received by from the secondary (transponder-based) radar system and attempts to predict the future location of each aircraft. The associating of primary and secondary radar returns from a single aircraft is referred to as collimation but is more often called the merge function. Collimation is the process of determining which primary radar target belongs with which secondary radar target, merging the appropriate position information, then displaying the combined target on the radar display. This enables the radar system to display aircraft identification and other appropriate information concerning that aircraft as well as to perform some computer calculations.

Once the merge system has collimated each target, its position is stored in the computer system for future computations. By storing the previous location of each aircraft, then calculating the distance each target travels between successive radar scans, the radar system can estimate the ground speed of the aircraft as well as its future location (assuming the aircraft does not change course or speed). This function, known as tracking, makes it possible for the computer to project aircraft flight paths and altitudes and warn the controller if aircraft are predicted to come too close to one another, or if one aircraft is flying at a dangerous altitude and is projected to approach an obstruction. These two tracking functions are known as conflict alert and minimum safe altitude warning. Future tracking enhancements to the computer system will enable the controller to use the computer to monitor and determine optimal flight profiles, arrange appropriate approach and departure sequencing and provide potential conflict evaluations on a real-time basis well ahead of the aircraft's current position.

Sensitivity Time Control

Since objects near the radar antenna tend to reflect more energy back to the antenna than objects farther away, the echoes on the PPI associated with these close targets will be much brighter than the others. Besides being distracting to the controller, these brighter targets persist for an excessive period, thereby cluttering the radar screen. To prevent this clutter, some means of equalizing the intensity of all of these targets must be provided.

Sensitivity time control (STC) circuitry is an electronic means of automatically controlling the receiver's sensitivity to equalize the display intensity of both nearby and distant targets. The circuitry reduces the receiver's sensitivity during the initial segment of the listening cycle, when strong echoes from nearby targets are received. After a few microseconds, the receiver sensitivity is gradually increased to normal to compensate for the reduced signal strength of echoes returning from distant targets. In most cases, the STC circuitry returns the receiver sensitivity to normal at approximately the same time that echoes reflected from objects located 20 miles away from the antenna are received, about 247 microseconds into the listening cycle. Therefore, any signal reflected from an object farther than 20 miles from the radar antenna is not attenuated by the STC circuitry.

Transmitter Controls

Although MTI and MTD can eliminate most nonmoving targets from the PPI, they are unable to eliminate most heavy precipitation. Raindrops, hail, and even snowflakes are excellent radar reflectors. Since most heavy precipitation has some inherent velocity, the moving target circuitry is unable to remove it from the PPI. During periods of widespread, heavy precipitation, the resultant clutter may completely mask actual aircraft targets. In an effort to remove as much of the echoes as possible, the FAA has equipped most radar transmitting systems with a means of switching from **linear polarization** (LP) to **circular polarization** (CP).

Polarization refers to the general orientation of the radar waves as they emanate from the radar antenna. The radar system normally operates with every radio wave polarized linearly, which means that the waves are parallel to each other. In this mode, the receiver makes no attempt to determine the polarization of the reflected signal. But during periods of heavy precipitation, the radar transmission can be changed so that the transmitted signal is polarized in two directions, one wave being polarized perpendicularly to the other. One wave is known as the vertical wave, and the other is the horizontal wave. When these waves are added electronically, the combination produces a vector that is a perfect circle, hence the name *circular polarization.*

The general principle of CP is that symmetrically shaped objects, such as raindrops, will reflect equal portions of the radar pulse's horizontal and vertical components, whereas asymmetrical objects, such as aircraft, will reflect uneven proportions of these two waves. When operating in the CP mode, the receiver measures the relative amount of vertical and horizontal polarization contained in each echo, subtracts one from the other, and uses the remaining signal to provide the echo on the PPI. The equal horizontal and vertical components of reflections from symmetrical objects will cancel each other out and no reflected energy will remain to be displayed. However, the signals from asymmetrical objects will always contain some energy after this electronic subtraction has occurred, and the echo will then be displayed on the PPI, albeit with less intensity.

When properly used, CP removes most of the unwanted precipitation returns, but it also reduces the display intensity of legitimate targets. Thus, CP should be used only when heavy-precipitation echoes threaten to overwhelm the rest of the targets on the radar display.

Display Controls

A typical radar system may have anywhere from 1 to 20 different PPIs operating off a single transmitter, antenna, and receiver. Each PPI display is equipped with controls that permit the controller to select or modify the display. These controls include range select, range mark interval and intensity, receiver gain, video map, and sweep decenter (see Figure 8-9).

Range Select

The **range selects switch** is used to select the range limits that should be displayed on the PPI. This control does *not* affect the operating range of the radar system—only the range to be displayed on that particular PPI. The selected range is measured in nautical miles from the center of the PPI to its edge (the radius). Airport surveillance radar systems used in approach and departure control facilities usually have a maximum range of about 100 nautical miles, with range select settings of 5, 10, 20, 40, 60, and 100 nautical miles. Air route surveillance radars used by ARTCC controllers usually have a maximum operating range of about 250 nautical miles, with range select settings of 50, 100, 150, 200, and 250 nautical miles.

Range Mark Interval and Intensity

Air traffic control radar displays also have various **range mark** intervals that can be superimposed on the PPI as concentric circles centered on the main bang. The controller can select both the intensity and the spacing of these range marks. Intensity is usually continuously variable from nonexistent to very bright, whereas spacing can be set to either 2, 5, 10, or 20 nautical-mile intervals.

Receiver Gain

Each PPI is also equipped with controls that vary the amplification of the radar reflections received by the radar system. The **normal video gain** control varies the intensity of the echoes being processed by the radar receiver. The **MTI/MTD video gain** control varies the intensity of the radar returns processed through the MTI/MTD circuitry. Through these two controls, maximum target intensity can be achieved while reducing the ground clutter as much as possible.

In most cases, MTI/MTD video gain is set fairly high, which amplifies the intensity of the moving targets displayed on the PPI. At the same time, the normal video gain is set fairly low, which reduces ground clutter while still effectively displaying slow-moving targets. If a distant target is difficult to detect, however, the normal video gain may be increased to assist in distinguishing the echo. Although this will also increase the displayed ground clutter, being able to distinguish the weak target outweighs increased ground clutter.

Video Map

Every PPI is also equipped with a **video map selector** and a **video map intensity** control. To properly use radar to separate and assist aircraft, the controller must know the position of each aircraft in relation to airports, navigation fixes, and airways. Radar maps are generated through an electromechanical system that superimposes the map directly on the PPI.

The video mapping unit uses a focused light source, a translucent map, and a photoelectric light cell to generate this map. Professionals from the National Oceanic and Atmospheric Administration draw each radar map precisely to scale. The maps are then reduced and etched on a small translucent disc, which is mounted in the video map generation unit and rotated in synchronization with the radar antenna. As the disc rotates, the light source is directed through the disc onto the photoelectric cell. Electronic pulses produced by this system are then transmitted to the PPI and reproduce the map directly on the luminescent screen.

As many as eight different maps may be available to each controller. The maps include symbology to indicate particular navaid positions, obstructions, airways, and intersection locations. Airport locations and the extended runway centerlines used for instrument approaches are also included on most **video maps** (see Figures 8-13 and 8-14). Sector boundaries and other important information defined in facility directives or letters of agreement are also contained in the maps. Every PPI display within a facility will usually have access to the video maps available at that facility. If the airspace is relatively simple, the facility may need only one video map that can be used at every radar position.

Most ATC facilities are of sufficient complexity that a specific number of different maps must be made available to each controller. Each of these maps is designed to be used by a different type of controller or for a particular airspace configuration. For example, an approach controller may use map 1 when arrivals are landing on a particular runway, but he or she may use map 2 when approaches are being conducted to another runway. Map 3 may be designed for the departure controller's use, whereas map 4 may contain specific information most useful to the local controller. Map 5 may include every known

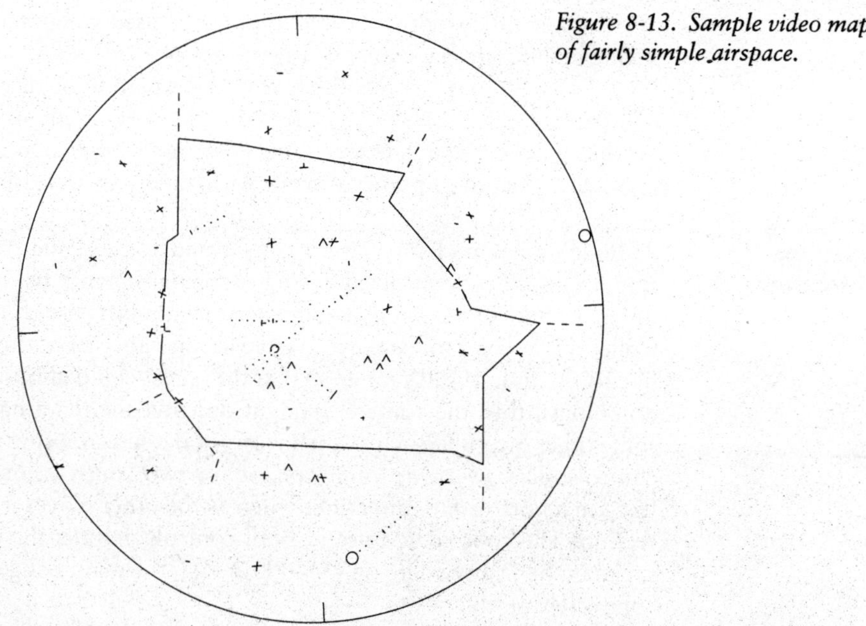

Figure 8-13. Sample video map of fairly simple airspace.

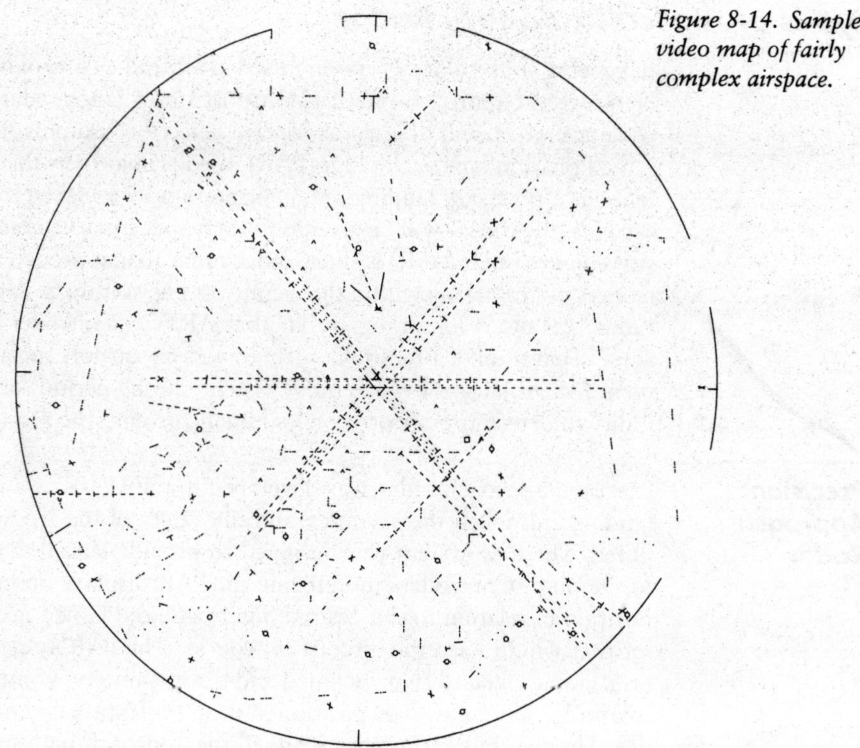

Figure 8-14. Sample video map of fairly complex airspace.

obstruction and airport location and may be used whenever a controller is assisting an aircraft experiencing an emergency.

Every PPI has a control that selects the map to be displayed and another that varies the map's working intensity. The working intensity is usually based on the controller's preference. Some controllers work best using a very bright map, whereas others prefer to work with an almost unreadably dim map.

Sweep Decenter

Although the main bang is normally located at the center of the PPI, this position may not always be ideal during particular operations. If, for instance, one PPI in a TRACON is being used to separate aircraft east of the airport, the controller might want to display the airspace directly east of the airport and might like to expand the PPI's range so that the sector could almost fill the display. To accomplish this, the controller might **decenter** the main bang and move it to some other location, such as the far left of the screen, using the **sweep decenter** controls. The sweep decenter consists of two controls: one moves the main bang in a north-south direction, whereas the other moves it in an east-west direction. The coordinated use of both controls permits the controller to move the main bang anywhere on the PPI. In fact, the main bang can even be moved completely off the screen.

Types of Air Traffic Control Radar

In general, four radar systems are used in air traffic control in the United States: (1) precision approach radar, (2) airport surveillance radar, (3) air route surveillance radar, and (4) airport surface detection equipment.

Precision approach radar (PAR), used primarily by the Department of Defense as a precision landing aid, is being rapidly replaced by the ILS and MLS systems (thus, PAR will be given only a cursory discussion in this text). **Airport surveillance radar (ASR)** is short-range radar that approach and departure controllers use primarily within the vicinity of busy airports. **Air route surveillance radar (ARSR)** is long-range radar that ARTCC controllers use to provide en route separation of aircraft. Control towers use **airport surface detection equipment (ASDE),** a short-range radar system, during periods of extremely low visibility to detect aircraft or vehicles moving around the airport.

Precision Approach Radar

Precision approach radar was developed in the 1940s as a precision approach landing aid, when the accuracy and the safety of the ILS were still being disputed. The new system was designed to provide lateral and vertical guidance to the pilot. Controllers monitoring the PAR displays observed each aircraft's position in relation to the desired flight path and issued instructions to the pilot that would keep the aircraft on course. The PAR system, which consisted of a mobile facility that included radio transmitters, controller displays, and two radar antennas, was positioned near the approach end of the runway in use. The two radar antennas scanned the approach path to that runway. One

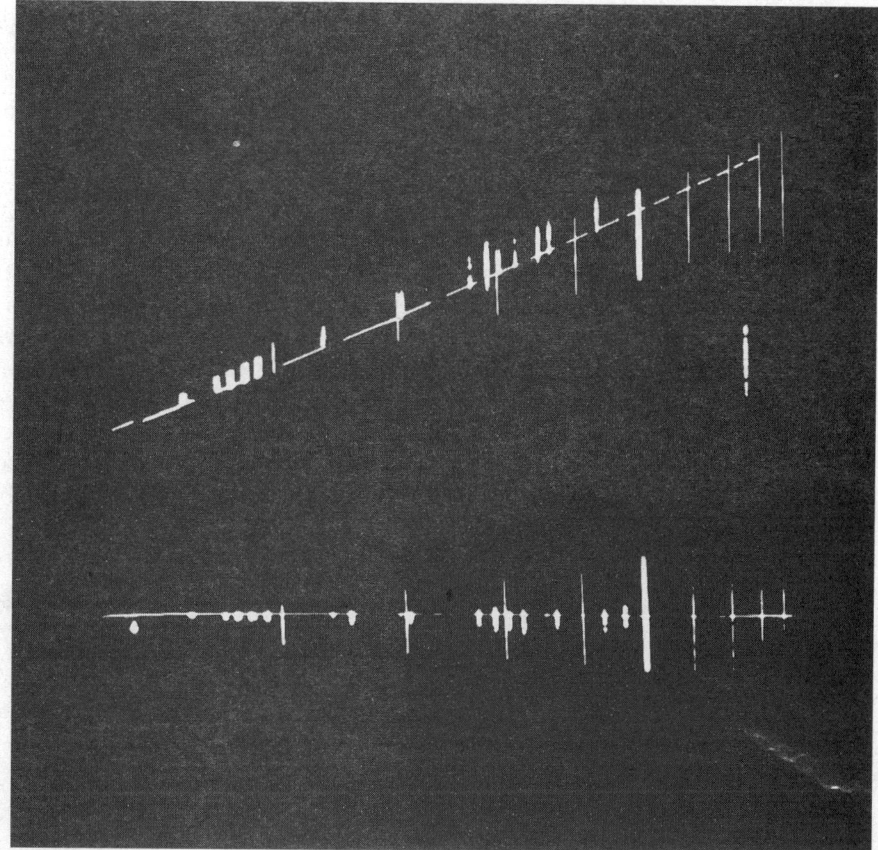

ITT Gilfillan Division

Figure 8-15. Precision approach radar display.

antenna scanned horizontally and displayed the aircraft's range and lateral position relative to the runway. The other antenna scanned vertically and displayed the aircraft's range and elevation. The PAR display included a video map that displayed the proper bearing and glidepath to the runway. Figure 8-15 shows a controller's PAR display, and Figure 8-16 provides a graphic presentation of the display. The controller monitored each aircraft's progress and advised the pilot to make the proper heading and rate-of-descent changes to keep the aircraft on the proper course.

The military preferred the PAR system as a precision approach aid since it was highly mobile and could be placed into operation at any temporary airport or landing field within hours. Many PAR installations at permanent airports were placed on rotating bases in the middle of the airport and could be turned in any direction to serve any of the runways (see Figure 8-17). While the military was installing and using PAR, the ILS was still undergoing development and testing and was still not reliable. Even when ILS had been perfected, it certainly was not a mobile system. The PAR installations, however, could be airlifted to remote sites and be operational in less than a day. In addition, any aircraft with an operable communications receiver could use a PAR approach; no special equipment was necessary. But to achieve the full benefits of ILS, aircraft had to carry expensive navigation receivers.

Precision approach radar was probably the best choice for military precision approaches because of the Defense Department's unique operational requirements. Since the military had to provide precision approaches at a limited number of airfields, the Defense Department could afford to install and operate a PAR installation at every military airfield. The PAR system proved to be very effective for the military and justified itself during the Berlin airlift when it guided aircraft to airports at a rate of one every 90 seconds.

Because of its unique requirements, however, the federal government (through the CAA) chose to implement the ILS across the continental United States. In retrospect, ILS was the correct choice for civilian aviation. Although the system was initially beset with a host of problems, it soon became reliable

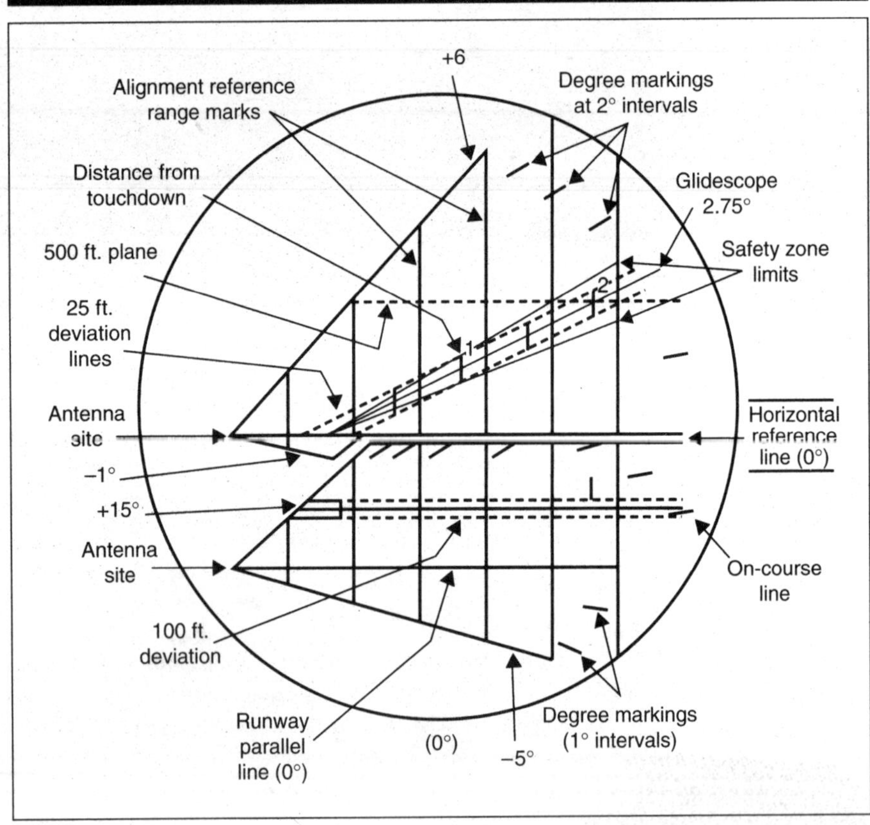

Figure 8-16. Graphic presentation of a PAR display.

ITT Gilfillan Division

Figure 8-17. Precision approach and automatic landing radar.

and accurate. As the base number of ILS installations increased, the airlines and personal aircraft owners began to install ILS receivers, and the price of the receivers decreased considerably. As pilots gained experience with the system, they soon accepted its accuracy and safety and began to lobby for additional installations. Eventually, ILSs were installed at many smaller airfields around the country. By 1986, more than 1,000 ILSs had been installed nationwide. The cost to install, staff, and operate that many PAR installations would have been prohibitive.

Since the late 1960s, the military services have operated PAR installations while equipping many of their aircraft with ILS receivers. Because most military aircraft and pilots use civilian airports occasionally, these aircraft must have the capability to conduct ILS approaches. In addition, most military airfields now have installed ILS transmitters. The PAR units at these fields are being decommissioned, although they are still kept in reserve and the controllers maintain proficiency in case a need arises to mobilize these facilities.

Airport Surveillance Radar

The primary short-range radar currently used by the FAA is airport surveillance radar. Most major civilian and military airports use ASR systems. Approach controllers mainly use ASR, which has a range of approximately 100 nautical miles, to separate local aircraft. The first ASRs used by the CAA were surplus military air defense radars. The first ASR that the CAA obtained for civilian air traffic control was the ASR-1 series. The FAA now uses the ASR-7, ASR-8, and ASR-9 series. ASR-9 radar is solid-state equipment that will be used as the backbone of the air traffic control system well into the twenty-first century.

Air Route Surveillance Radar

The CAA first purchased air route surveillance radar in 1956 to help the ARTCC controllers provide radar separation to en route aircraft. This long-range radar differs from ASR in that it transmits at a higher power level and at a lower pulse repetition frequency, permitting an effective range in excess of 250 nautical miles. The ARSR radar antennas are larger than the ASR antennas and revolve more slowly to give time for the distant radar echoes to return (see Figure 8-18). The first ARSRs ordered were the ARSR-1 series. The FAA now uses ARSR-2 and ARSR-3 systems and has acquired ARSR-4 systems. These radar systems will provide en route radar coverage well into the twenty-first century.

The U.S. Air Force has historically operated numerous radar installations in defense of the North American continent. In an attempt to decrease expenditures, increase operational efficiency, and reduce duplication of facilities, the Air Force and the FAA have agreed to jointly operate a number of long-range radar systems, known as **joint surveillance systems (JSSs).**

Michael Nolan

Figure 8-18. Air route surveillance radar primary and secondary radar system.

Norden Systems Division of United Technologies

Figure 8-19. Airport surface detection radar antenna and housing.

Joint surveillance systems use the same transmitter, antenna, and receiver for the Air Force and the FAA, but they are also equipped with an electronic splitter that sends duplicate radar information to military and civilian air traffic control facilities for processing. The FAA uses this information for air traffic control; the Air Force concentrates on air defense. Maintenance and operation costs are shared jointly by the FAA and the Department of Defense. This arrangement has proved to be one of the most cost-effective cooperative agreements between two federal agencies.

Airport Surface Detection Equipment

At airports where the surface visibility often makes it impossible to see each aircraft, a specialized radar system designed to locate and display the locations of moving, ground-based aircraft and vehicles has been designed. This short-range radar system is known as airport surface detection equipment (ASDE). Figure 8-19 shows the ASDE's radar antenna and housing.

The ASDE system provides surveillance of aircraft and airport service vehicles. At high-activity airports, radar monitoring of aircraft and ground vehicles is essential for safety during periods of reduced visibility. Aircraft and vehicle position information are reported even during periods of heavy snow and fog. The ASDE system is used whenever weather conditions preclude visual observation of the runways and taxiways. Because of its necessary sensitivity, a typical ASDE display is cluttered and difficult to interpret. Installation of the ASDE system is cost effective only at high-activity airports where reduced visibility is commonplace.

Air Traffic Control Radar Beacon System

One of the most significant developments in air traffic control technology has been the development of a secondary radar system known as the **air traffic control radar beacon system (ATCRBS)**, or secondary surveillance radar (SSR), or simply secondary radar. The ATCRBS was introduced in 1956 and its development achieved many of the goals set by the Project Beacon task force.

Primary radar efficiency and operation depends on a number of variables, including transmitter power, aircraft size and distance from the radar antenna, atmospheric conditions, and obstructions that may interfere with the transmitted radar signal. In addition, primary radar systems detect and display every aircraft within the range of the radar antenna, even if it is above or below the vertical limits of the controller's assigned sector. When using primary radar for air traffic control separation, a controller cannot positively determine each aircraft's altitude; instead he or she must depend on accurate altitude reports from the pilot—a time-consuming, inefficient, and potentially inaccurate means of verifying each aircraft's altitude. The ATCRBS can be used to alleviate many of these deficiencies.

It is difficult for a controller to positively identify a particular aircraft from primary surveillance radar. To determine an aircraft's identity, the controller must depend on the pilot-reported position. This procedure is fraught with potential hazards because many pilots who contact ATC facilities for assistance are unsure of their position. Thus, the effectiveness of using radar to help locate and identify these lost aircraft is reduced. Although there are other methods of identifying aircraft using primary radar, such as requiring the pilot to make a specific number of turns, the possibility of misidentifying an aircraft still remains. As long as positive radar identification of aircraft is this difficult, primary surveillance radar is an ineffective aircraft separation tool.

Using primary surveillance radar, the controller must initially identify each aircraft and maintain this identification while the aircraft is under his or her control. This is time consuming and mentally exhausting for the controller. In an attempt to maintain positive identification, air traffic controllers often used modern versions of the "shrimp boats" described in Chapter 1. Constructed of clear plastic, these modern shrimp boats carried the aircraft's identification and call sign penciled on their sides. As the aircraft progressed across the radar screen, the controller manually moved the shrimp boat, keeping it next to the radar echo. If the aircraft changed altitude, the old altitude was erased and the new altitude was penciled in. During periods of heavy traffic, an extra controller was usually assigned to each sector to move and update the shrimp boats.

The controller still had to maintain and update the aircraft's flight progress strip while trying to remember which target corresponded to which aircraft. During periods of heavy traffic or high stress, it was too easy to misidentify an aircraft, which could lead to a loss of aircraft separation, and perhaps to an aircraft accident.

Development of ATCRBS

Just before World War II, research had been conducted on a limited basis to try to alleviate some of the problems inherent in primary radar systems. The war helped to accelerate the development of a system that could differentiate between hostile and friendly targets on radar displays. Used extensively by the Royal Air Force in the air defense of the British Isles, this primitive radar identification system was known as **identification friend or foe (IFF).**

The IFF system used a ground-based transmitter (known as the interrogator) to broadcast a coded radio signal to the aircraft. The radio signal was composed of two pulses (known as **framing pulses**) separated by a short interval and collectively known as the **challenge pulse.** This challenge pulse could be sent in different **modes.** Each mode was identified by the interval of time that elapsed between each framing pulse. Every aircraft participating in the IFF program was equipped with a **transponder** that received the challenge signal, determined whether it was set to a mode that should be responded to, and if so returned a coded signal known as the **reply.**

The IFF transmitter could be set to operate in many available modes. Only aircraft whose transponders were set to the specific mode would reply to an interrogation. Each aircraft operating on a particular mode was assigned a unique code that could be used to determine that specific aircraft's identity. This combination of modes and codes was used to identify every aircraft appearing on the radar screen as either friendly or hostile.

The IFF system operated in conjunction with the primary radar system. If the IFF system determined that a target was "friendly" (by responding with the proper code in the appropriate mode), a designating symbol would be overlaid on the primary radar target displayed on the PPI. The aircraft's code could also be determined by interpreting the IFF symbol. If an aircraft did not respond in the proper mode or with the proper code, or if it did not reply to the IFF challenge at all, it was assumed to be an enemy aircraft. This system was probably one of the most important and unsung developments of World War II. Without the IFF system, it is highly unlikely that the Royal Air Force would have been able to defend the British mainland.

The air traffic control radar beacon system was developed using many of the same principles and the basic components of the IFF system. The ATCRBS uses two ground-based antennas to transmit a challenge to every aircraft using six different modes. Every aircraft equipped with a transponder that can be set to reply when challenged in any of these modes can generate a coded reply.

Table 8-2. Transponder Modes

Mode	Application	Framing Pulse Spacing (microseconds)	Characteristic
1	Military	3	3 μ sec
2	Military	5	5 μ sec
3/A	Military and civilian; known as mode 3 by military pilots and mode A by civilian pilots	8	8 μ sec
B	Civilian; primarily used in Europe	17	17 μ sec
C	Civilian; includes altitude encoding	21	21 μ sec
D	Civilian; not currently being used	25	25 μ sec

Ground-based computers can use this information to determine the aircraft's identity, predict flight paths, and provide other essential information to the controller.

One of the two secondary radar transmitting antennas is physically located directly on top of the primary radar antenna, rotating synchronously with it. The other ATCRBS transmitting antenna is placed in a fixed, vertical position next to the rotating antenna and is used for **side lobe suppression (SLS),** which will be discussed under "Secondary Radar System Deficiencies." To participate in the ATCRBS, each aircraft must be equipped with a transponder that can respond to any one of six modes and that can reply using one of 4,096 pilot-selectable codes.

The rotating antenna transmits short pulse pairs on a frequency of 1030 mHz. This interrogation signal, composed of two pulses known as P1 and P3, is transmitted sequentially using each of the six modes. (P2 is a pulse transmitted by the side lobe suppression antenna and will be discussed shortly.) Any aircraft transponder set to reply to one of these modes will reply on a frequency of 1090 mHz.

Each mode can be identified by measuring the time interval between the two pulses. The six different modes and their pulse intervals are shown in Table 8-2.

The aircraft's transponder will reply to an interrogation with two framing pulses that surround an additional set of pulses used to identify the four-digit code selected by the pilot. This entire series of reply pulses is known as a **pulse train.** The transponder uses the octal numbering system, which uses only the numbers 0 through 7, to transmit the code selected by the pilot. Each pulse within the code train is assigned a value based on the octal system. The ATCRBS receiver on the ground can decode the pulse train and use simple addition to determine the code selected by the pilot. Numeric beacon decoding systems and advanced computer processing systems can then use this information to identify each aircraft and perform other air traffic control functions.

ATCRBS Display

In its simplest method of operation, the ATCRBS causes a distinct slash, known as a **beacon slash,** to be created on the PPI display. This beacon slash directly overlays the primary target blip and identifies the target as a transponder-equipped aircraft, not a vehicle, train, or some other object. Using basic ATCRBS equipment, the controller can identify which aircraft is producing the slash in three ways. The controller can instruct the pilot to turn the transponder off or to the **Standby** position, which will cause the beacon slash to disappear. After a few seconds, the controller can then instruct the pilot to turn the transponder back on, which will cause the beacon slash to reappear. This method of identification is not very reliable because many situations may cause a beacon slash to temporarily disappear and then reappear. In most personal aircraft, the transponder antenna is located on the bottom of the fuselage. Any sustained turn by the aircraft *toward* the ATCRBS antenna on the ground will place the fuselage between the two antennas. The fuselage could shield the airborne antenna from the ground-based ATCRBS transmitter, and it will appear to the controller as if the aircraft's transponder is turned off. When the aircraft completes the turn and levels off, the transponder antenna will no longer be shielded by the fuselage and it will appear to the controller as if the transponder has been turned back on.

Two more positive means of aircraft identification require the pilot to either activate the Ident feature built into every transponder or select a particular code.

Ident Feature When the pilot depresses the **Ident** button, the transponder transmits a special reply pulse known as the **special identification pulse (SIP)** precisely 4.35 microseconds after the last framing pulse. This SIP is interpreted by the ATCRBS receiver and causes the beacon slash to become much wider. This double-width slash persists on the PPI for a couple of sweeps of the radar antenna. This procedure is a more positive means of aircraft identification because it is highly unlikely that a double-width beacon slash would appear on a PPI for any other reason.

Code Selection When the pilot selects a particular code on the transponder, it is known as **squawking.** Through the use of a selector panel, known as a **ten-channel selector,** the controller can program the ATCRBS receiver to produce an extra beacon slash for every aircraft squawking one of the selected codes. The resultant double beacon slash is often used to determine which controller has responsibility for the aircraft. In a typical facility, the approach controller might be allocated the use of transponder code 0400 and will assign this code to every inbound aircraft. The departure controller might then be allocated a different code, such as 4600, that will be assigned to every outbound aircraft. Through the selector panel, each controller can determine which aircraft is his or her responsibility. The approach controller programs the ten-channel selector so that only those aircraft squawking code 0400 are displayed with a double slash. The departure controller then programs the ten-channel selector so that only those aircraft squawking code 4600 appear as a double slash. Thus, each controller can identify which aircraft are arrivals and which are departures.

This primitive system of identification has, for the most part, been replaced by computerized systems that determine each aircraft's identity. However, this basic system has been retained at most ATC facilities as backup for the computer system.

Emergency Code Most ATCRBSs are also designed to sound an alarm and display a double-width double slash on the PPI whenever a transponder code of 7700 is received. Code 7700 is reserved for aircraft experiencing some type of emergency. The unmistakable display created on the PPI is called a **double bloomer** because of its shape and appearance.

Altitude Encoding When the interrogator transmits a challenge on mode C, any aircraft equipped with an altitude-encoding transponder transmits the aircraft's altitude using an additional series of pulses. Each pulse corresponds to a number that can be used to identify the aircraft's altitude. The numbering system used by altitude encoders is known as the **gray scale.** Once received, this altitude information can be processed by the computer system and displayed directly on the PPI. This altitude display can be used to eliminate many of the altitude reports commonly used by the controller when separating aircraft.

Secondary Radar System Deficiencies

Garbled Replies If two aircraft are located at the same azimuth from the radar antenna and are within 3.3 nautical miles of each other, it is highly likely that their transponder replies will overlap. The ATCRBS receiver may be unable to differentiate between the two replies, with unpredictable results. Portions of either aircraft's pulse train may not be processed by the system, causing either or both of the targets to disappear from the PPI. In rare instances, the replies may overlap in such a way that a false target between the two legitimate targets is produced. Such false targets may occur even if the two aircraft are separated vertically. Fortunately, this is a transient problem that can be somewhat controlled through improved receiver circuitry.

Interference Above the continental United States, most aircraft operating over 5,000 feet will be well within the range of two or more ATCRBS transmitters. Since every interrogator transmits on 1030 mHz, a transponder would have to reply to a multitude of interrogations in busy metropolitan areas. The aircraft transponder is unable to differentiate between the various interrogators and will reply to all, which can cause interference when the ground-based ATCRBS receivers detect transponders that are replying to other ground-based interrogators. The controller's PPI may become covered with small random dots, known as **fruit.** Fruit can be distracting and dangerous if it interferes with the controller's primary duty of separating air traffic. Fortunately, an electronic system of eliminating this interference is used by FAA radar systems. This feature, known as the **defruiter,** is normally operational at all times and removes most of the fruit from the radar display.

Side Lobe Suppression During normal operation, the rotating ATCRBS antenna also transmits low-powered pulses at angles of approximately 45° and 90° from the direction of the main transmission. These extraneous and unwanted transmissions, or **side lobes,** are common to radio transmissions. Side lobe transmissions can cause inaccurate aircraft azimuth determination if an airborne transponder mistakes one of these transmissions as the main ATCRBS challenge and generates a response. If the aircraft replies to both the main transmission and every one of the side lobes, multiple aircraft targets will be generated on the PPI. A close inspection of the PPI display will confirm that many of these targets are false replies, because no primary targets are being plotted in the same location. Nonetheless, during heavy traffic, the controller could easily become distracted by such false transponder targets. In addition, the multiple targets will confuse most of the computerized radar systems.

To eliminate this problem, airborne transponders have been designed to take advantage of the fact that side lobe transmissions are much lower in overall signal strength than the main transmission. Transponder circuitry cannot be designed to identify a side lobe simply by measuring its signal strength, since the received strength of each lobe is directly proportional to the aircraft's distance from the antenna. However, the aircraft transponder can compare the signal strength of the radar transmission with a reference signal transmitted from the radar site.

This reference signal is produced by a second ATCRBS transmitting antenna, known as the **side lobe suppression omnidirectional antenna,** located next to the rotating antenna. As previously discussed, the main ATCRBS antenna transmits two pulses—P1 and P3—at precise intervals with equal signal strength. The side lobe suppression (SLS) antenna transmits a pulse, P2, exactly 2 microseconds after the P1 pulse and at approximately the same power level. But the P2 pulse is transmitted equally in all directions (omnidirectionally). The aircraft transponder's SLS circuitry compares the relative signal strength of the P1 and the P2 transmissions. If they are both the same strength, the transponder concludes that it must have received a direct interrogation signal from the main lobe of the ATCRBS antenna and generates a suitable reply. But if the P1 pulse is received with lower signal strength than the P2 pulse, the transponder circuitry concludes that it has been challenged by a side lobe and will refuse to issue any response (see Figure 8-20).

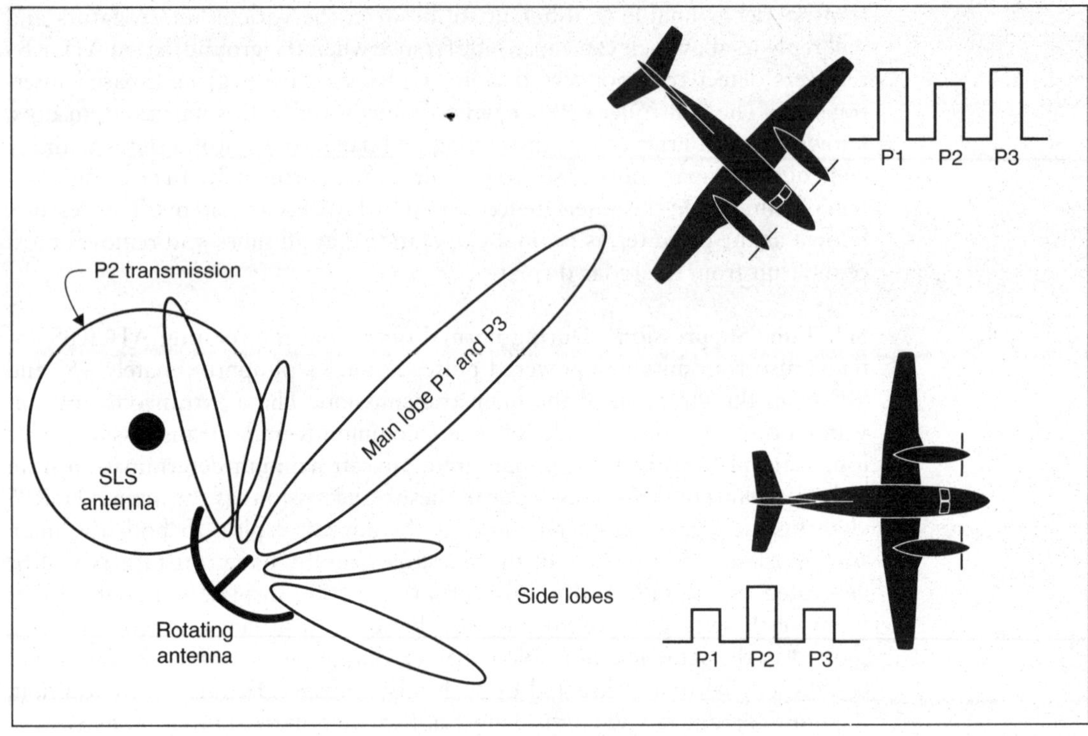

Figure 8-20. Example of main lobe and side lobe interrogation by the ATCRBS.

Computer Processing of ATCRBS

Although aircraft identification using secondary radar was a crude but workable system, it was a vast improvement over primary surveillance radar systems. But the controller still had to update flight progress strips that included each aircraft's identification, route of flight, and altitude and then correlate this flight information with the corresponding blip on the PPI.

Any misidentification of a particular aircraft could have disastrous results. During periods of intense traffic, when the controller was responsible for the separation of 20 or more aircraft, constantly maintaining the identity of each aircraft was virtually impossible. The clerical duties involved in updating the flight strips and the concentration required to correlate each target were overwhelming. In addition, the radar systems were unable to discriminate between IFR and VFR aircraft and could not even determine whether a particular IFR aircraft was within the controller's assigned sector. Every aircraft located within the range of the radar system was displayed on the controller's PPI. The controller then had to decipher this jumble of information and identify those aircraft within his or her area of control.

The Project Beacon task force recommended that a computerized beacon decoding system be designed that could decipher this information and assist the controller in maintaining positive identification of each aircraft. The system, as envisioned by the authors of the report, would process each aircraft's transponder replies, and then a computer would verify and correlate the information with flight plan information such as aircraft call sign, route of flight, and so on. This information would then be made immediately available to the controller. Some of the data, such as aircraft call sign and altitude, would be displayed directly on the PPI; the rest would be instantaneously available using auxiliary display equipment. The controller would no longer constantly need to refer to flight progress strips and shrimp boats for this information. The computer system would be programmed so that only the aircraft that were the controller's responsibility would be displayed. Even with this system, the controller would still need to update flight progress strips in case of computer malfunction, but this computerized beacon processing system would improve his or her ability to separate increasing amounts of traffic while also reducing the amount of concentration needed to accomplish the task.

One of the first semicomputerized ATCRBSs was a military beacon decoder system called **TPX-42,** essentially an advanced version of the identification friend or foe system developed in World War II. Known as a numeric beacon decoder system, TPX-42's primary capability is to decode the aircraft's transponder reply and display the four-digit transponder code next to the secondary radar target on the PPI. Using TPX-42, the controller still must update flight progress strips and use them to determine an aircraft's identity.

Typically, the controller assigns a different transponder code to every aircraft and writes this code on the appropriate flight progress strip. If the aircraft is on an IFR flight plan, the flight data processing computer in the ARTCC will already have automatically assigned a transponder code. Any VFR aircraft requesting radar services are usually assigned one of the codes allocated to the facility expressly for that purpose. By correlating the information contained on the flight progress strip with the four-digit transponder code displayed on the PPI, the controller can easily identify each aircraft.

If the aircraft is equipped with a mode C transponder (altitude-encoding), the TPX-42 system will display the aircraft's altitude directly below the four-digit transponder code. Additional TPX-42 capabilities include a stylized Ident feature and history dots. When the pilot presses the Ident button, the beacon slash becomes more prominent and the TPX-42 equipment creates a series of distinct shrinking circles centered over the aircraft's radar beacon slash. This shrinking set of concentric circles is prominent and easy to see even on a cluttered radar display (see Figure 8-21). The TPX-42 system is also designed to trail a series of dots behind each aircraft as it progresses across the radar scope. This feature makes it easier for the controller to determine the aircraft's direction of flight and ground speed.

The TPX-42 system was adapted for civilian use and was quickly and inexpensively installed at many low- to medium-activity airports. Although originally well suited for this environment, TPX-42 is not a very sophisticated system, is no longer considered a modern radar beacon system, and cannot be inexpensively upgraded. Further advances in ATCRBS technology have made

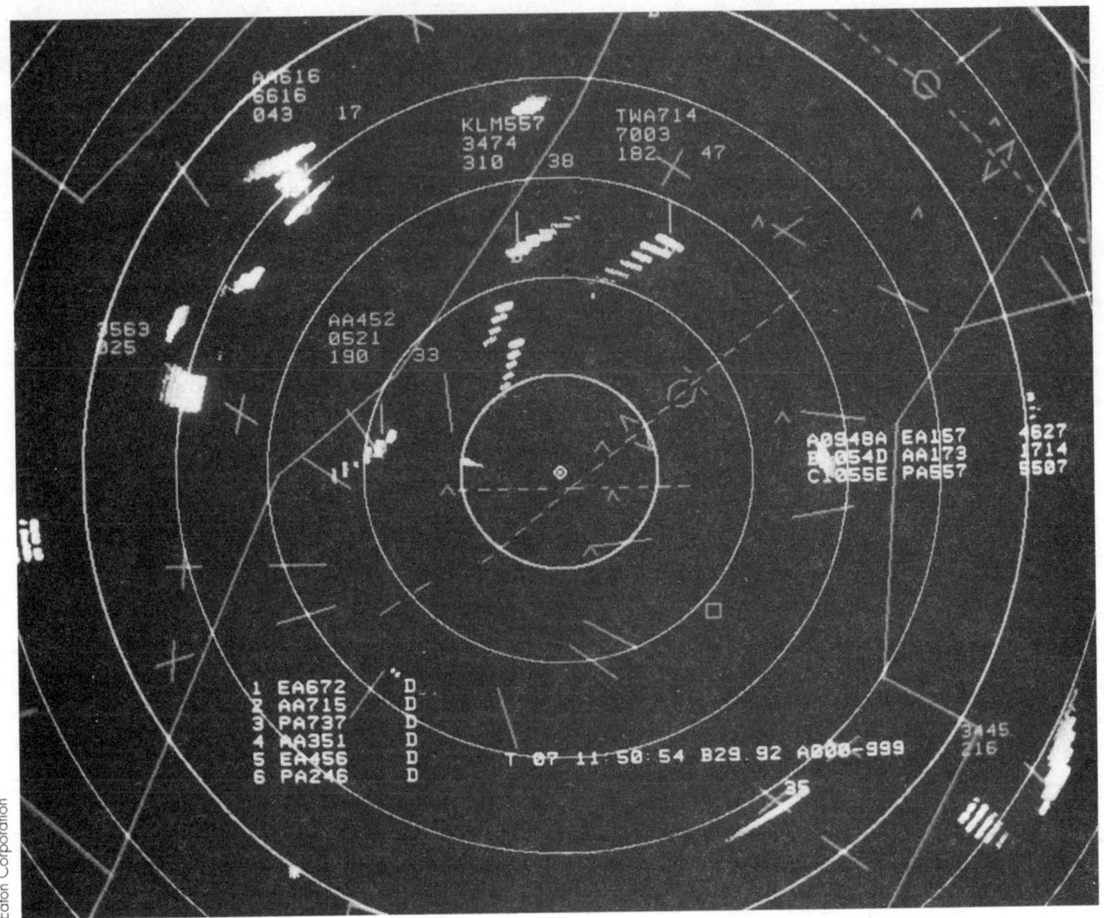

Eaton Corporation

Figure 8-21. TPX-42 display (with Programmable Indicator Data Processor enhancement).

TPX-42 obsolete; it is no longer being installed. Facilities still equipped with TPX-42 are scheduled to have the system decommissioned within the next few years, to be replaced by more modern equipment as soon as the FAA can procure it.

Computerized Radar Systems

The report issued by the Project Beacon task force focused the FAA's efforts on designing a totally new ATC computer system that would be expressly designed for civilian air traffic control. Initially, it appeared that a common system could be developed to serve both ARTCC and radar-equipped terminal approach control facilities. As conceived by the task force, this new system would identify each aircraft, predict its flight path and altitude, automatically pass this information to the next controller, and display aircraft information directly on the radar screen. The computer system would also be able to predict aircraft flight paths and notify the controller before a dangerous situation developed.

Initially, FAA research focused on developing a common computer system that could be used at every FAA air traffic control facility. But the en route centers and the terminal facilities had widely differing needs that could not be met using a single system. The ARTCC controllers needed equipment that would help them separate high-altitude, high-speed IFR aircraft; terminal controllers needed a system that would be more adaptable to local conditions and a mix of aircraft. Researchers realized that a single system could not meet the needs of both types of facilities. Ultimately, three different systems were developed. The flight plan data processing function was assigned to the ARTCCs and handled by a system previously described, known as the **flight data processing (FDP)** system. In addition, two computerized radar beacon processing systems were eventually developed: the **radar data processing (RDP)** system, which was destined for use in the ARTCCs, and the **automated radar terminal system (ARTS)**, which was designed for use in the terminal environment. Each system was designed to accommodate the requirements of a particular type of ATC facility.

Automated Radar Terminal System

The first prototype ARTS computer system, designated the ARTS-I, was installed in the Atlanta (Georgia) air traffic control tower in 1964. The system proved so successful that it was quickly expanded and installed in the New York Common IFR room (the predecessor to the current New York TRACON) in 1966. This modified ARTS-I was known as ARTS-IA. The ARTS-I and the ARTS-IA were essentially identical and provided tracking and identification capability for aircraft equipped with transponders. The ARTS was designed to identify each aircraft by matching its transponder code with flight plan data, provided either by the flight data processing computer located in the ARTCC or from controller entries made directly into the ARTS. Once the aircraft was identified, the ARTS computer system maintained constant identification and predicted the aircraft's future location. Since the ARTS-I was able to track only aircraft equipped with transponders, it was known as a **beacon tracking level (BTL)** system. A BTL system can track and provide identification and altitude information only on aircraft equipped with transponders.

Once the necessary flight plan data are entered into the ARTS computer, either manually or from the FDP computer located in the ARTCC, the BTL radar system can predict flight paths, initiate handoffs, and automatically provide the controller with continuous alphanumeric information on the radar screen. These alphanumeric data include an aircraft symbol and its associated **data block** (see Figure 8-22).

The computer-generated aircraft symbol is overlaid directly on the beacon target to indicate which controller has the responsibility for that aircraft.

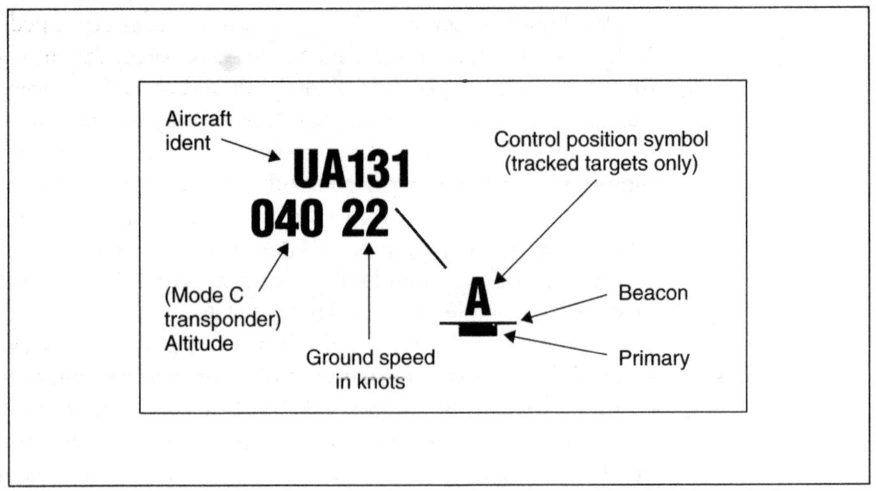

Figure 8-22. ARTS-III data block.

Different controller positions are usually assigned unique identifying symbols. For example, the east arrival controller might be assigned the letter E, whereas the west arrival controller might be assigned the letter W. The data block associated with each target includes the aircraft's call sign, the ARTS computer-assigned ID number, and the aircraft's altitude (if the aircraft is equipped with mode C). The controller uses the computer identification number to extract or enter flight data concerning that aircraft from the ARTS computer.

The ARTS-I computer system proved to be highly successful in relieving controllers of the tedious task of maintaining the correct association between radar targets and flight progress strips. The successful development and implementation of the ARTS system in Atlanta and New York led the FAA to award a contract for the installation of additional, enhanced ARTS systems at high-activity airports across the United States. This more advanced system is known as **ARTS-III.**

ARTS-III

The ARTS-III development contract was awarded in 1969, and by 1973 all of the needed systems were operational. As originally implemented, the ARTS-III was a BTL system that could track only transponder-equipped aircraft. In 1976 the FAA awarded an upgrade contract to add the capacity to every ARTS-III facility to track and identify aircraft that are not equipped with transponders. A computer system that can track aircraft using both primary and secondary surveillance radar echoes is known as a **radar and beacon tracking level (RBTL)** system (see Figure 8-23). This enhanced ARTS-III, known as **ARTS-IIIA,** has since become the FAA standard for high-activity airports. The ARTS-IIIA system includes the following components: a primary radar transmitter, antenna, and receiver; an ATCRBS transmitter, antenna, and receiver; a data acquisition

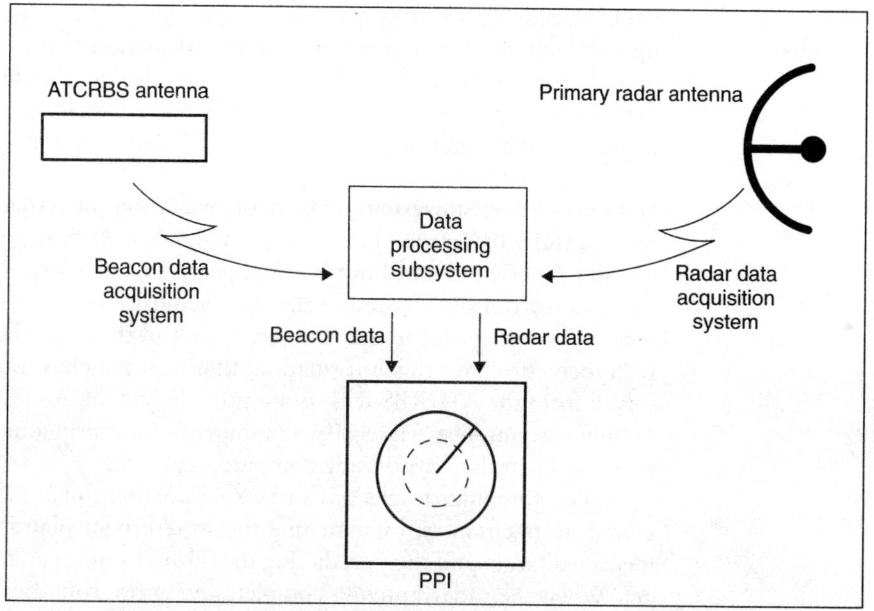

Figure 8-23. The primary and secondary radar systems transmit data to the data processing system of an ARTS-IIIA system. After the data processing system analyzes the data, they are sent to the PPI for display.

subsystem; a data processing subsystem; a data entry and display subsystem; and a continuous data recording subsystem.

Transmitter and Receiver The ARTS radar transmitter and receiver are standard FAA air traffic control radar systems. These systems include the FAA series of civilian radars, such as the ASR-6, ASR-7, ASR-8, and ASR-9 primary surveillance radars. They have a range of about 60 nautical miles and operate with a peak power of about 500 kilowatts.

ARTS also operates using standard secondary radar interrogators and receivers such as the ATCBI-5 and the ATCBI-6 secondary surveillance radar systems, which operate on a variety of modes, such as 1, 2, 3, A, and C. The secondary surveillance radar system used by ARTS also includes standard side lobe suppression circuitry.

Data Acquisition Subsystem The **data acquisition subsystem (DAS)** is a peripheral device that receives raw radar data from the primary surveillance radar system and beacon-derived information from the secondary surveillance system. The DAS then decodes this information, converts it to a digital format, and channels it to the data processing subsystem for further processing. The data acquisition system is actually composed of two different subsystems: the **radar data acquisition system (RDAS)** and the **beacon data acquisition system**

(BDAS). The RDAS subsystem digitizes primary radar information, converts it into a digital format, and transmits this information to the data processing system. The BDAS interprets the ATCRBS returns, correlates this information with those targets detected by the primary radar system, and then sends this information in a digital format to the data processing system.

Data Processing Subsystem The **data processing subsystem (DPS)** is the heart of the ARTS-IIIA radar processing system. The DPS is a high-speed, digital computer designed and built by the Sperry Univac Corporation to accept information from three sources: the data acquisition subsystem, the flight data processing system located at the ARTCC, and the data entry subsystem. The DPS then correlates this information; that is, it matches transponder codes received from the ATCRBS with those provided by the ARTCC's flight data processing system. The ARTS-IIIA computer is in continuous electronic contact with the flight data processing computer.

The computer **tracks** any targets whose flight plan information has been stored in its circuitry, by matching the stored flight plan information to each identified target and then predicting the future location and altitude of that target. When the radar antenna completes an entire rotation, the computer uses that information to look for a target at the aircraft's next predicted location and continues to process radar data received from that aircraft.

If the primary and the secondary radar returns from any aircraft are temporarily interrupted, the ARTS computer can still predict the track of the aircraft and advise the controller that radar contact has been lost. Aircraft in this predicament are in a **coast** mode. While in the coast mode, the ARTS computer predicts the aircraft's position and displays this information on the PPI. When the radar system reacquires the aircraft, the computer displays its exact position and initiates a new track.

Data Entry and Display Subsystem The **data entry and display subsystem (DEDS)** displays ARTS-derived information on the PPI. Also, the controller can use DEDS to input flight data into the computer. If necessary, the controller's entries are automatically sent to the FDP computer in the ARTCC. The DEDS is composed of two separate subsystems: the data display and the data entry sets. The data display system uses the same PPI used by the radar system to display the aircraft's data blocks and other pertinent information. It does not replace the primary targets and the ATCRBS beacon slashes but simply overlays alphanumeric information provided by the data processing system on the primary and secondary radar information. This system also displays information such as the current altimeter setting, ATIS code in use, and a list of projected inbound aircraft in seldom-used areas of the PPI.

The ARTS system only *overlays* information on the PPI; it does not eliminate the display of primary and secondary targets. Should the ARTS computer malfunction, the primary and secondary radar systems will continue to operate and display aircraft position information. Only the alphanumeric information provided by the ARTS computer system will be deleted from the radar display.

UNISYS Corporation

Figure 8-24. The data entry sets used on ARTS-III radar systems.

The **data entry sets (DES)** are the devices the controller uses to input flight data into the ARTS computer. This system is composed of an alphanumeric keyboard, a quick look selector, and a slew entry device (SED), sometimes referred to as a **trackball.** The keyboard sends function commands and flight data to the data processing subsystem (see Figure 8-24). The quick look selector permits the controller to select and display the full alphanumeric data blocks of aircraft under the control of other radar positions within the facility. The trackball is used to enter aircraft position information into the data processing system. When the controller manually rotates the trackball, a small symbol on the PPI moves in a corresponding direction. By placing the trackball symbol on top of an aircraft symbol and pressing one or two keys, the controller can obtain information about that aircraft and can more easily hand it off.

Continuous Data Recording Subsystem The continuous data recording subsystem (CDRS) is a magnetic-tape storage system that continuously records digital data that pass through the ARTS-III computer system. This information can be printed at a later date for data extraction, and events recorded by the system can be reconstructed for analysis. This feature is particularly useful when trying to locate lost aircraft, investigate accidents, or determine traffic patterns around an airport.

ARTS-IIIA Operational Characteristics

The ARTS-IIIA system is capable of accomplishing the following tasks: automatic track initiation, data block generation, automated handoffs, track drops, target coast, altitude filtering, conflict alert, minimum safe altitude warning, and special beacon code displays. Figure 8-25 is a graphical presentation of an ARTS-IIIA display.

Automatic Track Initiation Target information received by both the primary and the secondary surveillance radar systems is transmitted to the data processor and correlated with known flight plan and position information. Using this information, the data processor follows each target and predicts its future position. Each target is assigned to one of two groups. The first group consists of aircraft whose flight plan information has been derived from the FDP system in the ARTCC or from information entered by a controller; these aircraft are known as **associated tracks.** Any aircraft that has been assigned a transponder code by the ARTCC FDP system will be automatically tracked when the ARTS system receives the proper transponder code.

Any other aircraft being observed by either the primary or the secondary radar system and whose identity is unknown to the ARTS is known as an **unassociated track.** If the controller wants the ARTS computer to track that aircraft, he or she must enter the proper flight plan information into the ARTS computer and the computer will begin an associated track. Once the transponder code is received by the ARTS, the computer can track that aircraft.

Data Block Generation Any aircraft being tracked by the system will have an associated data block, which is generated by the ARTS computer and overlays the primary and secondary radar targets on the PPI. The data block displays the aircraft's call sign, computer identification number, ground speed, altitude (if mode C equipped), and any other information requested by the controller. Each controller may determine which data blocks should be displayed. An arrival controller, for instance, may not want to see the data blocks associated with departing aircraft. Whenever warranted, the controller may inhibit the data block display of any particular aircraft or any group of aircraft.

Automated Handoffs If the ARTS-IIIA-equipped facility borders another ARTS facility or an ARTCC, the ARTS equipment will permit the controller to perform **automated handoffs.** To initiate an automated handoff, the controller must slew the trackball symbol over the appropriate aircraft target and press the proper key. On the receiving controller's PPI, the data block for that aircraft will be flashing. If the receiving controller determines that the handoff can be safely accepted, he or she slews the trackball symbol over the aircraft's position symbol and then presses the appropriate key, which causes the data block to flash on the original controller's radar screen. At this point, the handoff has

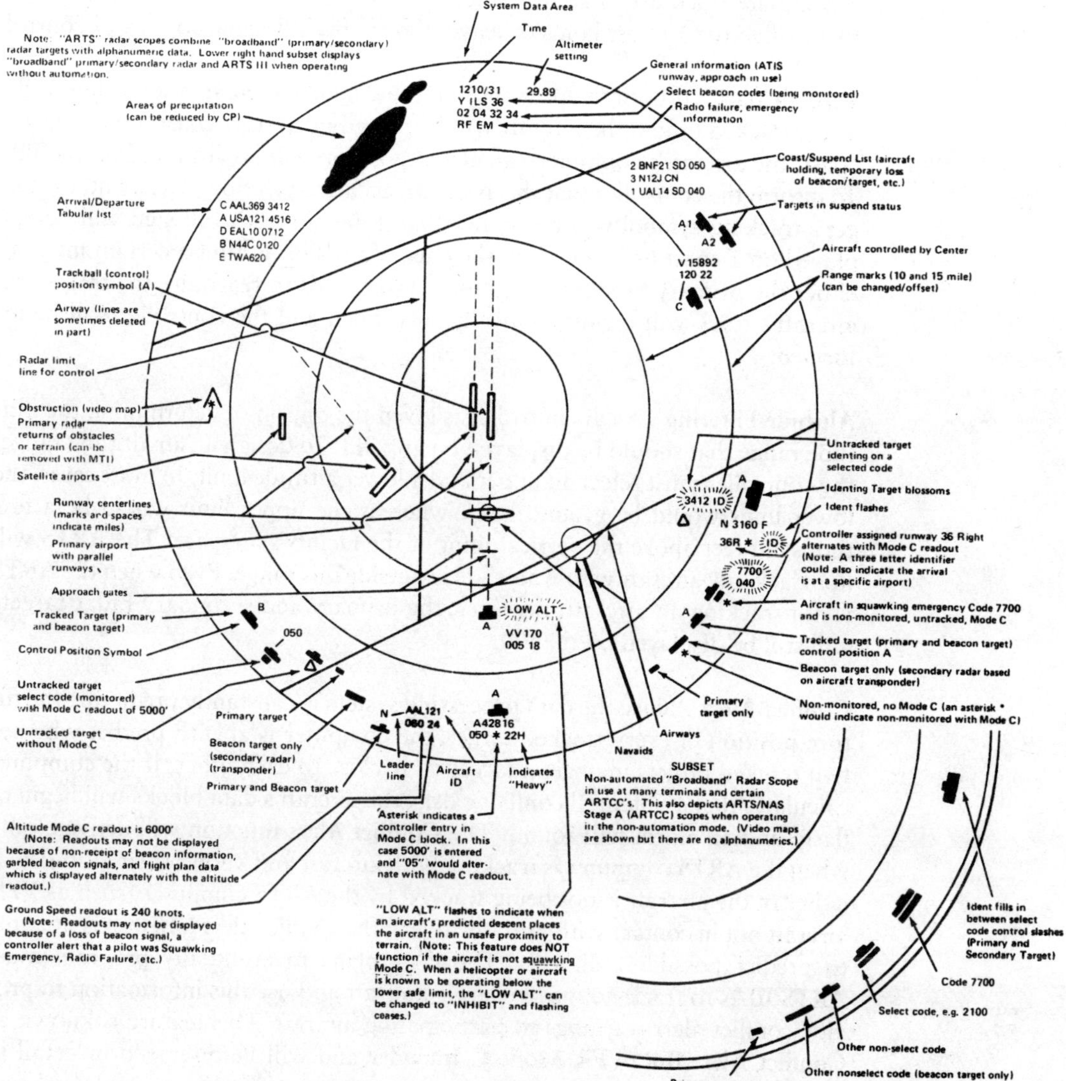

Figure 8-25. A graphic presentation of an ARTS-IIIA display.

been concluded, and the transferring controller can advise the pilot to contact the next controller on the appropriate frequency.

Track Drops If the controller or the computer determines that the aircraft no longer needs to be tracked, the data block will be removed from the display and the flight plan information will be deleted from the computer. The controller can initiate **track drops** at any time. The computer will automatically initiate them when the aircraft lands or leaves the facility's designated area of control.

Target Coast If both the primary and the secondary radar systems fail to detect a tracked target, the aircraft will be placed in a **target coast** mode. During coast, the aircraft's computer-calculated position will be displayed on the PPI. To inform the controller that this is not an accurate position placement, the target's tracking symbol will be changed, and the aircraft's call sign will be displayed on a **coast list** located on the edge of the PPI. If after a certain interval, either the primary or secondary radar system fails to reacquire the target, the aircraft's track will be automatically suspended and the controller will be informed.

Altitude Filtering Each controller is given the option of determining the altitude range that should be displayed on any PPI. To designate an altitude range, the controller must select an upper and a lower altitude limit. In most cases, the lower limit would be ground level, whereas the upper limit would be a few thousand feet above the vertical limit of the facility's airspace. The ARTS will not track any aircraft whose altitude is outside this range. Even when the ARTS is not tracking an aircraft, however, the primary and secondary radar targets will still be displayed on the PPI.

Conflict Alert Since the data processing system is constantly predicting the future position of every tracked target, the computer is able to predict whether two tracked targets may get unacceptably close to each other. If the computer calculates that a potential conflict exists, the aircraft's data blocks will begin to flash, and an alarm will sound. The **Conflict Alert** function will activate only when the ARTS computer is tracking both aircraft and knows their altitudes. If either of the aircraft is not being tracked by the ARTS computer (such as VFR aircraft not in contact with the controller), the conflict alert function is not able to predict possible collisions. The FAA plans to eventually program most ARTS-IIIAs to track nonparticipating aircraft and use this information to provide conflict alert warnings to participating aircraft. This feature is known as **Conflict Alert IFR/VFR Mode C Intruder** and will be discussed in detail in Chapter 12.

Inhibiting the operation of the conflict alert function is advantageous in certain areas within a facility's airspace, such as approach courses to parallel runways and areas located near the approach and departure ends of the runway. In each of these areas, a sufficient number of false alarms would routinely be produced and would defeat the purpose of conflict alert. The system would "cry wolf" so often that the controller might ignore an actual conflict alert.

Minimum Safe Altitude Warning The ARTS can also be programmed to warn a controller whenever a tracked aircraft appears to be descending too close to the ground or approaching an area of rising terrain. This feature, known as **minimum safe altitude warning (MSAW)**, operates by dividing the area served by the ARTS into 2-mile squares and assigning a terrain value to each square. If an aircraft's track is predicted to enter this area at an altitude below the minimum safe altitude or if the aircraft is predicted to eventually descend below this altitude, the aircraft's data block will begin to flash and an alarm will sound. At this point, the controller can advise the pilot of the impending hazard and recommend corrective action. The MSAW system can be inhibited from operating in areas where it might tend to produce false alarms, such as along the final approach course to the runway.

Special Beacon Code Displays The ARTS is designed to react to certain important transponder codes, including the emergency code, 7700; the radio failure code, 7600; and the hijack code, 7500. Any aircraft transmitting one of these special transponder codes will cause the aircraft's data block to flash and a special message to be displayed on the PPI.

Versions of ARTS-III

ARTS-III was developed as a programmable, modular system that could be easily modified and updated as conditions changed. ARTS-III has now evolved into four different versions: ARTS-III, ARTS-IIIA, New York TRACON ARTS-IIIA, and **en route ARTS (EARTS)**.

As mentioned previously, ARTS-IIIA expands the basic ARTS by having the ability to establish tracks on nontransponder-equipped aircraft and to automatically recover from some computer system failures. In addition, ARTS-IIIA permits utility or diagnostic programs to be run while the operational program is still in service. The FAA has converted all basic ARTS-III facilities to ARTS-IIIA facilities.

ARTS-IIIE is an enhanced ARTS that provides for multiple radar inputs and can drive more displays than the ARTS-IIIA. ARTS-IIIE can also send digital information to remote control towers located at outlying airports. ARTS-IIIE is installed at the New York TRACON and will become the standard installation at all large, consolidated TRACONs in the future.

EARTS was developed to serve the offshore ARTCCs in Anchorage, Alaska; Honolulu, Hawaii; and San Juan, Puerto Rico. The U.S. Air Force has also implemented a modified EARTS at Nellis Air Force Base, Nevada. A hybrid radar system that uses ARTS-type computers to drive ARTCC-type displays, EARTS was developed as an inexpensive way to computerize the operations of these three low-activity centers.

The EARTS accepts inputs from up to five different radar sites—either ASR or ARSR sites—and distributes this radar-derived information to the various display consoles in the center. The EARTS computer system evaluates the validity of this input and determines which data should be displayed to the controller. Thus, the EARTS computer can track aircraft that could not be detected by systems using only one radar site. This process of selecting, processing, and displaying radar-derived data is known as creating a **radar mosaic.**

The ability to create a radar mosaic is what differentiates EARTS from the standard ARTS. A standard ARTS PPI can display the radar returns from only a single radar antenna, whereas an EARTS display can mosaic the returns from numerous radar sites and present the controller with the most valid data from each site. In addition, EARTS facilities do not use standard PPI displays; instead they use standard ARTCC displays known as plan view displays (PVDs), which are necessary if the system is to provide a radar mosaic.

ARTS-II

As traffic increased nationwide in the late 1960s and early 1970s, the FAA realized that a system of computerized radar processing would have to be developed for low- to medium-activity facilities. The FAA determined that ARTS-III was inappropriate because it was too expensive to install and operate and required a team of computer programmers to maintain. TPX-42 was installed as a stop-gap measure at many of these airports, but it was intended to be used only until a more appropriate system could be developed. In 1974 the FAA contracted with the Burroughs Corporation (which has since merged with the Sperry Corporation to form UNISYS) to develop a low-cost ARTS radar processing system. To differentiate this system from the previous version of ARTS, it became known as **ARTS-II.** Initial installation of this system began in 1978.

The ARTS-II system was originally designed as a beacon tracking level system that did not provide many of the advanced programmable features of ARTS-IIIA, such as conflict alert and minimum safe altitude warning. The ARTS-II uses a minicomputer with a 256K word memory to process beacon returns and display alphanumeric data blocks on the controller's PPI. The primary difference between ARTS-II and ARTS-III is that ARTS-II *cannot* track nontransponder-equipped aircraft. ARTS-IIs require little maintenance or routine programming, yet they are less flexible and less expandable.

The FAA installed ARTS-II and its successor, ARTS-IIA, at over 100 airports across the United States. These systems are also in use in many foreign countries. The ARTS-IIA is an enhanced version that provides the controller with many of the ARTS-III features, such as conflict alert and minimum safe altitude warning. Although the ARTS-IIA system is somewhat limited in its programmability, it still provides sufficient computing power to track up to 256 aircraft at one time (see Figure 8-26).

Common ARTS

The replacement system for the aging ARTS system, known as STARS, will not become available until later this decade. Due to the inability of aging ARTS equipment to provide acceptable service, the FAA and the Lockheed-Martin Corporation developed an ARTS upgrade program known as common ARTS. Common ARTS is a replacement program for ARTS-II and ARTS-III equipment reaching the end of its service life. Common ARTS uses commercial off-the-shelf hardware and can be adapted to operate with both analog and digital airport surveillance radar systems.

The common ARTS system new software that is compatible with existing airport radar systems. Common ARTS is an expandable system that can take up to 15 inputs and drive as many as 200 individual display units. It uses commercial displays called ARTS Color Display (ACD) units to provide the controller with multicolor displays using a standard high resolution 20-inch video monitor. The ACD is similar to the color displays being installed in air route

FAA

Figure 8-26. ARTS-II radar equipment.

traffic control centers as part of the DSR program. All ARTS keyboard and trackball functions are displayed on the ACD; all the information previously displayed on the auxiliary ARTS displays can be displayed on the ACD itself. The controller can modify and move information using drop-down menus and windows similar to those used in personal computer graphical interfaces.

Common ARTS systems were installed at more than 140 sites including New York, Dallas/Fort Worth, Chicago, Denver, Minneapolis, Atlanta, and the Southern California Terminal Radar Approach Control facilities.

Radar Data Processing

The radar data processing (RDP) system used in the ARTCCs was designed at the same time as the flight data processing (FDP) system. Both systems make up the **National Airspace System stage A (NAS-A)**. The flight data portion of this automation project was known as phase one. Phase two involved the automation of the RDP system, which was implemented in all ARTCCs in the continental United States in 1974. The total National Airspace System, when finally completed, was considered the most complex computer system in existence in the world at that time.

During initial development of the RDP system, components of the prototype ARTS were modified to provide alphanumeric information to the controllers at the Indianapolis ARTCC. This prototype system was known as the **stored program alphanumeric (SPAN)** system. The development of SPAN was successful, but interfacing the ARTS-type equipment with the long-range radars used by the center was difficult. After experimenting with the SPAN system, the FAA decided that a different system should be installed at the ARTCCs, and the NAS-A system was developed.

Phase one of the National Airspace System program involved the installation of the FDP computer system, which was designed to provide automation capability to:

Accept and store flight plan information.

Print and distribute flight plan information in the form of flight progress strips.

Calculate and update flight plan data such as estimated time over specific intersections and estimated arrival time at the destination airport.

Transfer flight plan data automatically from one sector to the next within any particular ARTCC, from one ARTCC to the adjacent ARTCC, and from ARTCCs to FDIO-equipped control towers and TRACONs.

The second phase of the automation process was the completion of the RDP system, which provides for:

Radar input from multiple radar sites.

Radar mosaic capability.

Computer validation and selection of the most accurate data for display to the controller.

Automatic aircraft tracking.

Visual display of flight information, both on the radar display and on auxiliary cathode ray tube displays.

Automatic radar handoff capabilities.

Radar data processing as performed by the RDP system is a fairly complex process because every center may be equipped with many different primary and secondary radar systems that may be located hundreds of miles away. The data's accuracy must be verified and then converted into a format that can be transmitted long distances without inducing any errors. Once the data enter the main computer system, they must be sorted, merged with the data gathered from other radar sites, and then routed to as many as 100 different displays and input/output devices. In the RDP system, the FAA installed a **radar data acquisition and transfer (RDAT)** system at each of the remote radar sites. The RDAT system determines each aircraft's position using information received by the primary and secondary surveillance radar systems and routes this information to the **common digitizer (CD)**.

The common digitizer is a digital radar processor that receives data from the radar site, checks the information to determine its validity, and converts the target echoes into a digital format. The CD divides the radar's area of coverage into **range cells** of approximately 1 square mile. The CD determines the position of each aircraft and establishes which range cell the aircraft is located in.

As the radar completes each scan of the horizon, the CD determines each aircraft's altitude and transponder code and transmits the information to the appropriate ARTCC using a digital communications system.

Once received at the ARTCC, the radar data are electronically split, with identical data sent to the **central computer complex (CCC)**, which is the heart of the RDP system, and the **enhanced discrete address radar channel (EDARC)**, which is the backup computer system.

As originally designed, the CCC was an IBM 9020 computer. The IBM 9020s were initially installed in the early 1970s. As air traffic increased, however, this computer was unable to keep up and was replaced with IBM 3083 computers known as host computers. The host computer was replaced by the **host and oceanic computer system replacement (HOCSR)**, which is the primary device used in the RDP system and is also known as the CCC or the **prime channel.** The DARC system is known as the **backup channel.**

The HOCSR computer accepts the digital input transmitted from each radar site and correlates these data with the flight plan information stored in the FDP computer. The HOCSR then correlates the radar-derived position information with the appropriate fight plan information using the aircraft's transponder code. If more than one radar site is tracking a particular aircraft, the host computer determines which radar site is providing the most accurate data and then uses the information sent from that site. The unused data transmitted from other sites are temporarily stored.

For the RDP system to function properly, every aircraft within a particular ARTCC's area must be operating on a different transponder code. If two aircraft squawk the same code, the host computer will be unable to determine which is valid and will be unable to correlate properly. Since there is an insufficient number of available codes to issue a different transponder code to every aircraft, these codes are automatically assigned to each aircraft according to procedures contained in the **National Beacon Code Allocation Plan (NBCAP).**

The central computer complex correlates each aircraft's position and altitude with the flight data information (i.e., aircraft type, route of flight, navigation equipment) stored in the FDP computer. The host computer then plots the aircraft's current position, direction of flight, and ground speed and forwards this information to every affected control sector for display. The computer also calculates the aircraft's future position and altitude based on historical data. This track is very useful for predicting potential conflicts or for displaying the aircraft's predicted position whenever radar contact is temporarily interrupted.

The display information is routed to every sector workstation using a **computer display channel (CDC)**, which in turn routes the appropriate information to each controller's display (see Figure 8-27). The information displayed is not truly primary or secondary radar echoes; instead it is an electronic representation of radar-derived data.

The DSR display differs from the PPI display in that primary and secondary radar targets and ground clutter are not displayed. The HOCSR filters out this information and simply displays the information requested by the control-

FAA

Figure 8-27. Display system replacement system.

ler. In general, the only aircraft displayed are IFR or VFR aircraft operating within the vertical confines of the sector and VFR aircraft that do not have transponders or do not have altitude-encoding transponders. The VFR targets can be removed from the display whenever they begin to obscure IFR targets (see Figure 8-28).

The radar controller's workstation is equipped with a slew entry device (trackball) that simplifies the controller's request for data. Instead of typing in an aircraft's call sign when requesting data, the controller can move a symbol over the aircraft target on the display using the trackball and, with one or two key strokes, request specific information about that aircraft.

Enhanced Discrete Address Radar Channel

The **enhanced discrete address radar channel (EDARC,** although the new system is still commonly known as DARC) system operates as the backup for the RDP system. Before the development of DARC, whenever the prime channel malfunctioned, ARTCC controllers reverted to standard radar displays that did not display any alphanumerics. The displays were designed to convert to standard PPI screens and were hinged so that they could be pulled out to a horizontal position. As the controllers reidentified each aircraft, their identities were marked on plastic shrimp boats using grease pencils. The controller then moved the shrimp boats across the face of the display. The confusion that resulted whenever an unscheduled prime channel shutdown occurred usually immobilized the center for a time. If both the flight data processing and the radar data processing functions were lost simultaneously, the controllers would have to revert to manual handoffs and handwritten flight progress strips. In some cases, only partial computer failure would occur and only a few automated functions would be disabled.

To try to reduce the inevitable confusion that occurred whenever the prime channel malfunctioned, the FAA contracted with the Raytheon Corporation to design a backup system that would instantaneously take over and provide a display that was fairly similar to the one provided by the prime channel. This system was introduced into service in 1980 and is known as the discrete address radar channel.

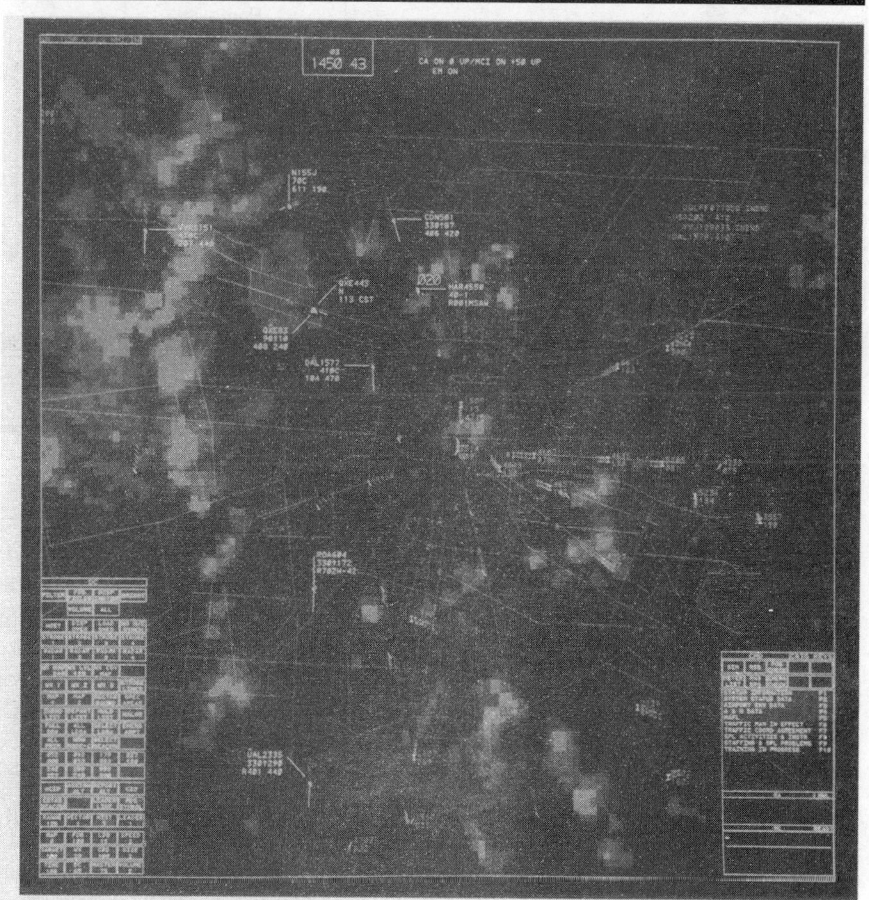

FAA

Figure 8-28. A DSR Display.

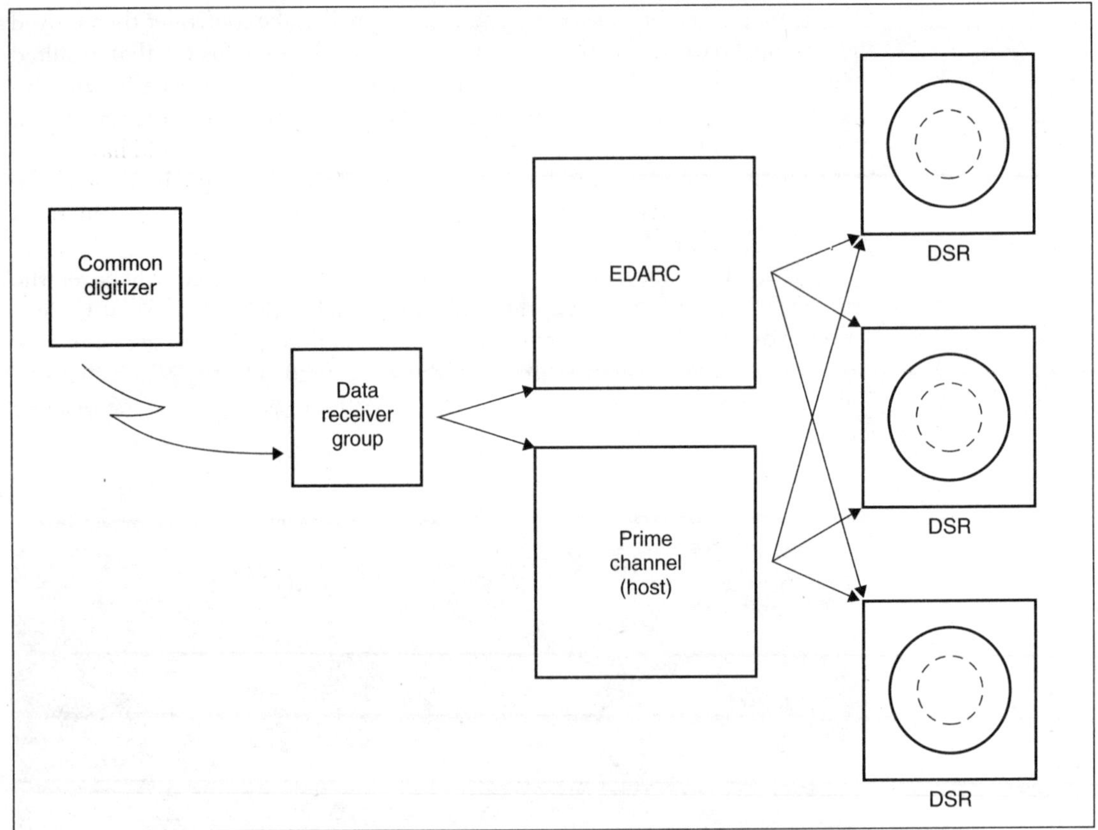

Figure 8-29. The digitized data transmitted by the common digitizer are split between the prime channel and the EDARC processors. Either display is available to every controller.

The EDARC system operates constantly in one of three modes: EDARC/NAS, EDARC, or EDARC/FDP. During EDARC/NAS operation, the EDARC system operates as a hot backup system in parallel with the prime channel. The prime channel provides both radar and flight data processing functions and the DARC operates strictly as a backup system. The DARC system processes the same information provided to the host computer. DARC and prime channel information are always available at each DSR and can be displayed at the flip of a switch (see Figure 8-29).

In case of full or partial failure of the host computer, the EDARC system can operate in one of the two remaining modes: EDARC/FDP or EDARC. If the entire CCC fails, resulting in a loss of both RDP and FDP functions, the EDARC system can take over and provide computer-generated display of traffic to each controller; this is the EDARC mode. In this mode, the EDARC display is fairly comparable to that offered by the prime channel. The EDARC system *cannot,* however, replace the flight data functions provided by the flight data processing system. The controllers must begin to handwrite and update flight progress strips; this is not as inconvenient as it first seems, however. In most cases, the prime channel failure will not last for a long time, and most of the control positions will already have preprinted flight progress strips available for each aircraft. In most cases the prime channel will be back on line before new flight strips actually have to be handwritten.

In many cases, only partial failure of the CCC may occur. Then the EDARC may take over the functions of the prime channel, while the host computer continues to operate the flight data processing functions; this mode is known as EDARC/FDP. While operating in this mode, the EDARC computers communicate with the host computer just as in prime channel operation; the host computer provides the flight data processing function and EDARC provides the radar data processing function. When the prime channel returns to full operation, the limited flight data stored in the EDARC computers can be sent back to the host computer, aiding in the reidentification of aircraft that should be tracked.

Center Radar ARTS Presentation

Center radar ARTS presentation (CENRAP) is a computer system used as a backup for airport surveillance radar. The program, utilized at airports equipped with ARTS systems, can provide ARTCC radar positioning information for display to the controllers via the ARTS. CENRAP can be utilized if the airport surveillance radar is unavailable. It normally is not as accurate as the main radar, since the ARTCC data may be derived from a radar site quite a distance from the main airport.

KEY TERMS

air route surveillance radar (ARSR)
air traffic control radar beacon system (ATCRBS)
airport surface detection equipment (ASDE)
airport surveillance radar (ASR)
altitude filtering
antenna
ARTS-II
ARTS-III
ARTS-IIIA
associated tracks
automated handoff
automated radar terminal system (ARTS)
backup channel
beacon data acquisition system (BDAS)
beacon slash
beacon tracking level (BTL)
blind speed
blip
boresight
center radar ARTS presentation (CENRAP)
central computer complex (CCC)
challenge pulse
circular polarization (CP)
clutter map
coast
coast list
common digitizer (CD)
computer display channel (CDC)
Conflict Alert
Conflict Alert IFR/VFR Mode C Intruder
continuous wave (CW) radar
data acquisition subsystem (DAS)
data block
data entry and display subsystem (DEDS)
data entry sets (DES)
data processing subsystem (DPS)
decenter
defruiter
discrete address radar channel (DARC)
double bloomer
duplexer
echo

enhanced discrete address radar channel (EDARC)
en route ARTS (EARTS)
feedhorn
flight data processing (FDP)
framing pulse
fruit
gain
Geo-Map
gray scale
ground clutter
history
host and oceanic computer system replacement (HOCSR)
Ident
identification friend or foe (IFF)
joint surveillance system (JSS)
linear polarization (LP)
main bang
minimum safe altitude warning (MSAW)
mode
moving target detection (MTD)
moving target indicator (MTI)
MTI gate
MTI/MTD video gain
National Airspace System stage A (NAS-A)
National Beacon Code Allocation Plan (NBCAP)
noise
normal video gain
plan position indicator (PPI)
precision approach radar (PAR)
PRF stagger
prime channel
pulse repetition frequency (PRF)
pulse repetition rate
pulse train
pulse-type radar
radar
radar and beacon tracking level (RBTL) system
radar cross section
radar data acquisition and transfer (RDAT)
radar data acquisition system (RDAS)
radar data processing (RDP)
radar mile
radar mosaic
radar scope
radial velocity
range cells
range mark
range select switch
receiver
receiver gain
reply
sensitivity time control (STC)
side lobe suppression (SLS)
side lobe suppression omnidirectional antenna
side lobes
special identification pulse (SIP)
squawking
Standby
stored program alphanumeric (SPAN)
sweep
sweep decenter
tangential track
target
target coast
target illumination
ten-channel selector
threshold
TPX-42
track
track drops
trackball
transmitter
transponder
unassociated track
video map
video map intensity
video map selector
waveguide

REVIEW QUESTIONS

1. What is the function of the radar transmitter, feedhorn, antenna, and receiver?
2. What is the difference between primary and secondary radar?
3. How is ground clutter removed from a radar display?
4. How is weather removed from a radar display?
5. What can controllers use radar for?
6. How does the use of radar make the air traffic control system more efficient?
7. What types of computerized radar systems does the FAA currently use and how do they differ?

9

Radar Separation

Checkpoints

After studying this chapter, you should be able to:

1. Identify the methods of primary radar identification.
2. Identify the methods of secondary radar identification.
3. Explain the differences between and the uses for handoffs and point outs.
4. Explain the difference between transfer of control and transfer of communications.
5. Explain the three methods of separating aircraft using radar.
6. Understand the use of radar for instrument approaches.
7. Explain the significance and the purpose of the approach gate.

Radar can be used more efficiently to separate aircraft than nonradar separation techniques and can also be used to provide additional ATC services to pilots. In general, radar is used by controllers to provide the following services to pilots:

Aircraft identification and location.

Aircraft separation.

Navigation assistance.

Instrument approaches.

Traffic advisories.

Unsafe condition alerts.

Aircraft Identification

Within certain limitations, radar can be used to easily locate and establish the identity of any aircraft, whether IFR or VFR, whose pilot is requesting air traffic control services. Since any particular radar system is capable of displaying hundreds of radar targets at any given time, the controller must be absolutely certain of a particular aircraft's identity prior to offering radar service to that pilot. Failure to identify a target when using radar creates an obvious safety hazard. If, for instance, a controller were to provide navigational assistance to the wrong aircraft, the instructions might cause the pilot to become disoriented or might cause the aircraft to crash into terrain or into another aircraft.

An aircraft must be positively identified using either the primary or the secondary radar system. Since each identification method has potential drawbacks, it is in the controller's and pilot's best interest to use multiple methods when identifying an aircraft using radar.

Primary Radar Identification

Primary radar identification methods are usually employed when the aircraft in question is not equipped with a transponder or when the transponder is inoperative or operating intermittently. Because primary radar identification methods are fairly imprecise, secondary surveillance radar identification techniques should be used whenever possible.

The easiest method for identifying an aircraft using primary surveillance radar is to observe an aircraft that has just departed from an airport. Since only one aircraft can depart from a runway at any given time, it can be safely assumed that a departing target observed within 1 nautical mile of the departure end of the runway is positively identified. This method of identification is not without its liabilities, however. The controller should be particularly cautious if the airport is constructed with parallel runways or intersecting runways whose departure ends are located close to each other, as two aircraft departing from separate runways at approximately the same time could be misidentified. The controller should also be aware of other traffic in the vicinity of the runway that might appear to be a departing aircraft. This traffic might include aircraft remaining in the traffic pattern or even large vehicles traveling near the airport.

A second method of identifying an aircraft using primary radar requires the pilot to report over a point whose exact location is known to the controller and displayed on the radar screen. Possible locations include airway intersections, prominent terrain features, or nearby cities or towns. If the controller observes a single target located over one of these landmarks, positive identification of that aircraft can be presumed. The controller should use caution when pilots report over highways, rivers, or large cities, because these are not precise locations. In addition, the controller should always consider that the pilot might be unsure of the aircraft's exact location when making initial contact with the controller. Highways, rivers, and towns can easily be misidentified by the pilot.

A third method of identifying a primary target using radar is to request that the pilot turn the aircraft to a particular heading and then observe which blip on the radar performs the proper turn. To ensure that this procedure results in identification of the correct aircraft, the controller must:

Issue the pilot a turn that differs by at least 30° from the current heading of the aircraft. Since pilots routinely make turns of 10° to 20° during the course of a flight, a 30° turn is sufficient to positively identify an aircraft. Because there is always the possibility of misidentification, controllers using this method should routinely request that the aircraft make a series of two or more turns before confirming the aircraft's identification.

Ensure that the aircraft is not located on or near any IFR or VFR routes where a turn similar to that issued might routinely be performed by aircraft.

Ensure that the unidentified aircraft is actually within the range of the radar system. Any turn issued to an aircraft not observed on the radar display might actually head that aircraft in a dangerous direction.

Ensure that only one aircraft is observed performing the assigned turns. Although prohibited by the FARs, it is not implausible for the pilot of one aircraft to perform turns issued to another aircraft. This could occur because of garbled radio communications or because the pilot mistakenly responds to a transmission meant for another aircraft.

Secondary Surveillance Radar Identification

More accurate methods of positively identifying aircraft can be provided by the secondary surveillance radar system. Because of their accuracy, in most cases it is preferable to use these methods to determine an aircraft's identity.

The most common method of secondary radar identification is use of the Ident feature included in the transponder. When the pilot presses the Ident button, the transponder transmits a **special identification pulse (SIP)** to the ATCRBS receiver on the ground, which interprets the SIP and causes the radar to display a double-width beacon slash on the PPI. Since the special identification pulse is the only method by which this double-width slash is normally produced, the use of the Ident feature provides accurate, positive identification of an aircraft. This method is not without its shortcomings, however, since overlap-

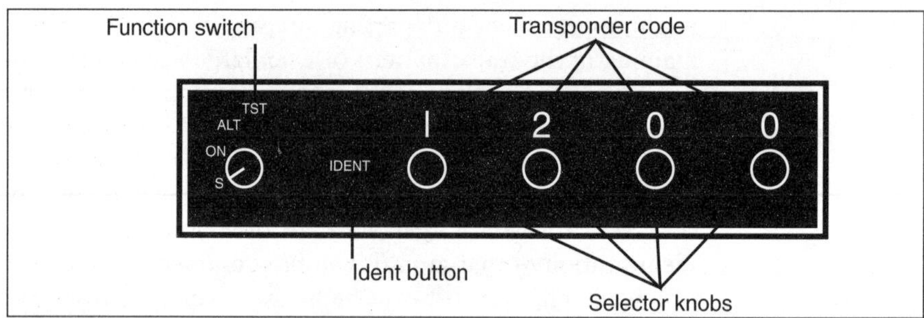

Figure 9-1. A light-aircraft transponder.

ping transponder returns from two aircraft can be interpreted by the ATCRBS receiver as an Ident. In addition, there is always the chance of miscommunication with a pilot, causing the pilot of an unintended aircraft to mistakenly send an Ident signal.

A second method of aircraft identification using the secondary surveillance system requires the pilot to switch the transponder from the On position to the Standby position (see Figure 9-1). When this is done, the transponder remains on but will no longer respond to interrogations from the ATCRBS transmitter on the ground. Once set to Standby, the transponder beacon slash on the PPI disappears, with only the primary radar target remaining. When the controller is certain that the beacon slash has disappeared from the radar display, the pilot is requested to return the transponder to the On position. Caution should be used when employing this technique, as many factors can cause the beacon slash to temporarily disappear from the PPI. For example, transient malfunctions in the ATCRBS equipment on the ground or on the aircraft may temporarily cause the beacon slash to disappear, and initiating a turn of the aircraft toward the radar antenna may temporarily shield the airborne transponder antenna, resulting in a loss of the beacon slash on the PPI. Therefore, whenever a controller uses this radar identification technique, the beacon slash must disappear for a sufficient time interval to ensure that the transponder has in fact been switched to the Standby position.

If the controller is using beacon decoding equipment that can display the aircraft's transponder code directly on the PPI, the controller can positively identify an aircraft by requesting that the pilot select a specific transponder code. This is accomplished by requesting that the pilot "squawk" a particular code (for example, "Falcon six two mike squawk two one four five"). Once the pilot has placed the assigned code into the transponder, it can be displayed directly on the PPI.

When using this technique, the controller must be aware of some of its possible limitations. If the pilot has not turned the transponder to the On position, it will not reply to the ATCRBS interrogation regardless of the code that has been selected. In addition, many transponders require a few minutes to warm up. If the pilot has just turned the transponder on, it may not reply to an interrogation for a number of minutes, even if the proper code has been selected.

If the controller is using a computerized secondary surveillance radar system that can generate a full data block on the PPI, the acquisition and display of this data block may be used to positively identify an aircraft. Once the controller has entered the aircraft's identification and transponder code into the computer, the display of the aircraft's data block on the PPI can occur only if the proper transponder code has been received by the radar system. Thus, it can be safely assumed that the acquisition and display of the aircraft's identity on the PPI is evidence that the aircraft is squawking the proper transponder code (see Figure 9-2).

All radar identification techniques depend on clear, concise communication between the controller and pilot. If the controller issues an improper instruction or if the wrong pilot reacts to an instruction, mistaken identification could result. That is why a combination of primary and secondary surveillance radar techniques should be used whenever a controller is identifying an aircraft.

Loss or Termination of Radar Contact Once an aircraft has been identified, the pilot is informed through the use of the phrase "radar contact." A controller's use of this phrase informs the pilot that radar identification has been estab-

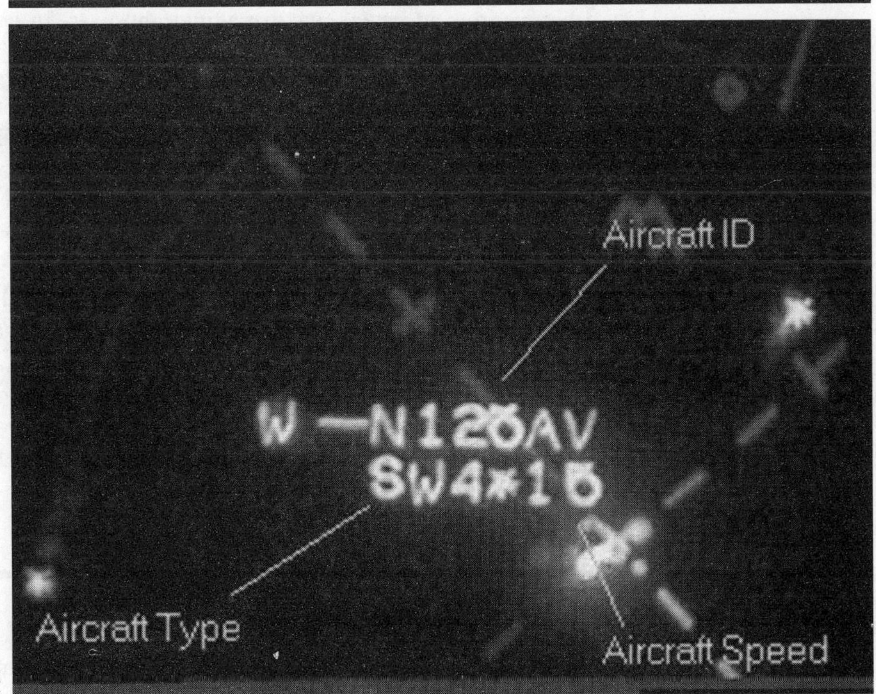

FAA

Figure 9-2. ARTS data tag.

lished and that radar services can now be provided. If radar identification is subsequently lost, the controller informs the pilot using the phrase "radar contact lost." This phrase advises the pilot that the controller can no longer identify the aircraft using radar and that radar services are no longer being provided. In addition, this phrase means that if the aircraft is operating on an IFR flight plan, the controller will separate the aircraft using nonradar separation methods. Before the controller can begin to offer radar services to an aircraft whose radar contact has been lost, he or she must reidentify the aircraft using one of the previously described methods.

If at any time during the flight either the controller or the pilot chooses to discontinue radar service to the aircraft, the pilot is informed using the phrase "radar service terminated." This phrase is most commonly used when a radar-equipped facility hands off an IFR aircraft to a nonradar facility or when a VFR aircraft reaches the outer limit of a facility's radar coverage area.

Altitude Verification Before using a mode C–generated altitude readout for aircraft separation, the controller must verify that the transponder onboard the aircraft is operating correctly and transmitting the proper altitude information. This verification can be accomplished in one of two ways:

The altitude displayed on the radar must vary by less than 300 feet from the pilot's verbally reported altitude.

The controller must observe a continuous altitude readout from an aircraft departing from an airport, and that altitude must vary by less than 300 feet from the airport elevation.

Invalid Mode C Operation Whenever the altitude readout from a transponder varies from that reported by the pilot, the controller must request that the pilot confirm the proper operation of both transponder and altimeter. The most likely cause of the problem is a malfunctioning altitude encoder, but it is also possible that the pilot misset the aircraft's altimeter, resulting in the aircraft actually flying at the wrong altitude. The phraseology to verify correct transponder operation is: "Cessna one mike lima, verify altitude and altimeter setting."

If the pilot reports that both the altimeter setting in use and the indicated altitude are correct, the mode C equipment on the aircraft is assumed to be malfunctioning and should be turned off. In this case, the controller should inform the pilot and request that the transponder be readjusted to operate on mode A (nonaltitude reporting): "Cessna one mike lima, stop altitude squawk, altitude differs by five hundred feet."

Transfer of Radar Identification

Once an aircraft has been initially radar identified by a controller, subsequent controllers need not repeat any of the radar identification procedures as long as the identification has not been terminated and has been continuously maintained and transferred to each subsequent controller. The transfer of radar identification from one controller to the next is extremely important, ensuring efficiency of the air traffic control system. If an aircraft's radar identification is lost just prior to or during a handoff, subsequent controllers must use nonradar techniques to separate that aircraft until its radar identification can be reestablished. Nonradar separation procedures inefficiently use existing airspace, and the reidentification of aircraft consumes an inordinate amount of the subsequent controller's time and concentration.

Because the transfer of radar identification is so important to the operation of the ATC system, the FAA has developed standardized procedures and terminology that should be used when transferring aircraft identification. The following standard definitions and terminology are excerpted from the FAA handbook:

Handoff	An action taken to transfer the radar identification of an aircraft from one controller to another if the aircraft will enter the receiving controller's airspace *and* radio communications with the aircraft will also be transferred.
Radar contact	The term used to inform the controller initiating the handoff that the aircraft has been identified and approval is granted for the aircraft to enter the receiving controller's airspace.
Point out	An action taken by a controller to transfer the radar identification of an aircraft to another controller if the aircraft will or may enter the airspace of another controller and radio communications with the aircraft will *not* be transferred.
Point out approved	The term used to inform the controller requesting the point out that the aircraft has been identified and that approval is granted for the aircraft to enter the receiving controller's airspace, without a transfer of radio communication.
Traffic	Term used to transfer radar identification of an aircraft to another controller for the purpose of coordinating another action. "Traffic" is normally issued in response to a handoff or a point out and is used to identify other aircraft that may become a factor in aircraft separation.
Traffic observed	Term used to inform the controller issuing traffic information that the traffic is identified and that any restrictions issued concerning that aircraft will be complied with.

Handoff Procedures

Positive radar identification must be transferred from one controller to the next whenever an aircraft traverses the boundary between air traffic control sectors. The first controller (known as the transferring controller) is responsible for ensuring that the following requirements are complied with:

The handoff must be concluded before the aircraft crosses the sector boundary. An aircraft is not permitted to cross a boundary between two sectors without the knowledge and the permission of the receiving controller.

During the handoff, the aircraft must be radar identified by the receiving controller. (This may be accomplished using any of the previously described methods.)

The **transfer of communication** must be accomplished before the aircraft crosses the sector boundary. This permits the receiving controller to be in radio contact with the pilot prior to the aircraft crossing the sector boundary. The receiving controller may not issue any clearance that will change the aircraft's route of flight or altitude while it is still within the transferring controller's sector.

The receiving controller's approval must be received before the aircraft crosses the sector boundary. Verbal communication is normally accomplished using intrafacility intercom or leased telephone circuits. Automated handoffs are accomplished using ARTS or NAS-A computer equipment.

Potential traffic conflicts must be resolved prior to the transfer of communication.

Both controllers must comply with the procedures specified in applicable letters of agreement or facility directives. These procedures include those for preferred routes and altitudes and radio frequency assignments.

Unless expressly negotiated between the involved controllers, the **transfer of control** occurs at the sector boundary.

Any restrictions issued by the receiving controller must be complied with.

Before accepting a handoff, the receiving controller must comply with the following rules:

The aircraft being transferred must be radar identified.

The controller must agree that the aircraft can be safely accepted and that separation must be provided.

Any applicable restrictions regarding that aircraft must be communicated to the transferring controller.

All of the procedures specified in the applicable letters of agreement or facility directives must be complied with.

The receiving controller is not permitted to change the altitude, heading, speed, or transponder code of the aircraft until it crosses the sector boundary. The transferring controller is still responsible for separating the aircraft until it has crossed the sector boundary and assumes that the aircraft will comply with the clearance that was in effect before the handoff occurred. If the receiving controller needs to alter the aircraft's route or altitude before it crosses the sector boundary, permission *must* be received from the transferring controller before the clearance is issued to the pilot.

Any of these conditions may be altered upon the consent of both controllers. This process of negotiating and granting permission for these changes is known as effecting **coordination.** Here is an example of handoff phraseology:

TRANSFERRING CONTROLLER:	Handoff, Delta two eleven, seven miles east of Kelly, seven thousand.
RECEIVING CONTROLLER:	Delta two eleven, radar contact.

Point Out Procedures

In instances in which an aircraft may cross a number of sector boundaries in a short time, it may not be efficient or practical to require that the pilot contact every responsible controller. Since some of the controllers may be in contact with the aircraft only briefly and may need to initiate a new handoff immediately after accepting the aircraft, it may be more efficient for the first controller to let the aircraft enter the second and the third controllers' airspace but to receive permission to transfer the aircraft's communication directly to the third controller. The second controller is advised of the aircraft's position and approves the aircraft's entry into the sector but waives the requirement to communicate with the aircraft. This sequence of events is known as a *point out.*

During a point out, the second controller remains responsible for the separation of the aircraft as it traverses through his or her sector but agrees not to communicate with the aircraft. The first controller is required to receive permission to traverse the second controller's airspace and must also hand off the aircraft to the third controller. This procedure is most advantageous when the aircraft will enter the fringes of the second controller's airspace and will remain there for only a short time.

In the example of a point out shown in Figure 9-3, the aircraft traverses sectors alpha, bravo, and charlie. Since the aircraft will be in bravo sector only

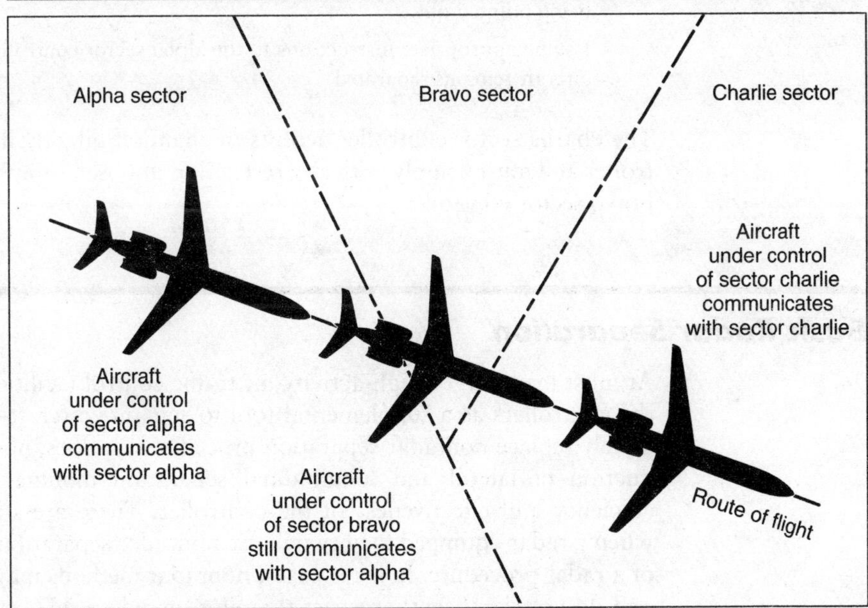

Figure 9-3. An example of a point out. The alpha sector controller points out the aircraft to the bravo sector while effecting a handoff with the charlie sector.

briefly, the transferring controller (controller alpha) chooses to point out the aircraft to the bravo controller while coordinating a handoff with the charlie sector controller. During a point out, the transferring controller (controller alpha) must adhere to the following rules:

> Permission from the bravo sector controller must be received before the aircraft enters bravo sector. The bravo controller is not obligated to approve the point out and may insist on communicating with the aircraft. If this occurs, the point out with the bravo sector controller becomes a handoff.
>
> The bravo sector controller may stipulate restrictions to be placed on the aircraft while it is in bravo sector.
>
> The bravo sector controller may also identify potentially conflicting traffic to the transferring controller (controller alpha), and the alpha sector controller must comply with any restrictions.
>
> The aircraft's altitude, heading, speed, or transponder code may not be altered while it is within the bravo sector.
>
> The alpha sector controller is responsible for initiating the handoff with controller charlie. This controller must advise the charlie sector controller that approval for the point out has been received from the bravo sector controller, and any restrictions placed on the aircraft by the bravo controller must be conveyed.

During a point out, the bravo sector controller is responsible for:

> Ensuring that every aircraft within bravo sector is separated from the aircraft being pointed out.
>
> Issuing appropriate instructions to the alpha sector controller to ensure that the aircraft remains separated.

The charlie sector controller accepts the handoff directly from the alpha controller and must comply with any restriction imposed by either the alpha or the bravo sector controller.

Basic Radar Separation

At most medium- or high-activity air traffic control facilities, radar is used by the controllers as a supplemental tool to separate aircraft—it does not completely replace nonradar separation procedures. It does, however, permit a reduction of lateral and longitudinal separation minima and increases the efficiency and effectiveness of the controller. There are still many occasions when a radar-equipped facility will use nonradar separation procedures in lieu of a radar procedure. In some cases, nonradar methods may be easier to apply and do not significantly restrict the pilot or reduce ATC system efficiency. In many areas, radar coverage does not extend as far as the FAA would like, and nonradar procedures are still used to separate aircraft. In addition, some FAA air traffic control facilities remain unequipped with radar and must rely on nonradar procedures.

But in most cases, use of radar increases ATC system efficiency, reduces controller workload, and enhances safety. When using radar, controllers can visualize the position of each aircraft, permitting most separation standards to be reduced. Radar also permits the controller to issue headings to pilots to more effectively use the airspace. In most cases, the routine use of holding patterns has been virtually eliminated through the use of radar. Lost aircraft can be assisted, pilots can be warned of nearby traffic, and instrument approaches can be conducted solely through the use of radar.

Radar is most commonly used by controllers to reduce the separation interval between aircraft participating in the air traffic control system. Controllers are not obligated to use radar separation procedures exclusively, however. If a nonradar separation method is more efficient or easier to apply in a particular situation, the controller is free to use nonradar procedures.

Separation Standards

Radar separation criteria are defined in much the same way as nonradar separation criteria. Separate procedures and criteria are used when applying vertical, lateral, longitudinal, or initial separation of aircraft. As when using nonradar procedures, the controller needs to apply only one of these methods of separation to any particular aircraft.

Vertical Separation Vertical separation procedures for controllers using radar are similar to those used by nonradar controllers. Aircraft at or below FL 290 must be separated by a minimum of 1,000 feet. Aircraft operating above FL 290 must be separated by a minimum of 2,000 feet. The exception occurs when two aircraft are either climbing or descending. The following aircraft can always be assigned the altitude that the previous aircraft has vacated if the pilot reports leaving that altitude or if the controller observes a valid mode C indication that the aircraft has passed through that altitude.

Longitudinal Separation After identifying the aircraft, the controller may use radar to reduce the required longitudinal separation. As long as the minimum longitudinal separation interval between each aircraft (usually 3 to 5 nautical miles) can be maintained, longitudinal separation is presumed to exist. When applying longitudinal separation using radar, the controller must measure the distance using the following reference points, which vary depending on the type of equipment operating on each aircraft:

> If neither aircraft is transponder equipped, the centers of the primary radar targets are used to measure the distance between targets. In no situation should the primary radar blips ever be permitted to overlap (see Figure 9-4).
>
> If both aircraft are transponder equipped, the distance between the targets is measured from the ends of the beacon control slashes.

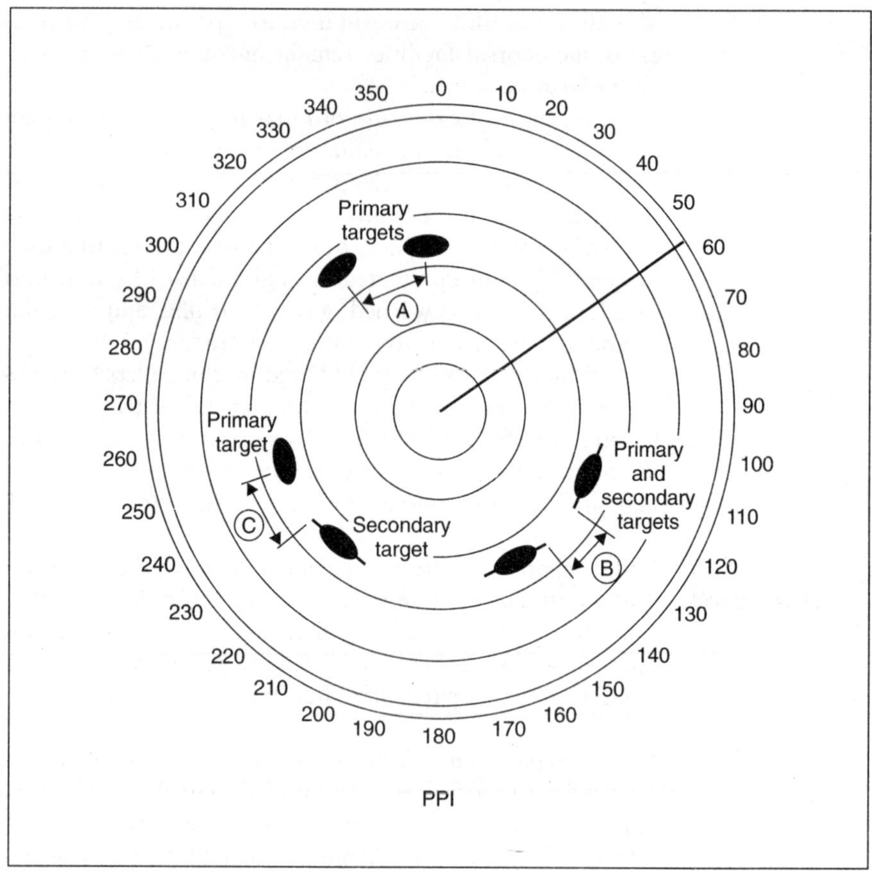

Figure 9-4. Target separation minima using a PPI display such as ARTS or STARS.

> If only one aircraft is transponder equipped, the distance between aircraft is measured from the center of the primary target to the end of the beacon slash.
>
> Controllers working at facilities equipped with all-digital displays, such as DSR, DARC, or EARTS, should measure the distance between the centers of the digitized targets. Under no circumstances should the targets be permitted to touch (see Figure 9-5).

The basic longitudinal separation minimum is 3 nautical miles, but because the width of the radar pulse increases as the pulse travels away from the antenna, distant targets appear much larger on a radar display than those located closer to the radar antenna. For this reason, the FAA has provided increased separation criteria for aircraft located more than 40 nautical miles from the radar antenna. The longitudinal separation standards for controllers using radar are (a) 3 nautical miles if both targets are located less than 40 nautical miles from

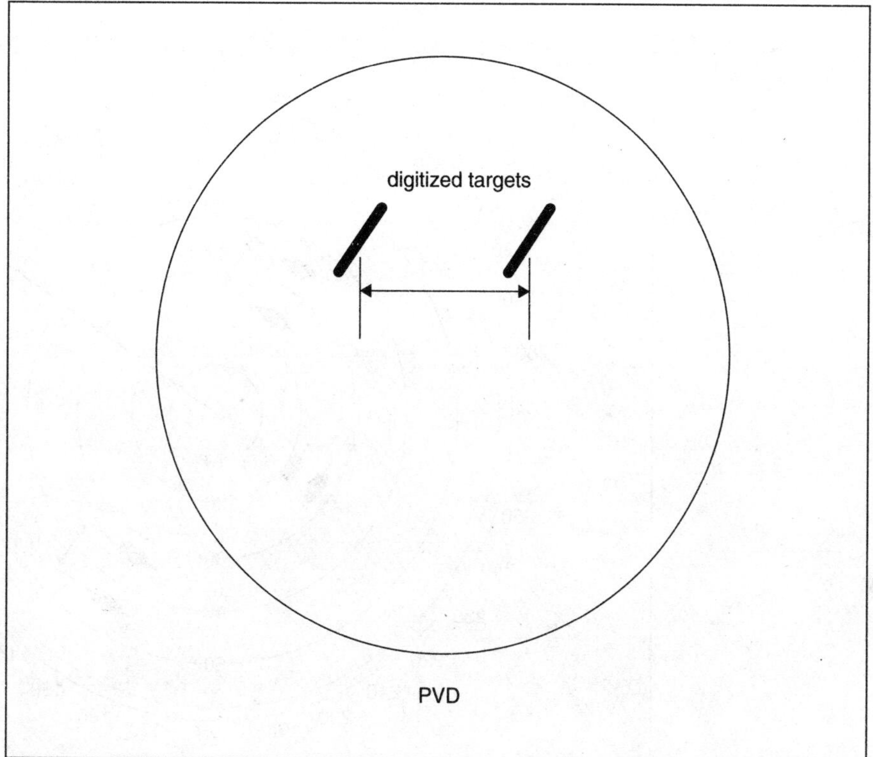

Figure 9-5. Target separation minima using a digital display (applicable to DARC, DSR, or EARTS displays).

the radar antenna, and (b) 5 nautical miles if either aircraft is 40 or more nautical miles from the radar antenna (see Figure 9-6).

Whenever a radar data processing system such as NAS-A, DARC, or EARTS (commonly known as a digital or mosaic radar system) is utilizing more than one radar site to create a radar mosaic, the controller is unable to determine which antenna is actually being used to locate any particular aircraft. Because of atmospheric conditions, terrain obstructions, and general system performance, it cannot always be assumed that the closest radar system is the one actually being used to determine the aircraft's position. For this reason, controllers using mosaic systems must always assume that each target is potentially 40 nautical miles or farther from the radar antenna and therefore must separate every aircraft by a minimum of 5 nautical miles.

These longitudinal separation criteria do not offer sufficient protection to a small aircraft behind or directly under and behind a larger aircraft. To reduce the chance that the smaller aircraft may encounter damaging wake turbulence, the FAA has mandated increased separation. The FAA handbook states that

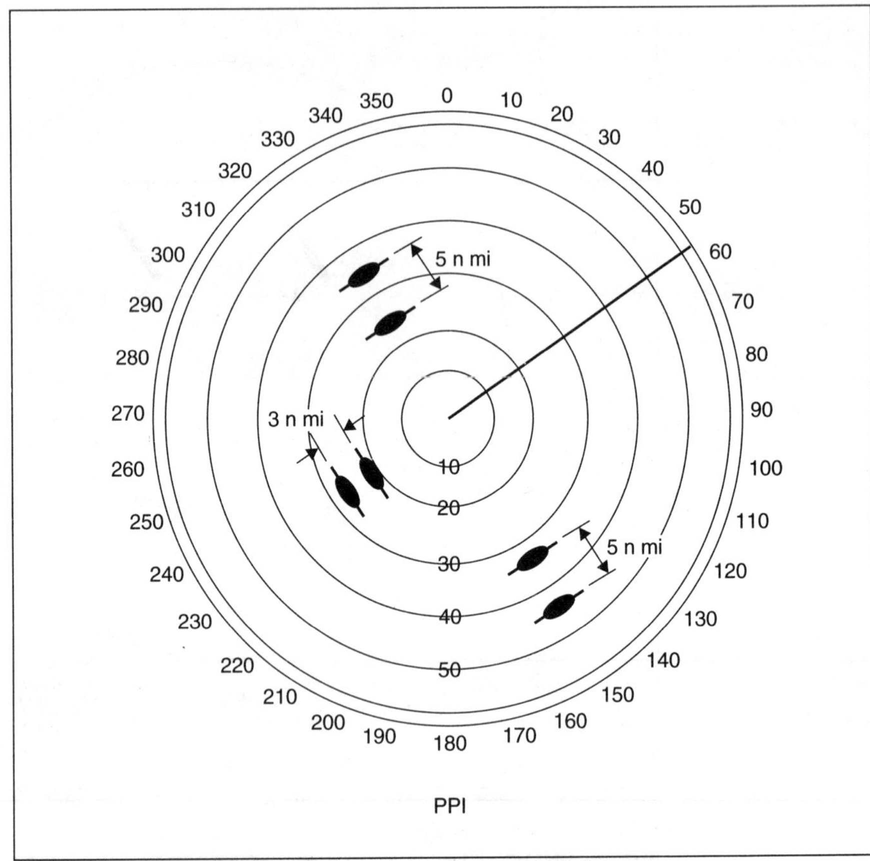

Figure 9-6. Longitudinal radar separation.

whenever a smaller aircraft is following a larger aircraft at the same altitude or 1,000 feet below it, the following longitudinal separation criteria must be used:

Four miles between a heavy aircraft following a heavy aircraft.

Five miles between a small aircraft following a heavy aircraft.

Five miles between a large aircraft following a heavy aircraft.

In addition, because of the increased risk of wake turbulence to small aircraft during the approach and landing phases of flight, the FAA has mandated increased longitudinal separation when a small aircraft is landing behind a larger aircraft (see Table 9-1). The handbook states that when a small aircraft is following a larger aircraft and is landing on the same runway, the following separation intervals must exist when the larger aircraft crosses the landing threshold:

Table 9-1. Wake Turbulence Minima (nautical miles of longitudinal separation)

En Route and Approach Separation

	Leading Aircraft		
Trailing Aircraft	*Small*	*Large*	*Heavy*
Small	3	3	5
Large	3	3	5
Heavy	3	3	4

Landing Separation

	Leading Aircraft		
Trailing Aircraft	*Small*	*Large*	*Heavy*
Small	3	4	6
Large	3	3	5
Heavy	3	3	4

A small aircraft following an aircraft classified as large must be separated by at least a 4-nautical-mile interval.

A small aircraft following an aircraft classified as heavy must be separated by at least a 6-nautical-mile interval.

Lateral Separation Lateral separation minima applied in a radar environment are similar to the longitudinal separation minima. One of the primary differences between the two is that wake turbulence avoidance is not a factor when utilizing lateral separation. The minimum lateral separation interval when using radar is:

Three nautical miles if both targets are located less than 40 nautical miles from the radar antenna.

Five nautical miles if either aircraft is 40 or more nautical miles from the radar antenna.

Whenever a controller is using a radar system that creates a radar mosaic, the controller must separate aircraft laterally by a minimum of 5 nautical miles. An exception to the basic lateral separation rule can be made whenever two aircraft are flying courses that diverge by at least 15° (see Figure 9-7). The FAA handbook states that lateral separation can be presumed to exist between these two aircraft when both of the following conditions exist:

Aircraft traveling in opposite directions have passed each other, if they are flying in the same direction, one aircraft has crossed the projected course of the other.

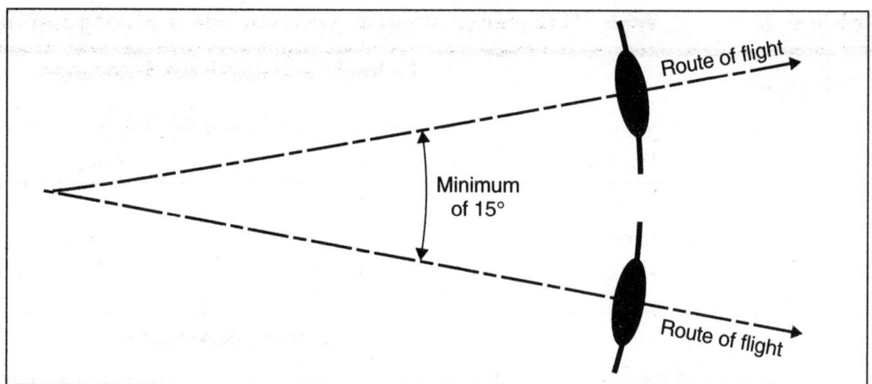

Figure 9-7. An example of lateral radar separation between two diverging targets.

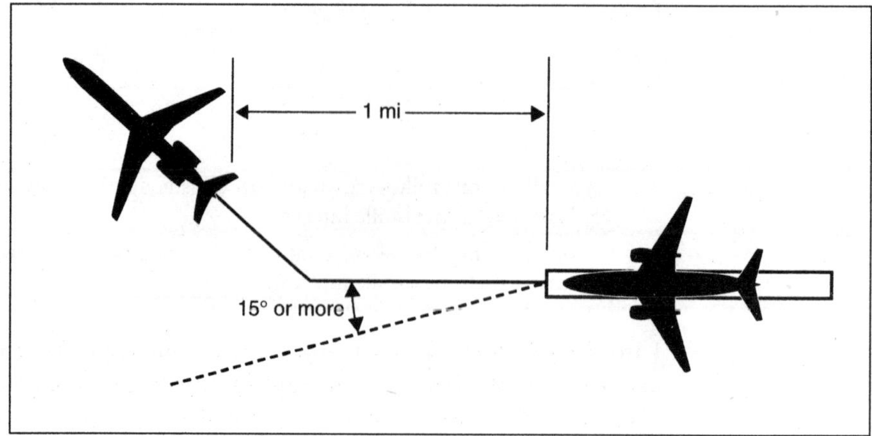

Figure 9-8. Aircraft departing from the same runway and diverging immediately after takeoff must be separated by at least 1 mile.

> Sufficient separation exists such that neither the primary targets nor the beacon control slashes touch each other.

Whenever these conditions exist, the controller may discontinue the use of either vertical or longitudinal separation.

Initial Separation of Departures Within a radar environment, lateral separation minima can be reduced when separating two aircraft departing from the same airport. These reduced minima can be applied only to aircraft whose courses will eventually diverge by at least 15°. (This differs from a nonradar environment where the course divergence requirement is a minimum of 45°.)

If both aircraft are departing from the same runway and their courses will diverge by at least 15° immediately after takeoff, a 1-mile separation interval must be maintained (see Figure 9-8). If the two aircraft will not diverge immediately after takeoff, the controller must apply longitudinal, vertical, or visual separation.

If two aircraft are departing from separate runways that do not intersect and both the runways' and the aircraft's courses diverge by at least 15°, simultaneous departures are authorized with no separation interval (see Figure 9-9).

If the runways intersect but diverge by at least 15° and the aircraft's courses will diverge by at least 15°, the following aircraft can be authorized to depart after the leading aircraft has crossed the runway intersection (see Figure 9-10).

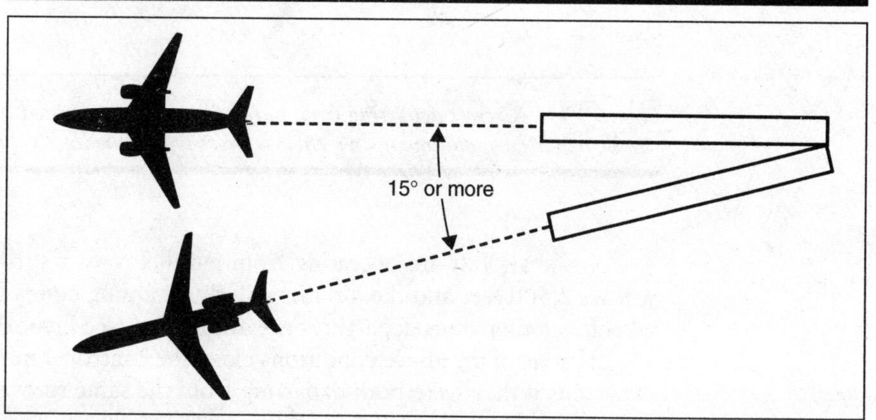

Figure 9-9. Aircraft departing from different, nonintersecting runways that diverge immediately after takeoff can be simultaneously departed.

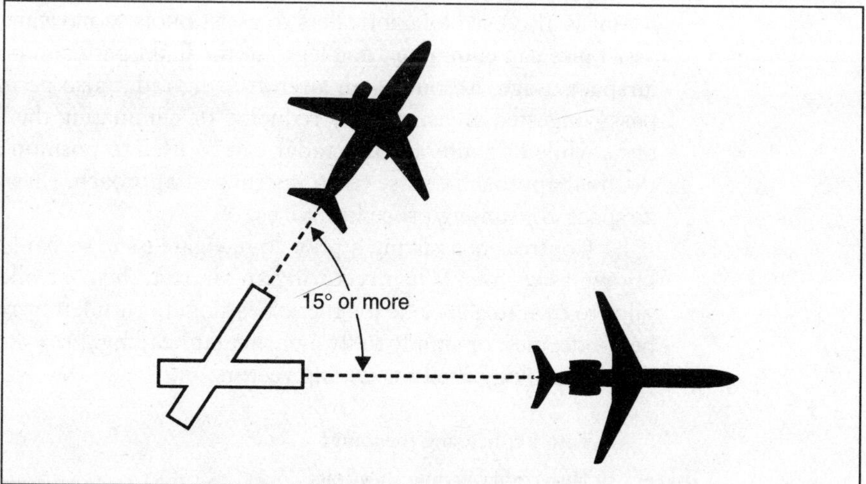

Figure 9-10. Aircraft departing from intersecting runways that diverge by at least 15° can be departed once the first aircraft has passed through the intersection.

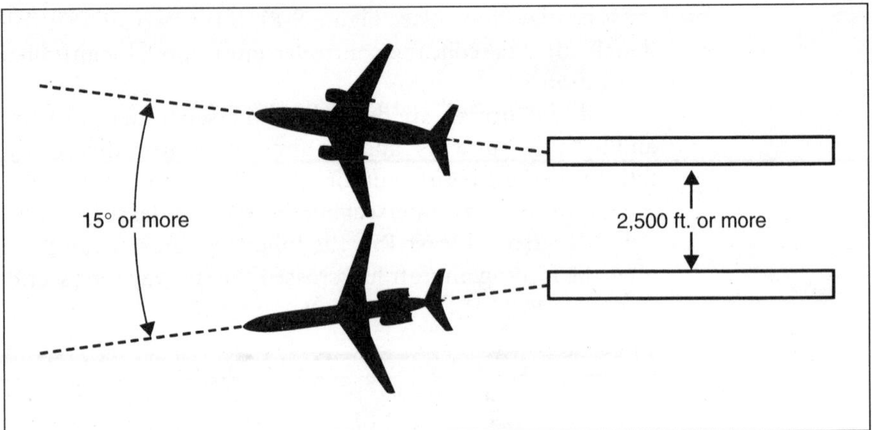

Figure 9-11. Aircraft departing from parallel runways separated by at least 2,500 feet can be departed simultaneously if their courses will immediately diverge by at least 15°.

If the aircraft are operating from parallel runways that are separated by at least 2,500 feet and the aircraft will fly diverging courses immediately after takeoff, simultaneous departures are authorized (see Figure 9-11).

If none of the above conditions exist, the controller must separate the two aircraft as if they were both departing from the same runway.

Radar-Assisted Navigation

Radar is also used by controllers to assist pilots to navigate their aircraft. This assistance can eliminate "dog legs" in the flight path and permit more efficient airspace usage. Rerouting an aircraft using radar also permits the pilot to bypass congested areas, thereby reducing or eliminating the use of holding patterns while en route. Finally, radar can be used to position aircraft directly on the final approach course of an instrument approach, eliminating the need for airspace-consuming procedure turns.

Controllers assisting a pilot to navigate issue verbal heading instructions known as vectors. When vectoring an aircraft, the controller must instruct the pilot to turn to a specific magnetic heading, to turn left or right a specific number of degrees, or simply to fly a particular heading. Here are some examples of the phraseology used for issuing vectors:

Turn left heading [heading]
Turn right heading [heading]
Fly heading [heading]
Fly present heading

Turn [number of degrees] left

Turn [number of degrees] right

Depart [fix] heading [heading]

The controller should be aware of a number of factors that can influence the phraseology used when issuing vectors. Since the winds aloft may significantly affect the ground track of the aircraft, and since the controller can only observe the ground track of an aircraft, potential crosswinds at the aircraft's cruising altitude must always be considered when assigning a heading. The controller must also be alert to the fact that the aircraft's heading indicator may be inaccurately set or might even be malfunctioning.

In consideration of these factors, the controller should instruct a pilot to turn in a specific direction only when the controller is positive of the aircraft's current heading. Since pilots are required by the FARs to turn in the direction requested by the controller even if it appears to be the "long way around," an incorrect direction of turn might produce unanticipated results. Although the "long way around" technique can be used to properly sequence an aircraft or to confine an aircraft to a specific area, controllers must use this technique with discretion. Pilots are trained to immediately comply with a controller's request, and they often will initiate the turn and request later confirmation if the vector seems inappropriate. In some situations, this request might be too late to prevent development of an unsafe condition. To prevent this situation, if the controller is unaware of an aircraft's current heading, the pilot should simply be instructed to "fly" a heading. Pilots interpret this instruction to mean that the aircraft should be turned in whichever direction results in the shortest turn.

Whenever a controller issues a vector to a pilot, it becomes an amendment to the aircraft's clearance. Since a vector is a change in the aircraft's route of flight, the controller is required to inform the pilot of the reason for the vector and at which point or time the pilot can be expected to resume normal navigation. This information can be used by the pilot if either the controller's or the aircraft's communications system should fail. The proper phraseology for vectoring an aircraft off its assigned route of flight is as follows:

N1234P, turn left heading three five zero, vector around traffic, expect to join victor niner in one five miles.

UAL211, fly heading two seven zero, when able, proceed direct to the Shelbyville VOR.

N321YT, turn right heading zero niner zero, vector for the ILS approach, expect a turn on to the final approach course in one five miles.

AAL321, depart the Lansing VOR heading zero eight zero, vector for the ILS runway two six left approach.

N555DM, turn two zero degrees left, vector around traffic, expect a vector direct to the Pullman VOR in three five miles.

VV678, turn right heading three five zero intercept victor three seventy-one.

Once the controller has issued a vector to the pilot, the controller is responsible for monitoring the progress of that aircraft until the pilot is able to reestablish normal navigation. If a possible misunderstanding concerning when the pilot should resume normal navigation could occur, the phraseology "resume normal navigation" should be used when the radar vector has terminated. To assist the pilot to reorient his or her aircraft at the conclusion of such a vector, the aircraft's current location should always precede this instruction. For example:

> N345MN, seven miles from the Knox VOR, resume normal navigation.
>
> UAL556, two zero miles east of the Northbrook VOR, when able proceed direct Badger.

While operating along a federal airway or an approved transition route or during the conduct of an instrument approach, the pilot must comply with the minimum altitudes provided on appropriate navigation charts. Once the aircraft has been vectored off one of these published routes, however, it becomes the controller's responsibility to ensure that the aircraft remains safely above terrain or local obstructions. To assist in this task, the FAA has developed **minimum vectoring altitudes** and has provided these altitudes to controllers at every radar-equipped facility (see Figure 9-12). The use of these altitudes is mandatory and provides each aircraft with standard IFR separation from any terrain or obstacle. In general, aircraft operating at minimum vectoring altitudes will remain at least 3 nautical miles laterally from or at least 1,000 feet above any obstruction.

Radar Arrivals and Approaches

Radar can also be used to expedite arrivals to the final approach course of an instrument approach. Instead of requiring each aircraft to conduct a lengthy procedure turn or enter a holding pattern before transitioning to the approach (two of the techniques used in a nonradar environment), a radar controller can vector each aircraft directly onto the final approach course. Since aircraft vectored onto the final approach course do not need—nor are they permitted by the FARs to perform—a procedure turn, vectoring reduces the separation interval between each aircraft while maximizing the use of the instrument approach. Radar vectors to the final approach course also permit a controller to more effectively manage the spacing of aircraft with dissimilar flight characteristics.

When vectoring an aircraft onto the final approach course, the controller must ensure that it is separated at all times. This task can be fairly difficult since the controller is usually required to sequence aircraft with different flight characteristics to the same final approach course. Even two identical aircraft may not fly the approach at the same speed because of aircraft loading characteristics, pilot preferences, or any number of other variables unknown to the controller.

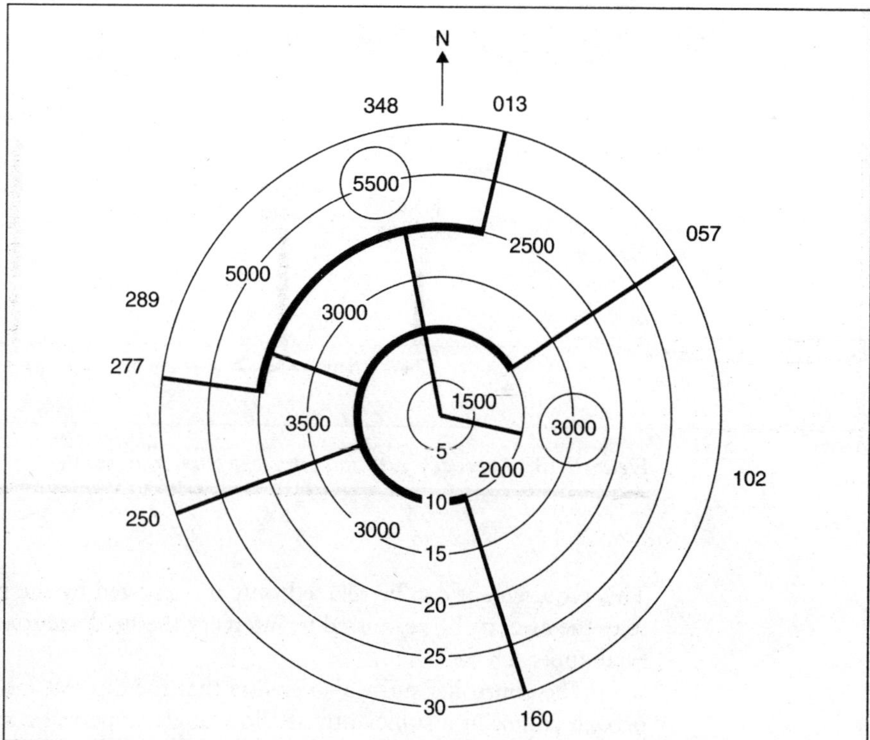

Figure 9-12. Sample minimum vectoring altitude chart.

Besides ensuring separation, the controller must ensure that the aircraft is positioned such that the pilot can make a safe and gradual transition to the final approach course. When the controller has assumed navigational responsibility while vectoring for the instrument approach, each aircraft must be vectored into the proper position and at an appropriate heading to ensure that the pilot can safely transition to the final approach course. To facilitate this transition, the FAA handbook specifies criteria that the controller must maintain during this procedure.

Approach Gate

The handbook specifies that an approach gate exists along every final approach course whenever radar vectoring to that instrument approach is in progress (see Figure 9-13). The approach gate is located either 1 nautical mile outside the final approach fix or 5 nautical miles from the end of the runway, whichever distance is greater. If the weather ceiling is lower than 500 feet above the minimum vectoring altitude or if the visibility at the airport is less than 3 nautical miles, every aircraft vectored to the final approach course must intercept the final approach course no less than 2 nautical miles outside the approach gate.

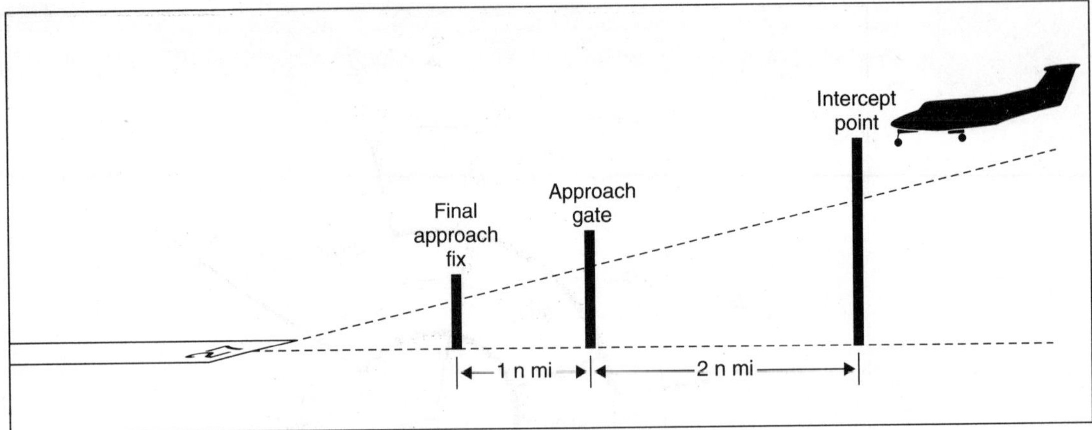

Figure 9-13. Approach gate and intercept point placement.

This requirement can be relaxed only if requested by the pilot, but in no case may the aircraft be permitted to intercept the final approach course inside the final approach fix.

The controller must also ensure that the aircraft intercepts the final approach course at a sufficiently shallow angle to permit a smooth transition to the final approach course. The FAA handbook specifies that if the aircraft will intercept the final approach course at a point less than 2 miles from the approach gate, the intercept angle should be less than or equal to 20°. If the aircraft will intercept the final approach course 2 miles or farther from the approach gate, the aircraft may intercept the final approach course at an angle no greater than 30° (see Figure 9-14).

Arrival Instructions

To provide for pilot preplanning during vectors to an instrument approach, the controller is required to inform the pilot of the aircraft's position and the altitude to be maintained during the vector. The following information must be issued to the pilot before the aircraft intercepts the final approach course:

The aircraft's position relative to a fix associated with the instrument approach. This fix is usually the final approach fix but may be any other navaid or intersection along the final approach course.

The pilot must be issued a vector that will cause the aircraft to intercept the final approach course at the proper point and at an allowable intercept angle.

The controller must give the pilot clearance to conduct the instrument approach.

If the aircraft is not on a published transition route to the final approach course, the controller must assign the aircraft an altitude that is no lower than the minimum vectoring altitude. The aircraft must remain at or above this altitude until it is established on a published segment of the instrument approach.

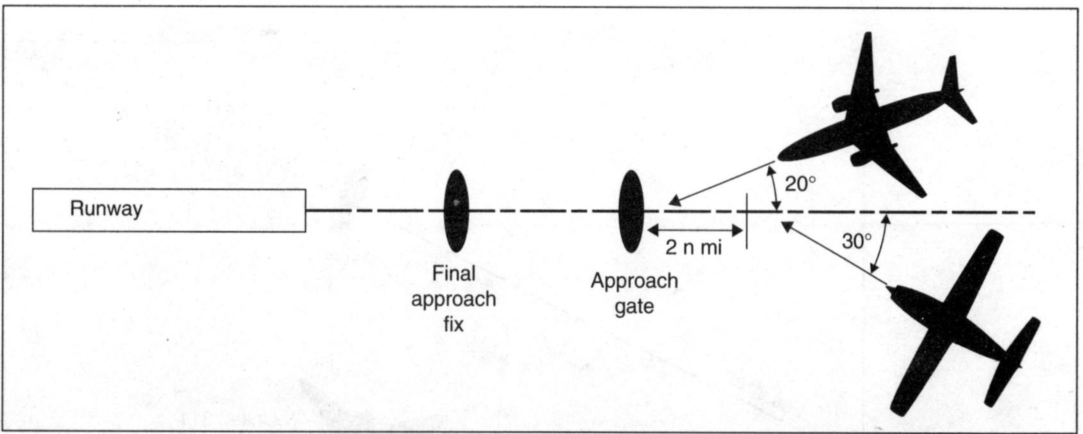

Figure 9-14. Proper intercept angle for aircraft intercepting the final approach course.

> The controller should issue instructions to contact the next controller, if appropriate, and the frequency to be used.

Here are examples of phraseology that would be used for the situations depicted in Figures 9-15 and 9-16.

> United three eleven, seven miles from Vagey, turn right heading zero two zero, intercept the final approach course at or above two thousand seven hundred, cleared for the ILS runway four approach. Monitor Minneapolis tower on one two six point seven, report the outer marker inbound.
>
> Cessna three five mike, one zero miles from Netts, turn right heading zero four zero, intercept the final approach course at or above three thousand. Cleared for the RNAV runway niner approach. Monitor Cleveland tower on one two zero point niner. Report Netts inbound.

ASR Approach

Radar can also be used by controllers as a navigational aid to conduct an instrument approach. During a radar-guided approach, the controller uses radar to monitor the aircraft's position relative to the runway centerline and provides instructions to the pilot to keep the aircraft on the centerline of the runway. This procedure, known as an **airport surveillance radar (ASR) approach,** can be used by pilots who are experiencing navigation receiver problems or who may be unable or unwilling to conduct any of the other instrument approaches at the airport. The minima for ASR approaches are published by the National Ocean Survey and other agencies and are normally included in the instrument approach procedures booklet utilized by pilots (see Figure 9-17).

During an ASR approach, the controller is responsible for advising the pilot of the aircraft's position relative to the runway centerline and then issuing vectors that keep the aircraft on the extended centerline. The controller is also

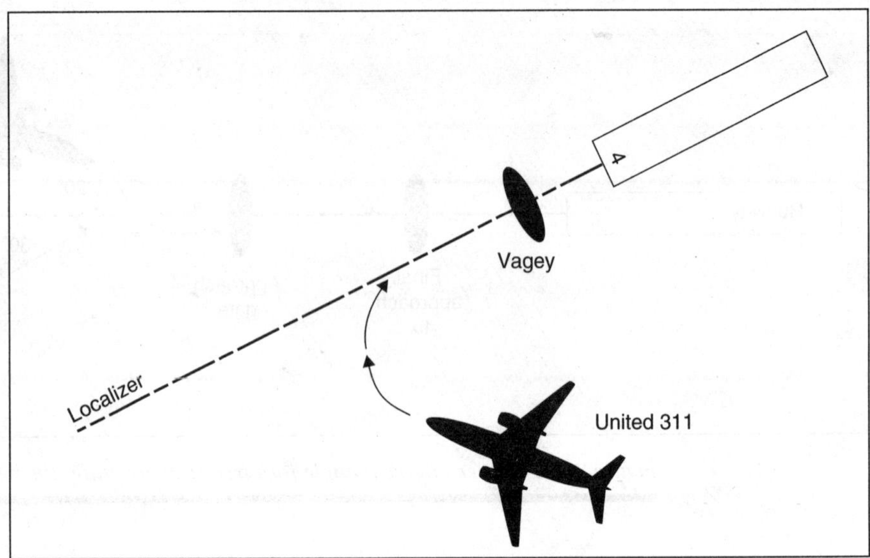

Figure 9-15. A sample turn onto the final approach course of an ILS approach.

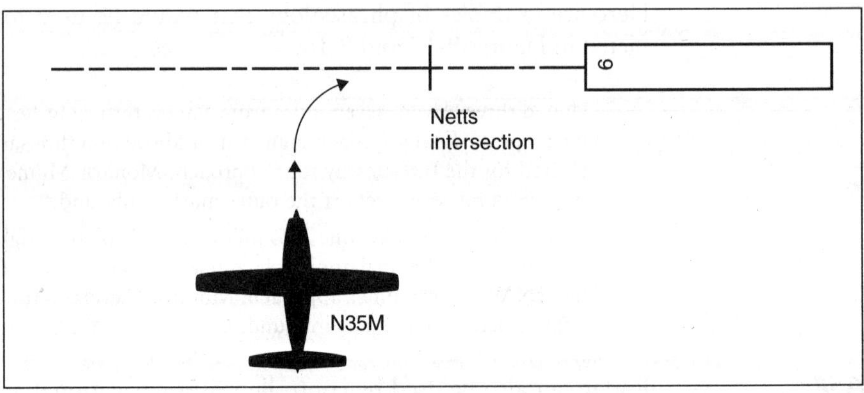

Figure 9-16. A sample turn onto the final approach course of an RNAV approach.

required to keep the pilot informed of the aircraft's distance from the approach end of the runway and may issue recommended altitudes if requested by the pilot, although it is up to the pilots to accurately monitor the aircraft's altitude.

Before starting an ASR approach, the controller must inform the pilot of the following:

The type of approach that will be conducted (a surveillance radar approach).

88266

CIVIL RADAR INSTRUMENT APPROACH MINIMUMS

ROCHESTER MUNI MN Amdt. 6, MAY 8, 1986 ELEV **1317**
RADAR—119.8 396.1

	RWY	GS/TCH/RPI	CAT	DH/ MDA-VIS	HAT/ HAA	CEIL-VIS	CAT	DH/ MDA-VIS	HAT/ HAA	CEIL-VIS
ASR	13		ABC	**1640**—½	354	(400—½)	D	**1640**—1	354	(400—1)
	31		ABC	**1660**/24	356	(400—½)	D	**1660**/50	356	(400—1)
	2		ABC	**1660**—1	343	(400—1)	D	**1660**—1¼	343	(400—1¼)
	20		ABC	**1660**—1	356	(400—1)	D	**1660**—1¼	356	(400—1¼)
CIRCLING			A	**1720**—1	403	(500—1)	B	**1780**—1	463	(500—1)
			C	**1780**—1½	463	(500—1½)	D	**1880**—2	563	(600—2)

Category D Rwy 13 visibility increased ¼ mile for inoperative MALSR.
Category D Rwy 31 visibility increased to RVR 6000 for inoperative SSALR.

SIOUX FALLS/JOE FOSS FIELD SD Amdt. 7, AUG 25, 1988 ELEV **1429**
RADAR—125.8 353.6

	RWY	GS/TCH/RPI	CAT	DH/ MDA-VIS	HAT/ HAA	CEIL-VIS	CAT	DH/ MDA-VIS	HAT/ HAA	CEIL-VIS
ASR	3		AB	**1920**/24	497	(500—½)	C	**1920**/40	497	(500—¾)
			D	**1920**/50	497	(500—1)	E	**1920**/60	497	(500—1¼)
	21		AB	**1920**/24	491	(500—½)	C	**1920**/40	491	(500—¾)
			D	**1920**/50	491	(500—1)	E	**1920**/60	491	(500—1¼)
	33		AB	**1920**—1	497	(500—1)	C	**1920**—1¼	497	(500—1¼)
			D	**1920**—1½	497	(500—1½)	E	**1920**—1¾	497	(500—1¾)
	15		AB	**1960**—1	534	(600—1)	C	**1960**—1½	534	(600—1½)
			D	**1960**—1¾	534	(600—1¾)	E	**1960**—2	534	(600—2)
CIRCLING			AB	**1960**—1	531	(600—1)	C	**1960**—1½	531	(600—1½)
			D	**1980**—2	551	(600—2)	E	**2080**—2¼	651	(700—2¼)

When control tower closed procedure not authorized.
Rwys 15 and 33 air carrier landing visibility reduction below ¾ mile for local conditions not authorized.
▼ ▲

NC-1

CIVIL RADAR INSTRUMENT APPROACH MINIMUMS

Figure 9-17. Civil radar (ASR) instrument approach minimums.

The location of the missed approach point (usually 1 mile from the approach end of the runway).

Lost communications procedures.

For example:

> Cherokee niner para alpha, this will be a surveillance approach to runway one zero, missed approach point one mile from end of runway. If no transmissions are received for one minute in the pattern or for one five seconds while on final, proceed VFR. If unable, maintain three thousand until established on the NDB runway two eight approach.

At least once before beginning the ASR approach, the controller must also inform the pilot of the aircraft's position and of when the pilot may expect to begin the descent to the published minimum descent altitude; the pilot need not acknowledge any further transmissions. When the aircraft reaches the final approach fix (usually 5 nautical miles from the approach end of the runway), the controller advises the pilot to descend to the MDA and issues instructions that will guide the aircraft to the end of the runway. For example:

> Beech five three november is one two miles from the airport on a left downwind, prepare to descend in seven miles.
>
> Beech five three november is five miles from end of runway, descend to your minimum descent altitude.

Once the pilot has initiated the descent, the controller monitors the aircraft's progress and issues course guidance information, the aircraft's position relative to the runway centerline, vectors that will return the aircraft to the runway centerline, and the aircraft's distance from the end of the runway. These transmissions are normally made at approximately 15-second intervals. If the pilot requests, the controller can also issue recommended altitudes, based on a standard descent rate of 300 feet per mile. The procedure for determining these recommended altitudes is found in FAA Order 7210.3, "Recommended Altitudes for Surveillance Approaches." Here is an example of phraseology for a complete ASR approach (depicted in Figure 9-18):

1. Mooney six one hotel, five miles from end of runway, fly heading zero niner zero, descend to your minimum descent altitude.
2. Mooney six one hotel, turn left heading zero seven zero, slightly right of course.
3. Mooney six one hotel, four miles from end of runway, fly heading zero seven zero, slightly right of course, correcting slowly, recommended altitude one thousand eight hundred.
4. Mooney six one hotel, on course, turn right heading zero eight zero.
5. Mooney six one hotel, three miles from end of runway, slightly left of course, turn right heading zero eight five, recommended altitude one thousand five hundred.

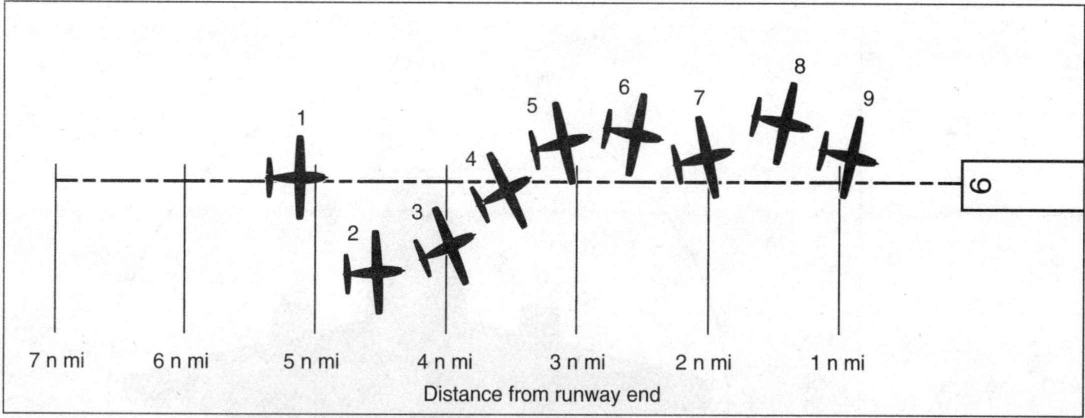

Figure 9-18. An aircraft conducting an ASR approach.

6. Mooney six one hotel, fly heading zero eight five, slightly left of course, correcting slowly.
7. Mooney six one hotel, two miles from end of runway, on course, turn left heading zero niner zero, recommended altitude one thousand two hundred.
8. Mooney six one hotel, drifting left of course, turn right heading one zero zero.
9. Mooney six one hotel, one mile from end of runway, slightly left of course, over missed approach point. Proceed visually, the tower has cleared you to land runway niner. If runway or approach lights not in sight, execute a missed approach and climb to three thousand feet heading zero niner zero.

Radar Traffic Information

Controllers can also use radar to assist pilots in avoiding potential conflicts with other aircraft by advising them of the relative position and altitude of any potentially conflicting traffic. This is known as providing traffic advisories. A controller may provide traffic advisories to any aircraft, whether IFR or VFR, on a workload-permitting basis. Priority is still given to providing separation to aircraft that are participating in the ATC system.

The controller should phrase the advisory in the following manner:

The azimuth of the conflicting traffic in terms of a 12-hour clock relative to the aircraft's ground track (see Figure 9-19).

The aircraft distance in nautical miles.

The direction in which the conflicting traffic is proceeding.

The altitude and type of aircraft if known.

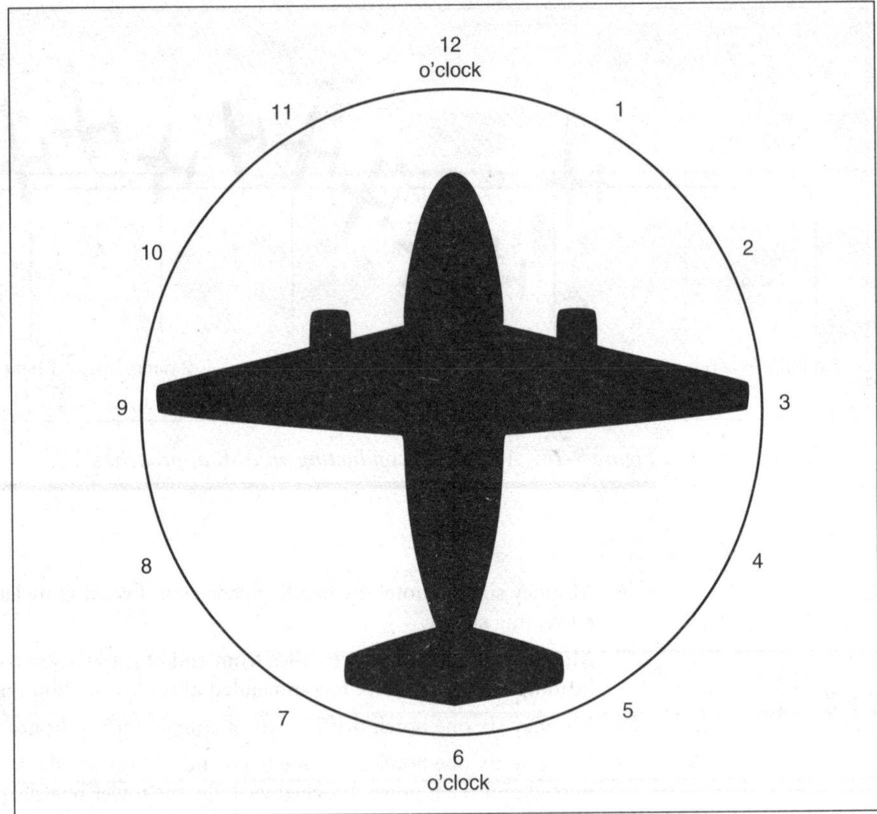

Figure 9-19. The use of the clock in issuing traffic advisories to an aircraft.

The following advisories would be used for the situations depicted in Figure 9-20:

> Eastern two eleven, traffic twelve o'clock, three miles, eastbound, type and altitude unknown.
>
> Eastern two eleven, traffic is a Lear at one o'clock, two miles, westbound at one two thousand.
>
> Eastern two eleven, traffic nine o'clock, five miles, southbound, a VFR military trainer, last reported altitude six thousand.

The controller should not issue the altitude of the conflicting traffic unless it is equipped with a mode C transponder and the controller has verified its accuracy. If the aircraft's altitude has not been verified, the pilot should be informed that the altitude indicated on the radar screen may not be accurate. Once the

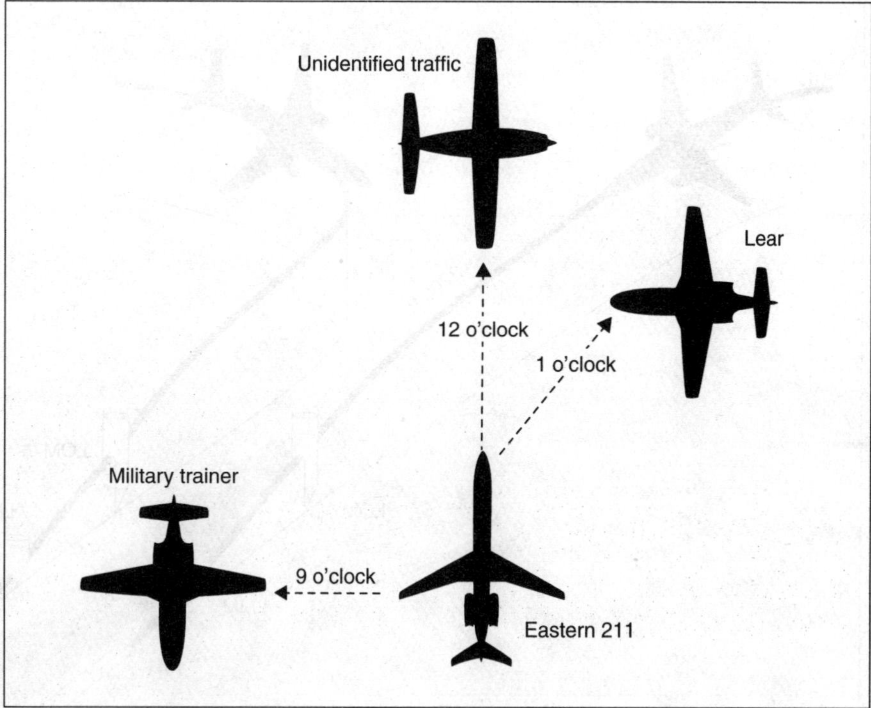

Figure 9-20. Sample traffic advisories.

conflicting traffic no longer poses a threat to the aircraft, the pilot should be so informed ("Eastern two eleven, traffic no longer a factor").

It is mandatory that the controller provide traffic advisories to certain types of aircraft if it appears that the two targets will merge. This is known as **merging-target procedures.** These procedures must be applied to the following aircraft:

Any aircraft operating at or above 10,000 feet MSL.

Turbojet aircraft regardless of altitude.

Presidential aircraft regardless of altitude.

Merging-target advisories can be discontinued if the controller is certain that both aircraft are separated by more than the vertical separation minima.

Both merging-target advisories and traffic advisories should be provided early enough that the pilot has ample time to locate the other aircraft or to ask the controller for a traffic avoidance vector. If the pilot requests such a vector, the controller should issue a heading that will prevent the targets from

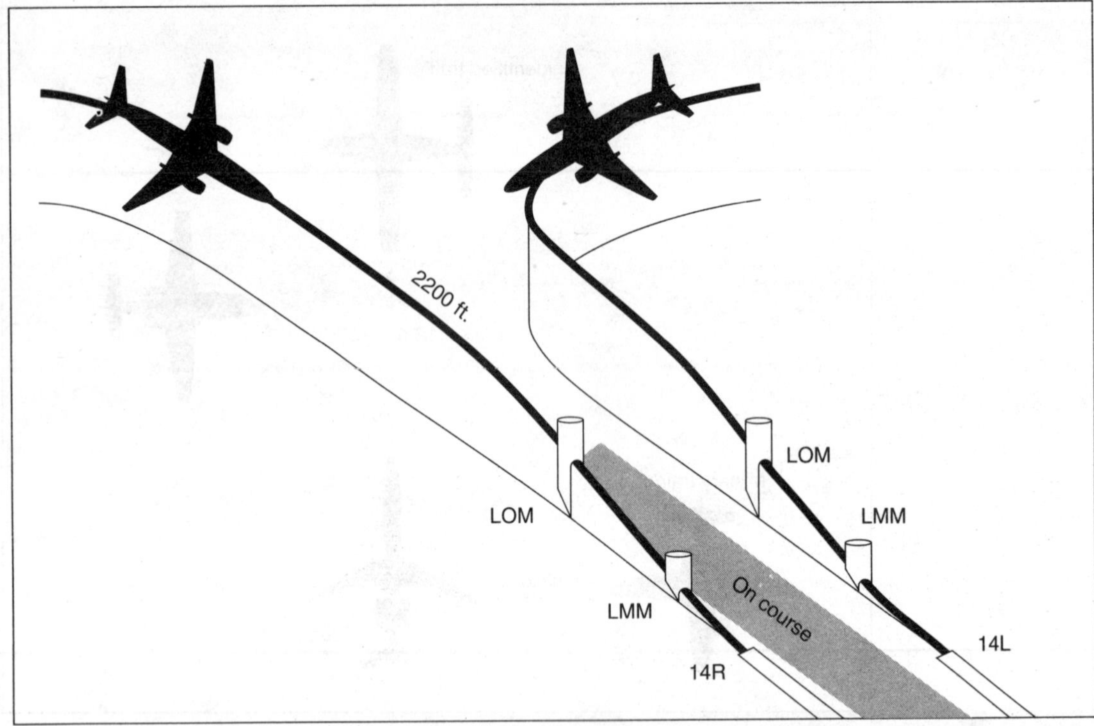

Figure 9-21. A situation in which two aircraft are heading toward parallel ILS approaches. In all likelihood, Conflict Alert will need to be inhibited in this area.

merging. If the controller is unable to do so, because of other conflicting traffic or procedural restrictions, the pilot should be informed of the reason.

Many of the computerized beacon processing systems used by the FAA are capable of projecting the flight paths of aircraft and can alert the controller in advance of certain potentially unsafe conditions. Once an alert has been sounded by the radar system, it is the controller's responsibility to resolve the situation or to advise the pilot to resolve the situation.

One of the most important of these safety systems is the **Conflict Alert** software program. Conflict Alert uses the tracking program already operational on these systems to predict when two tracked aircraft will approach each other within the vertical, lateral, or longitudinal separation minima. If Conflict Alert predicts that this condition might occur, the computer system alerts the controller, who can then evaluate the situation and initiate immediate corrective action if needed.

In certain terminal areas, however, Conflict Alert constantly predicts hazardous situations that do not exist. Since the computer software is unaware of the controller's or the pilot's planned actions, it can only predict an aircraft's future ground track based on its past history. This may cause false alerts to be routinely sounded at a busy approach control facility where the airspace configuration is necessarily complex.

For example, at a busy airport where two aircraft are being vectored for parallel runways, there is usually a point when both aircraft are in a position where, if they are not turned, they are likely to conflict (see Figure 9-21). This situation is perfectly acceptable since the controller is planning to turn each aircraft toward the appropriate runway. But because Conflict Alert has no way of knowing this, it sounds a warning each time two aircraft are in this position.

This routine sounding of an alert is distracting and potentially dangerous, because every time the alert is sounded the controller must verify whether an actual problem exists, diverting attention away from other traffic. In addition the routine sounding of a misleading alert will eventually cause controllers to disregard any alarm produced by the Conflict Alert system. For these reasons, FAA computer programmers inhibit Conflict Alert in areas where numerous false alarms are commonly generated. Inhibiting the Conflict Alert is not a decision made lightly. It is reached only after extensive coordination with the controllers, management, and computer system programmers. Additional information about Conflict Alert can be obtained from FAA Advisory Circulars 90-77 and 90-78.

Minimum Safe Altitude Warning

To assist air traffic controllers in detecting aircraft that are within or approaching an altitude that may be in close proximity to the ground or to obstructions, the FAA has implemented a computer software program known as **minimum safe altitude warning (MSAW)**. This program uses the mode C altitude encoder on the aircraft and the radar computer system tracking capabilities to predict whenever a tracked aircraft is within imminent danger of colliding with the ground.

To provide this service, FAA programmers have divided every ARTCC and radar approach control airspace into 2-mile squares known as **bins.** The highest obstacle within each bin is entered into a database that is instantaneously and continuously accessed by the radar processing system. During routine operation, the radar processing computer constantly compares the mode C–supplied altitude for every tracked aircraft against the information in the database. If the aircraft is less than 500 feet above the highest obstacle in the bin, the controller is alerted.

The MSAW software also predicts the aircraft's flight path for the next 30-second interval and calculates whether the aircraft will enter a bin below the minimum safe altitude if it continues on its present heading, altitude, or rate of climb or descent (see Figure 9-22). If the aircraft is predicted to enter a bin at an altitude lower than 300 feet above the highest obstacle, the controller is also alerted.

Because aircraft conducting instrument approaches must necessarily descend closer to the ground than the MSAW system permits, allowances must be made in these areas. The MSAW software relaxes the obstacle avoidance criteria but still monitors aircraft between the final approach fix and a point 2 miles from the approach end of the runway. The MSAW software is designed to predict both unreasonably low altitudes and excessive aircraft descent rates that might prove to be dangerous. Every aircraft within the approach area is monitored by the radar system and an alert is sounded if an aircraft descends 100 feet below the minimum altitude for that segment of the approach. In addition, the radar processing computer uses past altitude information to extrapolate the aircraft's current rate of descent. If it determines that the aircraft is currently above the minimum altitude for that segment but is predicted to descend 200 feet below the minimum altitude within the next 15 seconds, the controller is alerted. Because of differing aircraft types and approach minima for each runway, the MSAW software is inhibited within 2 nautical miles of the approach end of the runway.

Figure 9-22. Minimum safe altitude warning operation.

FAA

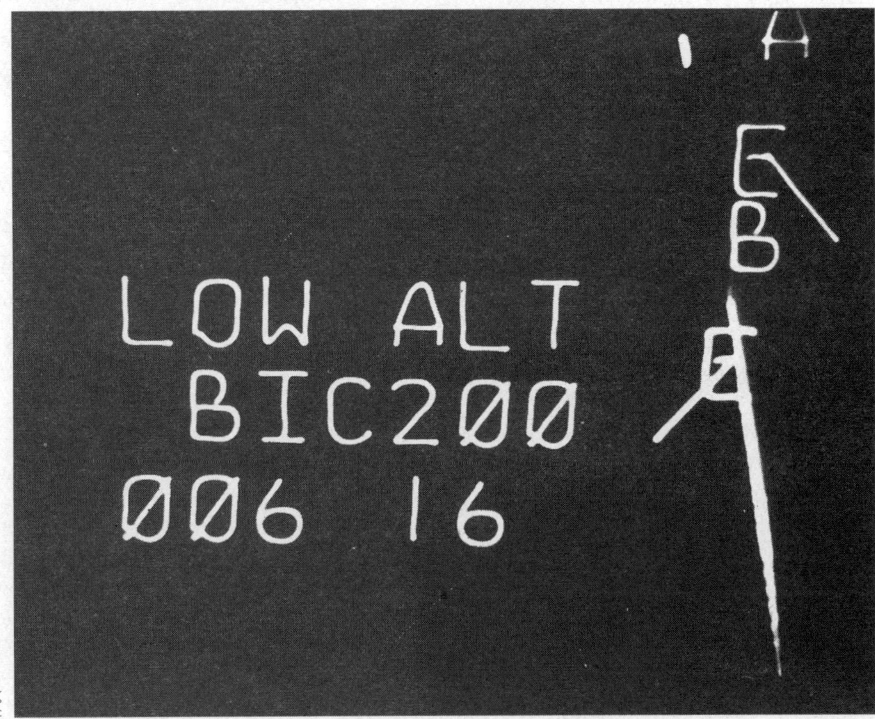

FAA

Figure 9-23. Low-altitude alert as shown on the controller's radar display.

Whenever an unsafe condition is predicted by the MSAW software, an alert is sounded and the letters "Low Alt" begin to flash in the aircraft's data block (see Figure 9-23). When this occurs, the controller must immediately evaluate the situation and, if appropriate, issue the pilot a verbal warning ("Cessna two papa alpha, low-altitude alert, check your altitude immediately, altimeter two niner eight six"). It is then up to the pilot to evaluate the situation and determine what actions are necessary to return the aircraft to the proper flight path.

KEY TERMS

airport surveillance radar (ASR) approach
approach gate
bins
Conflict Alert
coordination
handoff
merging-target procedures
minimum safe altitude warning (MSAW)
minimum vectoring altitudes
point out
point out approved
radar contact
special identification pulse (SIP)
traffic
traffic advisories
traffic observed
transfer of communication
transfer of control
vectors

REVIEW QUESTIONS

1. How are aircraft identified using radar?
2. How are aircraft separated using radar?
3. How is the transponder used in air traffic control?
4. How are handoffs accomplished?
5. How can radar be used as an instrument approach?

10

Operation in the National Airspace System

Checkpoints

After studying this chapter, you should be able to:

1. Describe the flow of flight plan information through the air traffic control system.
2. Describe the operation of the flight data processing system.
3. Describe the function of the Air Traffic Control System Command Center and traffic management units.
4. Explain the sectorization procedures used at a typical medium- and high-activity radar facility.
5. Understand the procedure used to assist lost or overdue aircraft.
6. Explain the uses of en route flight advisory service.

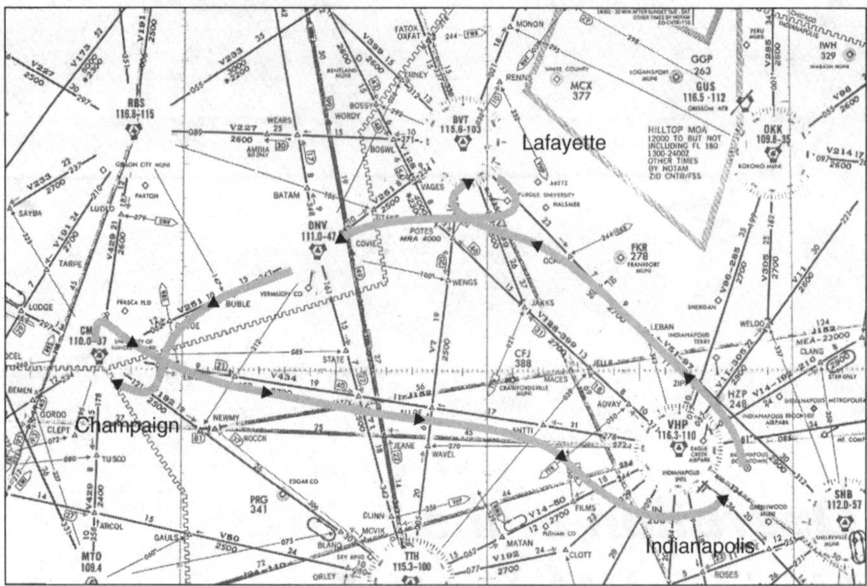

Figure 10-1. The route of flight of the simulated IFR flight.

A detailed description of flight through the air traffic control system is too complex to be completely described in an entire textbook, much less a single chapter. Thus this chapter attempts to summarize the process by presenting examples of how simple IFR and VFR flights are conducted in the air traffic control system, using two simulated flights. The first is a simulated IFR flight from Lafayette, Indiana, to Champaign, Illinois, then on to Indianapolis, Indiana, and back to Lafayette (see Figure 10-1). This route was chosen because it takes the aircraft from a nonradar approach control facility to a medium-activity radar-equipped tower, through an ARTCC to a high-activity radar-equipped tower, then back to the nonradar approach control facility.

The second example simulates a VFR flight from Lafayette, Indiana, to Champaign, Illinois. This route was chosen because it takes the aircraft from a small nonradar tower to an airport providing Class C airspace separation. After following these simulated flights, you should have a good idea of how the ATC system actually works.

Overview of an IFR Flight

Filing the Flight Plan

Pilots of personal and corporate aircraft usually contact a flight service station (FSS) to file a flight plan and receive weather briefings. An increasing number of these pilots, however, are beginning to use private weather-briefing firms and are now able to file flight plans directly through these organizations. Airline pilots usually have their flight plans filed for them by the airline itself. Military pilots usually file flight plans through their military operations office.

After contacting the flight service station, the pilot receives a thorough weather briefing that includes both current and forecast weather along the route of flight and learns of any adverse conditions affecting the ATC system or the destination airport that may influence the pilot's decision. These conditions include known or suspected ATC delays, navigation equipment outages, and any **notices to airmen (NOTAMs).** NOTAMs are entered into the FAA computer system by local flight service stations or at the **Flight Data Center (FDC)** in Washington, D.C. NOTAMs issued by local flight service stations concern local conditions such as airport or runway closures and unlit obstructions. NOTAMs issued by the Flight Data Center, known as FDC NOTAMs, concern en route navaid outages, changes to published instrument approach procedures, or any emergencies (see Figure 10-2).

Flight Plan Entry

Once the weather briefing is completed, the flight service specialist obtains the proper flight plan information from the pilot and enters it into the FSS computer. Within a short time, the appropriate information is sent to the flight data processing computer at the ARTCC with responsibility for the departure airport. The FDP computer checks the validity of this information and, if it is correct, stores it for later use.

Dallas-Love Field

FDC 7/1246 /DAL/FI/T DALLAS-LOVE FIELD, DALLAS, TX. RADAR-1 AMDT 24, S-13L MDA 1000 HAT 515 ALL CATS. VIS CAT C 5000, CAT D 6000. S-13R MDA 1000 HAT 524 ALL CATS. VIS CAT C 1 1/2, CAT D 1 3/4

FDC 7/1066 /DAL/FI/T DALLAS-LOVE FIELD, DALLAS, TX. ILS RWY 31R ORIG. CIRCLING VIS CAT B 1 1/4, CAT C 2 1/2, CAT D 2 3/4

FDC 7/1035 /DAL/FI/T DALLAS-LOVE FIELD, DALLAS, TX. ILS RWY 31L AMDT 15. S-I DME MINIMA MDA 1360 HAT 885 ALL CATS. VIS CATS A/B RVR 4000, CAT 2 1/4, CAT D 2 1/2. CIRCLING MDA 1400 HAA 913 ALL CATS. VIS CATS A/B 1 1/4, CAT C 2 3/4, CAT D 3. DME MINIMA: S-LOC 31R MDA 1340 HAA 853 ALL CATS. VIS CAT A 1/2, CAT B 3/4, CAT C 2, CAT D 2 1/4

FDC 7/811 /DAL/FI/T DALLAS-LOVE FIELD, DALLAS, TX. IFR TAKEOFF MINIMUMS RWYS 31L/R AND RWY 36 STANDARD. RWYS 13L/R AND RWY 18 1000-2 OR STANDARD WITH A MIN CLIMB OF 290 FT PER NM TO 1500.

FDC 6/1813 /DAL/FI/T DALLAS-LOVE FIELD, DALLAS, TX. ILS 31L AMDT 15, S-ILS 31L NA. S-LOC 31L. MDA 1600, HAT 1125 ALL CATS, CIRCLING 1600 HAA 1113 ALL CATS. ENTRA DME FIX RELOCATED LUE 4.1 DME. REASON: 889 AGL. 1339 AMSL CRANE 3.9 NM SE OF AIRPORT

Figure 10-2. A sample FDC NOTAM.

If the route information entered by the specialist at the FSS is incorrect or incomplete, an error message is returned to the FSS. It then becomes the FSS specialist's responsibility to correct the entry, ensuring that the route information is complete. At this point the abbreviation **FRC** (for **full route clearance**) is appended to the corrected flight data and then retransmitted to the ARTCC. The abbreviation FRC is then printed on the aircraft's flight progress strip, which alerts subsequent controllers that changes were made to the pilot's filed route of flight.

Once the information is received at the ARTCC, the FDP computer checks the route and determines whether any **preferred departure routes (PDRs), preferred arrival routes (PARs),** or preferred IFR routes apply to that particular aircraft. PDRs and PARs are determined by a letter of agreement or through facility directives (see Figure 10-3). If this particular flight will be affected by one of these types of routes, the preferred route will be printed on the flight progress strip bracketed by plus signs.

Flight Strip Printing

Thirty minutes prior to the aircraft's proposed departure time, the FDP computer causes a flight strip to be printed at the FDIO at the departure airport. If the departure airport is not served by an ATC facility, or if the facility is not FDIO equipped, the strip will be printed at the closest facility with FDIO. At this time, the FDP computer also assigns the aircraft a transponder code. Since the number of codes available is limited, this procedure is used to effectively ration transponder codes.

Assuming that the aircraft is departing from an airport equipped with FDIO (such as Indianapolis International), the flight strip will be printed at the clearance delivery position in the control tower. The clearance delivery controller is responsible for ensuring that the aircraft's altitude and route of flight conform to the appropriate letters of agreement and facility directives. The controller can then issue the clearance to the pilot (see Figure 10-4).

In most cases, facility directives specify that the aircraft be initially restricted to an altitude lower than that filed by the pilot. If the controlling facility has responsibility for the airspace extending up to 10,000 feet, for example, the clearance delivery controller must initially restrict the aircraft to this altitude, so that in case of temporary radio failure the aircraft does not leave the vertical confines of the facility's airspace before a handoff has been accomplished. At some facilities, additional constraints have been imposed on departing aircraft. It is not unusual to restrict an aircraft to an initial altitude of 3,000 to 6,000 feet. The advantages of this procedure will be explained shortly.

The clearance delivery controller must issue the pilot the clearance using one of two methods. If no changes were made to the pilot's requested route of flight, the controller can clear the pilot "as filed." This means that the route that the pilot filed with the flight service station is the same route as that contained in the clearance. An "as filed" clearance does not include the pilot's requested

PREFERRED IFR ROUTES

281

Terminals	Route	Effective Times (UTC)
MOLINE		
Fort Wayne	V156 BDF V144	0000–2359
South Bend	V156 BDF V144 MAPPS V156	0000–2359
ROCKFORD		
Fort Wayne	V128 SMARS V144	0000–2359
South Bend	V128 SMARS V144 MAPPS V156	0000–2359
SOUTH BEND		
Moline	V156 MAPPS V144 BDF V156 MZV	0000–2359
Rockford	V156 MAPPS V144 SMARS V128	0000–2359
SPRINGFIELD, IL		
Chicago Meigs	PNT MOTIF–STAR	0000–2359
Chicago Midway	PNT MOTIF–STAR	0000–2359
Chicago O'Hare	PNT V227 PLANO	0000–2359
Springfield, MO	V50 UIN V63 SGF V14	0000–2359
Tulsa	V50 UIN V63 SGF V14	0000–2359
TERRE HAUTE		
Kansas City	V50 CAP BQS–STAR	0000–2359
Springfield, MO	V7 PXV V190	0000–2359
Tulsa	V7 PXV V190 SGF V14	0000–2359

HIGH ALTITUDE

Terminals	Route	Effective Times (UTC)
CHICAGO (O'Hare—Midway)		
Atlanta	EON DNV TTH BWG RMG STAR	1100–0300
Baltimore	GIJ J146 J34 DJB J162 MGW EMI–STAR	
Boston	ELX CRL J554 JHW J82 ALB GDM GDM–STAR	
Bradley	ELX CRL J554 JHW J82 WILET RKA292 RKA RODA–STAR	
Charleston, WV	EON DNV VHP FLM	1200–0400
Cleveland	GIJ J146 PLAIN VWV WAKEM–STAR	
Columbus, OH	GIJ GIJ092036 FWA J178 APE	
Dallas/Fort Worth	RBS STL RZC FSM BYP	
Dayton	GIJ GIJ092036 FWA ROD DQN	
Denver	IOW DSM J10 LBF SAYGE–STAR	
	or	
	DBQ J94 ONL J114 SNY LANDR–STAR	
Detroit Metro–Wayne Co.	PETTY MKG POLAR–STAR	
Fort Lauderdale	GUIDO J73 TLH J43 PIE FORTL–STAR	
Fort Myers	GUIDO J73 TLH J43 PIE SRQ–STAR	
Greensboro	EON DNV VHP FLM HVQ PSK	1200–0400
Houston	RBS FAM J137 LIT J180	
Kennedy	ELX CRL J554 JHW J70 LVZ LENDY–STAR	
La Guardia	GIJ J146 MIP MIP–STAR	
Long Beach	MZV LMN J64 HEC KAYOH–STAR	1400–0100
Los Angeles	MZV LMN J64 HEC CIVET–STAR	1400–0100
	or	
	MZV J232 IRK J96 SLN J102 ALS J110 FMN J64 HEC CIVET–STAR	1400–0100
	or	
	MZV J18 SLN J102 ALS J110 FMN J64 HEC CIVET–STAR	1400–0100
Memphis	RBS FAM GQE–STAR	
Miami	GUIDO J73 TLH J43 PIE CYY–STAR	
Middletown	GIJ J146 J34 DJB J518 J512 HAR MDT	
Nashville	GUIDO J73 PXV CCT HEHAW–STAR	
New Orleans	RBS J71 J35	
Newark	ELX CRL J584 SLT FQM–STAR	

Figure 10-3. A sample set of preferred routes obtained from the Airport Facility Directory.

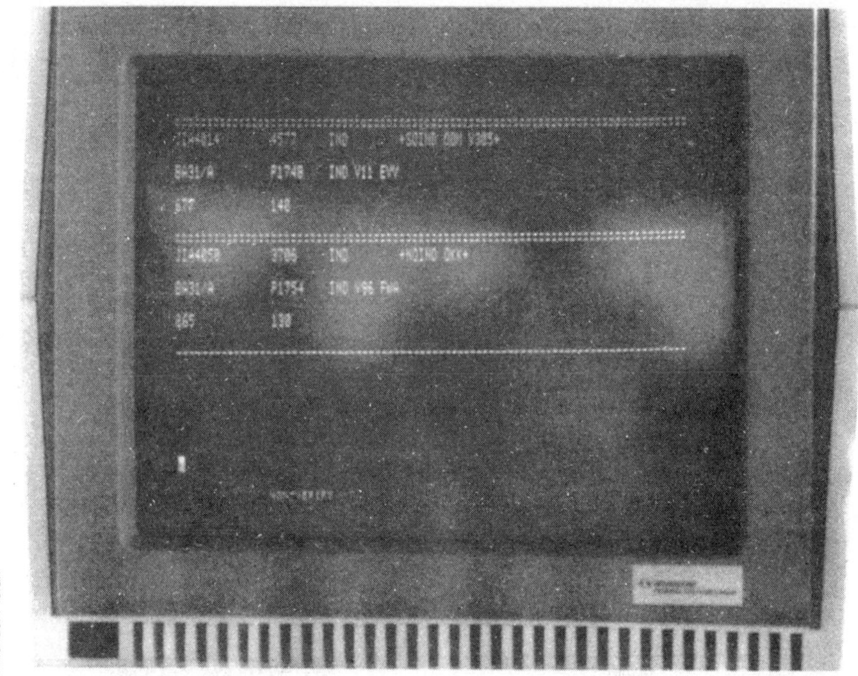

Michael Nolan

Figure 10-4. FDIO screen.

altitude. The altitude must always be stated by the controller when issuing a clearance to the pilot. The phraseology for an "as filed" clearance is:

> Cessna two five two mike november, cleared to Indianapolis International Airport as filed, climb and maintain one zero thousand, squawk three seven four one.

If the control tower is equipped with a departure control position, the clearance must also include the departure controller's frequency. In addition, if the facility directives specify that every departing aircraft should be temporarily restricted to a lower altitude, this restriction must be included as part of the original clearance. If a lower altitude is temporarily assigned, the pilot should be advised as to when the altitude filed in the flight plan might be expected:

> Cessna two five two mike november, cleared to Indianapolis International Airport as filed, climb and maintain five thousand, expect one two thousand one zero minutes after departure, departure control frequency one two three point seven five, squawk three seven four one.

If the aircraft is departing from an airport not served by a facility equipped with radar, the clearance must also include the first airway segment that the pilot has filed. This serves as a double check to ensure that the issued clearance is the same as that originally filed by the pilot:

> Cessna two five two mike november, cleared to Indianapolis International Airport as filed via victor ninety-seven, climb and maintain five thousand, expect one two thousand one zero minutes after departure, departure control frequency one two three point seven five, squawk three seven four one.

If a very small change has been made to the pilot's route of flight (such as the imposition of a PDR or a PAR), the phrase "cleared as filed" may still be used, but the changed portion of the route must be stated:

> Cessna two five two mike november, cleared to the Indianapolis International Airport via victor ninety-two south, Jetts intersection, then as filed, climb and maintain five thousand, expect one two thousand one zero minutes after departure, departure control frequency one two three point seven five, squawk three seven four one.

But if the route has been changed substantially, or if the abbreviation FRC is printed on the strip (signifying that the flight service specialist amended the clearance), the entire route of flight must be stated, preceded by the phrase "unable routing as requested," which alerts the pilot that a major change in the route of flight has been made:

> Cessna two five two mike november, unable routing as requested, cleared to Indianapolis International Airport via victor three ninety-nine, climb and maintain five thousand, expect one two thousand one zero minutes after departure, departure control frequency one two three point seven five, squawk three seven four one.

At terminal facilities not equipped with ARTS radar, after the pilot has verified the clearance, the clearance delivery controller enters the estimated departure time of the aircraft into the FDIO and passes the strip to the ground controller. At facilities equipped with ARTS radar systems, the ARTS computer will automatically send a departure message to the ARTCC upon receipt of the aircraft's transponder signal.

Ground Control

The ground controller is responsible for issuing a taxi clearance that will take the aircraft to the departure end of the appropriate runway. The ground controller is also responsible for any vehicles that must travel on the airport movement area. Taxi instructions are usually issued using a combination of some of the following clearances:

> Taxi runway two three.
>
> Taxi runway two seven left via the parallel and the outer taxiway.
>
> Taxi runway three three, follow the seven twenty-seven off your left.

> Taxi runway one two, pass behind the aircraft ahead and to your right on taxiway bravo.
>
> Runway five for departure, hold short of runway one zero, traffic landing on runway one zero.
>
> Runway two eight for departure, hold short of taxiway echo five.

If the aircraft must cross an active runway before reaching the departure runway, the ground controller must coordinate this crossing with the local controller. This is accomplished by asking the local controller for permission to cross the active runway at a certain location. The local controller may approve the request, deny it, or approve it subject to some restrictions:

GROUND CONTROL:	Cross runway two seven left at taxiway charlie?
LOCAL CONTROL:	Cross runway two seven left at taxiway charlie.

After the aircraft has crossed the runway, the ground controller must advise the local controller that the operation has been completed.

Local Control

It is the local controller's responsibility to safely sequence departing aircraft into the local traffic flow while still complying with any departure instructions issued by the departure controller. The local controller is not permitted to depart an IFR aircraft without the approval of the departure controller. This approval may be received specifically for each aircraft, or routine departure instructions may be specified in the facility directives.

Most radar-equipped facilities have devised a system that permits the local controller to depart an IFR aircraft without prior coordination with the departure controller. This method of operation requires that a specific block of airspace be reserved for departing aircraft; the local controller is authorized to depart aircraft into this area without prior coordination. The local controller still retains responsibility for the initial separation of IFR departures, however. When using this type of system, the approach controllers are responsible for keeping inbound aircraft separated from this departure area.

The departure area is usually the shape of a wide fan or a narrow corridor (see Figure 10-5). This wedge of airspace normally extends from the ground up to an altitude of 3,000 to 6,000 feet above the ground. As long as the clearance delivery controller has restricted the aircraft to the appropriate altitude and the local controller assigns a heading that will keep the aircraft within the confines of the departure area ("United five fifty-four, after departure turn left heading zero five zero, runway one two left, cleared for takeoff"), no prior coordination between the local and departure controllers is needed.

In most situations, the local controller pencils the aircraft's heading on the appropriate flight progress strip, then sends the strip via a tube that guides it directly to the appropriate departure controller's workstation. Once the local controller has departed the aircraft and resolved any conflicts with local traffic, the pilot is directed to contact the departure controller ("United five fifty-four,

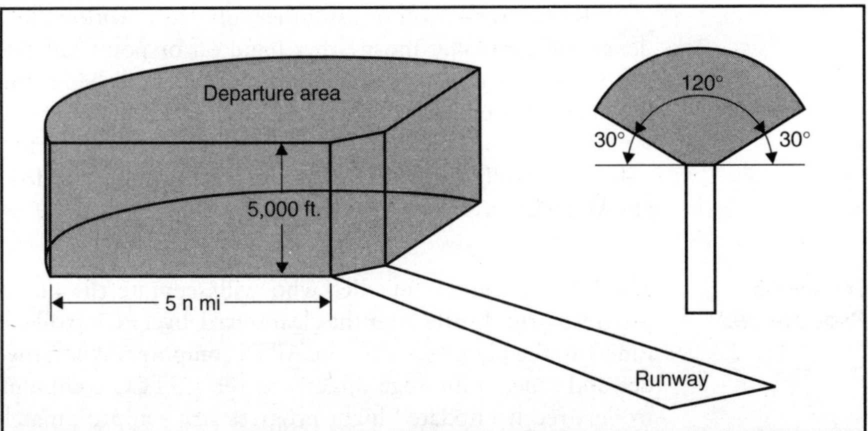

Figure 10-5. The departure fan used by the local controller to initially separate departures from arrivals.

contact departure"). Since the appropriate frequency was issued by the clearance delivery controller, the local controller is not required to repeat it.

Departure Control

Depending on the complexity of the facility, departure control may be operated by the approach controller, might be a separate control position, or might even be divided into a number of different subsectors. In any case, it is the departure controller's responsibility to separate this aircraft from all others while still complying with the appropriate facility directives and letters of agreement. Once the aircraft has been changed to the departure controller's frequency, this controller must radar identify the aircraft and verify the accuracy of the aircraft's mode C transponder, if the aircraft has one. At this time, if the facility is equipped with ARTS radar, the ARTS computer detects the aircraft's transponder transmission and automatically sends a departure message to the ARTCC computer.

Once the aircraft has been radar identified, the departure controller advises the pilot that radar contact has been established. If the pilot has not stated the altitude of the aircraft, the controller must ask the pilot for altitude verification before using the altitude readout for aircraft separation ("United five fifty-four, radar contact, verify at one thousand two hundred").

At this point the controller vectors the aircraft to join the route of flight while still complying with facility directives and letters of agreement. The controller also attempts to clear the aircraft to climb to the pilot's requested altitude as soon as is practical ("United five fifty-four, turn left heading two niner zero, join victor seven, climb and maintain one two thousand"). If this is not possible because of a lack of jurisdiction or traffic conflicts, the aircraft will normally be cleared to the altitude closest to that filed by the pilot.

If the aircraft will transit other subsectors within the terminal facility, the departure controller must either hand off or point out the aircraft to the appropriate controllers. Such handoffs are accomplished manually or through the use of automated procedures. If the aircraft is remaining at a fairly low altitude, it will usually be handed off to an adjoining terminal facility. But if the aircraft will fly at a sufficiently high altitude, it is normally handed off to the appropriate ARTCC.

En Route Procedures

The first en route controller who will separate the aircraft receives a flight progress strip shortly after the clearance delivery controller enters the departure time into the FDIO (or after the ARTS computer detects the aircraft's transponder and sends a message directly to the ARTCC computer). Subsequent controllers receive updated flight progress strips approximately 15 to 30 minutes before the aircraft enters each sector. The en route controllers use the information on the flight strip to prepare for the separation of that flight. Once the ARTCC radar system detects the aircraft's transponder signal, a data block containing the aircraft's call sign, altitude, and airspeed appears on the controller's PVD. At the point delineated in the appropriate letter of agreement, the departure controller hands off the aircraft to the ARTCC controller.

Once the en route controllers have accepted a handoff, they are responsible for separating that aircraft from all others within the sector. This may be somewhat difficult if the aircraft is sufficiently low and far enough away from an ARTCC radar site that it remains undetected by radar. In such cases, the aircraft will not appear on the ARTCC controllers' radar display and must be separated using nonradar procedures. The responsibility for this separation lies with the radar associate/nonradar controller. Once the aircraft is detected by radar, however, separation responsibility is that of the radar controller.

If the aircraft is operating below 18,000 feet MSL, it is separated by controllers responsible for low-altitude aircraft (known as low-sector controllers). But if the aircraft climbs to a higher altitude, it must be handed off to a high-altitude control sector. Once the aircraft reaches its assigned cruising altitude, it continues toward its destination, being handed off from controller to controller as it crosses sector boundaries.

Air Traffic Control System Command Center

The **Air Traffic Control System Command Center** (**ATCSCC**) in Washington works with **traffic management unit** (**TMU**) controllers in every ARTCC to coordinate the flow of aircraft through the center's airspace. Various factors may affect aircraft flying through the center, causing congestion, overloaded sectors, or potential aircraft delays. When widespread severe weather threatens to affect the ATC system, a **severe weather avoidance plan** (**SWAP**) is enacted by the ATCSCC to try to minimize the overall impact of such weather. Controllers may be required to reroute or even hold aircraft to ensure that efficient traffic flow is maintained and that no particular ARTCC control sector is inundated with traffic.

En Route Metering

As aircraft approach the destination airport, each successive controller begins to assign progressively lower altitudes to it. If the arrival airport is particularly busy, **en route metering** may be in effect. En route metering is an attempt by the FAA to try to match the inbound flow of traffic to the airport's **acceptance rate,** the calculated rate at which the airport can absorb traffic. If, for instance, calculations show that a particular airport can handle 60 aircraft operations in one hour, its theoretical acceptance rate is 1 per minute. A general rule of thumb is that a single runway can handle 30 arrivals per hour (one every 2 minutes) if the runway is being used for both arrivals and departures. If the runway is being used solely for arrivals, a 1-minute interval between aircraft can probably be maintained. This would permit the runway to handle 60 aircraft per hour.

If two aircraft are scheduled to arrive at the airport at the same time, one of the aircraft will have to be delayed for at least 1 minute. Such delays place a burden on the approach controller, since only a limited amount of airspace is available to maneuver aircraft. It becomes even more difficult to delay aircraft when more than two flights are scheduled to arrive at the same time. In this situation, the approach controller rapidly runs out of airspace in which to maneuver aircraft (a fairly common situation that occurs routinely wherever airlines operate hub-and-spoke scheduling systems).

The en route metering program calculates the airport's acceptance rate and determines the number of aircraft that can be handled in any given 5-minute period. If it is determined that the calculated airport acceptance rate will be exceeded, the en route metering software at the ARTCC begins to calculate appropriate delay strategies to temporarily reduce the number of aircraft inbound to the airport. The metering program prints out specific times that aircraft should cross en route fixes in order to delay each aircraft the required interval. It then becomes each ARTCC radar controller's responsibility to ensure that the aircraft cross these fixes at the appropriate times. This is usually accomplished by temporarily reducing each aircraft's speed. In most cases the metering fixes are approximately 50 to 100 miles from the destination airport (see Figure 10-6).

Approach Control

At radar-equipped terminal facilities, the approach controller is responsible for sequencing and separating inbound aircraft. At low-activity towers, this task may be delegated to only one controller. At high-activity airports, approach control duties may be assigned to up to five different types of controllers: (1) **feeder controllers,** whose responsibilities are to sequence arriving aircraft toward the final approach course; (2) a **final controller,** whose responsibility is to sequence aircraft on the instrument approach; (3) a **monitor controller,** who continuously monitors aircraft conducting parallel ILS approaches; (4) **satellite controllers,** who handle approaches and departures from low-activity airports located within the approach control's area of jurisdiction; and (5) **departure controllers,** who separate aircraft departing from the primary airport.

The remainder of this chapter is devoted to simulated IFR and VFR flights. As noted earlier, the particular routes were chosen because they illustrate how both radar- and nonradar-equipped facilities handle IFR and VFR flights.

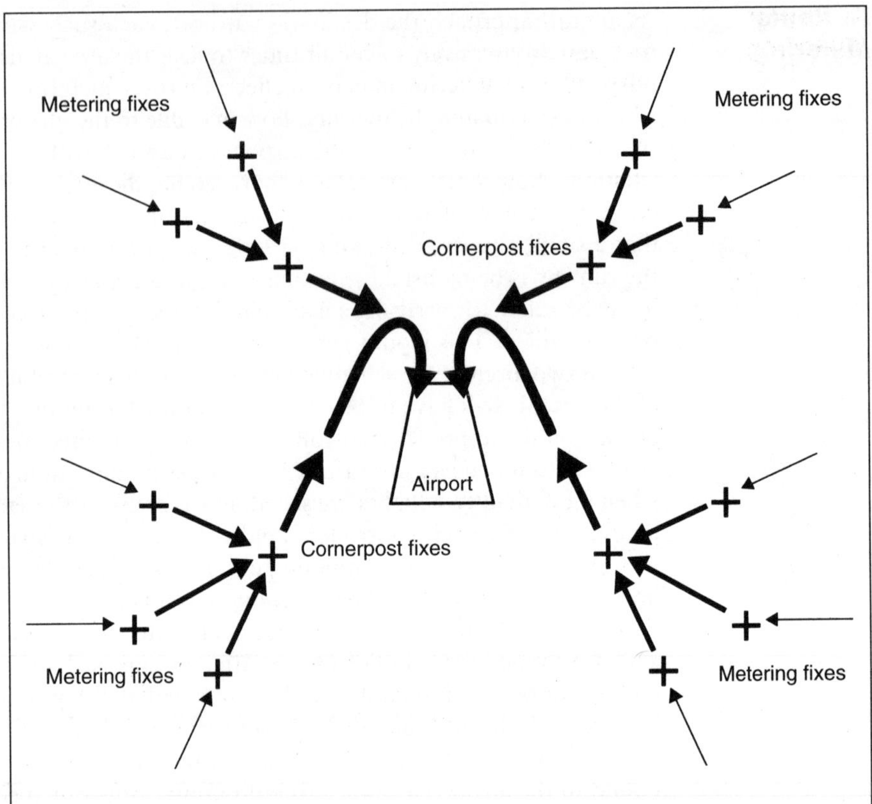

Figure 10-6. Traffic flow arriving in a major terminal. Aircraft first cross designated metering fixes before crossing the cornerpost fixes.

Examples of an IFR Flight

Lafayette to Champaign

Flight Plan Filing The pilot initially contacts a flight service station to receive a weather briefing and file an IFR flight plan. In this particular flight, the pilot would probably contact the **automated flight service station** (AFSS) in Terre Haute, Indiana. The flight service specialist at Terre Haute would conduct a weather briefing for the pilot and would then request the appropriate flight plan information (see Figure 10-7). In this example, we will assume that the pilot files a separate flight plan for each of the three legs of the flight. As the pilot passes along the appropriate information to the FSS specialist, it is entered directly into the FSS computer.

Once the briefing is finished, the FSS computer transmits the information to the flight data processing computer at the appropriate ARTCCs. Because both Lafayette and Indianapolis are within the boundaries of Indianapolis ARTCC,

U.S. DEPARTMENT OF TRANSPORTATION
FEDERAL AVIATION ADMINISTRATION

FLIGHT PLAN

(FAA USE ONLY) ☐ PILOT BRIEFING ☐ VNR ☐ STOPOVER | TIME STARTED | SPECIALIST INITIALS

1. TYPE	2. AIRCRAFT IDENTIFICATION	3. AIRCRAFT TYPE/ SPECIAL EQUIPMENT	4. TRUE AIRSPEED	5. DEPARTURE POINT	6. DEPARTURE TIME PROPOSED (Z)	6. DEPARTURE TIME ACTUAL (Z)	7. CRUISING ALTITUDE
☐ VFR ☒ IFR ☐ DVFR	N252MN	BE90/R	160	LAF	1500		50

8. ROUTE OF FLIGHT: V251

9. DESTINATION (Name of airport and city)	10. EST. TIME ENROUTE HOURS	10. EST. TIME ENROUTE MINUTES	11. REMARKS
CMI		30	

12. FUEL ON BOARD HOURS	12. FUEL ON BOARD MINUTES	13. ALTERNATE AIRPORT(S)	14. PILOT'S NAME, ADDRESS & TELEPHONE NUMBER & AIRCRAFT HOME BASE	15. NUMBER ABOARD
3	50	SPI	M. NOLAN LAF	2

17. DESTINATION CONTACT/TELEPHONE (OPTIONAL): B. NOLAN LAF

16. COLOR OF ARICRAFT: WHITE & GOLD

CIVIL AIRCRAFT PILOTS. FAR Part 91 requires you file an IFR flight plan to operate under instrument flight rules in controlled airspace. Failure to file could result in a civil penalty not to exceed $1,000 for each violation (Section 901 of the Federal Aviation Act of 1958, as amended). Filing of a VFR flight plan is recommended as a good operating practice. See also Part 99 for requirements concerning DVFR flight plans.

FAA Form 7233-1 (8-82) CLOSE VFR FLIGHT PLAN WITH ______________ FSS ON ARRIVAL

Figure 10-7. IFR flight plan for the Lafayette to Champaign segment of the simulated IFR flight.

the information for the first and third legs of the flight is transmitted to Indianapolis ARTCC. But because Champaign is located within Chicago Center's airspace, the flight plan information for the second leg is transmitted to the Chicago ARTCC computer.

The FDP computer examines the route information contained in the flight plan and verifies its accuracy. If it is in error, a message is sent to the Terre Haute flight service specialist, who must determine the routing error and make the required corrections. If any changes are made to the route of flight, the specialist causes the abbreviation FRC to be printed on the strip.

The FDP computer also checks the routing information in the flight plan to see whether any preferred arrival or departure routes apply to this flight. If so, they are automatically printed on the flight progress strip. A half hour prior to the pilot's estimated time of departure, the FDP computer activates the flight plan, issues the aircraft a transponder code, and causes a flight progress strip to be printed. Since the Lafayette control tower is not equipped with FDIO printers, the flight strip will be printed at the Indianapolis ARTCC sector directly above Lafayette approach control's airspace. This sector is known as the Purdue Low Sector.

The flight data controller at the Purdue Low Sector contacts the flight data controller at Lafayette tower and verbally passes along the information contained in the flight plan. This includes the aircraft call sign, aircraft type, transponder code, proposed departure time, requested altitude, destination airport, and route of flight. If the route is lengthy, normally only the segment of the route that concerns Lafayette tower will be passed along (see Figure 10-8).

The pilot next contacts the ground controller at Lafayette for taxi instructions and IFR clearance, and the controller issues appropriate taxi instructions. If the pilot has requested a route not approved by the letter of agreement between Lafayette and Champaign (see Figure 10-9), the controller will change the route of flight to conform to the LOA, then read the clearance to the pilot. The phraseology would be as follows:

> N252MN, cleared to the Champaign airport as filed via victor two fifty one, climb and maintain five thousand. Departure control frequency one two three point eight five. Squawk four two one two.

The ground controller passes the flight strip to the approach controller, who also handles the duties of the flight data and departure controllers. The ground controller verbally advises the local controller that N252MN is an IFR departure and that further coordination with the approach controller must be accomplished before N252MN can be released.

The approach controller then contacts the flight data controller at the Indianapolis ARTCC to request that N252MN be departed in the FDP computer. The flight data controller at the Purdue sector enters a departure message into the FDIO using the departure message format discussed in Chapter 6. Entering a departure message causes the appropriate flight plan information to be sent to the next ATC facility (Champaign) that will handle the aircraft. Since

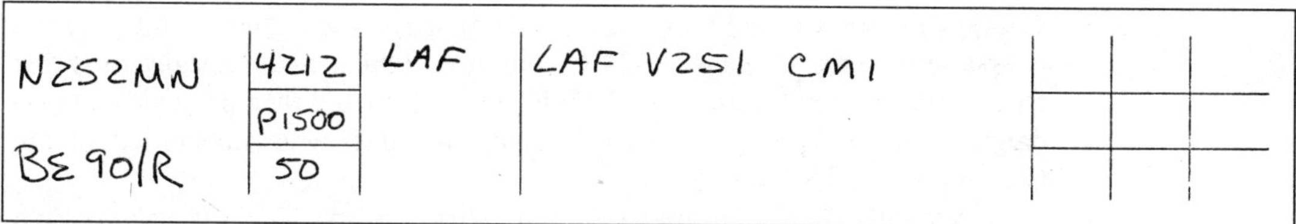

Figure 10-8. The flight progress strip that would be used for this segment of the flight.

CHAMPAIGN TOWER AND LAFAYETTE TOWER
LETTER OF AGREEMENT

SUBJECT: TOWER EN ROUTE PROCEDURE EFFECTIVE: January 1, 1989

PURPOSE: To establish procedures for providing tower enroute control service between Lafayette Tower and Champaign Tower.

CANCELLATION: Champaign Tower and Lafayette Tower Letter of Agreement dated January 1, 1984.

SCOPE: The procedures herein are for the purpose of conducting IFR operations between Champaign Tower and Lafayette Tower within the airspace allocated to each facility by their respective Center/Tower Letters of Agreement.

PROCEDURES:

Routing and Altitudes:

Champaign tower shall clear all aircraft landing in the Lafayette Approach Control area to the STAKS intersection via V251 at four thousand feet.

Lafayette Tower shall clear all aircraft landing within the Champaign Approach Control area at five thousand feet.

Coordination:

Coordination shall be effected prior to entry into the receiving fa cility's airspace.

An assumed departure time may be utilized provided the time is within three minutes of the actual departure time; otherwise the actual departure time shall be forwarded.

Frequency assignments:

Lafayette to Champaign - 121.35 mHz or 285.65 mHz

Champaign to Lafayette - 123.85 mHz or 393.00 mHz

Figure 10-9. Letter of agreement.

Champaign is not in Indianapolis Center's airspace, the flight plan information is first sent to the Chicago ARTCC computer, which automatically forwards this information to Champaign. A flight progress strip is then printed at Champaign, and the appropriate beacon code information is transmitted to the Champaign ARTS-II computer.

Next, the Lafayette approach controller contacts Champaign approach using the telephone. The flight data controller in the TRACON at Champaign usually answers. The Lafayette controller requests approval for N252MN to enter Champaign's airspace at 5,000 feet. The Lafayette controller requests the use of V251 and specifies that the transfer of control and communication will occur at the Staks intersection (which is the boundary between the two facilities). At this time, the Champaign flight data controller verbally communicates with the east arrival controller, who has responsibility for the airspace over Staks, and passes along Lafayette's request. Depending on traffic conditions (particularly at the Danville airport, which is almost directly under V251), the east arrival controller will approve the route and altitude or will request specific changes in either:

LAFAYETTE CONTROLLER:	APPREQ, N252MN, victor two fifty-one at Staks at five thousand.
CHAMPAIGN CONTROLLER:	N252MN at Staks at five thousand approved.

Lafayette Local Control The local controller must ensure proper runway separation between this aircraft and any other IFR or VFR aircraft landing or departing from Lafayette. In addition, the local controller must coordinate with the approach controller to ensure that N252MN is separated from other IFR aircraft within Lafayette's assigned airspace. When the pilots of N252MN advise the local controller that they are ready for takeoff, the local controller first coordinates the departure with the approach controller, who may request a slight change in route or altitude to separate N252MN from other IFR aircraft. If there is insufficient airspace to accommodate N252MN at this time, the departure controller may advise the local controller to **"hold for release."** This means that the aircraft must remain on the ground until sufficient airspace can be cleared to accommodate it. Once sufficient airspace is cleared for the aircraft, the approach controller advises the local controller, who can then clear N252MN for takeoff as soon as local traffic conditions permit.

Once approval has been received to depart the aircraft, the local controller sequences N252MN into the departure flow of aircraft, assigns a heading that will join the airway, and issues a clearance for takeoff. As soon as feasible after N252MN clears any traffic in the local area, the local controller advises N252MN to contact the departure controller:

> N252MN, turn right heading two seven zero, runway one zero, cleared for takeoff.
>
> N252MN, contact departure.

The approach controller at Lafayette also performs the duties of the departure controller and is responsible for separating N252MN while ensuring that the aircraft conforms with the clearance received from Champaign. To conform with the letter of agreement, N252MN must be established on V251 and level at 5,000 feet before crossing Staks ("N252MN, join victor two fifty-one, cross Staks at and maintain five thousand").

Once N252MN reports crossing Staks, Lafayette approach advises the pilot to contact Champaign approach ("N252MN, contact Champaign approach on one two one point three five").

Champaign Approach Control The Champaign Air Traffic Control Tower is an ARTS-II radar-equipped facility at a medium-activity airport. Champaign approach and departure control is housed in a TRACON room one floor below the tower cab. Champaign approach control is responsible for the airspace that extends from the surface up to and including 8,000 feet. Overlying Champaign are the Kansas City, Chicago, and Indianapolis ARTCCs. The Champaign tower Facility Directives specify that the approach control duties are normally assigned to two controllers. Each is given responsibility for a block of airspace extending from the surface of the Earth up to 8,000 feet. The dividing line between the two sectors is the extended centerline of runway 14R-32L (see Figure 10-10). In addition, a small block of airspace near the departure and arrival ends of each active runway is delegated to the local controller. The area at the approach end of the runway extends 3 nautical miles from the airport. The area at the departure end extends 5 miles from the end of the runway. This departure area extends from the surface of the Earth up to and including 3,000 feet MSL. Both the arrival and departure areas are used by the local controller for specific aircraft maneuvering and will be discussed later in this chapter. In this scenario, we will assume that the active runway at Champaign is runway 32 left. The controller assigned to the east sector is responsible for separating every other IFR aircraft from N252MN while vectoring it inbound for either a visual, an NDB, or an ILS approach to runway 32L. We will assume that the controller is vectoring for the NDB approach. The controller is required by the FAA handbook to maneuver the aircraft so that it can intercept the final approach course at a point at least 2 miles outside the approach gate. Since the final approach fix (Veals) is 5.5 nautical miles from the approach end of the runway, the approach gate is at 6.5 nautical miles. The controller must vector the aircraft to intercept the final approach course no closer than 8.5 nautical miles from the runway.

Once the east sector controller has approved N252MN's entry into Champaign's airspace, nonradar separation must be provided until N252MN can be radar identified. When the aircraft has established communication with the Champaign controller after crossing the Staks intersection, the controller will radar identify the aircraft and begin to use radar separation. This radar identification must be accomplished using one of the techniques discussed in Chapter 8. Most likely, the controller will use a combination of two or more of these methods, such as noting the reported location of the aircraft, observing

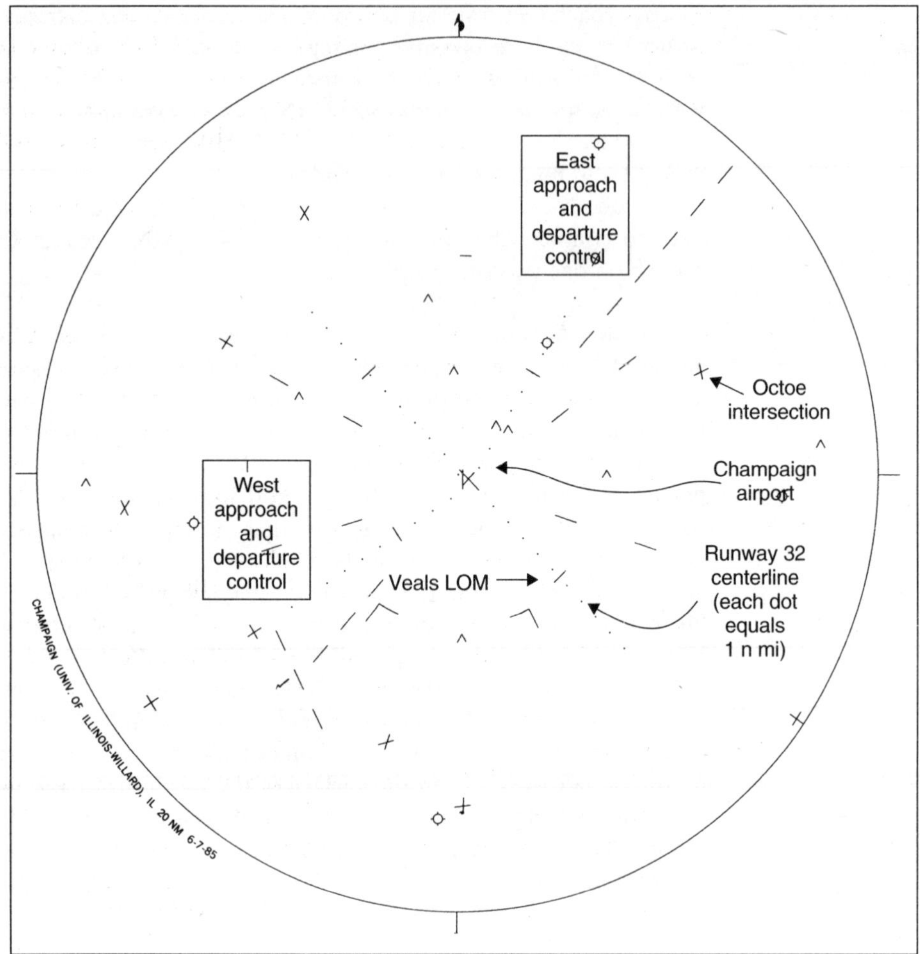

Figure 10-10. The radar map used by the Champaign approach controllers.

the acquisition of the aircraft's transponder signal by the ARTS-II system, and requesting that the pilot activate the Ident function of the transponder.

The controller advises the pilot when radar contact has been established and then begins to apply radar separation procedures and offer standard radar services. The controller also assigns a new altitude and informs the pilot of the NDB approach that is to be expected ("N252MN, radar contact three miles southwest of the Staks intersection, descend and maintain three thousand, expect vectors for the NDB runway three two left approach").

While vectoring N252MN for the NDB approach, the approach controller must accomplish the following:

> N252MN must remain separated from every other IFR aircraft within the east sector.
>
> Coordination with the west approach controller must be accomplished to determine the arrival sequence for N252MN, since both controllers are vectoring aircraft to runway 32L.
>
> Since an airport radar service area surrounds the Champaign airport, the east controller must separate N252MN from VFR aircraft while within the ARSA. While N252MN is outside the ARSA, the controller is required only to advise the pilot of any conflicting VFR traffic and attempt to provide resolution upon pilot request.

The east controller accomplishes these tasks by issuing vectors (and speed restrictions if needed) to N252MN ("N252MN, fly heading two one zero, vector for the NDB runway three two left approach"). Whenever one of these instructions is issued to the pilot, the controller should advise the pilot of the reasons for the instruction.

The controller must position the aircraft such that an easy transition can be made to the NDB 32L final approach course. This requires that the aircraft be positioned such that it will intercept the final approach course at least 2 miles outside of the approach gate, at an angle of less than 30° and at an altitude at or above the minimum vectoring altitude (see Figure 10-11). The pilot can then be cleared for the NDB approach.

As part of the approach clearance, the controller must advise the pilot of the aircraft's position relative to the final approach fix, the heading to fly to intercept the final approach course, the altitude to maintain until established on the final approach course, and the actual clearance to conduct the instrument approach ("N252MN, seven miles from Veals, turn right heading three zero zero, maintain two thousand six hundred until established on the final approach course, cleared NDB runway three two left approach. Monitor tower on one two zero point four, report Veals inbound"). Once the aircraft is established on the final approach course the pilot can be advised to monitor the local controller's frequency and report crossing the final approach fix (see Figure 10-12).

Local Control When the aircraft has crossed the final approach fix, it becomes the local controller's responsibility to sequence it into the local traffic flow. The local controller must ensure that no instructions are issued to the aircraft that might cause it to conflict with any inbound or outbound IFR aircraft being maneuvered by the approach controllers. The controller does so by using the 3-nautical-mile area near the approach end of the runway (see Figure 10-13). Champaign Facility Directives permit the local controller to maneuver N252MN once it enters this 3-mile area. The controller can maneuver N252MN to follow another aircraft or can clear N252MN to land on a different runway, such as runway 4, 22, or 32C. The one restriction on the use of this area is that the local controller is not permitted to turn N252MN back toward following IFR aircraft. Once N252MN has been properly sequenced, the local controller can

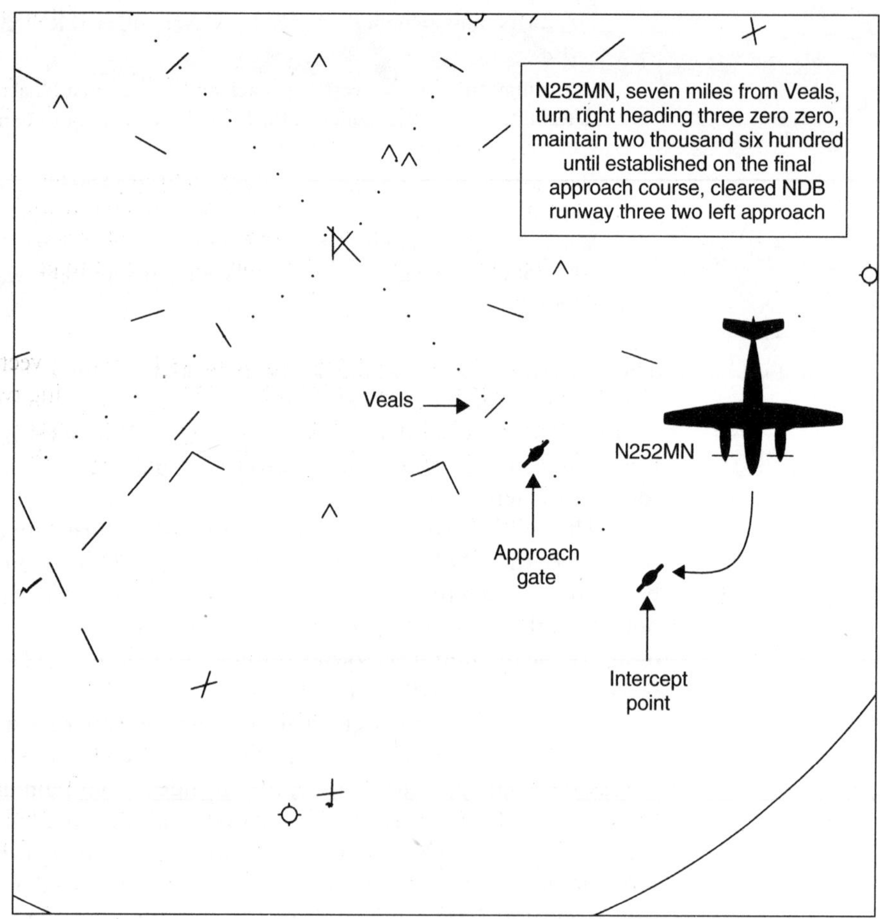

Figure 10-11. The approach gate and the intercept point on the NDB 32L approach at Champaign.

advise the pilots that the aircraft is cleared to land ("N252MN, cleared to land runway three two left").

After the aircraft has landed, the local controller advises the pilot to contact the ground controller for taxi instructions to the parking area ("N252MN, if able, turn right at the next intersection, contact ground on one two one point niner"). The Champaign ground controller then issues N252MN a clearance to taxi to the parking ramp ("N252MN, taxi to the ramp, transient parking is south of the tower").

If N252MN must cross an active runway before reaching the parking area, the ground controller must coordinate this crossing with the local controller. Once the aircraft is on the ground, the ARTS-II computer is programmed to terminate the aircraft's flight plan and the appropriate flight plan

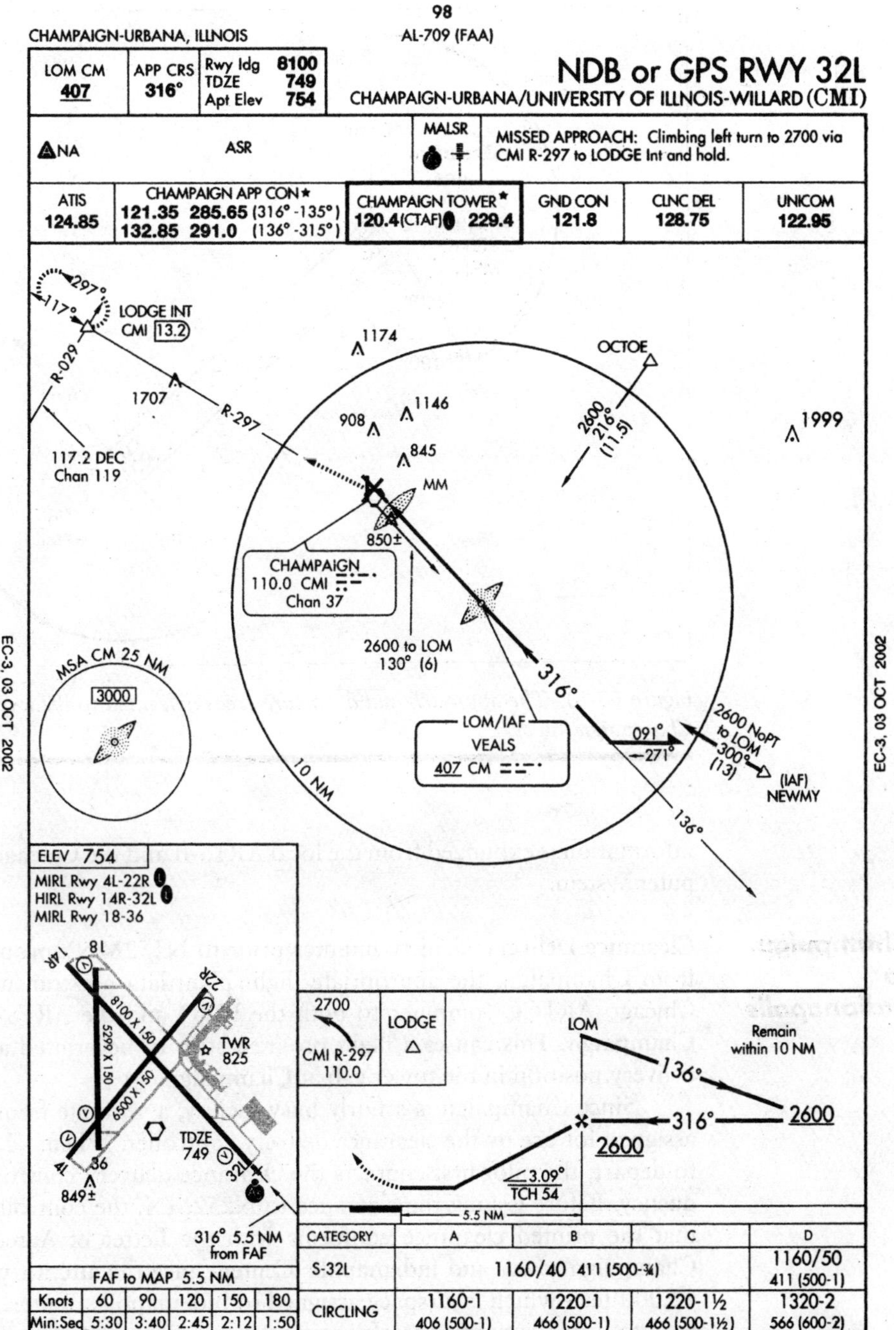

CATEGORY	A	B	C	D
S-32L	1160/40 411 (500-¾)			1160/50 411 (500-1)
CIRCLING	1160-1 406 (500-1)	1220-1 466 (500-1)	1220-1½ 466 (500-1½)	1320-2 566 (600-2)

FAF to MAP 5.5 NM					
Knots	60	90	120	150	180
Min:Sec	5:30	3:40	2:45	2:12	1:50

Figure 10-12. NDB runway 32L instrument approach chart.

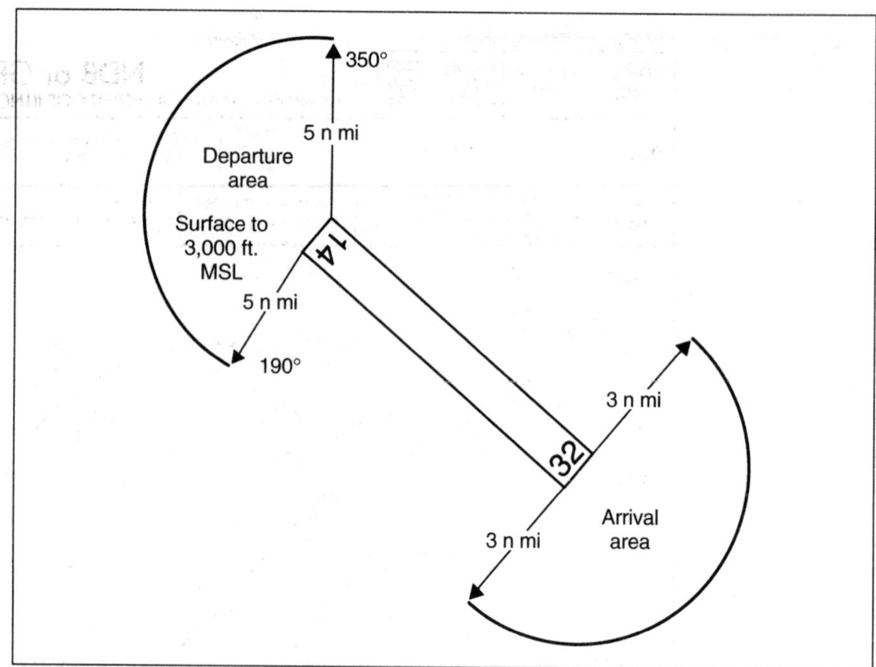

Figure 10-13. The approach and departure areas delegated to the local controller at Champaign tower.

information is expunged from the local ARTS-II and the Chicago ARTCC computer system.

Champaign to Indianapolis

Clearance Delivery Thirty minutes prior to N252MN's proposed departure from Champaign, the appropriate flight plan data are transmitted from the Chicago ARTCC computer to both the FDIO and the ARTS-II computer at Champaign. This causes a flight progress strip to be printed at the clearance delivery position in the tower cab at Champaign.

Since Champaign is a fairly busy facility, a separate frequency has been assigned for use by the clearance delivery controller. When N252MN is ready to depart, the pilot first contacts the clearance delivery controller on this frequency. Before issuing the clearance to N252MN, the controller must ensure that the printed clearance conforms with the Letter of Agreement between Champaign tower and Indianapolis Center, since the aircraft will be flying at 12,000 feet, which is airspace assigned to Indianapolis Center. If any changes need to be made to the aircraft's route or altitude to conform with the Letter of Agreement, they are made by the clearance delivery controller and entered into the FDIO.

The clearance delivery controller must also ensure that the clearance conforms with Champaign tower Facility Directives. In particular, at Champaign the Facility Directives state that every departing aircraft will initially be assigned an altitude of 3,000 feet MSL, the upper limit of the airspace located at the departure end of the runway, which has been delegated to the local controller. This area is commonly referred to as the departure fan. The clearance delivery controller is required to initially restrict every aircraft to this altitude and inform the pilot that the altitude filed in the flight plan can be expected later in the flight. ("N252MN, cleared to Indianapolis airport as filed, climb and maintain three thousand, expect one two thousand one zero minutes after departure. Departure frequency is one two one point three five. Squawk four one two one").

Since Champaign is equipped with an ARTS-II radar system, the clearance delivery controller does not need to send a departure message to the FDP computer at Chicago Center. Champaign's ARTS-II computer is connected to the FDP system at Chicago and will send a departure message upon receipt of N252MN's transponder code. When the clearance is issued by the clearance delivery controller, the flight progress strip is passed to the ground controller.

Ground Control When the pilots have received their clearance, they contact the ground controller for taxi instructions to the active runway, which in this example will remain runway 32L ("N252MN, turn left on the parallel taxiway, follow the DC-9, taxi to runway three two left"). If the aircraft must cross any active runways en route to 32L, the crossing must be coordinated with the local controller.

Local Control When the pilot is ready for takeoff, radio contact is established with the local controller. The local controller is responsible for sequencing N252MN into the local traffic flow while also providing initial IFR separation between departures. The Champaign Facility Directives specify that the local controller may assign aircraft departing runway 32L any heading between 190° clockwise to 350° ("N252MN, turn right heading three five zero, runway three two left, cleared for takeoff"). It is up to the approach controller to keep arriving aircraft clear of this area.

The local controller is permitted by the Facility Directives to depart IFR aircraft within this departure area without prior coordination with the departure controller as long as all of the following conditions are met:

> Every successive IFR departure must be assigned a heading that differs by at least 15° from that issued to the previous aircraft. If, for example, N252MN is assigned a heading of 350°, the next aircraft must be assigned a heading of 335° or less.
>
> Every departing aircraft must remain separated by at least a 1-nautical-mile interval.
>
> The local controller must ensure that every departing aircraft is separated from local pattern traffic before the pilot is advised to contact the departure controller.

Once the local controller has ensured that these conditions have been met, the pilot is informed to contact the departure controller ("N252MN, contact departure").

Departure Control When the ARTS-II computer detects N252MN's transponder transmissions, a departure message is automatically sent to the flight data processing computer at Chicago Center, causing a flight strip to be printed at every en route sector that N252MN will fly through. Since N252MN will be entering the airspace assigned to Indianapolis ARTCC, the appropriate flight plan information is transmitted to the FDP computer at Indianapolis Center. In addition, the flight plan information is passed along to the radar data processing computer at each center, which will begin to track N252MN upon receiving its transponder transmission.

At Champaign, the east and west approach controllers are also responsible for the separation of aircraft departing Champaign. Upon initial contact with N252MN, the east departure controller is required to perform the following tasks before any radar separation can be applied:

> The aircraft must be radar identified using one of the methods described in Chapter 9.
>
> If the aircraft is equipped with a mode C transponder, the controller must verify the accuracy of its operation.
>
> At Champaign, Facility Directives state that aircraft may not be turned until they are at least 500 feet above the altitude used by aircraft in the pattern conducting touch and goes. Since the pattern altitude is 1,500 feet MSL, the departure controller may not turn N252MN until the pilot has reported leaving or until a verified mode C readout indicates that the aircraft has left 2,000 feet MSL.
>
> Champaign's Facility Directives also state that the departure controller may turn an aircraft to a heading that will take it out of the departure fan but may not turn it back in toward the runway centerline. This procedure ensures separation between successive departures.

When all of these requirements have been met, the departure controller can advise N252MN that it has been radar identified and then vector the aircraft to join the proper airway (victor 434). As soon as traffic permits, the controller will authorize N252MN to climb higher than the 3,000 feet initial restriction ("N252MN, radar contact, turn right heading one two zero, join victor four thirty-four, climb and maintain eight thousand"). At no time may the aircraft be cleared to an altitude higher than 8,000 feet MSL, however, since that is the upper limit of Champaign's delegated airspace.

Once the aircraft has joined victor 434 and is climbing to or level at 8,000 feet, and every potential traffic conflict has been resolved, the Champaign departure controller can initiate a handoff with Indianapolis Center. This handoff must be accomplished before N252MN reaches the facility boundary (see Figure 10-14). If it cannot be accomplished before the aircraft reaches that point, the Champaign controller must ensure that N252MN remains within the confines of Champaign's airspace until a handoff or an alternative clearance can be coordinated.

Typically, the handoff with Indianapolis Center is accomplished somewhere around the Emtee intersection, which is 12 miles east of Champaign. The

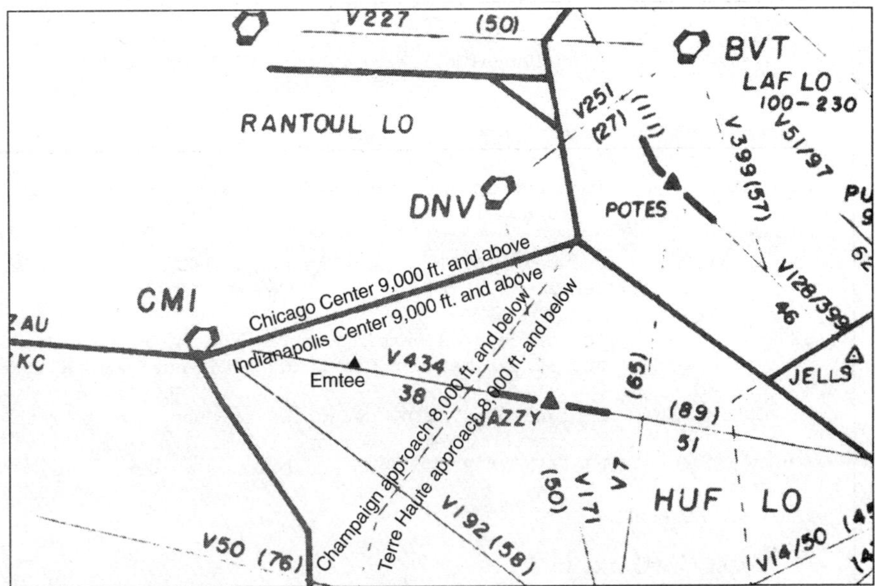

Figure 10-14. The airspace boundaries between the Champaign tower and Indianapolis Center.

Champaign controller initiates a handoff by slewing the ARTS-II track ball symbol over N252MN's radar blip and pressing the appropriate keys. This action automatically initiates the handoff process. Since Champaign is primarily located within Chicago Center's airspace, the electronic transmissions that perform this automated handoff actually travel from the Champaign TRACON to Indianapolis ARTCC via the Chicago Center computer.

At Indianapolis Center, at the Terre Haute Low Sector, N252MN's data block begins to flash on the controller's plan view display. If the Indianapolis Center radar controller decides to accept the handoff, the trackball symbol is slewed over N252MN's data block and the center controller presses the appropriate keys. N252MN's data block then ceases to flash on the center controller's PVD but starts to flash on the Champaign approach controller's radar screen, at which time the Champaign controller understands that the Indianapolis Center controller has accepted the handoff.

This type of handoff, accomplished solely through the computer, is known as an *automated handoff*. This procedure assumes that N252MN will comply with every restriction and procedure delineated in the letter of agreement between the Champaign tower and Indianapolis Center (see Figure 10-15). This letter of agreement specifies that every aircraft departing Champaign must be established on the airway printed on the flight progress strip and that potential traffic conflicts have been resolved by the Champaign controller. Any de-

CHAMPAIGN TOWER AND INDIANAPOLIS CENTER
LETTER OF AGREEMENT

SUBJECT: TERMINAL AREA CONTROL PROCEDURES EFFECTIVE: January 1, 1989

PURPOSE: To prescribe procedures to be used between Champaign ATCT and Indianapolis ARTCC.

CANCELLATION: Champaign ATCT and Indianapolis ARTCC Letter of Agreement dated January 1, 1984.

SCOPE: The procedures herein are for the purpose of conducting IFR operations between Champaign ATCT and Indianapolis ARTCC.

RESPONSIBILITY: Indianapolis ARTCC delegates to Champaign ATCT the authority and responsibility for control of IFR terminal and en route traffic at 8,000 feet and below.

ARRIVAL PROCEDURES:

CLEARANCE LIMIT

The destination airport shall be the clearance limit.

ROUTES

The filed route shall be the arrival route unless suspended by either facility. FDEP shall constitute approval and coordination.

ALTITUDES

Arrivals landing at any airport in Champaign ATCT's delegated airspace shall be cleared to cross the 24-mile DME arc of the Champaign VORTAC at 9,000 feet.

DEPARTURE PROCEDURES:

Champaign ATCT shall ensure that departures are handed off with at least 5 miles radar separation that is constant or increasing.

Aircraft filing for 9,000 feet or higher must be restricted to 8,000 feet until coordinated with Indianapolis ARTCC.

FREQUENCIES

Indianapolis ARTCC to Champaign - 121.35 mHz or 285.65 mHz
Champaign to Indianapolis ARTCC - 132.20 mHz or 307.10 mHz

Figure 10-15. Letter of agreement.

viation from these procedures requires verbal coordination between the two controllers.

The letter of agreement also specifies that Champaign must transfer the communication of N252MN before it reaches the boundary of Champaign's airspace ("N252MN, contact Indianapolis Center one three two point two"). This permits the ARTCC controller to be in radio contact with the aircraft before it actually enters Indianapolis Center's airspace.

If it becomes apparent that N252MN will cross the boundary between Champaign and the Terre Haute TRACON at an altitude below 8,000 feet MSL, permission for N252MN to enter Terre Haute's airspace must be received from the Terre Haute approach controller. It is assumed that this is the Champaign approach controller's responsibility, but once it becomes apparent that the aircraft may need to enter Terre Haute's airspace, the Champaign controller may no longer be communicating with the aircraft. This is a gray area in the procedures, and to be on the safe side, both the Champaign approach and the Indianapolis Center controllers should coordinate with Terre Haute if it appears that a point out may be necessary.

Indianapolis Center The Indianapolis Center controller is not required to re-identify the aircraft, since initial radar identification was accomplished by the Champaign departure controller and was subsequently transferred during the automated handoff. All that is required of the center controller is to verify the accuracy of the aircraft's mode C readout, issue the pilot any altitude changes, and issue the altimeter setting from the closest airport with a weather observer (which in this example is Terre Haute) ("N252MN, climb and maintain one two thousand, Terre Haute altimeter two niner eight four").

The Indianapolis ARTCC controller is then responsible for separating N252MN from other IFR aircraft within the Terre Haute Low Sector while also complying with internal Facility Directives that may affect that flight. The center controller must also sequence N252MN into the traffic flow for the Indianapolis airport. The controller must comply with the procedures described in the Indianapolis Center/Indianapolis Tower Letter of Agreement (see Figure 10-16). In particular, the Indianapolis Center controller must ensure that N252MN enters the Indianapolis approach control airspace either at or descending to 11,000 feet and enters over one of the designated arrival fixes. The Antti intersection, which is 21 miles west of the Indianapolis VOR on victor 434, is one such fix.

Indianapolis Approach Control Procedures Indianapolis TRACON controllers procedurally separate inbound and outbound aircraft using a modification of a "box" system of procedural separation. In a typical box configuration, the letter of agreement describes a box that is drawn around the affected TRACON's airspace. Each corner of the box (known as a cornerpost) is delineated by an intersection or navaid.

At Indianapolis, the cornerposts are delineated by the Jells and Antti intersections to the northwest, Clang to the northeast, the Shelbyville (SHB) VOR

INDIANAPOLIS TOWER AND INDIANAPOLIS CENTER
LETTER OF AGREEMENT

SUBJECT: TERMINAL AREA CONTROL PROCEDURES EFFECTIVE: January 1, 1989

PURPOSE: To prescribe procedures to be used between Indianapolis ATCT and Indianapolis ARTCC.

CANCELLATION: Indianapolis ATCT and Indianapolis ARTCC Letter of Agreement dated January 1, 1984.

SCOPE: The procedures herein are for the purpose of conducting IFR operations between Indianapolis ATCT and Indianapolis ARTCC.

RESPONSIBILITY: Indianapolis ARTCC delegates to Indianapolis ATCT the authority and responsibility for control of IFR terminal and en route traffic at 10,000 feet and below.

ARRIVAL PROCEDURES:

CLEARANCE LIMIT

The destination airport shall be the arrival airport. Indianapolis ARTCC shall clear arrivals via the metering fixes depicted on attachment #1.

ROUTES

The filed route shall be the arrival route unless suspended by either facility. FDEP shall constitute approval and coordination.

ALTITUDES

Arrivals landing at any airport in Indianapolis ATCT's delegated airspace shall be cleared over one of the arrival fixes either level at or descending to 11,000 feet. Indianapolis ATCT may descend these aircraft below 11,000 feet once the transfer of communication has been accomplished.

DEPARTURE PROCEDURES:

Indianapolis ATCT shall ensure that departures are handed off with at least 5 miles radar separation that is constant or increasing.

Aircraft filing for 11,000 feet or higher must be restricted to 10,000 feet until coordinated with Indianapolis ARTCC.

Indianapolis ATCT shall ensure that departures cross one of the four departure gates shown on attachment #1 as NOIND, EAIND, SOIND, and WEIND.

FREQUENCIES

Indianapolis ARTCC to Indianapolis ATCT - 121.35 mHz or 285.65 mHz
Indianapolis ATCT to Indianapolis ARTCC - 132.20 mHz or 307.10 mHz

Figure 10-16. Letter of agreement.

at the southeast, and the Kelly intersection to the southwest. Where box systems are used, the letter of agreement specifies that every inbound IFR aircraft must enter the approach control's airspace at one of the cornerposts. These areas are known as **arrival gates** (see Figure 10-17). The letters of agreement also specify that departures must remain clear of the cornerposts and depart the area through the sides of the box. The sides of the box are known as **departure gates.**

Handoff to Indianapolis Approach Approximately 30 minutes before N252MN enters Indianapolis approach control's airspace, the appropriate flight plan information is transmitted to the ARTS-III computer at Indianapolis tower. Once N252MN is detected by the secondary radar system at Indianapolis tower, the ARTS-III computer initiates a computer track. As N252MN nears the boundary between Indianapolis ARTCC and Indianapolis approach, the ARTCC

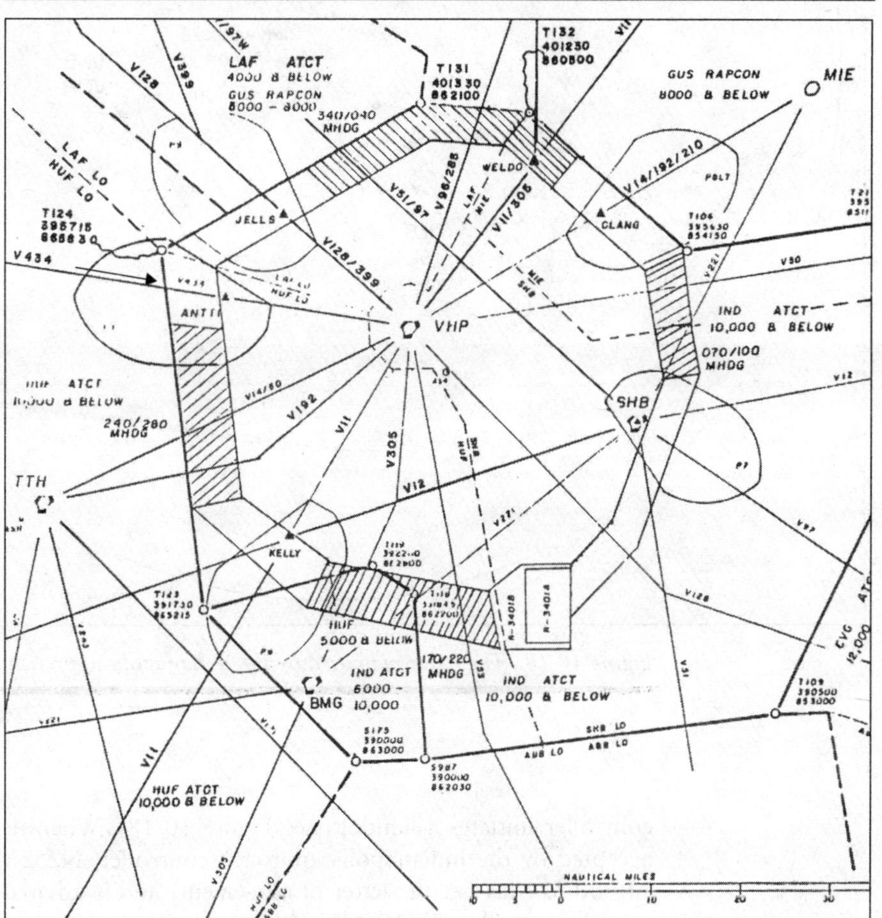

Figure 10-17. The arrival gates used by Indianapolis approach control.

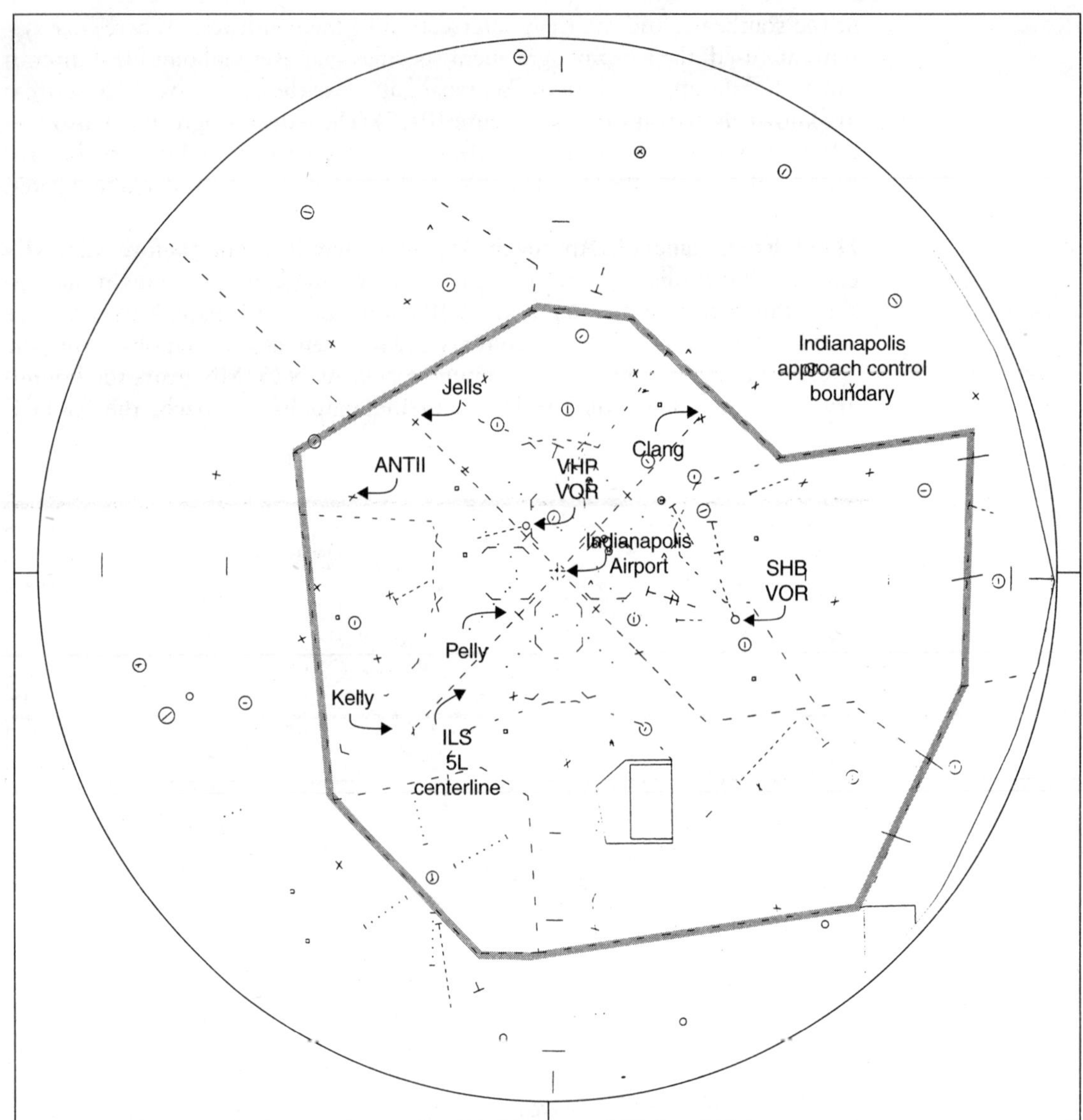

Figure 10-18. The radar map used by the Indianapolis approach controllers.

controller initiates a handoff (see Figure 10-18). When the handoff has been accepted by the Indianapolis approach controller, N252MN is descended to 11,000 feet (as per the letter of agreement) and is advised to contact the approach controller ("N252MN, descend and maintain one one thousand, contact Indianapolis approach control on one two one point one").

Indianapolis Approach Control Since Indianapolis tower is located at a high-activity airport, it is not possible to simply divide the approach control airspace into two sectors as is done at Champaign. The Facility Directives at Indianapolis specify that as many as six different controllers may be assigned approach and departure control responsibilities, corresponding to six control sectors (see Figure 10-19). Two of these sectors are designated as arrival sectors (and are

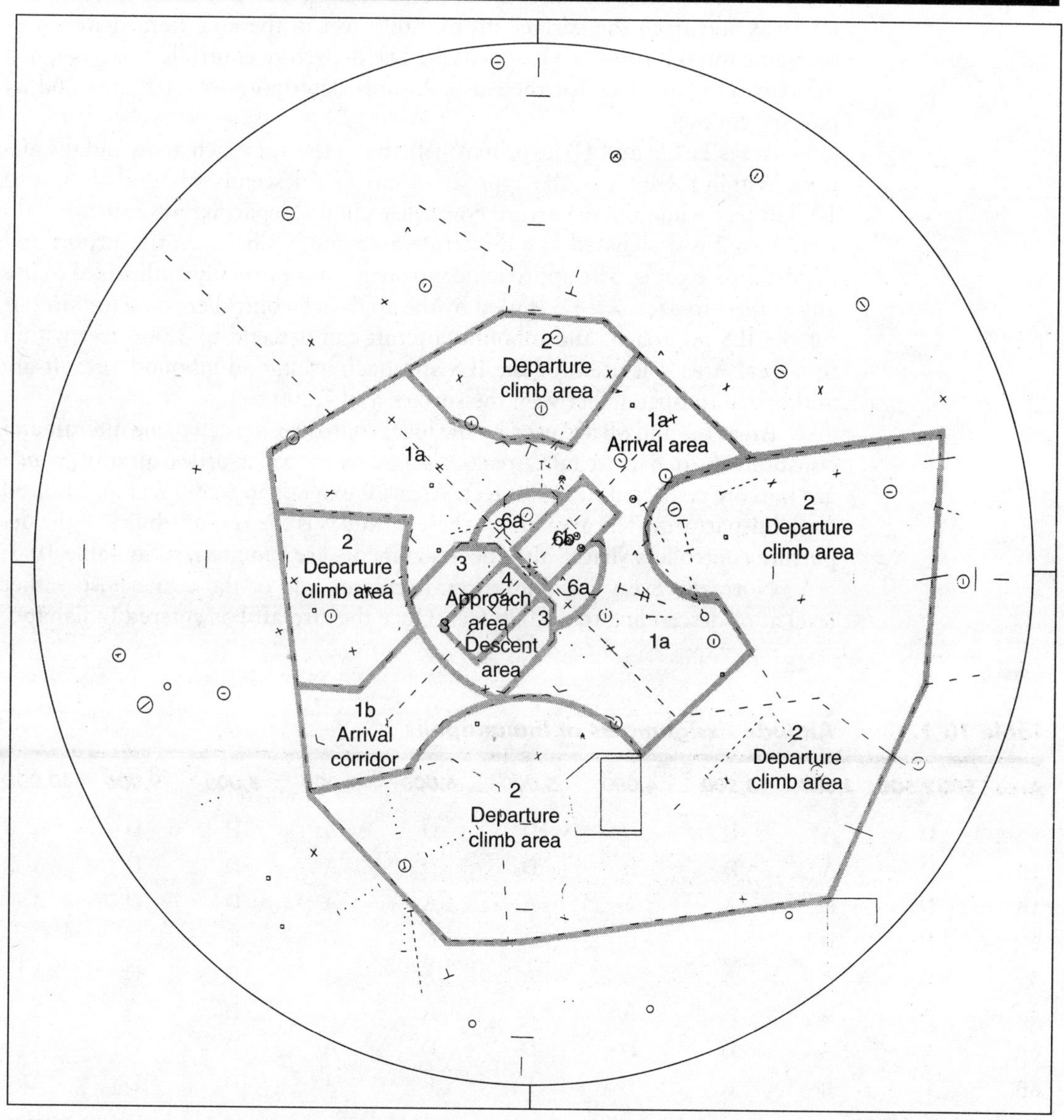

Figure 10-19. The arrival and departure areas as specified in the Indianapolis Facility Directives.

known as east arrival and west arrival), whereas the other four are departure sectors (known as north, south, east, and west departure).

In this scenario, we will assume that runway 5L is the primary runway in use at Indianapolis International Airport. In this runway configuration, the two arrival controllers are assigned the airspace on both sides of runway 5L. Each departure controller is assigned a 90° segment of airspace delineated by the extended centerlines of runway 5L-23R and 14-32. In general, the arrival controllers are delegated the airspace at 3,000 feet, 7,000 feet, and 10,000 feet in all areas and from the surface up to 7,000 feet in the area immediately surrounding the ILS runway 5L approach. The departure controllers are assigned the remaining airspace for their use. A short description of each area and its purpose follows.

Areas 1, 1A, and 1B lie primarily between the approach gates and the airport. Within these areas, the approach controller descends inbound aircraft to 10,000 feet while the departure controller climbs departing aircraft to 9,000 feet. Area 2 is designated as a departure area and lies between the airport and the departure gates. The approach controller is not normally authorized to use any of this airspace. Area 3 is used by the approach controller to vector aircraft for the ILS approach, and inbound aircraft can descend to 3,000 feet within this area. Area 4 is used for the ILS approach itself, and inbound aircraft are authorized to operate between the surface and 7,000 feet.

Areas 6A and 6B are used by the local controller for departing aircraft and constitute the departure fan. Area 6A is used by propeller-driven aircraft, which are initially restricted to 2,500 feet. Area 6B extends to 6,000 feet and is used for jet departures. The area above these altitudes is the responsibility of the departure controllers. These altitude assignments are summarized in Table 10-1.

Normally every inbound aircraft crosses one of the cornerposts either level at or descending to 11,000 feet. Once the aircraft has entered Indianapo-

Table 10-1. Altitude Assignments at Indianapolis

Area	SFC-2,500	3,000	3,500	4,000	5,000	6,000	7,000	8,000	9,000	10,000
1	D	A*	D	D	D	D	A	D	D	A
1A	D	A*	D	D	D	D	A	D	D	A
1B	D	D	A	A	A	A	A	D	D	A
2	D	D	D	D	D	D	D	D	D	D
3	D	A	A	A	A	A	A	D	D	A
4	A	A	A	A	A	A	A	D	D	A
6A	L	A	D	D	D	D	A	D	D	A
6B	L	L	L	L	L	L	A	D	D	A

A = Arrival controller; D = Departure controller; L = Local controller; * = VFR altitudes.

lis TRACON's assigned airspace, the arrival controller is permitted to descent the aircraft to 10,000 feet.

Every aircraft inbound to Indianapolis is vectored toward the airport and sequenced behind other inbound aircraft. Once the aircraft is within about 15 nautical miles of the airport (area 1), the arrival controller is authorized to descend the aircraft to 7,000 feet if coordination has been accomplished with the appropriate departure controller ("N252MN, descend and maintain seven thousand, vector for the ILS runway five left approach").

When N252MN is within about 7 miles of the airport, it has entered area 3, where the arrival controller may utilize the airspace extending from 7,000 feet to 3,000 feet MSL. At this point the aircraft is normally turned outbound, parallel to the ILS final approach course. Once in this position, traffic permitting, the aircraft will be descended to 3,000 feet in preparation for the ILS approach ("N252MN, turn right heading two two zero, descend and maintain three thousand").

As the aircraft flies this extended downwind pattern, the west arrival controller must coordinate with the east arrival controller to establish N252MN's sequence. When the sequence has been agreed on, the west controller will vector N252MN to follow the preceding aircraft, and if needed, the controller may issue the aircraft a speed adjustment that will maintain proper separation ("N252MN, reduce speed to one seven zero").

When N252MN is in the proper position, adequately separated from both preceding and following aircraft, the controller will turn the aircraft to intercept the runway 5L localizer (see Figure 10-20). Once the aircraft is turned inbound, it enters area 4 and can be authorized to conduct the ILS runway 5L approach ("N252MN, seven miles from Pully, turn left heading zero seven zero, intercept the final approach course at or above two thousand eight hundred, cleared for the ILS runway five left approach. Monitor tower on one two zero point niner and report Pully inbound").

At all times while being vectored for the ILS approach, N252MN is procedurally separated from departing aircraft. The only coordination involved between the approach and departure controllers occurs when the arrival controller descends N252MN from 10,000 feet to 7,000 feet. The two arrival controllers must coordinate the sequence of inbound aircraft with each other since each controller is vectoring for the same runway.

One of the limitations on the arrival controller is that very little airspace is assigned for maneuvering aircraft. The arrival controllers must keep each aircraft within the confines of areas 1B, 3, and 4 while descending the aircraft to 3,000 feet MSL. Because of this lack of airspace, the arrival controllers at Indianapolis become highly skilled at predicting the future flight paths of aircraft and judiciously use speed adjustments to safely sequence arrival aircraft while still confining these aircraft to the specified airspace.

Local Control At Indianapolis tower, the local controller is responsible for sequencing N252MN into the departure flow of traffic but has little flexibility

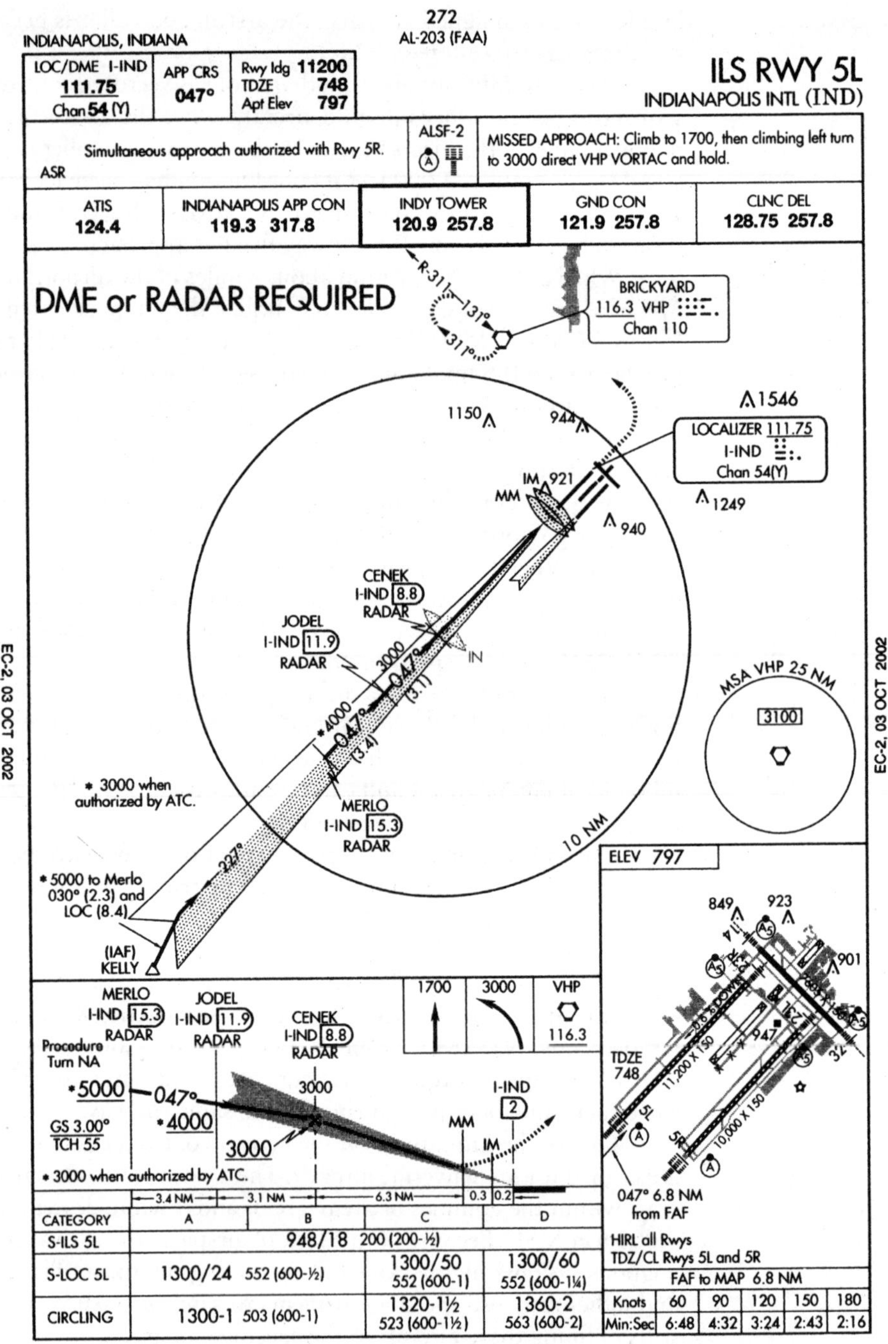

Figure 10-20. The ILS runway 5L approach at Indianapolis.

to maneuver the aircraft without coordinating with the controllers in the TRACON. If circumstances require, the local controller can clear N252MN to land on the parallel runway, runway 5R, but may not assign N252MN to any other runway without coordinating with the controllers in the TRACON ("N252MN, cleared to land runway five left. Traffic is a Cessna ahead and to your right landing runway four right").

Once N252MN has landed, the local controller advises the pilot to contact the ground controller, who clears the aircraft to taxi to the parking ramp.

Indianapolis to Lafayette

Flight Data and Clearance Delivery Except when the facility is very busy, the flight data and the clearance delivery positions at Indianapolis tower are combined into one operating position. Approximately 30 minutes before N252MN's departure to Lafayette, the FDP computer at Indianapolis Center transmits the appropriate flight plan information to the FDIO and the ARTS-III radar computer at the Indianapolis tower. When the flight progress strip for N252MN is printed, the flight data controller must verify the accuracy of information contained in the flight progress strip and must initiate any changes that will make N252MN conform to the Lafayette Tower/Indianapolis Tower Letter of Agreement (see Figure 10-21).

The pilots of N252MN contact the clearance delivery controller to receive their IFR clearance. The controller issues the clearance as amended but initially restricts the aircraft's altitude to 2,500 feet, because at Indianapolis tower Facility Directives require that all propeller-driven aircraft initially be assigned an altitude of 2,500 feet, whereas pure jet aircraft are restricted to an initial altitude of 6,000 feet. These altitudes confine each aircraft to the local controller's departure fan. The clearance delivery controller issues a clearance that restricts N252MN to 2,500 feet, advising the pilots that they can expect their requested altitude (4,000 feet) 10 minutes after departure. In addition, the clearance delivery controller issues the appropriate departure control frequency ("N252MN, cleared to the Ockel intersection via victor ninety-seven, climb and maintain two thousand five hundred, expect four thousand one zero minutes after departure. Departure frequency one two one point seven five. Squawk four one five one").

The clearance delivery controller then contacts the Lafayette approach controller using telephone equipment. Since Lafayette is a nonradar and non-FDIO facility, Indianapolis tower is required by letter of agreement to forward the appropriate flight plan information concerning N252MN and must also coordinate N252MN's entry into Lafayette's airspace. The Indianapolis controller advises the Lafayette controller of N252MN's route and altitude and its estimated departure time from Indianapolis. The Lafayette controller considers this request and then approves or disapproves it. As part of the approval, the Lafayette controller restricts the aircraft to an intersection located just within the boundary between Indianapolis and Lafayette (Ockel). Restricting N252MN to this intersection alleviates the need for the Lafayette controller to reserve a large block of airspace for N252MN:

INDIANAPOLIS TOWER AND LAFAYETTE TOWER
LETTER OF AGREEMENT

SUBJECT: TOWER EN ROUTE PROCEDURE EFFECTIVE: January 1, 1989

PURPOSE: To establish procedures for providing tower en route control service between Lafayette Tower and Indianapolis Tower.

CANCELLATION: Indianapolis Tower and Lafayette Tower Letter of Agreement dated January 1, 1984.

SCOPE: The procedures herein are for the purpose of conducting IFR operations between Indianapolis Tower and Lafayette Tower within the airspace delegated to each facility during Lafayette Tower operational hours.

PROCEDURES:

Routing and Altitudes:

Indianapolis Tower shall clear all aircraft landing in the Lafayette Approach Control area to the OCKEL intersection via V97 at four thousand feet.

Lafayette Tower shall clear all aircraft landing within the Indianapolis Approach Control area to the JELLS intersection via V399 at five thousand feet.

Coordination:

Coordination shall be effected prior to entry into the receiving facility's airspace.

An assumed departure time may be utilized provided the time is within three minutes of the actual departure time; otherwise the actual departure time shall be forwarded.

Frequency assignments:

Lafayette to Indianapolis - 124.65 mHz or 317.80 mHz
Indianapolis to Lafayette - 123.85 mHz or 393.00 mHz

Figure 10-21. Letter of agreement.

INDIANAPOLIS CONTROLLER: APPREQ, N252MN, victor ninety-seven at four thousand, departing Indianapolis at one five three seven.

LAFAYETTE CONTROLLER: N252MN, Ockel via victor ninety-seven at four thousand approved.

When N252MN's entry into Lafayette's airspace has been approved, the clearance delivery controller at Indianapolis circles the altitude approved by

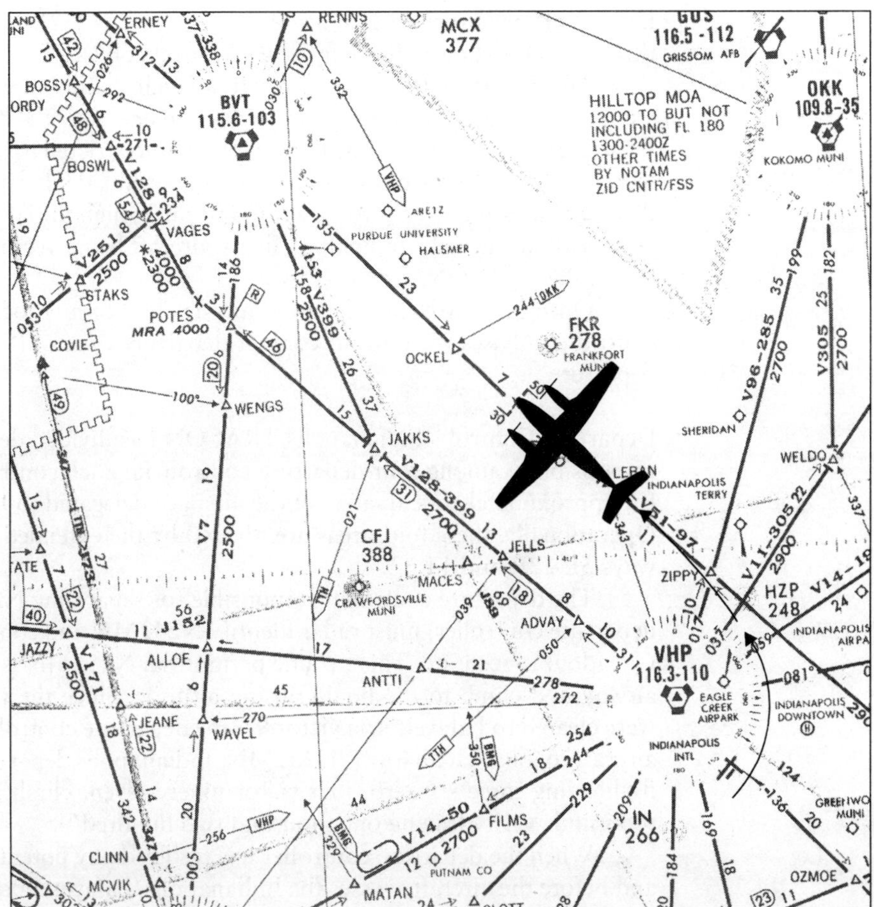

Figure 10-22. The flight path of N252MN as it returns to Lafayette.

Lafayette and places the estimated time of departure on the flight progress strip. This strip is then passed to the ground controller (see Figure 10-22). When the pilot contacts ground control on 121.90 mHz, the ground controller issues the aircraft the appropriate taxi instructions and passes the strip to the local controller.

Local Control The local controller is responsible for sequencing N252MN into the local traffic flow while providing initial separation between departures. At Indianapolis, this is accomplished by assigning a departure fan to the local controller that varies in dimension depending on the type of aircraft departing. Assuming that runway 5L is in use at Indianapolis, the departure fan is shaped such that the local controller may assign propeller-driven aircraft initial headings that range between 320° and 140°. The upper limit of the departure fan for propeller-driven aircraft is 2,500 feet ("N252MN, turn left heading three six zero, runway five left, cleared for takeoff").

The departure fan for jet aircraft is much smaller in width, and the local controller may issue jet aircraft only an initial heading of 50° (runway heading). Jet aircraft are permitted, however, to climb unrestricted to an altitude of 6,000 feet. The variable size of the departure fan permits the local controller to issue diverging headings to slower aircraft and fairly high initial altitudes to jet aircraft without severely impinging on any aircraft being vectored by the controllers in the TRACON.

Once N252MN has departed and is clear of any local traffic, the local controller advises the pilot to contact departure control ("N252MN, contact departure").

Departure Control Indianapolis TRACON has divided departure separation responsibility among four departure controllers. Each controller is responsible for approximately one quarter of the airspace delegated to Indianapolis tower. Operationally, these four areas are divided by the extended centerlines of runways 5L-23R and 14-32.

The departure controller responsible for separating N252MN (the north departure controller) must radar identify N252MN and confirm that the mode C readout is accurate. Then, traffic permitting, N252MN is cleared to join the airway and climb to the final cruising altitude. Since the pilots of N252MN were cleared to Lafayette via victor 97, the departure controller must vector the aircraft to join that airway ("N252MN, Indianapolis departure, radar contact, fly heading three six zero, join victor ninety-seven, climb and maintain four thousand. Verify leaving one thousand two hundred").

When the departure controller has resolved any potential traffic conflicts, and before the aircraft crosses the Indianapolis/Lafayette boundary, the departure controller advises the pilots of N252MN that their radar service is being terminated and that they should contact Lafayette approach control. This transfer of communication normally occurs somewhere near the Leban intersection ("N252MN, radar service terminated three miles southeast of the Leban intersection, contact Lafayette approach on one two three point eight five").

Lafayette Approach Control Upon initial contact with the Lafayette approach controller, N252MN is advised which approach to expect at Lafayette and is then cleared past Ockel or cleared to enter a holding pattern at Ockel. If N252MN must enter a holding pattern, the Lafayette controller must inform Indianapolis approach, since the Ockel holding pattern overlaps Indianapolis's airspace. But in most cases, N252MN will be cleared past Ockel to the Earle outer marker on the ILS runway 10 approach (see Figure 10-23) into Lafayette ("N252MN is now cleared to the Earle outer compass locator via Ockel direct. Hold west of Earle on the localizer. Maintain four thousand. Expect further clearance at one six one zero. Time now one five five three").

Whenever traffic permits, N252MN will be cleared for the ILS approach. If more than one aircraft is inbound to Lafayette, however, each aircraft will

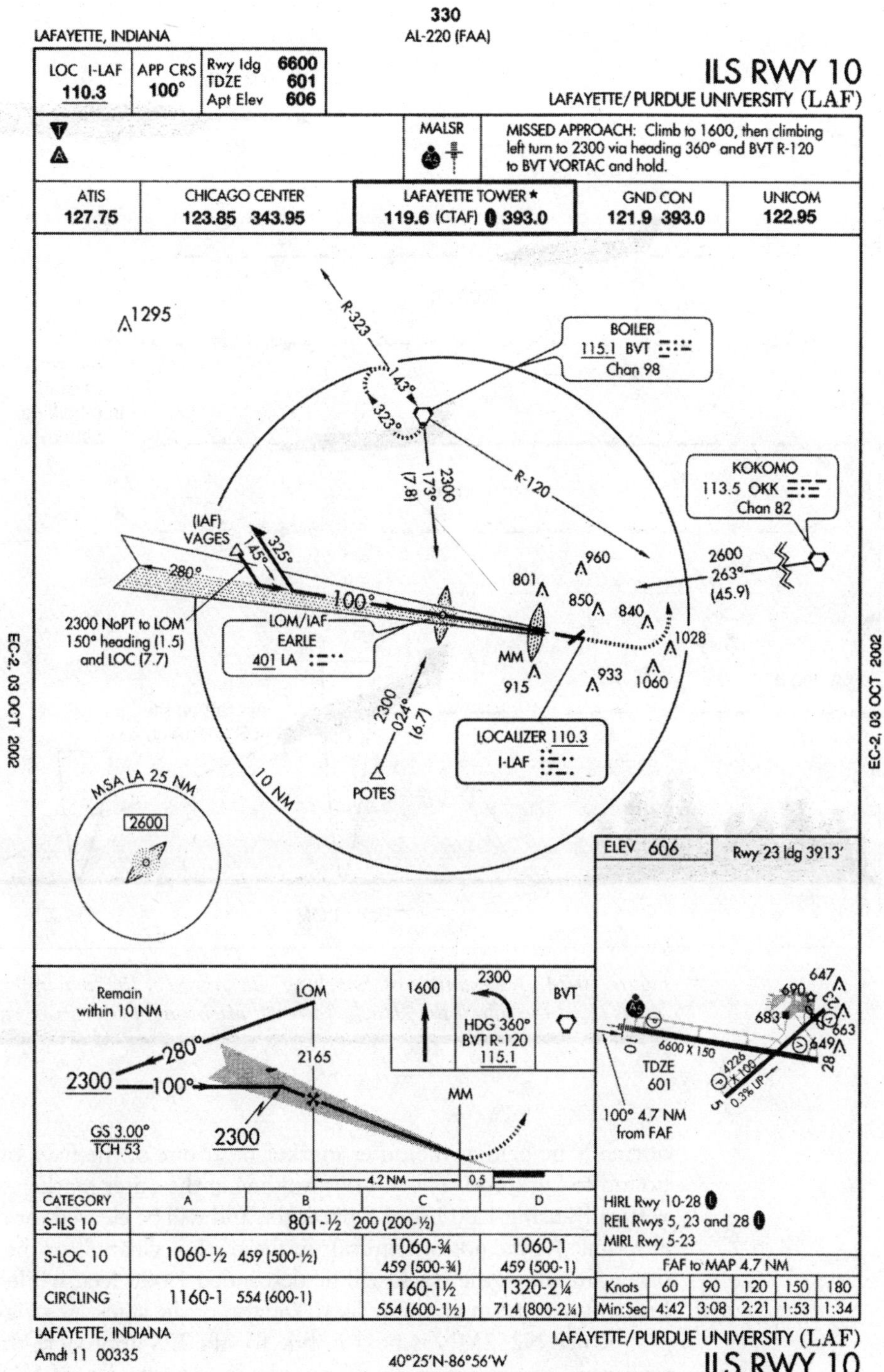

CATEGORY	A	B	C	D
S-ILS 10	801-½ 200 (200-½)			
S-LOC 10	1060-½ 459 (500-½)		1060-¾ 459 (500-¾)	1060-1 459 (500-1)
CIRCLING	1160-1 554 (600-1)		1160-1½ 554 (600-1½)	1320-2¼ 714 (800-2¼)

FAF to MAP 4.7 NM					
Knots	60	90	120	150	180
Min:Sec	4:42	3:08	2:21	1:53	1:34

Figure 10-23. ILS runway 10 approach at Lafayette/Purdue University airport.

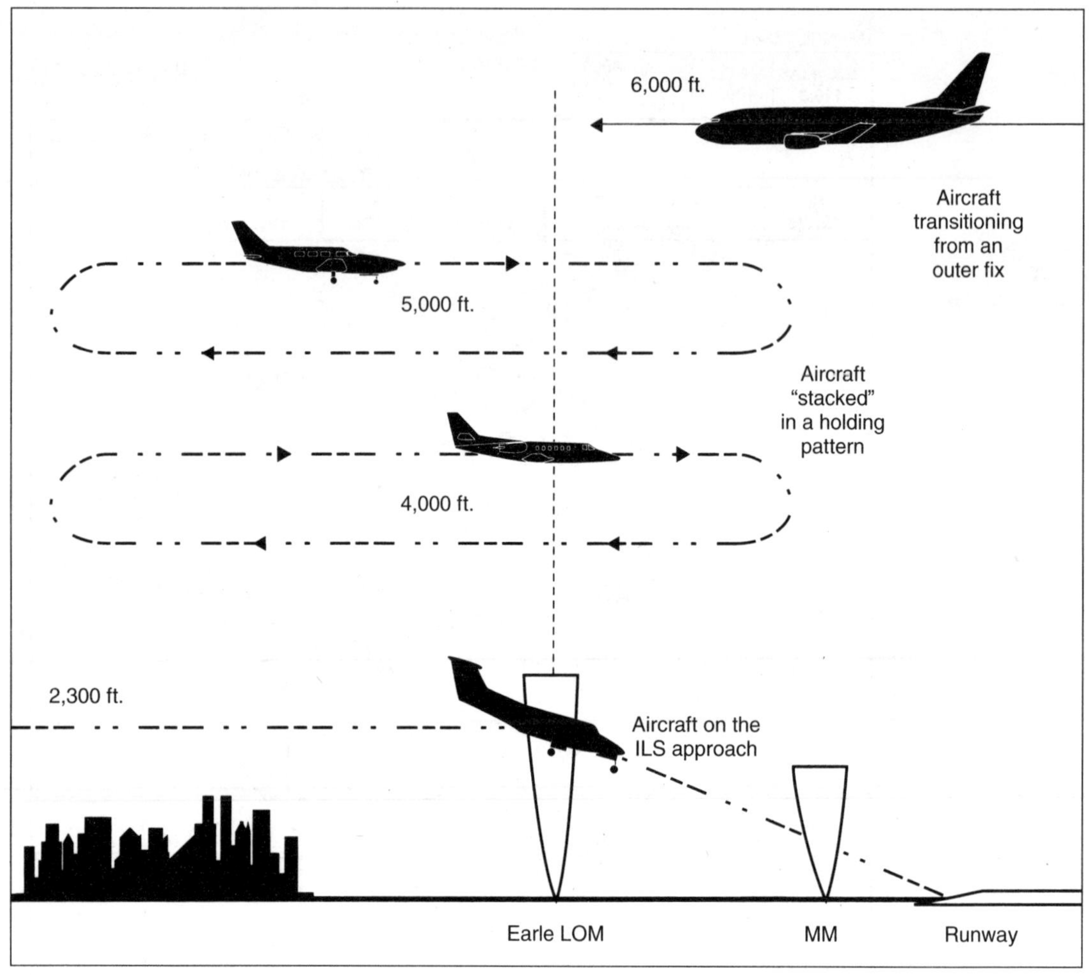

Figure 10-24. An example of "stacking" aircraft over the final approach fix. As room opens up at the top of the "stack," aircraft are brought over from various outer fixes.

normally be held at the outer marker or at one of the fixes on the airways inbound to Lafayette. The aircraft holding at the outer marker will be "stacked" vertically, using 1,000-foot separation, and will be cleared for the approach one at a time. As the bottom aircraft in the stack is cleared for the approach, every other aircraft in the stack will be descended 1,000 feet, while a new aircraft is brought over from an outer fix to the top of the stack (see Figure 10-24).

Once N252MN is next in line for the ILS approach, the approach controller at Lafayette verbally advises the local controller of N252MN's position and the type of approach being conducted. The aircraft can then be cleared for the approach ("N252MN, cleared for the ILS runway one zero approach. Monitor tower one one niner point six, report Earle inbound").

Lafayette Tower Once the aircraft crosses the outer marker, it is the local controller's responsibility to sequence N252MN into the inbound flow of traffic ("N252MN, cleared to land runway one zero. Traffic is a Piper Cherokee on short final runway two three").

When N252MN has landed, the pilots contact the ground controller, who will issue a clearance to taxi to the parking ramp.

Example of a VFR Flight

Lafayette to Champaign

Pilots flying VFR are not required to file a flight plan but are encouraged by the FAA to do so. The flight plan itself is not directly transmitted to air traffic control facilities but is instead used primarily to assist in the location of lost aircraft.

Pilots who contact a flight service station for a VFR flight will receive essentially the same weather briefing information as that given to an IFR pilot but will have the briefing specifically arranged for them. Since Lafayette is within the Terre Haute AFSS's area of responsibility, the pilot of N252MN would normally call the Terre Haute Flight Service Station for this weather briefing. At the conclusion of the weather briefing, the FSS specialist asks whether the pilot wishes to file a flight plan. Assuming that the pilot does, the briefer enters the appropriate information into the FSS computer (see Figure 10-25).

Ground Control When N252MN is ready to depart Lafayette, the pilot first contacts the Lafayette ground controller and receives taxi clearance ("N252MN, taxi to runway one zero").

After taxiing to the active runway, the pilot contacts the local controller for departure instructions. The controller's responsibility to VFR pilots is to provide appropriate runway separation to each aircraft ("N252MN, Lafayette tower, turn right on course, runway one zero cleared for takeoff").

Once N252MN is airborne and clear of the Lafayette Class D airspace, the pilots contact the Terre Haute Flight Service Station to "activate" their VFR flight plan. A VFR flight plan is *not* activated automatically; it is up to the pilots to initiate contact with the appropriate ATC facility (usually a flight service station) to activate the flight plan.

The pilot can contact the FSS in various ways. The first method is to use a **remote communications outlet (RCO)** to the flight service station. An RCO permits pilots to communicate with distant flight service stations using a single frequency. The radio transmitter and receiver are located at an airport distant from the flight service station (in this case at the Lafayette Airport) but are connected to it by telephone communications equipment. The FSS specialist at

Form Approved: OMB No. 2120-0026

U.S. DEPARTMENT OF TRANSPORTATION FEDERAL AVIATION ADMINISTRATION **FLIGHT PLAN**	(FAA USE ONLY) ☐ PILOT BRIEFING ☐ VNR ☐ STOPOVER	TIME STARTED	SPECIALIST INITIALS

1. TYPE	2. AIRCRAFT IDENTIFICATION	3. AIRCRAFT TYPE/ SPECIAL EQUIPMENT	4. TRUE AIRSPEED	5. DEPARTURE POINT	6. DEPARTURE TIME PROPOSED (Z)	6. DEPARTURE TIME ACTUAL (Z)	7. CRUISING ALTITUDE
X VFR / IFR / DVFR	N252MN	BE90/R	160 KTS	LAF	1500		65

8. ROUTE OF FLIGHT

D→

9. DESTINATION (Name of airport and city)	10. EST. TIME ENROUTE HOURS	10. EST. TIME ENROUTE MINUTES	11. REMARKS
CMI		35	

12. FUEL ON BOARD HOURS	12. FUEL ON BOARD MINUTES	13. ALTERNATE AIRPORT(S)	14. PILOT'S NAME, ADDRESS & TELEPHONE NUMBER & AIRCRAFT HOME BASE	15. NUMBER ABOARD
3	30		M. NOLAN LAF	2
			17. DESTINATION CONTACT/TELEPHONE (OPTIONAL)	

16. COLOR OF AIRCRAFT	
BLACK & WHITE	CIVIL AIRCRAFT PILOTS. FAR Part 91 requires you file an IFR flight plan to operate under instrument flight rules in controlled airspace. Failure to file could result in a civil penalty not to exceed $1,000 for each violation (Section 901 of the Federal Aviation Act of 1958, as amended). Filing of a VFR flight plan is recommended as a good operating practice. See also Part 99 for requirements concerning DVFR flight plans.

FAA Form 7233-1 (8-82) CLOSE VFR FLIGHT PLAN WITH ____________ FSS ON ARRIVAL

Figure 10-25. Sample VFR flight plan for N252MN's simulated VFR flight from Lafayette to Champaign.

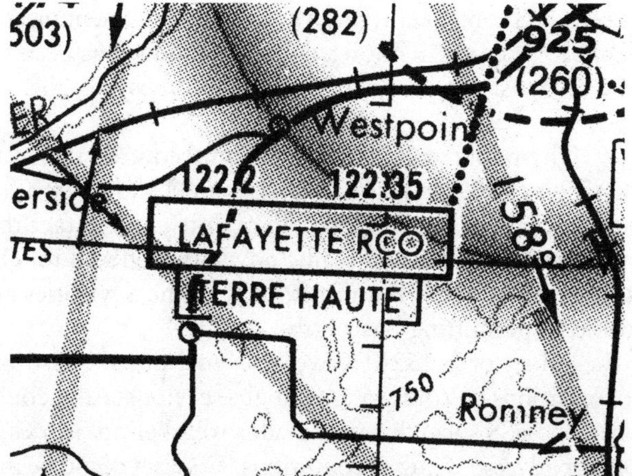

Figure 10-26. A remote communications outlet as depicted on a sectional chart.

Terre Haute can communicate with aircraft on the ground or within the immediate vicinity of Lafayette using the Lafayette RCO. Remote communications outlet frequencies are printed on VFR navigation charts (see Figure 10-26).

Another method of communicating with a flight service station requires the pilot to transmit on one frequency and receive the reply from the flight service specialist on another frequency assigned to a navigation aid, usually a VOR. These facilities can also be found on VFR navigation charts; the appropriate transmitting frequency is next to the navaid followed by the letter R (which indicates FSS receive only). The receiver is remoted to the flight service station via telephone equipment. The FSS specialist in turn communicates with the pilot by transmitting on the VOR frequency, which does not impair the operation of the VOR.

A third method of communicating with an FSS specialist requires that the aircraft be within range of the FSS itself. Besides having their own discrete frequencies, almost all FSSs have the capability of communicating using 122.2 mHz. If pilots are unsure of local FSS frequencies, they can almost always establish communication on 122.2 mHz.

At Lafayette, the pilots of N252MN would probably contact Terre Haute FSS on 122.35 mHz using a remote communications outlet. To activate their flight plan, the pilots must advise the FSS specialist of their departure time from Lafayette. The FSS specialist then enters the departure information into the FSS computer. This causes the following information to be sent to the flight service station with responsibility for Champaign, which in this case is the St. Louis FSS:

Aircraft identification.
Aircraft type.
Destination.
Estimated time of arrival (ETA) at Champaign.

The St. Louis FSS computer returns an acknowledgment message to Terre Haute Flight Service. Once the acknowledgment message has been sent, N252MN becomes the responsibility of St. Louis if it becomes overdue.

En Route The pilots of N252MN are not required to establish contact with any ATC facility while en route to Champaign. If the aircraft is within range of a radar-equipped facility, however, the pilots can contact that facility and request **radar traffic advisories.** Traffic advisories offered to VFR aircraft are the same as those offered to IFR aircraft; VFR traffic advisories are offered to pilots on a workload-permitting basis only.

If the pilots of N252MN were to encounter questionable or changing weather conditions en route, a local flight service station could offer them some assistance. The FSS specialist could offer weather advisories, forecasts, and pilot reports of adverse weather conditions. Contact could be made with the flight service station through an RCO or direct communications with an FSS using 122.2 mHz, or the pilot could request **en route flight advisory service (EFAS).** EFAS is a weather advisory service provided by certain flight service stations to en route VFR or IFR aircraft (see Figure 10-27). At these specially equipped stations, an individual controller is on duty to provide timely weather information to en route aircraft. EFAS is not intended to be used for filing or opening flight plans; it is designed to be used by pilots as a weather exchange service only. The EFAS specialist has all of the most pertinent weather information available, including real-time weather radar provided by the National Weather Service (see Figure 10-28). In addition, the EFAS specialist constantly solicits weather information from area pilots and controllers. The EFAS specialist is thus able to provide timely weather and safety-related information to those pilots who need it most. EFAS operates using a common frequency of 122.0 mHz. Pilots who desire to contact the EFAS specialist should broadcast their position relative to the nearest VOR using this frequency. Since every EFAS specialist monitors 122.0, the controller with the appropriate jurisdiction will answer the pilot and provide the required information.

EFAS has been enormously successful, and therein lies its only major problem. Since EFAS operates on a common frequency nationwide, it is possible for high-flying aircraft to interfere with EFAS transmissions over a number of states at one time. To alleviate this problem, the FAA is beginning a program of establishing discrete frequencies for aircraft utilizing EFAS above 18,000 feet MSL. **High-altitude EFAS,** as it is known, will provide a separate frequency for use by aircraft operating at or above FL 180 within each ARTCC area.

Champaign Approach Control Once N252MN is within Champaign approach control's area of radar coverage (about 40 nautical miles), the pilots can contact the Champaign TRACON for radar traffic advisories. Although contact is not mandatory at this distance, it is recommended in order to enhance safety around busy terminal areas. The pilots of N252MN are required to contact Champaign approach prior to entering the Champaign Class C airspace, however (see Figure 10-29).

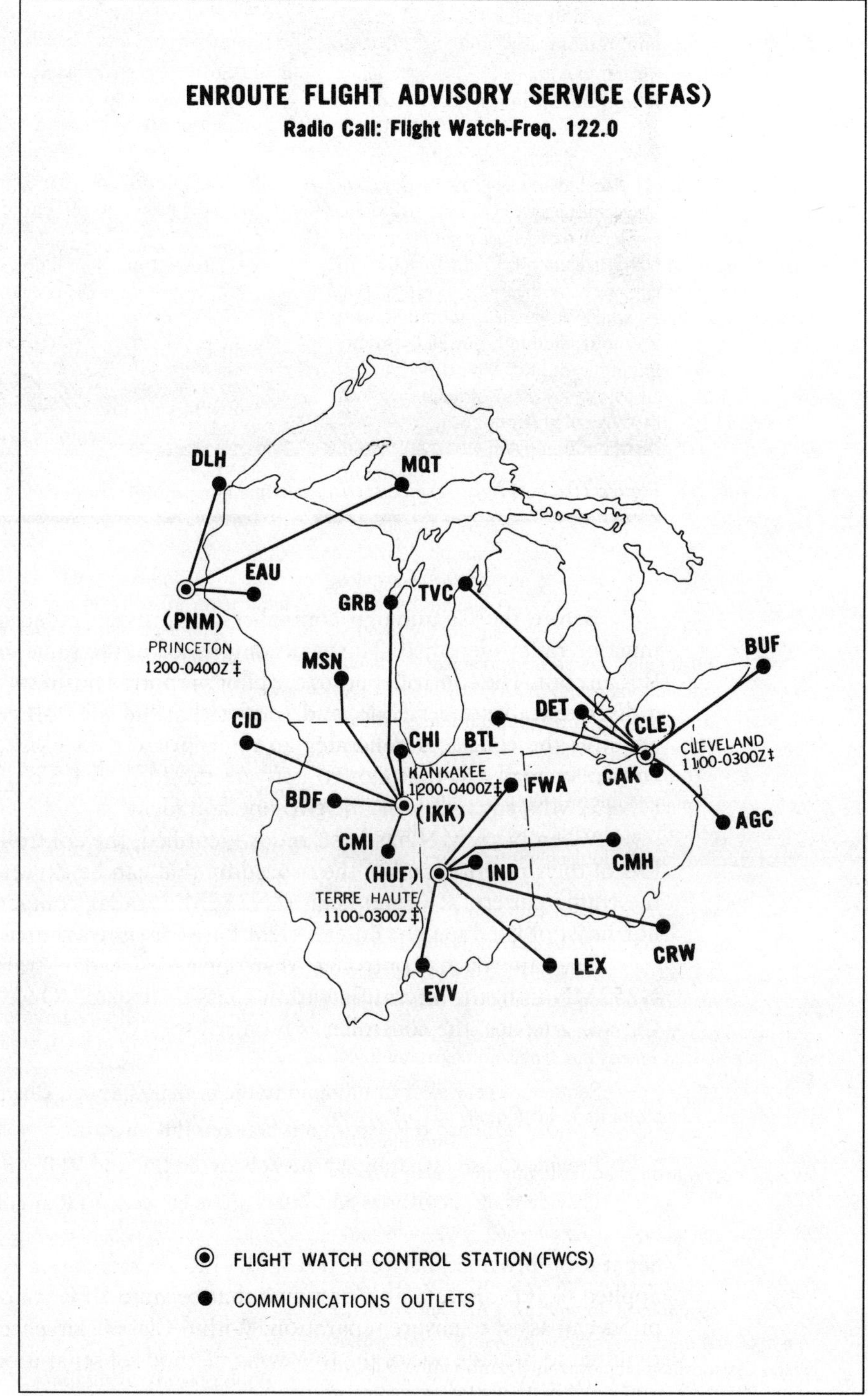

Figure 10-27. En route flight advisory service.

FAA

Figure 10-28. Flight service station controller.

Before the Champaign controller can provide radar service, N252MN must be radar identified. This is accomplished in the same manner as with an IFR aircraft. The controller notes the pilot's reported position, assigns N252MN a discrete transponder code, and verifies that the ARTS-II computer properly acquires the code and generates an appropriate data block. In addition, the controller may ask the pilot to activate the Ident feature of the transponder ("N252MN, squawk four one two one and ident").

When N252MN has been radar identified, the controller advises the pilots of their position and of the procedure that can be expected when entering the traffic pattern at Champaign ("N252MN, radar contact two seven miles northeast of Champaign. Enter a right base for runway three two left").

The approach controller then provides radar traffic advisories to N252MN until the aircraft is within Class C airspace. Once N252MN enters Class C airspace, the controller is required to:

Sequence every aircraft inbound to the primary airport (Champaign).

Provide standard IFR separation between IFR aircraft.

Provide Class C separation criteria between IFR and VFR aircraft.

Provide traffic advisories and safety alerts between VFR aircraft.

Separation provided between IFR and VFR aircraft is not as stringent as that applied to IFR aircraft. It is assumed that because VFR conditions exist, both pilots can assist to ensure separation. Within Class C airspace, the controller is required to provide one of the following methods of separation between an IFR and a VFR aircraft:

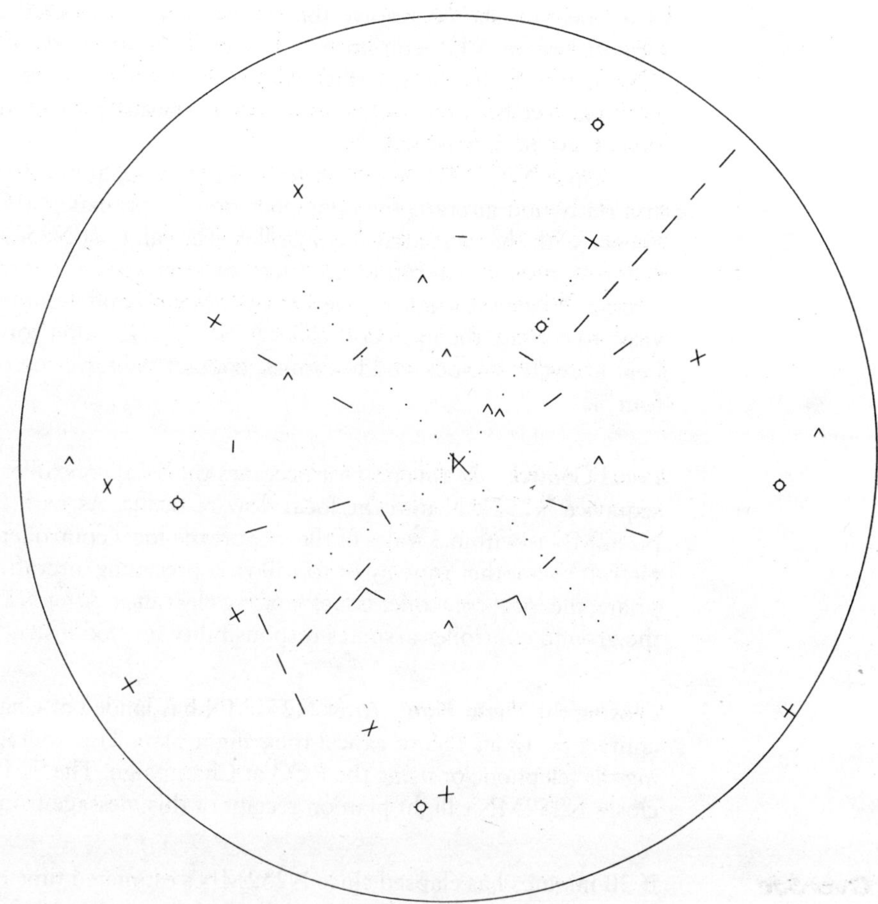

Figure 10-29. Champaign approach control area.

Visual separation.

A 500-foot vertical separation.

Lateral or longitudinal **conflict resolution.**

When providing conflict resolution, the controller must ensure that the displayed radar targets do not touch each other. In addition, both aircraft must be issued the applicable traffic advisories concerning the other aircraft.

A radar controller is not required to separate two VFR aircraft but must offer traffic advisories and safety alerts. A **safety alert** is defined by the *Aeronautical Information Manual* as a condition in which, in the controller's judgment, the aircraft are in unsafe proximity. Whenever this condition arises, the controller must issue a traffic advisory and offer the pilots an alternate course of action that should resolve the situation. It is expected that because both aircraft are in VFR conditions, they will assist in the conflict resolution ("N252MN, traffic alert, traffic twelve o'clock and one mile, eastbound at three thousand five hundred. Advise you turn left heading two four zero or climb to four thousand immediately").

Once N252MN has been appropriately separated from other inbound and outbound aircraft, the controller must coordinate N252MN's arrival sequence with the west arrival controller. The pilots of N252MN are instructed to follow another aircraft or are vectored to ensure proper spacing behind that aircraft. When the sequence has been established and ensured, the pilots are advised to contact the local controller ("N252MN, traffic you are following is a Lear at twelve o'clock and five miles, contact the tower on one two zero point four").

Local Control At this point it becomes the local controller's responsibility to sequence N252MN into the local flow of traffic. As with IFR arrivals, when N252MN is within 3 miles of the airport the local controller can maneuver the aircraft to another runway or to follow a preceding aircraft. When it is appropriate, the local controller issues landing clearance. After N252MN has landed, the ground controller assumes responsibility for taxi instructions.

Closing the Flight Plan After N252MN has landed at Champaign, the pilots contact St. Louis FSS to cancel their flight plan. This contact can be made using the telephone or using the RCO at Champaign. The St. Louis FSS specialist closes N252MN's flight plan on receipt of this message from the pilot.

Overdue Aircraft

If 30 minutes has elapsed since N252MN's estimated time of arrival at Champaign and the St. Louis FSS specialist has not received N252MN's flight plan cancellation, N252MN is considered overdue. Once an aircraft is classified as overdue, **search and rescue (SAR)** procedures are instigated.

During search and rescue operations, the destination FSS is responsible for initiating every attempt to locate the aircraft. The first action that the St. Louis controller takes is to send a **QALQ message** to every FAA facility at an airport where N252MN may have landed. In addition, the QALQ message is sent to the departure FSS (Terre Haute) and to every ARTCC within the area. A QALQ message is a request for information concerning the overdue aircraft. Any facility that receives a QALQ must briefly check with every controller and examine recent flight strips to determine whether any contact has been made with the overdue aircraft. Each of these facilities is required to answer the QALQ request, even if no contact has been made with N252MN.

On receipt of a QALQ message, the departure FSS transmits all the pertinent flight plan information concerning N252MN to the St. Louis controller. This information is also transmitted to every facility that might have had contact with N252MN.

Information Request If the replies to the QALQ request are all negative, meaning that no FAA facility in the nearby area has located N252MN, St. Louis FSS transmits an **information request (INREQ)** to the departure FSS, to every flight watch FSS along N252MN's route of flight, to other FSSs or ARTCCs along N252MN's planned route of flight, and to the **Rescue Coordination Center (RCC)** with responsibility for the area N252MN would have been flying through. In this example, the appropriate RCC is Langley Air Force Base in Virginia.

On receipt of an INREQ message, every facility begins a check of facility records to determine whether radio contact was made with N252MN. Every FAA facility along N252MN's route of flight, such as flight service stations, towers, and ARTCCs, is also contacted to determine whether communication with N252MN occurred. At the conclusion of these checks, a reply message is transmitted to St. Louis FSS describing the results of the search.

Alert Notice If the replies to the INREQ are negative, the St. Louis FSS specialist transmits an **alert notice (ALNOT)** to every FAA facility within 50 miles of N252MN's proposed route of flight. These facilities then conduct a communications search of every airport within their immediate vicinity. In most cases, the airport manager or operator is telephoned, and this individual conducts a visual search of the airport property. If no one can be contacted at the airport, local law enforcement personnel are requested to check for N252MN at the airport. In addition, flight service stations within this area transmit a request over the appropriate frequencies asking every airborne aircraft to monitor the emergency frequency (121.5 mHz or 243.0 mHz) and listen for emergency communications or a transmission from the **emergency locator transmitter (ELT)** on board N252MN.

If an hour has elapsed since the original ALNOT transmission, the St. Louis FSS contacts the Rescue Coordination Center and provides all the pertinent information about that flight to the RCC officer. If N252MN has not been located by this time, the U.S. Air Force assumes complete responsibility for locating N252MN and may initiate a ground and air search for the aircraft, utilizing the Civil Air Patrol.

KEY TERMS

acceptance rate
Air Traffic Control System Command Center (ATCSCC)
alert notice (ALNOT)
arrival gates
automated flight service station (AFSS)
conflict resolution
departure controller
departure gates
emergency locator transmitter (ELT)
en route flight advisory service (EFAS)
en route metering
feeder controllers
final controller
Flight Data Center (FDC)
full route clearance (FRC)
high-altitude EFAS
hold for release
information request (INREQ)
monitor controller
notices to airmen (NOTAMs)

preferred arrival routes (PARs)
preferred departure routes (PDRs)
QALQ message
radar traffic advisories
remote communications outlet (RCO)
Rescue Coordination Center (RCC)
safety alert
satellite controller
search and rescue (SAR)
severe weather avoidance plan (SWAP)
traffic management unit (TMU)

REVIEW QUESTIONS

1. What air traffic control services are routinely offered to IFR pilots but not VFR pilots?
2. How are flight data transmitted to the appropriate facilities?
3. How does a nonautomated ATC facility differ from an automated facility when coordinating IFR aircraft?
4. What types of ATC services are available to pilots?
5. How are lost and overdue aircraft assisted?

11

Oceanic and International Air Traffic Control

Checkpoints

After studying this chapter, you should be able to:

1. Describe the function of ICAO in international air traffic control.
2. Describe reduced separation vertical minima.
3. Explain oceanic air traffic control minima.
4. Describe operating procedures and separation minima within minima navigation performance specification airspace.

In this chapter, the air traffic control procedures used in areas adjacent to the United States are explained, as are the procedures used by the FAA to separate aircraft in international airspace. Because few controllers will ever have the opportunity to control traffic in international airspace, this discussion is kept brief. Emphasis is placed on those areas where FAA controllers are most likely to control international air traffic, specifically Canadian and North Atlantic airspace. Since Canada and the United States share one of the longest borders in the world, many United States controllers may have an opportunity to control Canadian air traffic. The North Atlantic airspace is also covered since it contains some of the busiest and densest airspace in the world.

International Air Traffic Control

The United States is a member of the International Civil Aviation Organization (ICAO) and therefore must abide by its requirements and regulations. One of ICAO's original goals was to standardize the world's aviation systems through the development and dissemination of suggested procedures for aviation regulatory agencies. The standards developed by ICAO are known as **International Standards and Recommended Practices** and are classified individually as **ICAO Annexes** to the original **Convention on International Civil Aviation** held in Chicago in 1944.

Every member of ICAO has agreed to generally abide by these Annexes unless they must be modified to meet national needs. The adoption of these procedures has permitted pilots to fly around the world using the same language (English), the same navigation aids (VOR, ILS, NDB, and MLS), and the same procedures. Without the ICAO Annexes, every country would be free to develop its own navigation systems, to use its own method for numbering airways and runways, and to use its native language (or languages) for air traffic control. Through the perseverance of ICAO and the cooperation of its members, international air travel is just about as easy as travel within the United States.

The ICAO Annexes were adopted in 1944 and have been continuously modified, adapted, and expanded since that time. Small changes to the original Annexes are amendments that are made when needed, whereas major changes are made during **Air Navigation Conferences** held every few years in Montreal, Canada, the permanent headquarters of ICAO. At Air Navigation Conferences, the member nations of ICAO meet to discuss, develop, and ratify major additions and changes to the ICAO International Standards and Recommended Practices.

Up to this point ICAO has approved 18 Annexes and 3 **Procedures for Air Navigation Services (PANS)** that are to be used by member nations when developing and operating their aviation systems. These Annexes are:

Annex 1	Personnel Licensing
Annex 2	Rules of the Air

Annex 3	Meteorology
Annex 4	Aeronautical Charts
Annex 5	Units of Measurement in Air-Ground Communications
Annex 6	Operation of Aircraft
Annex 7	Aircraft Nationality and Registration Marks
Annex 8	Airworthiness of Aircraft
Annex 9	Facilitation
Annex 10	Aeronautical Telecommunications
Annex 11	Air Traffic Services
Annex 12	Search and Rescue
Annex 13	Aircraft Accident Inquiry
Annex 14	Aerodromes
Annex 15	Aeronautical Information Services
Annex 16	Aircraft Noise
Annex 17	Security
Annex 18	Safe Transport of Dangerous Goods by Air
PANS	Aircraft Operations Rules of the Air and Air Traffic Services ICAO Abbreviations and Codes

ICAO requires that every country publish manuals describing its ATC system and any differences from ICAO standards. In the United States, these publications are the ***International Flight Information Manual*** (IFIM) and the ***Aeronautical Information Publication*** (AIP), published by the FAA. In general, the United States conforms to most of the recommendations made by ICAO for the operation of air traffic control systems. A few differences should be noted, however.

ICAO recommends three types of aircraft operations: VFR, IFR, and controlled VFR (CVFR). Controlled VFR flights are separated by controllers as if they are IFR, but the pilots are not IFR rated and must remain in VFR conditions. Controlled VFR is not used in the United States. ICAO also recommends phraseology not normally used in the United States. American pilots and controllers pronounce decimal points as "point," whereas ICAO recommends that it be pronounced "decimal." U.S. en route facilities are known as ARTCCs, whereas ICAO phraseology refers to such facilities as Area Control Centers. Other than these few minor differences, the United States ATC system conforms to ICAO standards.

Canadian Air Traffic Control

Because Canada and the United States share one of the longest national borders in the world, the two nations' air traffic control systems interact considerably. This interaction has led to the development of a highly coordinated ATC system in which both countries have agreed to assist each other in many areas. For example, in cases of small American airports fairly close to Canadian ATC facilities, the FAA has delegated control for that airspace to the Canadian government. Likewise, parcels of Canadian airspace have been delegated to American ATC facilities.

For this governmental cooperation to come about, both countries' air traffic control systems had to be compatible with each other. Through discussions and agreements between the FAA and **Transport Canada** (the Canadian governmental authority charged with the development, regulation, and operation of that nation's air traffic control system), both ATC systems have developed similarly. The procedures used by Transport Canada are in many respects similar to those used by American controllers. They are so similar in fact, that in 1985 the United States and Canada signed an agreement recognizing the inherent safety of each other's ATC system. More important, the agreement permits the controllers of one country, when authorized to separate aircraft flying over the other, to use the procedures developed by the home country to separate those aircraft. In other words, in areas where Canada has authorized the FAA to separate aircraft within Canadian airspace, the FAA controllers may use FAA procedures. In U.S. airspace that the FAA has delegated to Canada, the Canadian controllers may use Canadian procedures to separate American aircraft.

International Airspace

The ICAO agreements specify that every nation will control its own sovereign airspace but will permit ICAO to determine who shall provide air traffic control service within international airspace. Since ICAO is only a voluntary regulatory body and does not provide any direct air traffic control service, international ATC has been delegated to those member nations willing to accept this responsibility. ICAO has assigned a fairly large area of international airspace to the United States within which the FAA provides air traffic control service. Because this airspace does not actually belong to the United States, the rules and regulations applicable to U.S. pilots and controllers do not always apply in this airspace. The appropriate rules of operation are found in ICAO Annex 2 (Rules of the Air) and Procedures for Air Navigation Services. These procedures are considered as supplements to the FAA handbook and are used when offering international air traffic control services.

U.S. pilots and controllers are required by the FARs to conform with Annex 2 when operating in international airspace. FAR 91.1 states, "When over the high seas, comply with Annex 2 (Rules of the Air) to the Convention on International Civil Aviation." Most foreign aircraft operators are also required by their government regulations to conform with Annex 2.

ICAO has divided the airspace of the world into **flight information regions (FIRs)**. These regions identify which country controls the airspace and determines which procedures are to be used. For the purpose of en route ATC, normally one major air traffic control facility is identified with each FIR. In the United States, this would be the air route traffic control center. In the rest of the world, these facilities are known as area control centers (ACC).

The boundaries of each FIR normally follow the geopolitical boundary of the underlying country, but in some cases individual nations have agreed to grant control authority of their airspace to area control centers located in other countries. In oceanic airspace, certain countries have agreed to provide ATC service outside of their national boundaries. Although the countries do not have political ownership of the airspace, the aeronautical regulations of most ICAO member countries require that their pilots abide by the rules and procedures used by the ACC controlling the oceanic airspace.

Air traffic control procedures common to the United States may very often be lacking in some airspace around the world. Limited facilities and differing national priorities often require less emphasis on providing ATC as sophisticated as we are familiar with. In many areas of the world, much of the upper airspace is designated class G airspace, a classification unused in the United States. Within class G airspace, both IFR and VFR flights are permitted and aircraft will receive flight information, if available. This is often limited to reports of known aircraft and weather conditions. In many areas of the world, pilots provide much of their own separation. Using a common radio frequency, pilots operating in class G airspace transmit their location and altitude at regular time intervals. When approaching common fixes or navaids, the pilots transmit their location, altitude, and estimated crossing times. If two aircraft seem to be in close proximity to one another, the pilots are responsible for informing each other and altering their routes and/or altitudes accordingly. Procedures such as these are most often used in less developed nations. Even in these areas, more sophisticated equipment and procedures common to the United States are utilized in the areas surrounding major metropolitan areas.

Airport Identifiers

ICAO has an established method of issuing unique identifiers for every commercial airport in the world. This format is different than that used by the **International Air Transport Association (IATA)** to identify airports. IATA uses a three-letter format, whereas ICAO uses a four-letter system. IATA's coding format is used primarily by travel agents and airline personnel. The ICAO format is used exclusively in air traffic control.

The ICAO system breaks down the four-letter code into three segments. The first segment identifies which area of the world (the aeronautical fixed service routing area or AFSRA) the airport is located in. The second segment identifies the specific country, and the third segment identifies the particular airport.

For example, the ICAO airport code of EHAM is assigned to the Amsterdam-Schiphol airport. The letter E in the airport code identifies the airport as being located in northern Europe. The H specifies the Netherlands, and the AM is the code assigned to the Amsterdam-Schiphol airport. In contrast to this system, the IATA code for Schiphol used by airlines and travel agents is AMS.

The only exceptions to this coding scheme are the United States and China. These countries have so many commercial airports that they have unique first letters assigned just to their country: Z for China and K for the United States (although airports in Alaska and Hawaii are identified with the Pacific Ocean letter P). The second, third, and fourth letters are assigned to specific airports in these countries. Over 10,000 airport codes are in use worldwide. The first letter of the codes for AFSRA are:

Code Letter	***Region***
A	Southwest Pacific
B	North Atlantic
C	Canada
D	North Africa
E	Northern Europe
F	South Africa
G	West Africa
H	Central Africa
K	United States of America
L	Southern Europe
M	Central America
N	Pacific Islands
O	Middle East
P	Hawaii and Alaska
R	Eastern Asia and Japan
S	South America
T	Caribbean and Virgin Islands
U	Northern Europe and Russia
V	South Asia
W	Southeast Asia
Z	China

Some selected second letters of the codes include:

Country	***Code Letter***
Alaska	A
Bolivia	L
Brazil	B
Canada	Y
Chile	C
Columbia	K
Egypt	E

France	F
Germany	D
Greece	G
Hawaii	H
Iceland	I
India	E
Ireland	I
Italy	I
Japan	J
Korea	K
Mexico	M
Netherlands	H
New Zealand	Z
Peru	P
Philippines	P
Russia	H
Saudi Arabia	E
South Africa	A
Spain	E
Ukraine	K
Virgin Islands	I

A sample of the airport codes used by the ICAO and IATA follows. A more complete list is included in Appendix C.

ICAO Code	***IATA Code***	***City***	***Country***	***Airport***
CYMX	YMX	Montreal	Canada	Mirabel International
EDDF	FRA	Frankfurt	Germany	Rhein Main
EGLL	LHR	London	England	Heathrow
KATL	ATL	Atlanta	Georgia	Hartsfield International
PAFA	FAI	Fairbanks	Alaska	International
PHNL	HNL	Honolulu	Hawaii	International
ZBAA	BJS	Beijing	China	Capital

European Air Traffic Control

Air traffic control in Europe is similar to the United States. The only difference is that in Europe, each country provides its own ATC system. The result is as if each state in the United States had its own ATC system. In an effort to better coordinate Europe's air traffic system, Eurocontrol was formed.

Eurocontrol, the European Organization for the Safety of Air Navigation, has 31 member states: Germany, Belgium, France, Luxembourg, Netherlands, Albania, the United Kingdom, Ireland, Portugal, Greece, Turkey, Malta, Cyprus, Hungary, Switzerland, Austria, Norway, Denmark, Slovenia, Sweden, the Czech Republic, Italy, Romania, the Slovak Republic, Spain, Croatia, Bulgaria, Monaco, Moldova, Slovenia, and the former Yugoslav republic of Macedonia.

Eurocontrol was founded in 1960 for overseeing air traffic control in the upper airspace of member states; but due to political considerations, it was never able to operate as planned. Today Eurocontrol has as its most important goal the development of a coherent and coordinated air traffic control system in Europe. Its primary objectives are to manage the implementation of the European Air Traffic Control Harmonization and Integration Program (EATCHIP), as well as a series of associated concepts and future strategies. Eurocontrol also operates the central flow management unit (CFMU) that operates similar to the ATCSCC in Washington, D.C. An additional mission is to carry out research and development work aimed at increasing air traffic control capacity in Europe.

Eurocontrol's Maastricht Upper Area Control Center was Europe's first international air traffic control center. The center was founded by Eurocontrol at the request of Belgium, Luxembourg, Netherlands, and Germany. It provides air traffic control services in the upper airspace (24,500 feet) of one of the busiest regions of Europe: the airspace over Belgium, Luxembourg, Netherlands, and part of Germany.

Atlantic Ocean Air Traffic Control

The most highly congested international airspace controlled by the FAA is over the **North Atlantic Region (NAR)**. The high traffic in this airspace becomes congested because of the time zone differences between North America and Europe. Most of the traffic across the North Atlantic between 8:00 A.M. and 3:00 P.M. EST (1300Z and 2000Z) is westbound. Aircraft leave Europe in the morning and arrive in North America in the early afternoon. Eastbound traffic is most concentrated between 8:00 P.M. and 3:00 A.M. EST (0100Z and 0800Z), leaving North America in the evening and arriving in Europe in the early morning. Because of this highly directional and concentrated traffic flow, special procedures have been developed for this airspace.

The North Atlantic airspace is delegated to seven air traffic control facilities. The FAA facilities are the New York and San Juan ARTCCs, and the remainder of the airspace is divided among area control centers in Greenland, Newfoundland, Great Britain, Iceland, and the Azores.

ICAO standards specify that all of the airspace at or above 5,500 feet MSL within this area is controlled airspace. The airspace below 5,500 feet MSL is uncontrolled. ICAO also specifies that the transition level over the North Atlantic begins at 5,500 feet MSL. This means that pilots flying at or above 5,500 feet are required to adjust their altimeters to standard pressure (29.92 Hg). Since these altitudes are no longer true altitudes, they are known as flight levels. Unlike in the United States, however, the lowest flight level over the Atlantic is FL 55 (compared to FL 180 over the continental United States).

This difference in transition altitudes (5,500 feet versus 18,000 feet) can create some separation difficulties as aircraft enter or leave U.S. airspace. An aircraft arriving from Europe at FL 170 would have its altimeter set to 29.92. Once the aircraft enters United States airspace, however, the pilot would readjust the altimeter to the local station pressure (let us assume that it is 29.82). When the altimeter has been readjusted, it would indicate that the aircraft is level at 16,900 feet MSL, not at 17,000. Although the pilot would certainly begin an immediate climb to 17,000, if an aircraft had been directly below at 16,000, a temporary loss of separation would already have occurred. Thus, controllers must be careful when aircraft pass horizontally through a transition level while in level flight.

North Atlantic Separation

Aircraft operating over the North Atlantic are separated using nonradar techniques, since radar service is not available over most of this route. This nonradar separation must be expanded from that used within the United States because of a number of factors that may affect aircraft in flight. The separation intervals must necessarily be increased, since radio communication is difficult to maintain over this distance. In addition, since aircraft cannot be directly observed and position determination is less accurate over the Atlantic, the separation interval between aircraft is increased. All of these factors combine to decrease the capacity of the North Atlantic airways.

Primarily two sets of airways are used by flights over the Atlantic Ocean. The first is a series of one-way airways at fairly low altitudes, which are commonly used by single- or multi-engine propeller-driven aircraft. Aircraft operating on these routes are normally within range of VHF communications facilities and can use VOR or NDB navigation facilities. Most of these airways are designed such that flight over water is reduced to a minimum. The other set of airways is a flexible system of changing airways primarily used by airline, military, and business jet operators. These airways can be used only by aircraft equipped with accurate area navigation equipment. Within this airspace, known as **minimum navigation performance specifications airspace (MNPSA or MNPS airspace)**, only those aircraft that are properly equipped and certified may operate. Within this airspace, separation intervals are reduced and the airspace is utilized more efficiently. MNPS airspace exists primarily above FL 275; it is explained in detail later in this chapter.

Within non-MNPS airspace, aircraft must still be separated using vertical, lateral, or longitudinal separation techniques. The application of these techniques is similar to the methods used by domestic air traffic controllers, but the separation interval has been proportionally increased.

Vertical Separation Vertical separation is applied to oceanic aircraft in exactly the same manner as it is applied to domestic aircraft. Aircraft operating up to and including FL 290 are separated by a 1,000-foot vertical interval. Aircraft operating above FL 290 are separated by a 2,000-foot vertical interval. The only difference between domestic and oceanic vertical separation occurs with supersonic and high-altitude military aircraft. Because of the high airspeeds involved and the inherent inaccuracies of barometric altimeters at high altitudes, supersonic aircraft operating above FL 450 must be separated from nonsupersonic aircraft by a 4,000-foot vertical interval. Military aircraft operating above FL 600 must be separated by a 5,000-foot vertical interval.

Lateral Separation Since the distance between navaids on the North Atlantic airways is much greater than on domestic routes, the airways are wider than those within the continental United States. Because aircraft operating on oceanic routes cannot be observed on radar, increased lateral separation must be used between aircraft operating on parallel routes at the same altitude. When providing lateral separation within oceanic airspace, the controller must still assign aircraft routes whose protected airspaces do not overlap, but the widths of these routes will vary depending on the type of aircraft using the route, the length of the oceanic route, and the aircraft's altitude.

Supersonic aircraft operating above FL 275 anywhere over the North Atlantic must be laterally separated by 60 nautical miles. Aircraft operating between North America and Bermuda at any altitude must be separated by 90 nautical miles. Any other aircraft operating over the North Atlantic must be laterally separated by at least 120 nautical miles.

Longitudinal Separation Longitudinal separation can also be applied to oceanic aircraft, but great care must be taken to ensure that the following aircraft never overtakes the leading aircraft. This is accomplished by ensuring that the faster aircraft is the leading aircraft or, if this is not possible, by assigning the following aircraft a particular airspeed to ensure that it will not overtake the leading aircraft.

Within North Atlantic airspace, a 10-minute longitudinal separation interval must be maintained between supersonic aircraft; a 20-minute separation interval must be maintained between nonsupersonic, turbojet-powered aircraft; and a 30-minute separation interval must be maintained between all other aircraft.

Reduced Oceanic Separation

A number of procedural initiatives are currently underway to increase capacity in the oceanic airspace while maintaining or improving safety. More than 40 years ago, when separation standards were first developed, procedures required a 1,000-foot minimum vertical separation between IFR aircraft below FL 290 and 2,000-foot separation above FL 290. The adoption of 2,000-foot separation above FL 290 reflected the belief that altimeters in use at that time were less accurate at higher altitudes.

During the 1980s, an effort was conducted under the auspices of the ICAO to assess the feasibility of reducing the vertical separation minimum

(VSM) above flight level 290 from 2,000 to 1,000 feet. Drawing on input from the same contributors, the panel drafted ICAO guidance material for world-wide and regional application of **reduced vertical separation minimum (RVSM)** in 1991. Following ICAO approval of the draft material, the North Atlantic Region began coordination efforts to develop plans and programs for implementing RVSM. Aircraft seeking airworthiness approval for RVSM operations must be equipped with two independent altitude measurement systems, one approved secondary surveillance radar altitude reporting transponder, one altitude alert system, and one automatic altitude control system.

The FAA and other participating states have begun a phased implementation of RVSM in the North Atlantic. The goal of RVSM is to reduce the vertical separation between FL 290 to FL 410 from the current 2,000-foot minimum to 1,000-foot minimum. Operational trials of RVSM began in the North Atlantic airspace from FL 330 to FL 370, inclusive, in March 1997. Expansion of the RVSM to FL 310 and FL 390 was accomplished in April 1998. Additional phases lead to full implementation from FL 290 to FL 410. Implementation of RVSM has resulted in almost doubling the available oceanic tracks across the North Atlantic within the relevant altitudes.

The current oceanic ATC system uses filed flight plans and position reports to track an aircraft's progress and ensure that separation is maintained. The progress of an aircraft is monitored by ATC using position reports sent by the aircraft over high frequency (HF) radio. Position reports are infrequent (approximately one report per hour), and the accuracy of these reports depends on the accuracy of the onboard navigation system and timing standard aboard the aircraft. HF communication is subject to interference, disruption, and delay because it involves the use of radio operators to relay messages between pilots and controllers. These deficiencies in communications and surveillance have necessitated large horizontal separation minima. As a result of improved navigational capabilities made possible by technologies such as GPS, Traffic Alert and Collision Avoidance System (TCAS), and controller to pilot data link communications (CPDLC), oceanic minimum horizontal separation standards, both lateral and longitudinal, should be able to be reduced. In the Pacific, lateral separation standards will be reduced from the current 60 nautical miles to 100 nautical miles depending on location, to 50 nautical miles in specific oceanic areas. Where traffic density permits, longitudinal separation in the Pacific will be reduced from the current time-based standard of 15 minutes to 50 nautical miles. The reduced lateral separation minima will allow increased capacity during peak hours. The reduced longitudinal minima will provide increased opportunities for altitude changes to achieve optimum altitudes, thus saving fuel.

InTrail Climb and InTrail Descent

InTrail Climb (ITC) and InTrail Descent (ITD) procedures have been developed that enable a trailing aircraft in an oceanic environment to climb or descend through the altitude of a leading aircraft to a more desirable cruising altitude. Using ITC or ITD, an aircraft flying behind and 2,000 feet above or below an aircraft along the same oceanic route may request to climb or descend through the altitude of the leading aircraft as long as the distance between them is at least 15 nautical miles and the ground speed closure rate is 20 knots or less. The pilot wanting to ascend or descend uses the TCAS traffic display to positively identify and determine the distance to the leading aircraft. The trailing aircraft initiates the procedure, coordinates with the leading aircraft, and obtains climb or descent clearance from ATC. During this maneuver, ATC maintains responsibility for separation. Standard nonradar spacing criteria are applied by ATC after the procedure is completed.

ITC and ITD are the first procedures to utilize the display of traffic information on the flight deck to assist air traffic controllers in monitoring and reducing aircraft spacing requirements. ITC and ITD procedures reduce the typical nonradar in-trail distance necessary to approve a climb or descent from 10 minutes (approximately 100 nautical miles) to a minimum of 15 nautical miles.

MNPS Airspace Operations

As a result of increased traffic demand, time zone restrictions, and aircraft performance characteristics, most of the North Atlantic aircraft operations occur within a fairly small block of airspace. This block extends from the northeastern United States to Great Britain and from about FL 285 to FL 420. Because of the constraints placed on controllers when separating aircraft within this area, it can become highly congested at peak operating times. To maximize the usage of this airspace, a system of flexible, organized tracks has been developed that replaces the typical airway structure used for air traffic control (see Figures 11-1 and 11-2). These tracks exist in MNPS airspace.

MNPS airspace lies between the North Pole and the 27th parallel and between FL 285 and FL 420. It is located within several flight information regions controlled by the New York ARTCC and by the Shanwick, Gander, Sondestrom, and Santa Maria **Oceanic Area Control Centers (OACC).**

In an attempt to maximize the use of this limited airspace, international agreements have reduced the separation interval between aircraft operating on these tracks. In return, increased accuracy and reliability of onboard aircraft navigational systems is required. The International Air Transport Association (IATA), in cooperation with ICAO and its member nations, has developed this organized track system and the associated aircraft equipment requirements. Both are described in the *North Atlantic MNPS Airspace Operations Manual.*

Aircraft operating on these tracks are required by the air regulations of their home nation to have the appropriate equipment described in the minimum navigation performance specifications published by the IATA. These specifications require that every aircraft be equipped with two independent, long-range area navigation systems, such as inertial navigation or GNSS, and appropriate high-frequency communication equipment. (High-frequency communication is required since VHF is line of sight and cannot be utilized during most of the oceanic flight.)

Typically, the organized tracks are developed approximately 24 hours before they are actually to be used. Track development takes into consideration

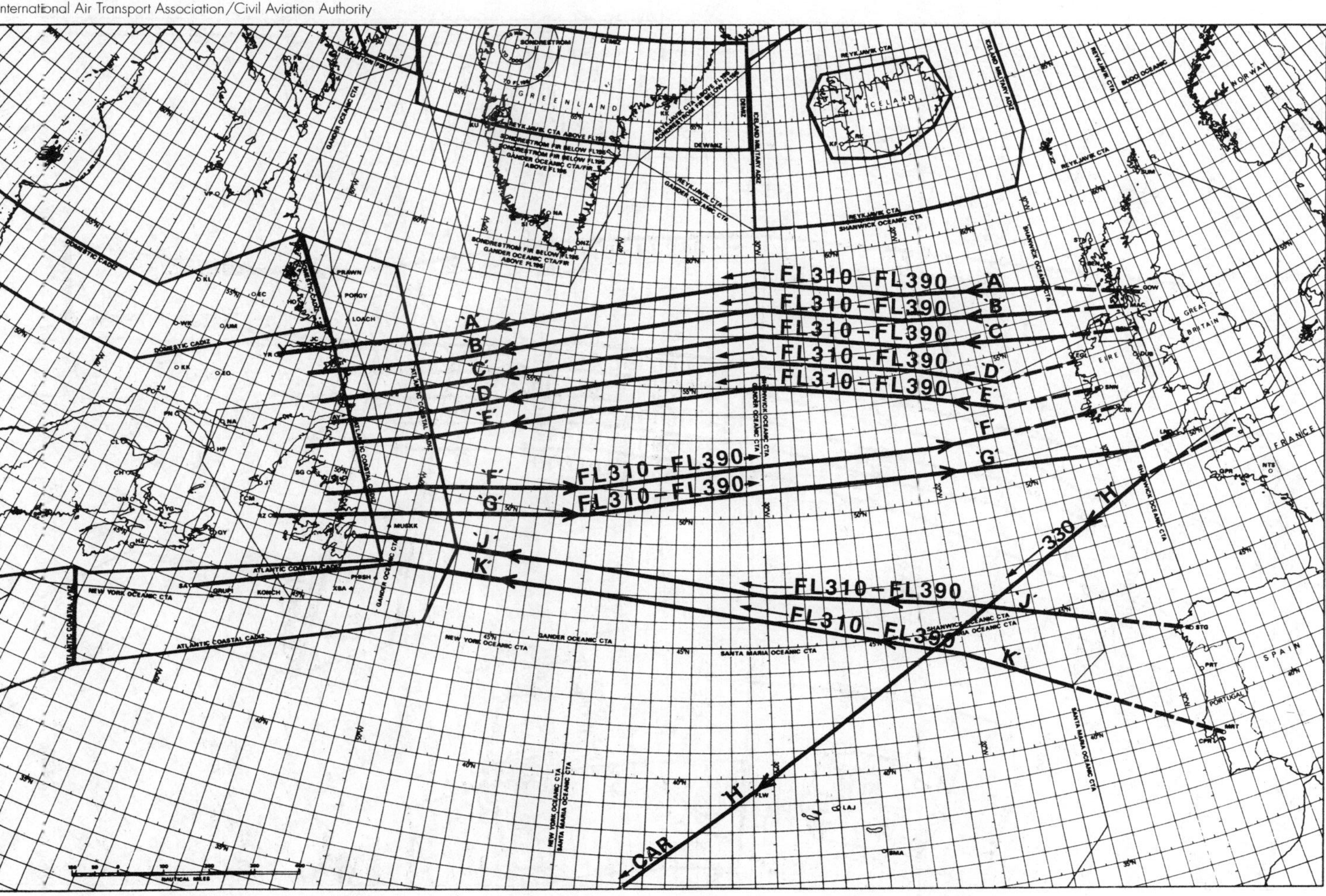

Figure 11-1. Typical westbound traffic flow and tracks over the North Atlantic Region.

International Air Transport Association/Civil Aviation Authority

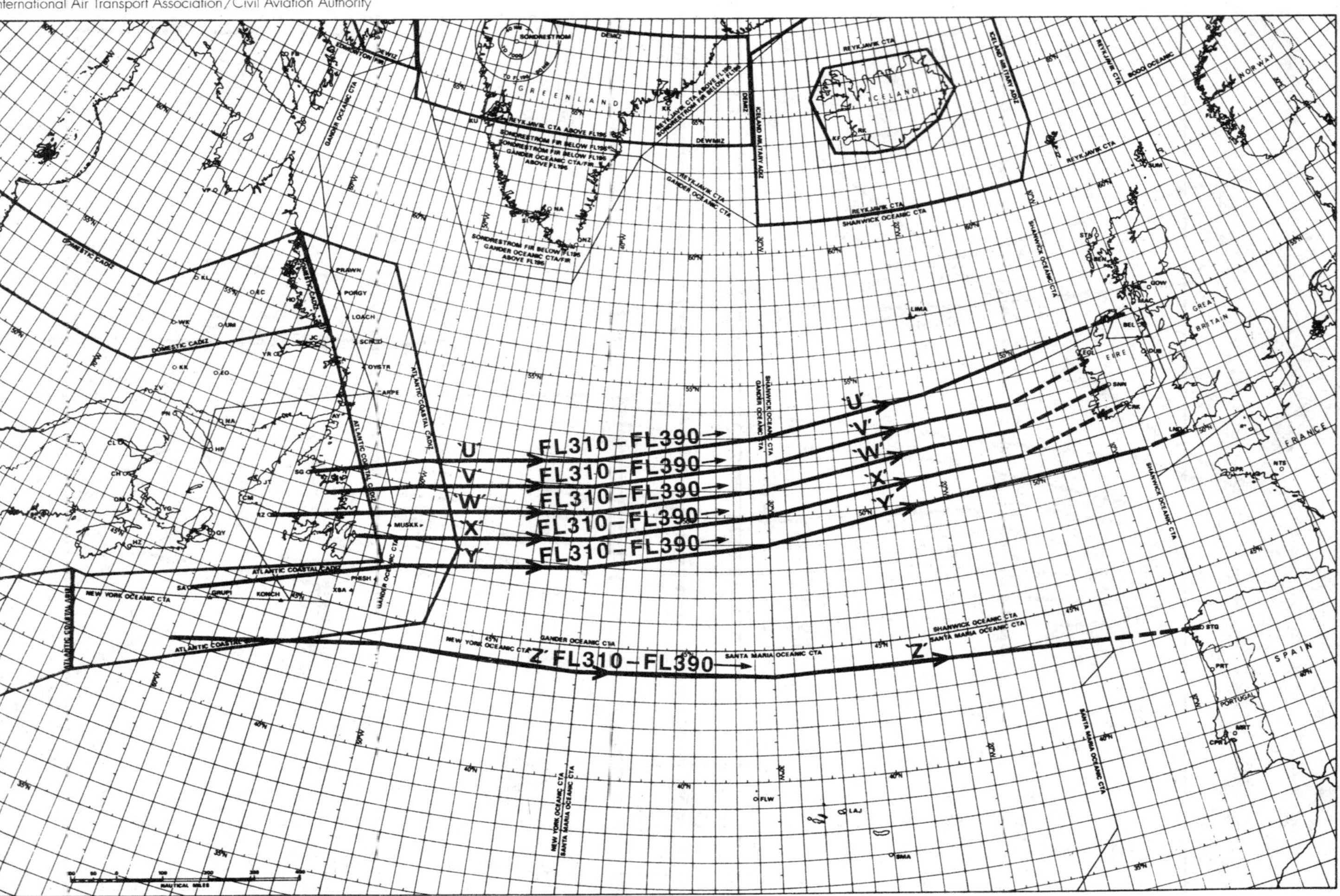

Figure 11-2. Typical eastbound traffic flow and tracks over the North Atlantic Region.

the winds aloft and the weather that may be encountered en route, the anticipated number of aircraft that will be traveling in each direction, and the impact the tracks will have on adjacent and adjoining ATC facilities.

When the factors have been determined, the organized track system for the next day is provided to potential ATC system users and the ATC facilities themselves. In most cases, two track systems are published. The first is primarily designed for westbound traffic and is effective from 6:00 A.M. to 5:00 P.M. EST (1100Z to 2200Z). The second track system is designed primarily for eastbound traffic and is in effect from 7:00 P.M. to 4:00 A.M. EST (0000Z to 0900Z). The time interval between these two track systems is used to clear any late aircraft from the system before the tracks are reversed.

Since the actual locations of the tracks change daily, a fairly simple system of naming each track has been developed. This naming scheme informs the pilots of each track location and whether it is eastbound or westbound. The early-morning, westbound tracks are labeled A (Alpha) through K (Kilo). The northernmost track is Alpha and the southernmost is Kilo. The late-afternoon, eastbound tracks are Uniform through Zulu, with the northernmost track being Uniform. Position reports are normally made at every point where the track crosses a meridian at intervals of 10° of longitude.

The following is a sample MNPS track message:

```
ATC DELAYS AND ADVISORIES
ATCSCC ADVZY 015 DCC 03/27/98 ZBW NATS
ZBW NORTH ATLANTIC ADVISORY FOR 03/27/98 2000Z 03/28/98
0400Z
AIRCRAFT DEPARTING JFK PLEASE FILE THE FOLLOWING ROUTES
TO MINIMIZE DEPARTURE
DELAYS DESTINED TO EUROPE:
TRACK U/ JFK. GREK12. MARTN..EBONY.N95B.CYMON.TRAKU
TRACK V/ JFK.MERIT2.PUT.WITCH..ALLEX.N79B.YQX.TRAKV
TRACK W/ JFK.BETTE2.ACK..TUSKY.N63B. VIXUN.TRAKW
TRACKX/ JFK.BETTE2.ACK..BRADD.N53B.YYT.TRAKX
TRACK Y/ JFK.HAPIE2 YAHOO. .KANNI.N45A.COLOR.TRAKY
ANY QUESTIONS CALL ZBW TMU AT 6038867666
NAT1/2 TRACKS FLS 310/390 INCLUSIVE
MAR 28/0100Z TO MAR 20/0800Z

U CYMON 51/50 52/40 53/30 54/20 54/15 BABAN
EAST LVLS 310 320 330 340 350 360 370 380 390
WEST LVLS NIL
EUR RTS WEST NIL
NAR N95B N97B
V YQX 50/50 51/40 52/30 53/20 53/15 BURAK
EAST LVLS 310 320 330 340 350 360 370 380 390
WEST LVLS NIL
EUR RTS WEST NIL
```

```
NAR N79B N83B
W VIXUN 49/50 50/40 51/30 52/20 52/15 DOLIP
EAST LVLS 310 320 330 340 350 360 370 380 390
WEST LVLS NIL
EUR RTS WEST NIL
NAR N63B N67B
X YYT 48/50 49/40 50/30 51/20 51/15 GIPER
EAST LVLS 310 320 330 340 350 360 370 380 390
WEST LVLS NIL
EUR RTS WEST NIL
NAR N53B N55A
Y COLOR 57/50 48/40 49/30 50/20 50/15 KENUK
EAST LVLS 310 320 330 340 350 360 370 380 390
WEST LVLS NIL
EUR RTS WEST NIL
NAR N45A N49A
NAT2/2 TRACKS FLS 310/390 INCLUSIVE
MAR 28/0100Z TO MAR 28/0800Z
Z BDA 33/60 35/50 39/40 43/30 47/20 OMOKO GUNSO
EAST LVLS 330 350 370
WEST LVLS NIL
EUR RTS WEST NIL
NAR NIL
REMARKS
1. SEE INTERNATIONAL NOTAMS A0666/98 AND C0507/98 REGARD-
ING REVISED CLEARANCE DELIVERY PROCEDURE FOR EASTBOUND
AIRCRAFT EXITING THE GANDER DOMESTIC FIR.
2. TRACK MESSAGE IDENTIFICATION 087.
3. MNPS AIRSPACE EXTENDS FROM FL285 TO FL420. OPERATORS
ARE REMINDED THAT SPECIFIC MNPS APPROVAL IS REQUIRED TO
FLY IN THIS AIRSPACE.
4.40 PERCENT OF GROSS NAVIGATIONAL ERRORS OCCUR AFTER A
REROUTE. ALWAYS CARRY OUT WAYPOINT CROSS CHECKS.
NAT1/2 TRACKS FLS 310/390 INCLUSIVE
MARCH 27/1130Z TO MARCH 27/1800Z
A 55/10 57/20 59/30 59/40 58/50 PORGY HO
EAST LVLS NIL
WEST LVLS 310 320 330 340 350 360 370 380 390
EUR RTS WEST NIL
NAR N284B N292C N294C N298G N302C N304E N306C N308E N312A
B 54/15 55/20 56/30 56/40 54/50 CARPE REDBY
EAST LVLS NIL
WEST LVLS 310 320 330 340 350 360 370 380 390
EUR RTS WEST BABAN
NAR N204B N206C N210C
```

```
C 53/15 54/20 55/30 55/40 53/50 YAY
EAST LVLS NIL
WEST LVLS 310 320 330 340 350 360 370 380 390
EUR RTS WEST BURAK
NAR N184B N188B N192B
D 52/15 53/20 54/30 54/40 52/50 DOTTY
EAST LVLS NIL
WEST LVLS 310 320 330 340 350 360 370 380 390
EUR RTS WEST DOLIP
NAR N162B N164B
E 51/15 52/20 53/30 53/40 51/50 CYMON
EAST LVLS NIL
WEST LVLS 310 320 330 340 350 360 370 380 390
EUR RTS WEST GIPER
NAR N142B N148B
F 50/15 51/20 52/30 52/40 50/50 YQX
EAST LVLS NIL
WEST LVLS 310 320 330 340 350 360 370 380 390
EUR RTS WEST KENUK
NAR N124B N130C
```

Since controllers have no radar image of the aircraft, they instead build up a picture of traffic using position reports sent from the aircraft, mainly using high frequency voice radio. These position reports usually follow this format: airline name, flight number, current position, time and flight level, next position, estimated time at next position, followed by the next position after that. Usually the reports will be a mixture of coordinates and reporting points. An example would be:

> Delta 112, 54 North 20 West at 0634, Flight Level 350, estimating 53 North 15 West at 0656, BURAK next.

This would be written by the controller on a flight strip in the following format:

> DAL112 54N20W 0634 F350 53N15W 0656 BURAK

Using these reports in conjunction with the flight progress strips and maps depicting the route tracks, controllers ensure that there is always separation between all aircraft.

MNPS Airspace Separation

Aircraft operating along the organized track system are separated vertically as previously described in this chapter. The primary difference between normal oceanic separation and MNPSA separation is the reduction of both lateral and longitudinal separation.

Lateral Separation Because of the navigation equipment accuracy required and because each aircraft carries a redundant navigational system, the lateral separation interval within MNPS airspace can be reduced to 60 nautical miles. This can be accomplished only as long as both aircrafts' navigational systems remain operational. If one of them fails, the appropriate ATC facility will begin to separate the aircraft using 120 nautical miles of lateral separation. This can be fairly difficult, given the traffic and communications constraints and considering that the controller is using flight progress strips to separate the aircraft.

Longitudinal Separation Within MNPS airspace, longitudinal separation intervals can be significantly reduced over those used in normal oceanic airspace. If the pilots are capable of determining and maintaining a particular mach airspeed, the separation interval can be reduced even further. Mach is a means of measuring airspeed as a percentage of the speed of sound and is a much more reliable method of determining a high-altitude aircraft's airspeed.

If the leading aircraft operates at the same airspeed or is faster than the following aircraft and both aircraft's mach speed cannot be determined, a 10-minute longitudinal separation interval is sufficient. But if each aircraft's mach number can be determined, longitudinal separation can be reduced to the following values:

If the leading aircraft is .02 mach faster than the following aircraft, longitudinal separation can be reduced to 9 minutes.

If the leading aircraft is .03 mach faster than the following aircraft, longitudinal separation can be reduced to 8 minutes.

If the leading aircraft is .04 mach faster than the following aircraft, longitudinal separation can be reduced to 7 minutes.

If the leading aircraft is .05 mach faster than the following aircraft, longitudinal separation can be reduced to 6 minutes.

If the leading aircraft is .06 mach faster than the following aircraft, longitudinal separation can be reduced to 5 minutes.

KEY TERMS

Aeronautical Information Publication (AIP)
Air Navigation Conferences
controlled VFR (CVFR)
Convention on International Civil Aviation
flight information regions (FIRs)
ICAO Annexes
International Air Transport Association (IATA)
International Flight Information Manual (IFIM)
International Standards and Recommended Practices
mach
minimum navigation performance specifications (MNPS) airspace
North Atlantic Region (NAR)
Oceanic Area Control Centers (OACC)
Procedures for Air Navigation Services (PANS)
reduced vertical separation minimum (RSVM)
tracks
Transport Canada

REVIEW QUESTIONS

1. What are some differences between domestic and oceanic separation standards?
2. Who grants authority for the separation of oceanic aircraft?
3. What are the requirements for operating in MNPS airspace?
4. Under what conditions can longitudinal separation be reduced within MNPS airspace?
5. What is the purpose of the North Atlantic organized track system?

12

The Future of the National Airspace System

Checkpoints

After studying this chapter, you should be able to:

1. Identify the problems that constrain current air traffic control systems.
2. Describe the operation of the future air traffic control system.
3. Describe the improvements planned for the FAA's communications systems.
4. Describe the improvements planned for the nation's navigation system.
5. Describe the FAA's plan for automating air traffic management.
6. Explain the concept of "free flight."
7. Describe the function of a "conflict probe."

The U.S. air traffic control system has developed in spurts over the last 40 years. Intervals of relative calm have been repeatedly interrupted by periods of major change and system improvement. Most of these improvements have been made in response to changes in public opinion that resulted from major aircraft accidents or incidents and that forced the FAA and its predecessor, the CAA, to develop a series of air traffic control system research and development plans. The SC-31 and the Project Beacon task force reports were among the more well-known plans.

The current ATC system was developed over time to meet system users' needs with modern technology, and it has worked remarkably well. What it lacked, however, was long-term, goal-oriented ATC research and development programs. Instead of trying to predict the future and then developing a system designed specifically for the future, the FAA and the CAA were usually trying to catch up with improved technology. This has not entirely been the FAA's fault; in most instances, the appropriation and procurement process and unforeseeable changes in the aviation industry—such as deregulation, the PATCO strike, and hub-and-spoke airline systems—have been responsible.

As the 1970s came to a close, the FAA found itself unprepared for major changes in the industry that would dramatically affect the air traffic control system. Major hardware components had reached their operational capacity and were becoming increasingly antiquated and obsolete. The computer system installed in the ARTCCs was routinely reaching capacity and was increasingly prone to failures and breakdowns. Many terminal controllers were using outdated equipment that was increasingly expensive to operate and maintain. Some of this equipment was so old that the data plates still bore the insignia of the CAA—an agency that had been replaced 25 years earlier.

Automated Air Traffic Control

Air traffic is forecast to grow at a rate of over 10 percent per year over the next decade. As a result, the primary objective of the FAA is to introduce automation and improved reliability into the ATC system.

If the FAA continues to separate traffic as it does now, a significant number of new controllers will need to be hired and trained. In theory, this growing demand might be met by hiring additional controllers and managers, but this approach would be prohibitively expensive and would not improve the system's overall efficiency. As the ATC system operates now, every controller is responsible for separating aircraft within a certain block of airspace. Using current technology and procedures, each controller can separate a finite number of aircraft. Employing additional controllers might permit the FAA to reduce the size of every sector, thereby reducing the load on each controller, but system capacity would be only marginally increased, and the amount of coordination necessary to operate the system would increase monumentally. Whereas the

ATC system of the 1950s was drowning in paperwork, the FAA system of the twenty-first century would smother in coordination.

The FAA's proposed solution is to increase every controller's productivity, thereby increasing the number of aircraft that can be separated. The FAA believes that by using sophisticated computer hardware and software, every controller can become less involved in the mechanics of separating aircraft and can become more of a traffic manager or monitor. The computer system envisioned by the FAA would help the controller determine whether potential aircraft conflicts exist and would even be able to automatically propose alternative resolutions to the controller.

Procedural Separation Standards

The Current System In the current air traffic control system, the pilot determines the flight's objectives and decides how those objectives can best be met. These objectives include the destination airport, route of flight, proposed altitude, cruising airspeed, time of departure, climb and descent profiles, and speed schedules. The flight plan, however, conveys only a limited number of these objectives to the controller; parameters such as speed scheduling and climb and descent profiles are not transmitted.

The controller can query the pilot about these parameters or can make gross determinations of the aircraft's flight profile by interpreting the flight track and altitude information displayed on the radar scope. Using this limited information, the controller is responsible for separating participating aircraft. The task is further complicated by procedural restrictions that may cause rerouting of aircraft using preselected routes or altitudes. In addition, at any time during the flight the pilot may request a change in altitude or route or both. Then, the controller must predict the consequences of the request and ensure that both aircraft and procedural separation are maintained.

Although the current ATC system attempts to satisfy each pilot's request for a specific route or altitude, procedural restrictions are used to ensure positive aircraft separation and an efficient, orderly flow of traffic. As the number of aircraft participating in the ATC system increases, however, additional route and altitude restrictions must be instituted to reduce the potential conflict between converging streams of traffic. In reality, the imposition of these procedural restrictions separates potential, not real, traffic; thus, aircraft may often be denied the use of "empty airspace." Ironically, this may be the very airspace that the pilot had originally requested but was not cleared to use because of an ATC-imposed procedural restriction. The routine use of these restrictions results in increased fuel use, increased flight times, loss of flexibility, and, occasionally, reduced traffic flow.

Great care must also be taken not to overload the controller with the task of separating these aircraft. Procedural restrictions are commonly used to separate areas of traffic flow from each other, but they are seldom necessary for the actual separation of aircraft. The routine imposition of procedural restrictions reduces the controller's workload, thereby decreasing the potential for a loss of separation. Up to a point, the use of procedural separation actually increases the amount of traffic that can be separated by each controller because he or she is not required to constantly predict the flight track of each aircraft (and every ensuing potential conflict). Unfortunately, this system also requires that every aircraft remain within the procedurally prescribed routes and altitudes. Once these routes become saturated, no additional aircraft can be accepted by the controller, even if sufficient airspace exists elsewhere.

Procedural restrictions tend to keep aircraft at inefficient altitudes or to result in circuitous routes. These procedures are prearranged among ATC facilities to ensure that potentially conflicting traffic flows are always separated, either laterally or vertically, allowing for a small range of individual deviations and eliminating the need to coordinate individual clearances with nearby sectors or facilities. Since the limiting factor that leads to the imposition of these procedural restrictions is the controller's capacity to coordinate clearances and predict separation conflicts, and not airspace saturation, an automated process would greatly reduce the need for rigid procedural restrictions on system capacity.

ATC Modernization

The capacity of today's National Airspace System (NAS) is constrained by rules, procedures, and technologies that require pilots and air traffic controllers to conduct operations within narrow, often inefficient, guidelines. As air traffic continues to grow, system inefficiencies and their associated costs are compounded. Due to the constraints in today's ATC system, users cannot reduce their operational costs by flying preferred routes or by receiving timely departure/arrival slots. The current ATC system does not permit pilots to choose either routes or altitudes on any real-time basis. Most traffic management programs "manage" aircraft by delaying or rerouting them. This system of traffic management wastes fuel and increases the time that aircraft are in flight, thereby artificially clogging up the system. Although this air traffic control system provides a high level of safety, it was not designed to handle the volume of traffic already present or the approximately 10 percent per year increase predicted for the next century.

In response to these limitations, the FAA and the aviation industry are working together on two major interdependent air traffic control initiatives—free flight and ATC modernization. According to the FAA, **free flight** is defined as "a concept for safe and efficient flight operating capability under instrument flight rules (IFR) in which the operators have the freedom to select their path and speed in real time. Air traffic restrictions are imposed only to ensure separation, to preclude exceeding airport capacity, to prevent unauthorized flight through special use airspace (SUA), and to ensure the safety of flight. Restrictions are limited in extent and duration to correct the identified problem. Any activity that removes restrictions represents a move toward free flight."

In theory, a free-flight ATC system would let the pilots choose their own routes, altitudes, and airspeeds and modify them in real time as they see fit. Air traffic control would only intervene if necessary to prevent the loss of separation. This system is a far cry from today's ATC system in which pilots are placed on specific routes at predetermined altitudes. Much needs to be accomplished before free flight can become a reality. In addition, the current system needs to be maintained in some form until the transition to free flight can be accomplished.

Current ATC Initiatives

In recent years, initiatives have been developed by the FAA to permit more flexibility while still managing the flow of traffic. One set of programs attempts to match traffic flow to the controller's ability to monitor and separate traffic. These programs are known as traffic flow management programs. The goal of these traffic flow management projects is to increase overall system efficiency without reducing minimum safety standards. Most recent efforts toward increasing ATC system efficiency center on smoothing out the peaks and valleys in traffic flows and matching these flows to theoretical maximum values. Other initiatives are being tested that attempt to help the controller detect and solve potential traffic conflicts. These are generically known as conflict prediction or conflict probe programs. Both of these type of initiatives must be perfected before free flight can become a reality.

As operational experience has been gained in the ATC system over the last few decades, the FAA, while taking into account dynamic changes in weather and traffic flow patterns, has calculated theoretical maximum traffic values for various components of the ATC system. These components include airways, airports, final approach routes, specific intersections, and air traffic control sectors. Virtually all of the traffic management initiatives of the FAA attempt to dynamically match traffic demand with the theoretical maximum traffic values for these components. The three major programs currently in use are the departure delay, **en route metering (ERM)**, and **en route sector loading (ELOD) programs.**

Departure Delay Program

The **departure delay program** involves matching actual airport demand with a calculated airport acceptance rate (AAR). AAR is a dynamic variable that considers weather conditions, available runways, noise abatement routes, and traffic flow patterns to determine the maximum number of aircraft that can land at an airport during any given time period. The Air Traffic Control System Command Center (ATCSCC) shown in Figure 12-1, in cooperation with traffic management units (TMUs) located at every ARTCC and busy terminals, continually calculates airport acceptance rates for major airports and determines whether predicted airport demand will exceed that value for any given time period. If an airport overload appears likely, the ATCSCC delays selected aircraft that are still on the ground in an attempt to match future aircraft arrivals to the airport's predicted acceptance rate.

This program cannot be made highly automated as airport acceptance rates can change dramatically and depend on unforeseen and unpredictable variables, such as rapidly changing weather conditions and runway closures.

FAA

Figure 12-1. Air Traffic Control System Command Center located in Herndon, Virginia.

The FAA admits that the matching of demand to capacity is more of an art than a science—computer-assisted art, but art nonetheless.

En Route Metering Program

The departure delay program is unable to make short-term corrections to traffic flows, since it can only affect aircraft that have not yet departed. In any case, ATCSCC-imposed ground delays only grossly affect the actual arrival times of aircraft. Unknown variables such as unforecast winds aloft, aircraft loading, and pilot selection of airspeed will likely change the ultimate arrival time of each aircraft.

The en route metering program attempts to fine-tune this process in real time. As aircraft proceed toward their destination, the air traffic control surveillance system begins to calculate actual arrival times at the airport. If the airport acceptance rate is predicted to be exceeded for a given time period, the metering program calculates appropriate delays for each inbound aircraft to effect proper spacing. The en route metering program calculates time and distance from each aircraft to the airport and issues times for each aircraft to cross a predetermined fix. It is left up to the controller to determine the means of delaying each aircraft. Route changes may be employed, as may speed restrictions. The overall goal of the metering program is to ensure that aircraft arrive at the airport in a timed sequence so that holding patterns do not have to be utilized.

En Route Sector Loading Program

The en route sector loading (ELOD) program is a dynamic computer program operated within each air traffic control center. ELOD constantly predicts future traffic within each sector and compares this figure with a theoretically defined maximum traffic density for the sector. Whenever an overload is predicted, controllers determine the nature and cause of the overload, determine whether it is a transient condition or an immediate problem, and manually initiate aircraft rerouting or traffic metering if it is deemed necessary.

Procedural Changes

The principal philosophical change required for the transition to free flight is a shift from the concept of air traffic *control* to air traffic *management.* ATM differs from ATC in several ways: the increased extent of collaboration between users and air traffic managers, greater flexibility for users to make decisions to meet their unique operational goals, and the replacement of broad restrictions with user-determined limits and flight restrictions only when required.

The procedural changes required for free flight correspond directly to the change in philosophy from ATC to ATM. Under the current air traffic system, aircraft are frequently restricted to ATC-preferred routes, which may not be the routes preferred by the pilot or airline. Air traffic controllers direct pilots to change their direction, speed, or altitude to avoid adverse weather or traffic congestion. In contrast, free flight will grant pilots substantial discretion in determining their routes. Many decisions will be collaborative, taking advantage of the best information available to the pilot and air traffic manager to ensure safe, efficient flights.

National Route Program

In a preliminary move toward free flight using current technology, the FAA has instituted the **National Route Program (NRP)**. The NRP gives airlines and pilots increased flexibility in choosing their routes. Under the NRP, flights are not limited to preferred routes but may proceed direct from departure to destination (subject to route limitations within a 200-nautical-mile radius of takeoff or landing). The NRP has been expanded in phases, with each phase lowering the base altitude included in the program. NRP operations are currently authorized at all flight levels at or above FL 290 across the contiguous United States. Future efforts will focus on expansion of the NRP to altitudes below FL 290.

In conjunction with the National Route Program, the FAA is trying to increase flexibility by eliminating preferred routes whenever and wherever possible. The air traffic control system currently has close to 2,000 preferred routes. During any given day, pilots using the low-altitude victor airway system add approximately 125,000 miles of extra distance to their flight plans as a result of preferred routes. Although the FAA cannot eliminate all preferred routes, since they help ensure aircraft separation, a recent audit by the FAA indicates that at least 100 to 150 of these routes could be eliminated without negatively affecting system operations.

CNS Improvements

The FAA is increasingly experiencing problems with its older ATC systems, which have exceeded their expected life cycle. The technology in many systems is antiquated, and replacement parts are increasingly difficult to acquire. In many cases, the FAA must hand make its own parts or cannibalize parts from decommissioned systems to repair operational radars. Aging air traffic control systems also experience lower reliability rates and increased maintenance cost. This equipment is so old that many maintenance technicians are retiring before the systems they have serviced their entire careers can be replaced.

To achieve free flight, the FAA will need to replace much of the equipment, computers, and software currently in use in the ATC system. As part of this process, the Radio Technical Commission for Aeronautics (RTCA) has drafted a preliminary report describing a path to free flight. The RTCA committee report recommends that the FAA adopt a plan that would make free flight the mode of operation of the National Airspace System by the year 2010. The report also specifies many of the CNS/ATM improvements that must be made before free flight can be implemented. Some of these improvements include:

- A new, all-digital communications system must be developed that will permit the transmission of both voice and data.
- GNSS must become the navigation system standard. This will reduce the navigational error inherent in all air traffic control separation equations.
- **Automatic dependent surveillance (ADS)** will become the primary means of transmitting aircraft positional data to ATC ground stations.
- New displays in aircraft and ATC facilities must be developed to improve pilot and controller situation awareness.
- Computer-based decision support systems must be developed to assist or take over some of the routine controller separation duties.

Communications System Changes

Voice radio is the only electronic method for controller-pilot communications that exists today. These radio communications are the essential link for controllers to provide safety or flight instructions. Given the global nature of aviation, aircraft must be able to interface with air traffic control systems anywhere in the world. Today's air traffic control system is overburdened by ever-increasing levels of air traffic and the limitations imposed by congested voice communications.

In domestic airspace, flight information is typically transmitted and received using very high frequency (VHF) and ultra-high frequency (UHF) voice radio. As the number of aircraft operations has grown and the demand for communications has continued to rise, frequency congestion has become an increasing problem. This congestion increases controller and pilot workload, creates delays, and increases the likelihood of missed or misinterpreted information. The airspace in the immediate vicinities of Los Angeles, Chicago, New York,

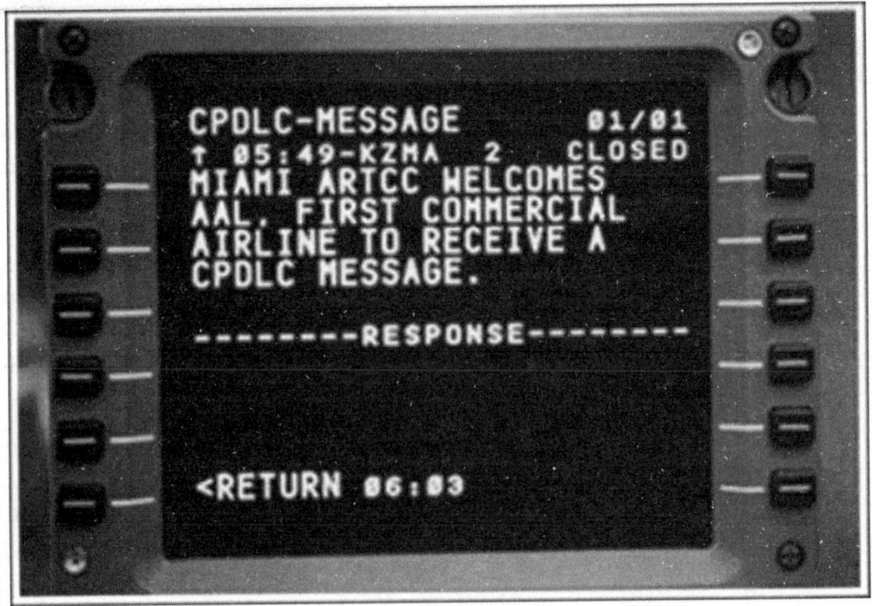

FAA

Figure 12-2. First CPDLC message used in air traffic control.

and Atlanta is already out of available channels. By the year 2007, the FAA predicts that it will be unable to provide any additional radio channel assignments nationwide.

Communication between aircraft and ground facilities in the future will rely less on voice radio communication and more on electronic data transmitted to and from the flight deck via data link technology. Analog radios will be replaced by digital equipment for both voice and data.

Between now and 2003, the FAA will add digital communication capabilities through the expanded use of data link services. The term "data link" refers to the overall system for entering, processing, transmitting, and displaying information. Data link technology is designed to transmit and receive air-ground voice, alphanumeric, and graphic information.

The first operational data link system, called **controller pilot data link communications (CPDLC),** was installed and became operational at the Miami ARTCC in 2002 (see Figure 12-2). To test this new technology, selected sectors at Miami ARTCC are equipped to transmit messages to a group of specially equipped airline and USAF aircraft. The new system has been used to digitally transmit text messages from the ground to the aircraft and back. The first phase of the CPDLC program (called Build 1) includes the capability to send four types of messages: initial air traffic control aircraft contact, altimeter setting, communications transfer to the next sector, and selected text messages from an approved menu. These messages are digitally displayed on a computer screen both at the controller workstation and in the cockpit of the selected aircraft. After experience has been gained with the new system, and more aircraft and sectors have been equipped, the FAA plans to enhance the capabilities of the system to be able to transmit additional requests and information including heading, altitude and speed assignments, pilot-initiated altitude requests, and clearance amendments.

The Next-Generation Air/Ground Communication System (NEXCOM) is proposed to provide air-ground communications in the future ATC system. These radios will have both digital voice and data capability and will be designed to operate with the existing analog system. Thus, aging ground radios can be replaced and later converted to digital operations without the need for an intermediate radio. NEXCOM radios will use a new modulation technique, such as time division multiple access, code division multiple access, or carrier sense multiple access, that is software programmable. This new radio system will be based on the VHF digital link standards defined by ICAO and will be backward compatible with the current radio system. Information that can be exchanged using the new system will include aircraft information, weather, traffic and approach information, and routine pilot-controller communications.

Navigation System Changes

The accuracy, coverage, and capabilities of current navigation systems vary considerably. The current airway structure is very rigid, and many approach and landing aids must be adapted to the geographic location and their own technical limitations. Many times pilots are restricted to single airway routes and approach paths. In general, the en route systems are not accurate enough for use as approach aids, so additional systems must be installed at airports, increasing system cost and complexity.

The navigation system of the future will rely on satellite-based GNS systems such as GPS and its international counterparts. In the United States, it is proposed that GPS become the standard for en route and approach navigation. The satellite navigation system currently used in the United States needs to become more accurate and have greater integrity, however, if it is to become the sole source of en route and approach navigation. The GPS system needs some form of accuracy augmentation to allow for more precise positioning information required for instrument approaches.

The wide area augmentation system (WAAS) described in Chapter 2 is an augmentation of GPS that includes integrity broadcasts, differential corrections, and additional ranging signals. It is being developed to provide the accuracy, integrity, availability, and continuity required to provide enhanced navigational signals for both en route and Category I precision approaches. WAAS will consist of a network of wide area ground reference stations that receive and monitor GPS signals. Data from these reference stations will be transmitted to master stations (see Figure 12-3), where the signal from each satellite will be assessed and corrections to the signal will be calculated. This correction signal will be rebroadcast to all aircraft via the WAAS satellites. WAAS should be come available around the year 2005.

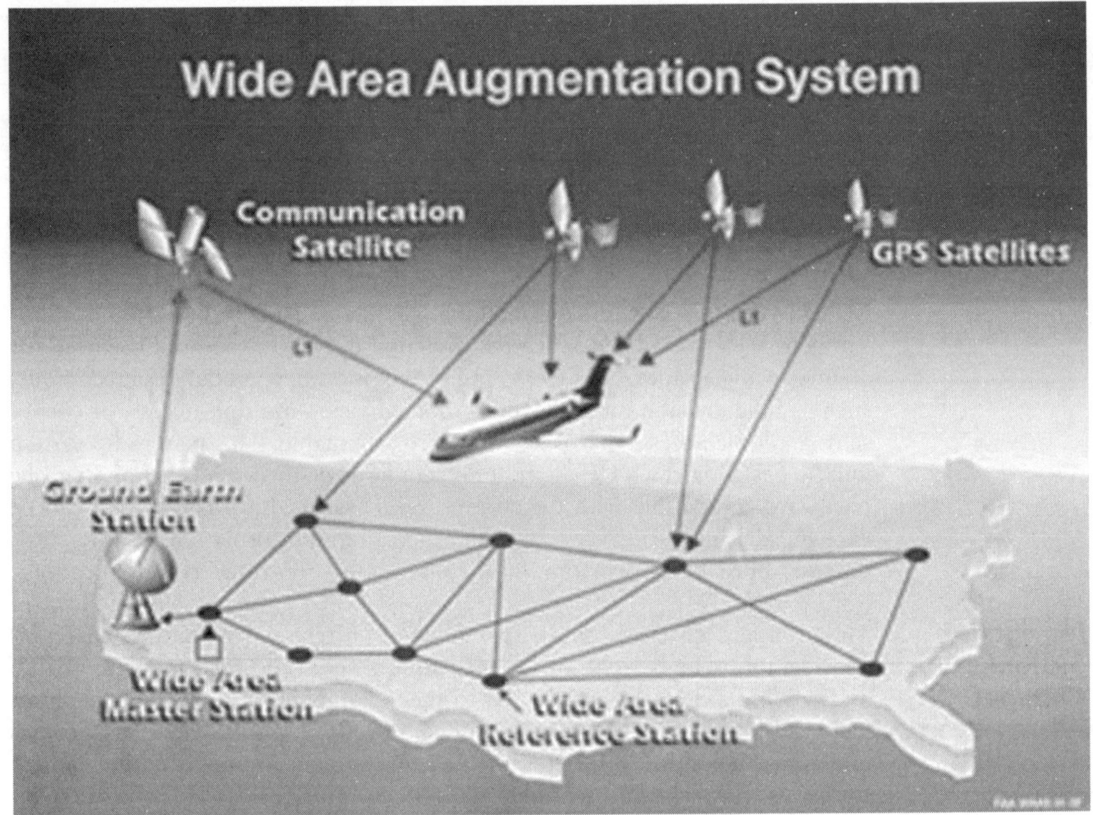

Figure 12-3. Wide area augmentation system. Courtesy Federal Aviation Administration.

WAAS will provide the accuracy sufficient for the needs of en route aircraft and can provide navigational guidance for nonprecision approaches. WAAS may be able to provide the equivalent of Category I ILS approaches. But since the WAAS corrective signal is transmitted over a very wide area, the signal must be averaged; therefore, WAAS may have difficulty providing consistent Category I criteria and will most likely never be accurate enough to replace Category II or III instrument landing systems.

The local area augmentation system (LAAS) will be needed for Category II and III precision landing minima. LAAS relies on a precisely surveyed ground station located near the airport that will precisely measure any GPS errors and will provide corrective information for just that airport or terminal area. LAAS will transmit a highly accurate corrective signal to aircraft within an operational radius of about 25 to 30 nautical miles. This signal should be so precise that Category II and III approaches will become widely available.

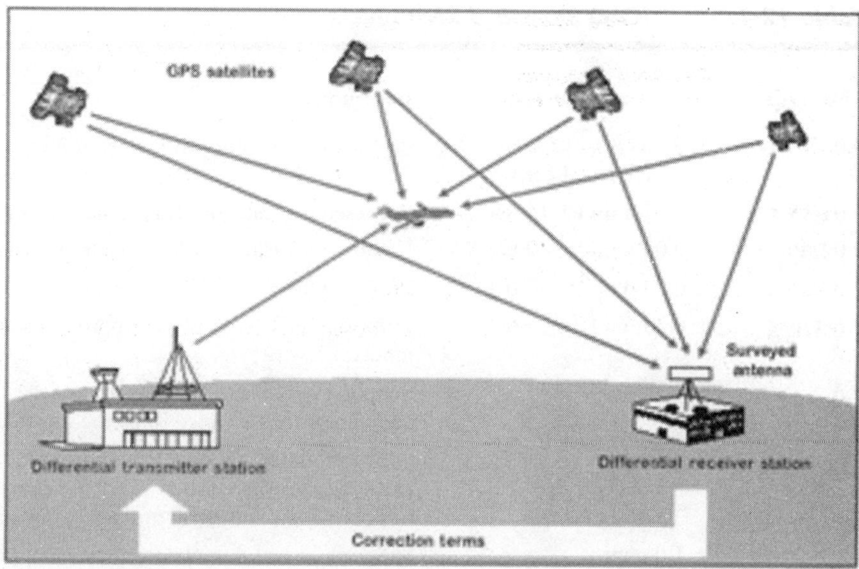

Figure 12-4. Local area augmentation system. Courtesy Federal Aviation Administration.

A single LAAS (as shown in Figure 12-4) will be able to provide service for multiple runways so long as they are located within about 25 nautical miles of the LAAS. LAAS can also provide terminal navigation, airport surface navigation, and guided missed approach and departure procedures. LAAS should become available around the year 2005.

Required Navigation Performance

The FAA is transitioning to a new measure of navigational accuracy known as required navigation performance (RNP). RNP changes the national airspace system from a ground-based, fixed navaid system to one in which aircraft operators can select which technologies they wish to utilize for en route and terminal phases of flight. Aircraft operators will be able to install various independent navigational sensors for their aircraft, choosing from VOR, VORTAC, DME, TACAN, GNSS, LORAN, and ILS navigation systems as they see fit. In lieu of designing airspace and airport procedures around specific navigation aids, RNP will define the navigation performance required for aircraft to operate within any particular airspace. RNP defines the navigation performance accuracy (RNP level) that needs to be achieved for operation within the defined airspace or for a particular procedure. RNP level is based on the minimum acceptable accuracy that can be achieved at least 95 percent of the time.

As RNP certified aircraft fly through the national airspace system, the aircraft navigation system will automatically choose the most appropriate and

Table 12-1. ICAO Standard RNP Levels

RNP Type	Required Accuracy (95% Containment)	Description
0.003/z	± 0.003 n mi [± z ft. (where z ≤ 14 ft.)]	Planned for Category III Precision Approach and Landing
0.01/15	± 0.01 n mi [± 15 ft.]	Proposed for Category II Precision Approach to 100 ft. DH
0.02/40	± 0.02 n mi [± 40 ft.]	Proposed for Category I Precision Approach to 200 ft. DH
0.03/50	± 0.03 n mi [± 50 ft.]	Proposed for RNAV/VNAV approaches
0.3/125	± 0.3 n mi [± 125 ft.]	Proposed for RNAV/VNAV approaches using barometric inputs
0.3	± 0.3 n mi	Supports initial/intermediate approach, 2D RNAV approach, and departure; expected to be the most common application
0.5	± 0.5 n mi	Supports initial/intermediate approach and departure; only expected to be used where RNP 0.3 cannot be achieved due to poor navaid infrastructure and/or obstacles
1	± 1.0 n mi	Supports arrival, initial/intermediate approach and departure; also envisaged as supporting the most efficient route structure
4	± 4.0 n mi	Supports normal en route routes and airways in domestic airspace
5	± 5.0 n mi	Meets European RNAV standards
10	± 10.0 n mi	For use in oceanic and remote areas
12.6	± 12.6 n mi	Supports limited optimized routing in areas with a reduced level of navigation facilities
20	± 20.0 n mi	Used in early oceanic RNP operations

accurate set of navigation sensors to utilize and will continuously notify the pilot of the accuracy of the system (using RNP level). This data will be displayed in real time on the appropriate instruments and cockpit displays.

The RNP level is a statement of the **total navigation system error (TSE)** allowable within a particular airspace. TSE takes into account avionics system errors and computation errors. The TSE must not exceed the specified RNP value for 95 percent of the flight. RNP is a measure of total cross-track and along-track errors relative to a preplanned flight track. For example, an aircraft operating with at least RNP-4 capability would be capable of maintaining a position within 4 nautical miles of the route centerline 95 percent of the time. An aircraft capable of maintaining RNP-0.3 will be able to navigate within 0.3 nautical miles of the course centerline. RNP levels currently approved for worldwide use are detailed in Table 12-1.

RNP procedures do not normally specify that an operator equip the aircraft with a specific type of navigation equipment, but instead require that the aircraft have the ability to maintain a specified navigational accuracy during the conduct of the flight or the procedure.

Table 12-2. FAA-Approved RNP Levels

RNP Level	Typical Application
0.3	Used for instrument approaches. May provide equivalent to ILS Category I
1.0	Used for departure and terminal operations
2.0	En route standard

RNP has been accepted by the FAA as the future means of defining en route and approach accuracy. The FAA currently defines a subset of fewer RNP levels than the ICAO, but in general the two systems of categorization are compatible (see Table 12-2). As the RNP concept matures, it is likely these two identification systems will merge into one worldwide standard.

RNP levels are beginning to be assigned to existing routes and procedures based on the use of a specific navigational system or sensor such as VORTAC, DME, or GPS. In the future, new RNP-specific routes and procedures will be defined using RNP level as the only defining criteria. When using these new procedures, unless otherwise noted, the required minimum RNP level will be depicted on the chart and it will be the pilot's responsibility to ensure that the aircraft navigation system can meet those requirements. If it cannot, then another route or approach procedure must to be selected by the pilot. For example, a departure procedure may describe specific navigational equipment to be used (such as VORTAC), and include the phrase "or RNP-1.0." An instrument approach procedure will be described by referring to the specific navigation aid needed (such as ILS) and the phrase "or RNP-0.3." In either of these cases, either a conventionally equipped aircraft or an RNP-certified aircraft able to maintain the required RNP level could fly the procedure. As the RNP system matures, new routes and procedures will be developed that will require RNP certification. These procedures will not list a conventional navigation aid, but will simply stipulate the RNP level necessary if they are to used.

It is expected that RNP-capable aviation systems will require very little pilot selection of specific navaids or systems. The various sensors installed on the aircraft will automatically detect and utilize the most accurate available system within range and display this information to the pilots. This type of navigation is already commonplace in flight management systems (FMSs) found on most modern corporate and airline aircraft. When utilizing these systems, the pilot is not required to physically select any particular navigation aid. The aviation suite on board the aircraft detects that most accurate available navigation system or systems, determines the aircraft's location, and displays the aircraft on a moving-map indicator. Ancillary information, such as which navigation aid is being used and the current accuracy level (RNP level), is also displayed on the screen.

When operating on RNP-approved routes or procedures, it will become the pilot's responsibility to understand his or her particular aircraft's capabilities and advise the controller if an equipment failure or other malfunction might cause the aircraft to lose its ability to continue operating in the designated RNP airspace. If a pilot determines that a specified RNP level cannot be achieved or maintained, it becomes his or her responsibility to advise the controller and determine acceptable alternative routes and procedures to continue the flight. The FAA, in cooperation with ICAO, has already approved RNP for some oceanic operations. RNP-10 routes have been established in the northern Pacific (NOPAC) operating area. This action has increased capacity and efficiency by reducing the distance between assigned flight tracks to 50-nautical-mile lateral separation. In addition, the FAA has assisted and approved avionics installations and flight operations in areas of Europe, where some routes have already been designated as RNP-5 routes.

The FAA has developed a transition plan for the evolution of navigation to an RNP environment. In the first phase, a limited set of RNP benefits, such as the ability to operate on existing routes and procedures, will be offered to existing aircraft that are currently equipped with RNP-compatible navigation systems. This phase has already begun and will last until 2005. During the second phase, from 2005 to 2015, the FAA will develop independent RNP routes that will only be available to RNP-certified aircraft. During this phase, the standard airways and approach procedures will be maintained. The third phase, beginning in 2015, will mark the beginning of an all-RNP era of flight. By this point in the development, RNP capability will be required of all aircraft operating in the national airspace system.

Navigation Security

After experiencing inadvertent jamming of GPS signals around the United States on multiple occasions, and after conducting a formal review of the events on and following September 11, 2001, the FAA has re-examined its transition strategy from a ground-based to a space-based aviation navigation system.

Previously, most FAA planning documents described a transition from the existing navigation system to one based entirely on GNSS. However, recent events have identified problems with this strategy and required the FAA to develop a new plan.

Occasional GPS interference was once regarded by the FAA as an accidental, somewhat improbable and most likely localized event. It is now considered more likely to occur deliberately and over a larger geographical area for a longer period of time. The FAA has had to focus future strategies on developing contingency plans for deliberate and longer-term GPS disruption. The goal of the FAA is to provide a backup system that can safely recover aircraft in the air and sustain commercial flight operations for the length of the disruption.

In 2002, the FAA released a navigation and landing transition plan that outlines a strategy for transitioning from the current to the future navigation system while minimizing the potential effects of both deliberate and inadvertent jamming of future air navigation transmitters.

Previous FAA plans described the decommissioning of the ground-based navaid system once GNSS became the primary source of navigation data. There are currently a little more than 1,000 VOR stations in operation across the

United States. The FAA now plans to maintain a limited navigation system of about 470 strategically placed VORs. By relocating existing VORs to more optimal locations, the FAA will develop a system of uninterrupted navigational coverage for all aircraft operating at altitudes of 5,000 feet MSL and above. In addition, all existing DMEs and TACAN stations would remain operational. DME can provide very accurate inputs into flight management systems, while TACANs are necessary for military homeland defense.

There are currently over 1,200 ILS installations at airports across the United States. As GPS-based procedures become commonplace, total ILS installations will be reduced with the proviso that at least one system always remain as a basic backup at every airport currently equipped with ILS. To provide all-weather capability, the approximately 100 Category II and III ILS installations currently installed would remain operational.

The FAA is currently conducting an extensive review to consider whether the LORAN-C network, which is currently operating but scheduled for decommissioning, can be upgraded and utilized as a backup for the GPS system. With some modification, LORAN could most likely be used to provide acceptable en route and nonprecision instrument approach capability to most airports. LORAN still has many limitations, however, that make it unable to act as a complete and reliable backup for the GPS system. Technical studies and modification still must be conducted to determine whether LORAN-C can become the sole backup system for GPS.

Surveillance Systems

Knowing the position and intended path of aircraft relative to others is necessary to ensure safe separation. The accuracy and certainty with which aircraft positions can be tracked determines the procedures and spacing used in ATC. Improved aircraft surveillance can improve the overall efficiency of the nation's airspace by reducing separation requirements. To reduce separation standards while still providing free flight requires the ability to accurately and reliably locate and track aircraft with greater precision and a faster update rate than is used today.

Separation methods employed today include visual positioning confirmation, radar positioning, and pilot position reports transmitted via voice radio. Visual separation is common in both general aviation and commercial air transport operations, though its use is limited to clear weather conditions. Radar permits air traffic controllers to monitor aircraft movements and more easily apply separation criteria. If radar is not available, pilot position reports are utilized.

Surveillance coverage and accuracy will be enhanced in the future by incorporating GNSS-derived positioning information with information provided by radar and/or pilot reports. This information can be translated into 4-D (position plus time) and made available to both pilots and controllers.

A new system known as automatic dependent surveillance (ADS) has been proposed to both augment and eventually replace ground-based surveillance radar systems. Unlike radar, which tracks aircraft using interrogating radio signals, ADS is based on aircraft-derived positioning and transmits reports based on onboard navigational instruments (most likely GNSS). ADS relies on data link technologies to transmit this information. There are two forms of ADS currently under development: ADS-Address (ADS-A) and ADS-Broadcast (ADS-B). The ADS-A system exchanges information between a specific aircraft and ATC on request. The ADS-B system broadcasts information periodically to all aircraft in the immediate vicinity and all air traffic management facilities within a specified area.

The primary objective of ADS-A and ADS-B technology is to improve surveillance coverage, particularly in areas having poor or no radar coverage. When ADS-equipped aircraft are within radar coverage, their positions will be verified by radar reports, providing independent and redundant surveillance. In areas not covered by radar, ADS will provide the controllers with "radarlike" computer displays of aircraft location. This will permit separation requirements in areas with inadequate or nonexistent radar coverage to be reduced.

At the airport, ADS will be used to assist controllers and pilots to maintain positive identification of taxiing aircraft and help prevent runway incursions. Currently, ground-based traffic can only be monitored through the use of airport surface detection equipment (ASDE) radar. The current model in use by the FAA, ASDE-3, provides radar surveillance of aircraft operating on the airport surface. When combined with a computer-based conflict prediction system, it can alert controllers to potential and actual surface conflicts. ASDE is an expensive system to install, with less than 25 systems slated for U.S. installation. ASDE is viewed as a short-term solution to preventing runway incursions, with the hope that ADS-B will become commonplace near the year 2010.

The FAA currently plans to utilize surveillance radars to provide independent airspace coverage for the next decade. Once ADS becomes commonplace and reliable, surveillance radars may be decommissioned. Until that time, radar will still be the primary form of aircraft surveillance. The focus of the FAA in this area is to upgrade the existing radar system to distribute radar data in a way that will enable the FAA to increase its effective airspace coverage using fewer radars. Obsolete radar systems will be replaced with newer digital systems. The information provided by these systems can be transmitted to multiple computer sites, with the data redistributed throughout the ATC system. Each ATC facility will no longer be required to be tied into a single radar site. Radar data can be distributed systemwide, even to facilities that do not have current radar capability.

To provide en route surveillance capability, all of the older ARSR-1 and -2 radars will be decommissioned and replaced, if necessary, by digitally enhanced ARSR-3 or newly acquired ARSR-4. In the terminal environment, older, analog ASR-7 systems will be replaced with the new digital output ASR-11 radars (see Figure 12-5). Many of the existing ASR-8 radars will be given an upgrade that includes digital output, turning them into ASR8-Ds. The most recent addition to air traffic control, the all-digital ASR-9 radars, will be installed at the larger airports.

FAA

Figure 12-5. ASR-11 digital airport surveillance radar.

These upgrades are planned to be complete around the year 2004. ADS-B will become more commonplace after that point, and surveillance radar will be operated in parallel with ADS from the year 2008 through 2012.

Air Traffic Management

Manual air traffic control procedures need to be augmented with computer-based decision support systems if the ATC system is to become more efficient and capable. Current FAA hardware and software systems are generally incompatible with one another, and implementing any standardized system throughout the FAA is next to impossible. Before any widespread decision support systems can be installed, the FAA must upgrade and standardize the hardware used in the air traffic control system. Only after this is accomplished can uniform and effective software be developed to assist controllers in their air traffic management tasks.

Hardware

Air traffic controllers currently use a combination of procedures and automated systems to separate traffic. The decision support systems in use today provide only limited assistance to air traffic controllers. Most routine decisions are made based on the training, experience, and judgment of the individual controllers, who must follow a set of narrowly defined air traffic procedures. As the volume of air traffic increases and as procedures allow greater pilot discretion, the efficient management and monitoring of air traffic will require the use of more advanced decision support systems.

The previously described programs ensure the efficient and effective collection, transfer, and display of information. Decision support systems need to be designed to improve the effectiveness of tasks such as flight planning, traffic sequencing, conflict checking, and conflict resolution. These new systems will enable controllers to simultaneously provide greater flexibility, reduce delays in congested airspace, and enhance overall safety. These systems will need new computer hardware and software installed at every major air traffic control facility. The FAA currently plans to replace outdated air traffic control computers at all the ARTCCs and approximately 225 FAA and Department of Defense radar approach control sites across the country.

The current automation systems available through either the ARTSs located at approach controls or the host computer at the ARTCCs have exceeded their expected life cycle. Numerous hardware components are no longer commercially available, and the antiquated software is expensive to maintain. The current system capacity cannot be economically expanded to meet projected traffic levels, because the hardware architecture has reached its limits. The hardware and software architecture will also delay introducing some new functional enhancements designed to reduce controller workload, improve safety, and increase throughput capacity at major airports. In 1994, due to the failure of the advanced automation system, the FAA restructured its acquisition into more manageable segments. One of these segments is called the **standard terminal automation replacement system (STARS).**

STARS STARS is a new automation system that uses a modern, commercial open architecture that solves existing capacity problems and can grow to meet future needs (see Figure 12-6). STARS is structured to acquire commercially available systems that meet FAA requirements with minimal development. This would theoretically allow a quick acquisition and deployment.

The STARS program will deploy new controller workstations and computer hardware to replace existing ARTS-IIA, ARTS-IIE, ARTS-IIIA, and ARTS-IIIE systems. Over 200 systems for terminal radar approach control facilities and large consolidated TRACONs will be installed. This will involve the installation of over 1,700 individual controller workstations. STARS is designed as an all-digital air traffic control system composed of commercial off-the-shelf hardware and commercially available software with ATC-specific software enhancements. STARS will interface with existing ATC systems to obtain the data necessary to perform terminal air traffic control automation processing.

FAA

Figure 12-6. Standard Terminal Automation Replacement System used in a terminal ATC facility.

STARS is a joint FAA/Department of Defense program to replace ARTS and other older technology systems at over 200 FAA and DOD terminal radar approach control facilities and associated towers. STARS is designed to be used to provide air traffic control to aircraft operating within terminal areas.

The full STARS installation includes 20-by-20-inch full-color displays with associated input devices, computers, and networks. STARS will provide the controller with aircraft positional information in a full-color format, with weather overlay, flight data, and other ATC-related information directly on the display console. STARS utilizes commercial workstation computers interconnected by modern local area networks to provide the controllers with modern color displays that provide controller-modifiable computer windows and graphics. The STARS system is capable of tracking up to 1,350 airborne aircraft simultaneously within a terminal area. STARS can accept inputs from up to 16 short- and long-range radars and can supply information to more than 100 controller positions and 20 remote tower locations.

Since its inception, the STARS program has encountered extensive delays and political controversy. Originally slated to begin operation in Boston in 1998, STARS has only been deployed in preliminary form at two low-activity FAA TRACON sites, El Paso and Syracuse. STARS has a history of escalating in cost. The original STARS program was estimated to cost about $400 million. By 1993, the estimated cost for deployment was $654 million. In 1996, when the actual deployment contract was awarded, the estimated cost had escalated to over $900 million. By 1998, after addressing multiple "human factor" issues that came up in testing, the total cost for implementation was estimated to be about $1.4 billion.

An additional problem is that STARS requires digital input to track air traffic and weather systems. The ASR-9 is the only terminal digital radar currently operated by the FAA and is only in place at 57 TRACONs. STARS deployment depends on the successful development and installation of the next-generation terminal radar planned by the FAA, the ASR-11. The ASR-11 program has also experienced setbacks that are expected to delay delivery. At the request of Congress, the FAA has developed a contingency plan in case the ASR-11 program suffers major delays.

The El Paso and Syracuse installations are currently hampered by the lack of specific peripheral equipment needed for STARS operation. At these two locations, the STARS system has been adapted to utilize much of the existing ARTS support equipment and software; however, adapting the system has hampered its operation. Until all STARS-specific hardware and peripherals can be developed and installed, full and complete STARS testing and implementation will be delayed. In the meantime, the FAA has plans to install and evaluate the limited STARS system at 11 other facilities: Albany, Albuquerque, Birmingham, Bradley International/Windsor Locks, Connecticut, Cleveland, Des Moines, Detroit, Las Vegas, Memphis, Omaha, and Providence.

As part of this project, the FAA is simultaneously developing a low-cost STARS display (know as STARS-Lite) that can be placed in airport control towers. STARS-Lite uses a remote link similar to a computer modem to access TRACON data to provide data to a control tower operating at a satellite airport. STARS-Lite can be used by tower controllers to separate air traffic and will replace the D-BRITE radar systems currently used at some remote tower locations.

En Route Automation Modernization The host and oceanic computer system replacement (HOCSR) program will provide adequate operation of the ARTCC radar system for the next couple of years but is inadequate for the planned software enhancements that will enable free flight. Before free flight can become a reality, the computer system at the ARTCCs will need to be replaced. The FAA plans to develop the **en route automation modernization (ERAM)** computer system to replace the existing air traffic control automation system utilized at the centers.

ERAM will replace both the host computer system software and hardware and the direct-access radar channel backup system. Still in its conceptual stage, ERAM will be designed to provide a system that will be both modular and expandable. Based on commercial off-the-shelf hardware and custom software, ERAM will be designed with new surveillance and flight data processing capabilities, will be able to provide enhanced traffic flow management metering and sequencing, and will be easily adaptable to any ARTCC site. The ERAM system will also provide integrated monitoring and control capabilities, data recording and playback, data reduction and analysis systems, and new development tools and utilities. It is planned that ERAM will provide the ARTCC controller with increased automation availability and reliability, thereby reducing delays, minimizing traffic restrictions, and improving NAS efficiency.

Once ERAM becomes operational, all the software being developed by the FAA to enhance the ATC system will be easily installed.

The ERAM system will receive digital surveillance aircraft positioning reports from external sensors such as radar and ADS-A and ADS-B. The surveillance data processor provides automatic tracking, conflict alert, and minimum safe altitude warning. The flight data processor subsystem will replace the current FDP system and will be responsible for exchanging and processing flight planning information between and within facilities. The flight data processor will also compute a four-dimensional route of flight for every aircraft and make it available to air traffic management decision support systems and other advanced planning applications. The flight data processor will establish the association between tracks and flight plans for formatting and sending appropriate information for display.

Software

Once the ERAM system has been installed, new decision support software can be installed, tested, and utilized by air traffic controllers. Some preliminary systems have already been tested and installed. The **Collaborative Decision Making (CDM)** program is currently operating and was established by the FAA in an effort to improve traffic flow management by establishing closer collaboration between the FAA and the airlines.

All aircraft currently flying under IFR flight plans in U.S. airspace are currently tracked by ATC computer systems at the ARTCCs. These data are electronically transmitted to the ATCSCC in Washington, D.C., and are then correlated and verified by the ATCSCC computers. Flights essential to national security or law enforcement are normally removed from the system, and then the data are made available to others in the aviation community. Using this information, the FAA and system users can determine actual and projected traffic flows at major airports, monitor flight progress, and make strategic scheduling decisions.

The data are currently available to the traffic management units located at every ARTCC and most large TRACONs. Many airlines utilize these data also. The information is now available to interested systems users through private information vendors. These vendors take the raw data provided by the FAA and present the data to users using customized software. Using this information, individuals can visualize traffic flow into and out of airports, predict landings and departures, and even track specific flights.

Individual corporations, such as Dimensions International and UNISYS, have developed software that will permit groups of users to utilize these data (see Figures 12-7 and 12-8). Large corporations or even individuals can gain access to the flight information needed to manage their aviation operations. This

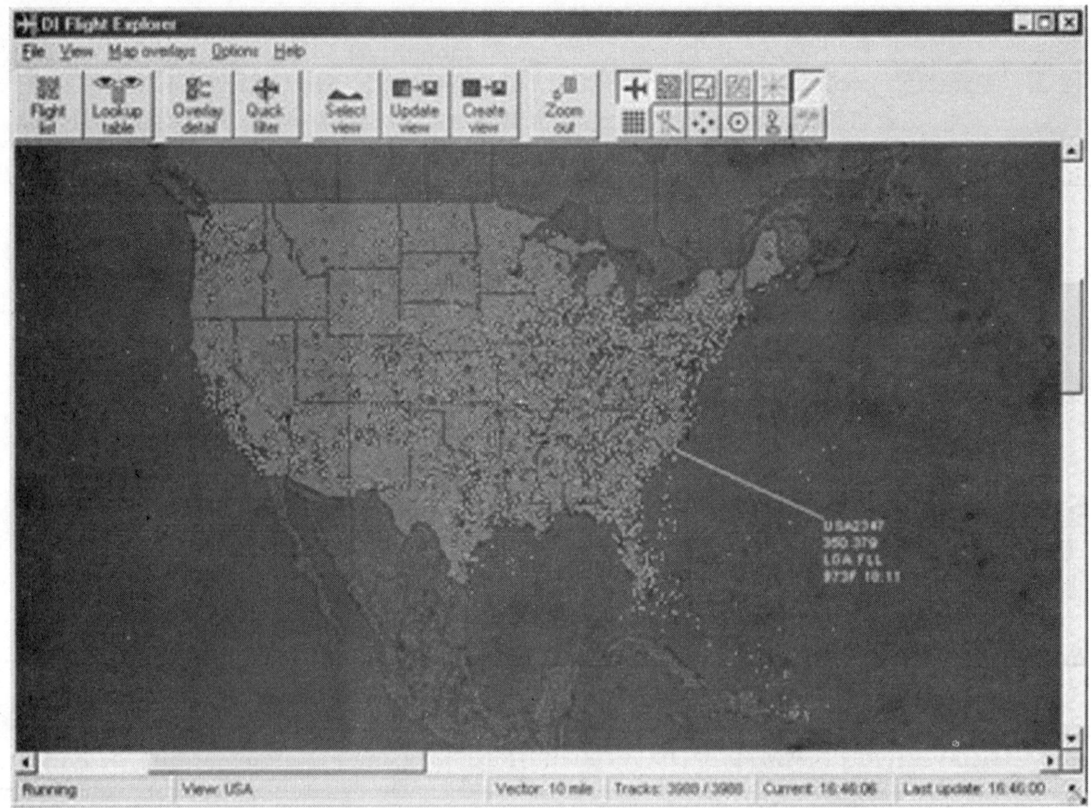

Dimensions International

Figure 12-7. Dimensions International aircraft situation display.

type of data will become indispensable to airport managers, small airlines, corporate flight operations, and even fixed base operators.

As the ATC system becomes increasingly congested, these collaborative programs will become more important to system users. Systems such as these may help eliminate or reduce the scope and duration of ground delays and allow ATC users more flexibility in responding to airport arrival constraints. By using these and other software programs to establish accurate pictures of real-time schedule information, system users will be better able to determine actual and projected traffic flow demands at major airports (see Figure 12-9).

The Center TRACON Automation System

The **Center TRACON Automation System** (CTAS) is a system developed by NASA that will provide users with airspace capacity improvement, delay reduc-

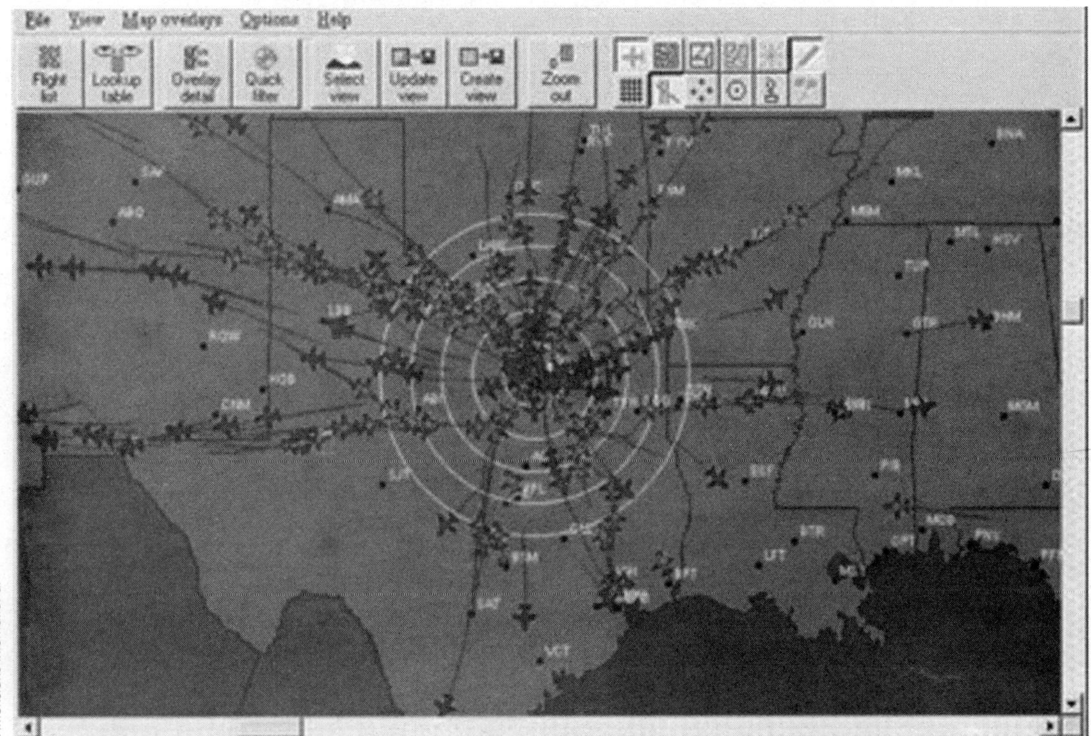

Dimensions International

Figure 12-8. Traffic flow in and out of Dallas/Fort Worth International airport.

tions, and fuel savings benefits by applying computer automation to assist controllers in efficiently descending, sequencing, and spacing arriving aircraft. CTAS provides four major functions: traffic management advisor, descent advisor, final approach spacing tool, and departure path expediter.

Traffic Management Advisor

The **traffic management advisor** (TMA) schedules arrival traffic to runways and generates a landing sequence as far out as 200 miles from the airport. A graphically depicted linear clock, showing time in the future as a vertical time line, is used to display aircraft identification tags alongside scheduled and estimated times of arrival. This provides traffic managers with a visual reference of the time interval between consecutive aircraft and the relative position of all aircraft (see Figures 12-10 and 12-11).

Descent Advisor

The **descent advisor** (DA) provides descent points, along with speed, heading, and altitude advisories. It also helps ARTCC controllers deliver aircraft to the TRACON feeder fixes on time and with minimum fuel consumption. Optimal, conflict-free descent profiles and routings are continually computed based on traffic, aircraft performance, weather, and airport/airspace configurations.

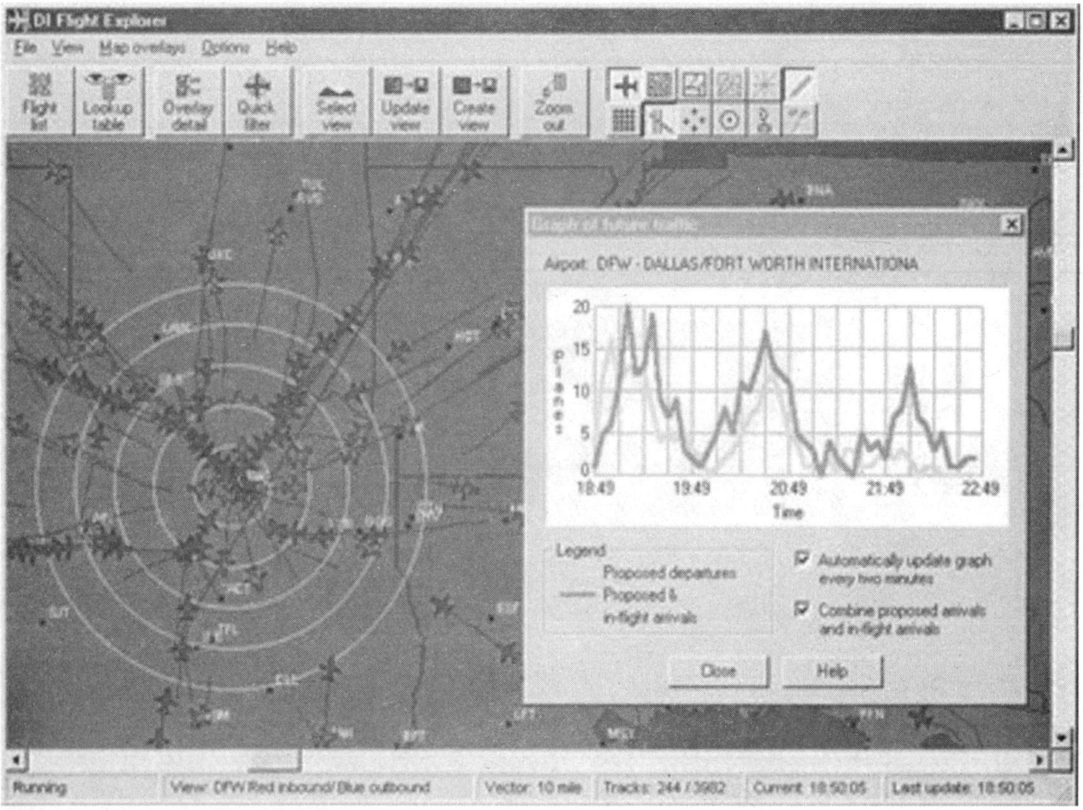

Dimensions International

Figure 12-9. Graph of aircraft inbound to and departing from Dallas/Fort Worth International airport.

Figure 12-10. Traffic management advisor time line graph.

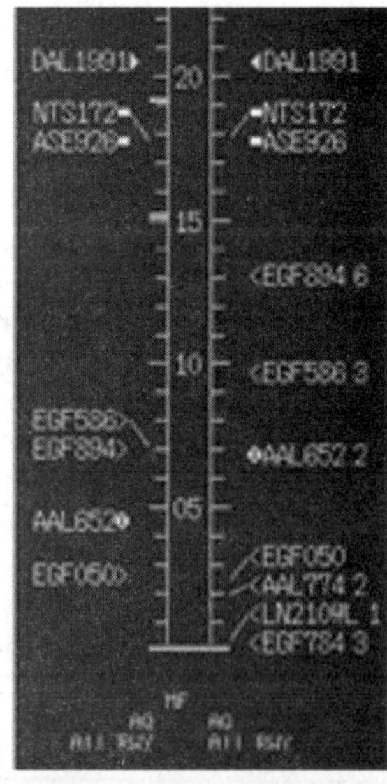

FAA

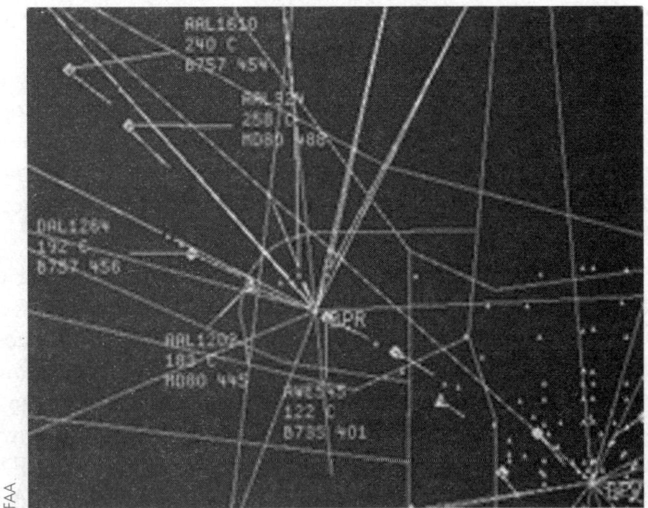

Figure 12-11. Traffic management advisor graphical display of aircraft tracks.

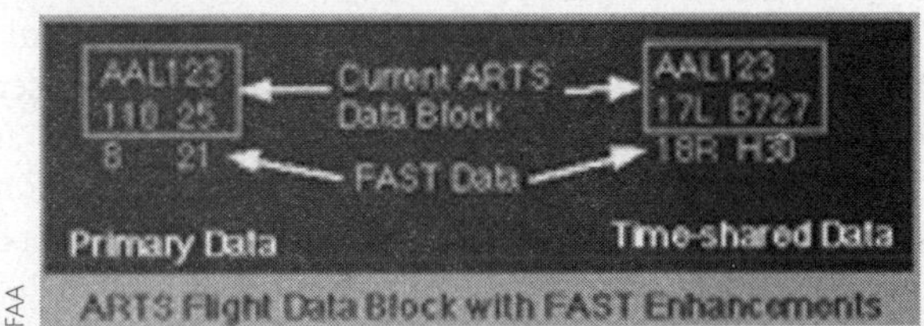

Figure 12-12. ARTS flight data block with FAST enhancements.

Final Approach Spacing Tool

The final approach spacing tool (FAST) assigns landing runway and sequence numbers in conjunction with precise speed and turn advisories to assist approach controllers in efficiently spacing aircraft to the runway. Advisories are continually provided that update changes that have accumulated since the previous advisories were displayed (see Figure 12-12).

Expedite Departure Path

The expedite departure path (EDP) program provides climb profiles and routings for departure traffic. This enables the controller to sequence departures from satellite airports with aircraft operating from the primary airport.

CTAS has been installed as a prototype system and testing is in process. But before free flight can become a reality, many of these proposed CNS/ATM improvements must be tested on a limited basis using real pilots, controllers, and aircraft. This is planned by the FAA in a program known as Safe Flight 21.

Safe Flight 21

Safe Flight 21 is an FAA-sponsored program to explore and demonstrate the use of ADS-B and other related technologies for eventual use in the free flight concept of operations. Safe Flight 21 is composed of two ongoing demonstration projects: the Ohio River Valley and the Capstone projects.

Ohio River Valley Project

The Ohio River Valley project tests ADS-B avionics using commercial cargo aircraft in the Ohio River Valley. These tests have taken place in terminal areas with significant cargo operations, including Memphis, Tennessee; Wilmington, Ohio; Louisville, Kentucky; Scott Air Force Base, Illinois; and Nashville, Tennessee. The Ohio River Valley project is co-sponsored by the Cargo Airline Association and the FAA. The Cargo Airline Association purchased, equipped, and is maintaining the avionics for the test aircraft. The FAA purchased, installed, and is maintaining ground systems at the five sites. A ground broadcast was installed at one site that receives data from the other sites and depicts ADS-B targets merged with radar targets. These merged ADS-B and radar targets are transmitted to participating aircraft. This enables the pilots to see both targets on a cockpit display, along with selected broadcast information such as weather maps, special use airspace status, and wind-shear alerts.

The Ohio River Valley project has been assessed in a series of operational evaluations. The first evaluation took place in July 1999 at the Wilmington site. It concentrated on measuring the improvement in the test aircraft's ability to make approaches in low visibility conditions and pilots' enhanced ability to see and avoid adjacent traffic. Cargo carriers, the FAA, NASA, the military, and academia participated in this initial evaluation. During the demonstration, aircraft equipped with ADS-B enabled pilots to consistently maintain close separation.

The second operational evaluation took place in October 2000 at the Louisville site, with some 20 aircraft participating. Cargo Airline Association members provided eight aircraft and NASA, the U.S. Navy, and the Aircraft Owners and Pilots Association provided additional aircraft for the trials. This evaluation measured the pilots' and controllers' abilities to maintain approach spacing, departure spacing/clearance, runway and final approach occupancy awareness, and airport surface situational awareness through the use of shared ADS-B data.

The third evaluation took place at the Memphis site in May 2001. It followed up on the successes of the previous trials and demonstrated additional capabilities of the ADS-B technology such as enhanced presentation of surface targets to controllers and enhanced airport surface management.

Capstone Project

The Capstone project is being conducted in the western portion of mainland Alaska. Up to 200 selected commercial, government, and private aircraft will be equipped, on a voluntary basis, with avionics furnished by the FAA. Participating aircraft will be equipped with an IFR-certified GPS navigation receiver, an ADS-B transmitter/receiver, and a multifunction color display with moving map display function and terrain advisory capabilities.

A network of data link ground stations will be installed at up to 12 existing FAA and joint-use facilities in the test area and connected via existing communications systems to FAA air traffic control facilities, service providers, and aircraft operator bases. A ground broadcast server and a gateway processor will be installed at the Anchorage ARTCC to receive data from each remote site and interface with the existing **micro en route automated radar tracking system (Micro-EARTS).** The Micro-EARTS will be programmed to depict ADS-B targets on one or more air traffic controller displays merged with radar-derived targets. The merged data will be transmitted to participating aircraft via ADS-B. This will enable the pilots to see both ADS-B and radar targets on the multifunction display. Lessons learned from both of these projects will be applied to future free flight operations.

Conflict Probes

Most of the overload in the current ATC system is, in reality, overloaded controllers. They must try to constantly predict future traffic positions and decide whether potential conflicts exist, then develop error-free conflict resolutions. All of this must be accomplished in real time using 20- to 50-year-old technology. All of the previously mentioned programs give the controller the tools to better manage or space aircraft. They do very little in the way of assisting the controller with detecting and resolving potential traffic conflicts.

Most of the traffic flow management problems in this country are not a result of a lack of airspace, but an inability of the controller to make accurate and confident separation predictions. As traffic density increases, controllers must apply additional procedural forms of separation. In most cases, the airspace involved has not reached its maximum traffic saturation, but the procedures in place cause temporary, transitory overloads of the controller's ability to visualize the airspace assigned to each aircraft. Until the controller's capacity to predict potential traffic conflicts can be improved, procedural restrictions will remain in place, resulting in increased traffic delays.

Almost all of the FAA traffic flow management programs involve the restricting of aircraft to effect procedural separation. In most areas of the country, if the controller were able to accurately predict potential separation conflicts, and if the airspace assigned to each aircraft could be reduced, procedural restrictions might be significantly reduced, thereby reducing aircraft delays. Two experimental programs are being tested that are designed to assist the controller with this task. These systems are generically known as **conflict probes.**

Conflict probes provide controllers with the ability to look ahead to identify potential separation conflicts with greater precision and accuracy. By estimating current position and predicted flight paths, conflict probes can check for potential loss of separation at current and future times. Depending on the type of software being utilized, conflict probes can be triggered automatically or manually.

User Request Evaluation Tool

The **user request evaluation tool** (URET) is a computer program that probes for potential conflicts of selected flight paths. When requested by the controller, URET can project the position of a selected aircraft into the future and determine whether a potential conflict might exist with another selected aircraft. In its present configuration, URET is selected by the controller concerning specific aircraft. It does not operate in the background continuously monitoring traffic conflicts.

URET processes real-time flight plan and track data from the host computer through a one-way interface. These data are combined with site adaptation, key aircraft performance parameters, and winds and temperatures from the National Weather Service to build four-dimensional flight profiles, or trajectories, for predeparture, inbound, and IFR flights. URET also adapts itself to the observed behavior of aircraft, dynamically adjusting predicted speeds, climb rates, and descent rates based on the performance of each individual flight as it is tracked through en route airspace (see Figure 12-13).

URET uses the predicted trajectories to continuously detect potential aircraft problems up to 20 minutes into the future and to provide a strategic alert

FAA

Figure 12-13. URET computer terminal.

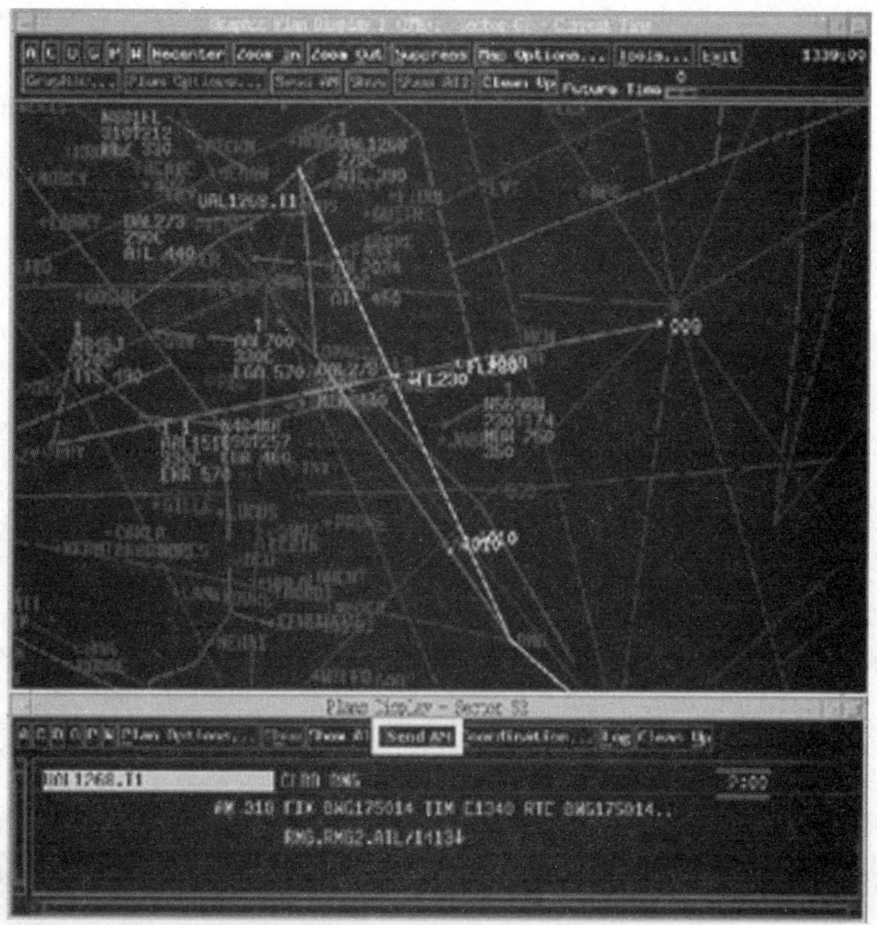

FAA

Figure 12-14. Graphical URET display.

to the appropriate sector. In addition, trajectories are the basis for the system's trial planning capability. Trial planning allows the controller to check a desired flight plan amendment for potential conflicts before a clearance is issued. This "what if" checking provides better capabilities for handling pilot requests.

URET is presented to the controllers in both a textual and graphical format. The text system displays aircraft lists and manages the presentation of flight data (call signs, routes, and altitudes), trial flight plans, and conflict information for the sector. Clearance language is also generated for trial plans. The graphical display provides the capability to view aircraft routes and altitudes, predicted conflicts, and results of trial plan resolutions. A point-and-click interface similar to today's personal computers enables quick entry and conflict checking of trial route, altitude, or speed changes (see Figure 12-14).

Free-Flight Operational Concept

Under the free-flight concept, pilots operating under instrument flight rules will be able to select their aircraft's course, speed, and altitude in real time. The ATC system will only intervene when necessary to provide positive aircraft separation. Currently, a pilot establishes a flight plan, which is essentially a contract, with air traffic control. This plan requires the aircraft to fly along a specific route and altitude. Any deviation from the contracted route or altitude must be renegotiated with an air traffic controller. For example, if a thunderstorm is encountered along the flight path, the pilot must notify the air traffic controller of the need to change course, and the controller must approve such a course change. Potential traffic conflicts are detected by the controller who unilaterally changes the contract, either by issuing a new route and/or a new altitude.

Under the free-flight concept, the pilot will be free to choose the aircraft's route, speed, and altitude and to make changes as necessary without requiring the controller's preapproval. This pilot's flexibility would only be restricted to ensure separation when traffic density precluded free flight, to prevent unauthorized entry into special use airspace, and to ensure safety of flight.

The free-flight concept is based on the development and identification of two airspace zones. These are known as the protected zone and the alert zone. The physical dimensions of each zone are dynamically calculated based on the aircraft's speed and performance characteristics and on the CNS equipment available (see Figure 12-15).

Under the free-flight concept, the air traffic management system must ensure that the protected zone, the one closest to the aircraft, can never overlap the protected zone of another aircraft. The alert zone extends well beyond the protected zone and, on contact with another aircraft's alert zone, the pilots, an air traffic controller, or decision support system software will determine what correction is required to ensure that the aircraft's protected zones do not touch. In principle, until the alert zones of two aircraft touch, the pilots can maneuver the aircraft freely. If the alert zones of two aircraft do touch, course corrections, altitude changes, or some other flight restriction will be issued to ensure separation. Most of these commands will be sent via data link or voice.

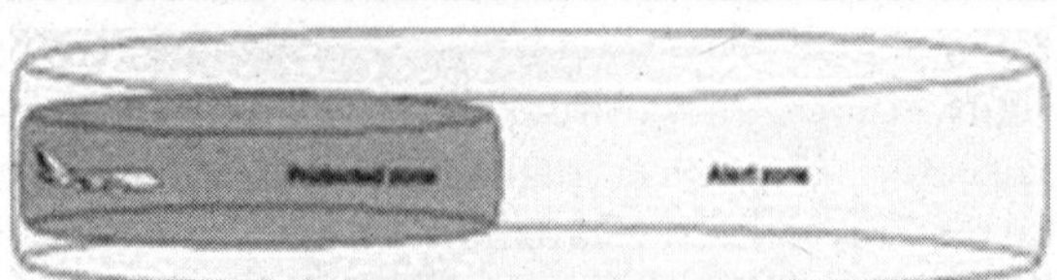

Figure 12-15. Free-flight zones. Courtesy Federal Aviation Administration.

Implementation Timetable

The FAA has proposed that free flight become the dominant form of air traffic management by the year 2010. But before it can become a reality, considerable work must be accomplished by the FAA and the user community. The current CNS system must be upgraded. FAA hardware and software must be installed and developed. Conflict probes must be developed, tested, and validated. More importantly, the users of a free-flight system, the pilots and controllers, must become comfortable with permitting some of the decision-making process to be delegated to computers and related air traffic management systems. Free flight will also require more of a partnership between users and the FAA than the system in use today.

Other Improvements to the ATC System

Weather is the single largest contributor to delay in the ATC system and is a major factor in aircraft safety incidents and accidents. Accurate forecasts and timely, accurate dissemination of weather information are critical to the safety of the National Airspace System. Many of the problems with today's weather systems can be attributed to limitations in the accuracy, predictability, analysis, transmission, coordination, and display of weather data. The FAA plans to develop technologies that will improve the dissemination of consistent, common, and timely aviation weather information in graphical format to all users of the aviation system.

Sources of aviation weather information that should be made available to pilots and controllers come from many types of atmospheric sensors, including automated weather observing systems, low-level wind-shear alert systems, terminal Doppler weather radar (TDWR), next-generation weather radar, and equipment operated by the National Weather Service and commercial vendors. This information will be displayed using enhanced graphics on new display systems in ATC facilities and in the aircraft cockpit.

Weather and Radar Processor

The Weather and Radar Processor (WARP) system will collect, process, and disseminate **next-generation radar (NEXRAD)** and other weather information to controllers, traffic management specialists, area supervisors, pilots, and meteorologists. WARP will provide a mosaic of multiple NEXRAD images to the DSR and STARS equipment. This will improve the quality of weather information available to air traffic controllers, thereby reducing accidents and air traffic delays. WARP will also provide the center weather service unit meteorologists with automated workstations that will greatly enhance their ability to analyze rapidly changing, potentially hazardous weather conditions. WARP will ensure that the latest and best information is provided to all system users within the en route environment.

Integrated Terminal Weather System

The Integrated Terminal Weather System (ITWS) is a fully automated weather prediction system that will give air traffic personnel and pilots enhanced information on weather hazards in the airspace within 60 nautical miles of an airport. When fully implemented, ITWS will have the capability to generate predictions of weather phenomena such as microbursts, gust fronts, storm cell movements, and runway winds up to 10 minutes in advance. Additionally, the system will display weather data in tower cabs, terminal radar approach control facilities, and their associated air route traffic control centers to facilitate coordination among air traffic control personnel. This system is a step toward avoiding delays caused by localized, hazardous weather and will increase the margin of safety.

In addition, a system is being developed that adds a wind-shear processor to the existing airport surveillance radar (ASR-9) to detect wind shear. This ASR-9 enhancement will provide products similar to TDWR at reduced cost but at somewhat lower performance levels. This enhancement is intended for airports that have an ASR-9, do not have TDWR, and need improved wind-shear detection capability.

Large Consolidated TRACONs

Metropolitan areas have experienced increased air traffic demand that makes the existing terminal airspace structure inefficient. This inefficiency contributes to delays, inefficient routings, complex air traffic control procedures, and airspace saturation. Consolidating two or more adjacent ATC facilities often makes it possible to redesign the airspace and make it more efficient (see Figure 12-16).

At least nine terminal areas have potential for facility consolidation. Before it can be accomplished, FAA efficiency and technical studies are conducted to determine their feasibility. The nine areas that the FAA has investigated or is investigating for TRACON consolidation (and the facilities that have been or might be included in the new facility) include:

Chicago, Illinois (Chicago, Milwaukee, Rockford, and Peoria)

Southern California (Los Angeles, Coast, Burbank, Ontario, and San Diego terminal radar approach controls)

Denver, Colorado (Denver, Colorado Springs, Pueblo, and Grand Junction)

Dallas/Fort Worth, Texas

Atlanta, Georgia (Atlanta, Macon, and Columbus)

Northern California (Oakland, Sacramento, Stockton, Monterey, and Travis AFB)

Washington/Baltimore terminal area (Dulles, National, Baltimore/Washington, and Andrews AFB)

Central Florida terminal area (Jacksonville, Orlando, Daytona Beach, Tampa, and Patrick AFB)

New York City

FAA

Figure 12-16. Potomac TRACON serves northern Virginia, Washington, D.C., and southern Maryland.

Precision Runway Monitor

During instrument meteorological conditions, airports with parallel runways spaced less than 4,300 feet apart cannot conduct independent simultaneous instrument approaches due to equipment limitations. This results in decreased capacity during periods of poor visibility.

The **Precision Runway Monitor (PRM)** system has been designed to help alleviate this problem. The PRM system combines a high-update-rate radar with computer displays that enable controllers to monitor aircraft on independent IFR approaches to dual and triple parallel runways spaced less than 4,300 feet apart. The PRM system requires installation of a nonrotating, phased-array-type secondary radar system that can update the controller with aircraft position two or three times per second. The electronically scanned antenna system provides a faster update rate than conventional radars, because it uses a computer-controlled electronic scanning sensor beam. Possible navigational transgressions can be detected at an early stage, and the controller can maneuver nearby aircraft away from the blundering aircraft. A separate monitor display is set up in the TRACON, and it must be staffed when independent parallel approaches are in use. This approach is expensive and can only be accomplished at a limited number of airports, but it will enable these selected airports to increase capacity, reduce delays, and save fuel during reduced visibility.

KEY TERMS

aircraft situation display
airport acceptance rate
Center TRACON Automation System (CTAS)
Collaborative Decision Making (CDM)
conflict probe
controller pilot data link communications (CPDLC)
data link
departure delay program
descent advisor (DA)
en route automation modernization (ERAM)
en route metering program (ERM)
en route sector loading program (ELOD)
expedite departure path (EDP)
final approach spacing tool (FAST)
free flight
micro en route automated radar tracking (micro-EARTS)
National Route Program (NRP)
next-generation radar (NEXRAD)
Precision Runway Monitor (PRM)
prediction/resolution advisory too (PRAT)
Safe Flight 21
standard terminal automation replacement system (STARS)
total navigation system error (TSE)
traffic management advisor (TMA)
user request evaluation tool (URET)

REVIEW QUESTIONS

1. What improvements are planned to the ATC system?
2. How will these changes affect pilots and controllers?
3. What changes will be made to the management of the ATC system?
4. What changes will be made to the aviation navigation system?

13

The Federal Aviation Administration

Checkpoints

After studying this chapter, you should be able to:

1. Identify the relationship of the FAA to other federal agencies.
2. Describe the general structure of the FAA and where air traffic control fits into that structure.
3. Describe the regional structure of the FAA.
4. Describe the process of becoming a controller for the FAA.
5. Identify the various screening programs applied to prospective air traffic controllers.
6. Describe the FAA controller training process.

The FAA is a diverse branch of the federal government that can trace its roots back 50 years to the Bureau of Air Commerce. The predecessors of the FAA have been parts of larger organizations (such as the Commerce Department) as well as completely separate divisions of the government (the Federal Aviation Agency). Since 1967, however, the Federal Aviation Administration has been the largest part of a cabinet-level agency, the **Department of Transportation (DOT)**. The FAA administrator reports directly to the secretary of transportation, as do the heads of the Federal Highway Administration, the National Highway Traffic Safety Administration, the Federal Railroad Administration, the Urban Mass Transportation Administration, the Coast Guard, and the St. Lawrence Seaway Development Corporation. Major policy changes in aviation are either approved or initiated by the secretary of transportation, who is the official transportation spokesperson for the executive branch of the U.S. government. All FAA funding requests are submitted to the **Office of the Secretary of Transportation (OST)**, which then makes an official budget request to the president.

Administrative Structure

The FAA's mission is to promote and regulate aviation in the United States. To perform this mission, the FAA has been divided into several principal lines of business, the Washington headquarters of the FAA, the Technical Center in Atlantic City, the Aeronautical Center in Oklahoma City, nine regional offices, and hundreds of operational facilities across the country (see Figure 13-1).

The Washington headquarters houses the administrators, deputy administrator, staff offices, and seven associate administrators. The FAA administrator and deputy administrator are directly responsible to the secretary of transportation for the operation and functioning of the FAA. The FAA administrator is selected to serve a 5-year term, and historically the administrator has been an FAA outsider. The administrator's position is not a career position, but a politically appointed office. The administrator is usually a person with a high level of aviation knowledge and an extensive aviation background.

To ensure civilian control of aviation, the Federal Aviation Act of 1958 requires that the FAA administrator be a civilian. Therefore, military aviators interested in this position are required to resign their commissions before they can be appointed administrator.

The actual working structure of the FAA has been modified by almost every administrator. The description of the FAA structure contained in this chapter was accurate as of 1998, but because of the rapid technological and political change in the country and in the aviation industry, the structure of the FAA will most likely change again before too long.

To fulfill its mission, the FAA maintains a work force of over 40,000 employees organized along seven principal lines of business. These are Regulation

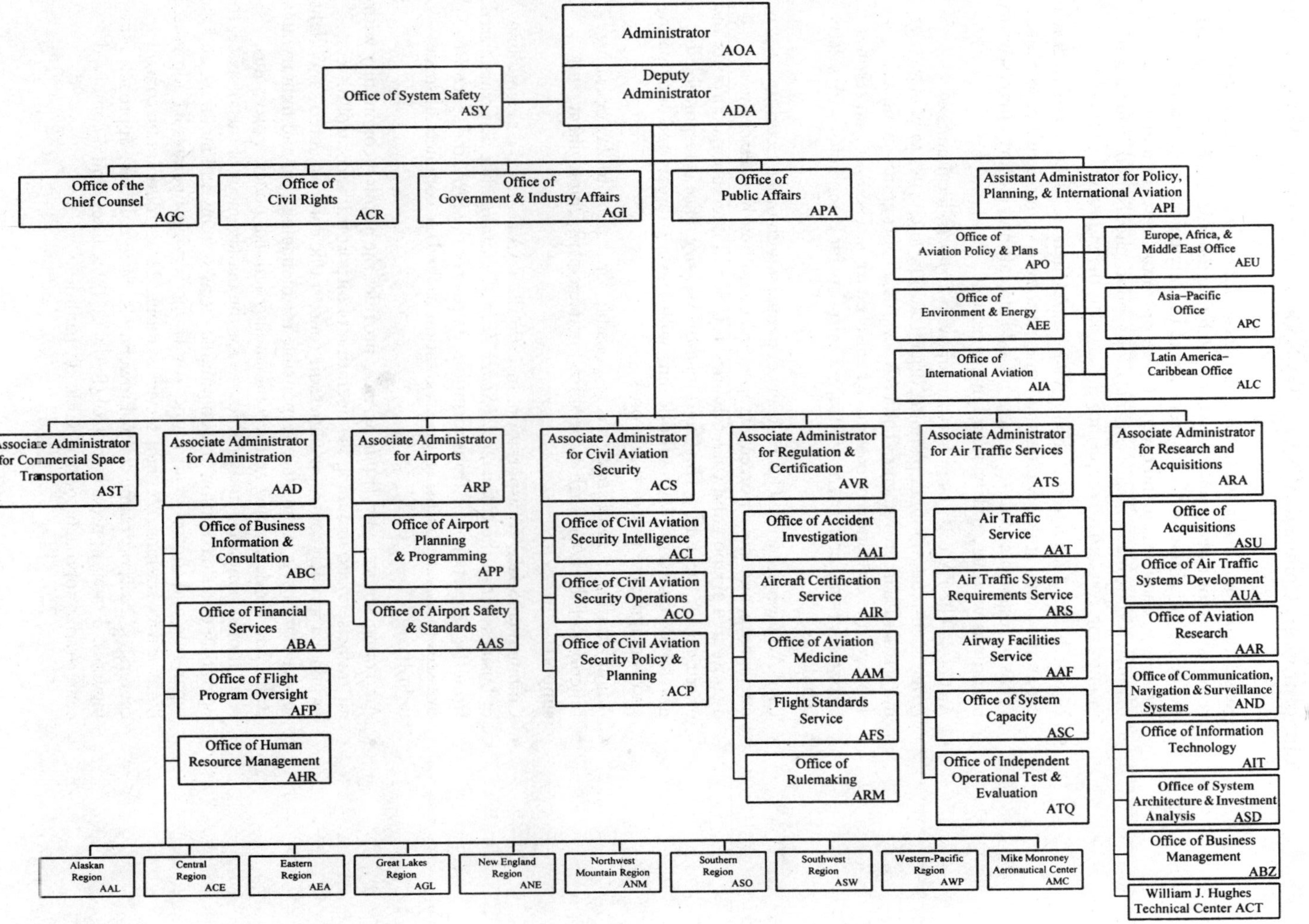

Figure 13-1. FAA structure.

and Certification, Research and Acquisitions, Airports, Civil Aviation Security, Administration, Commercial Space Transportation, and Air Traffic Services.

- Regulation and Certification. The FAA's *Regulation and Certification* office is responsible for three primary elements: defining the rules and policies governing the design, manufacture, and operation of U.S. civil aircraft; certificating and overseeing commercial and cargo aircraft operators, aircraft and avionics manufacturers, maintenance organizations, pilot schools, and airmen; and monitoring regulatory compliance throughout the industry.
- Research and Acquisitions. The FAA's *Research and Acquisitions* office develops programs aimed at developing and implementing ATC modernization plans. Areas of research and development include aircraft and airport safety, security, hazardous weather, and aviation human factors.
- Airports. The FAA works in partnership with *airports,* local units of government, metropolitan planning organizations, and states to plan airport infrastructure improvements. The FAA awards airport grants-in-aid and approves the collection of passenger facility charges. The FAA also certifies airports serving air carriers and inspects airports for compliance with safety standards.
- Civil Aviation Security. The FAA, in cooperation with other security, intelligence, and law enforcement agencies, continuously assesses threats against civil aviation and vulnerabilities at U.S. and international airports and at critical FAA facilities. The *Civil Aviation Security* office is responsible for developing countermeasures for dealing with existing threats and anticipating and managing the risk of future ones.
- Administration. *Administration* is responsible for the day-to-day operating functions of the FAA, such as personnel management, employment, and support.
- Commercial Space Transportation. In 1996 the FAA acquired a seventh line of business when the *Commercial Space Transportation* office was transferred from the Department of Transportation. As part of the FAA, the space office oversees the safety, licensing, and regulation of the U.S. commercial space industry.
- Air Traffic Services. *Air Traffic Services* provides the aviation community with air traffic services, including the operation of airport traffic control towers, terminal approach control facilities, air route traffic control centers, and flight service stations. It is the office responsible for managing civil and military air traffic and for developing and recommending national policies, programs, regulations, standards, and procedures for the management of the airspace, the operation of air navigation and communications systems and facilities, and the separation and control of air traffic. Air Traffic Services operates 61 automated flight service stations, 29 flight service stations, 21 air route traffic control centers, 352 airport traffic control towers, 185 terminal radar approach control facilities, 2 radar approach control (RAPCON) facilities, and 3 combined center/radar approach control (CERAP) facilities.

FAA Regions

When the FAA was formed in 1958, most of the important policy decisions were made and implemented by the Washington headquarters staff, with little regional input. FAA facilities in the field were often left to implement these staff decisions. This lack of input to the decision-making process prevented the FAA from developing and implementing programs that could be tailored to regional and local needs. In an attempt to return responsibility for implementing FAA policies to those individuals who had more direct contact with the flying public, in 1961 FAA Administrator Najeeb Halaby began to shift many of the day-to-day operations away from Washington, D.C., to the FAA regional headquarters. His plan was to use Washington to develop national policy and standards and let the regional administrators decide how to best implement those policies. In Halaby's judgment, decentralization promised increased flexibility and efficiency and would provide better service to the aviation public by increasing local input into FAA decisions while reducing decision-making time.

The FAA still operates as a decentralized agency much like that which existed when Halaby retired from the FAA administrator's position in 1965. Although the number and locations of the regional offices have changed since that time, they are still responsible for carrying out most of the FAA's policies. There are currently nine FAA **regional offices (ROs)** across the country (Figure 13-2)

Figure 13-2. FAA regional boundaries and regional office locations.

with one additional office located in Belgium (see Table 13-1). Every FAA facility in the field is assigned to one of these regional offices. Each region has regional directors, assistant directors, and departments that roughly correspond to those at FAA headquarters.

Getting Hired by the FAA

A potential controller must take a number of steps and pass a number of tests before he or she can be considered for employment by the FAA. The air traffic control profession is a demanding one, and the FAA must ensure that every potential controller has the ability, education, and desire to perform the job safely and efficiently. Thousands of individuals begin the process of becoming an FAA controller, but few can pass all of the tests and the training required to become part of this high-paying profession.

Three steps must be taken before an applicant can be hired as an air traffic controller by the FAA: preemployment testing, the FAA Academy training program, and on-the-job training at an FAA facility.

Preemployment Testing

Before becoming employed by the FAA, prospective controllers must successfully pass an aptitude test, a general employment interview, a medical examination, and a security investigation.

On the basis of extensive studies, the FAA has determined that the skills and abilities necessary to perform the duties of an air traffic controller begin to decline at a relatively young age. In recognition of this fact, Congress passed Public Law 92-297 in 1972; it requires that controllers be no older than 30 when first employed by the FAA either in an air traffic control tower or in an ARTCC. This age requirement does not apply to flight service specialist positions, however.

Aptitude Test Prospective controllers must begin by passing a comprehensive ATC aptitude test known as the **Air Traffic Selection and Training (AT-SAT)** test. The AT-SAT test is designed to determine whether the applicant has the basic mental skills needed to become an air traffic controller. These abilities include spatial awareness, ability to project in time and space, and ability to mentally form 3-D pictures from 2-D information.

Interview Applicants selected by the FAA to continue the employment process are asked to report to a nearby air traffic control facility for a personal employment interview. The interviewer looks for evidence that the applicant has the personal characteristics (such as motivation, practical intelligence, and speaking ability) required of an air traffic controller. Applicants who successfully complete this phase of the application process are scheduled for the medical examination.

Table 13-1. FAA Regional Offices

Region	Area Served	Location
Alaskan	Alaska	Anchorage, Alaska
Central	Iowa Kansas Missouri Nebraska	Kansas City, Missouri
Eastern	Delaware Maryland New Jersey New York Pennsylvania Virginia Washington, D.C. West Virginia	New York, New York
Great Lakes	Illinois Indiana Michigan Minnesota North Dakota Ohio South Dakota Wisconsin	Chicago, Illinois
New England	Connecticut Maine Massachusetts New Hampshire Rhode Island Vermont	Boston, Massachusetts
Northwest Mountain	Colorado Idaho Montana Oregon Utah Washington Wyoming	Seattle, Washington
Southern	Alabama Florida Georgia Kentucky Mississippi North Carolina South Carolina Tennessee	Atlanta, Georgia

(*continued*)

Table 13-1. (Continued)

Region	Area Served	Location
Southwest	Arkansas Louisiana New Mexico Oklahoma Texas	Fort Worth, Texas
Western Pacific	Arizona California Hawaii Nevada	Los Angeles, California
Europe, Africa, and the Middle East		Brussels, Belgium

Medical Examination Those who pass the AT-SAT and do well in an interview are scheduled for a medical exam. The medical examination includes physical and eye exams, blood chemistry tests, an audiogram, a psychological test, and a drug screening. The physical examination is similar to the second-class medical exam required of commercial pilots. Hearing and visual acuity are checked. Air traffic controllers must have 20/20 vision or vision correctable to 20/20 in each eye. In addition, ATC applicants must not have any medical history of psychosis, neurosis, or substance dependence (including alcohol, narcotic, or nonnarcotic drug dependence).

Every applicant for employment as an air traffic controller is required to provide a urine sample that will be screened for illegal drugs. The presence of these drugs disqualifies the applicant from future employment with the FAA. The drug screening and the medical exam are repeated continually during the career of an air traffic controller.

Security Investigation On successful completion of the medical examination, the FAA conducts a detailed security investigation of the applicant. This investigation includes inquiries to former employers and educational institutions and a review of any appropriate FBI, military, and police files. If the investigation reveals any questionable information concerning the conduct, reliability, character, trustworthiness, or loyalty of the applicant, an offer of employment with the FAA will normally not be extended. It takes 45 to 180 days to complete this investigation.

FAA Academy Training

On successful completion of the above steps, the applicant may be hired as a conditional employee of the FAA. Every newly hired FAA controller is required to successfully complete the controller training program conducted at the FAA Academy in Oklahoma City, Oklahoma (see Figure 13-3). The program is designed to evaluate both the academic and practical skills of the controller.

Michael Nolan

Figure 13-3. The Aeronautical Center Headquarters building in Oklahoma City.

The training program at the FAA academy is based on the concept of "train for success." The aforementioned screening program is designed to ensure that every student selected for training at the academy has demonstrated the skills necessary to become an air traffic controller.

The training at the academy focuses on the basic knowledge and skills required of an air traffic controller. Students learn about different aircraft types and operations and about the administrative and operational structure of the FAA. Air traffic control rules, regulations, and operational procedures are stressed.

Practical experience through simulation is gained in control tower operation and nonradar and radar separation. Students utilize computer-based training systems and computer-controlled control tower and radar simulators as an integral part of this unique training program.

The Academy training program is approximately 15 weeks in length and is composed of three parts: **aviation academics, part-task training,** and **skills building.** Conditional employees who have previous air traffic control experience, such as ex-military controllers, may be permitted to skip either the first or the first and second portions of the academy training program. Since they require less training, these applicants will be given hiring priority by the FAA. It

AIR TRAFFIC - SELECTION AND TRAINING (AT-SAT)

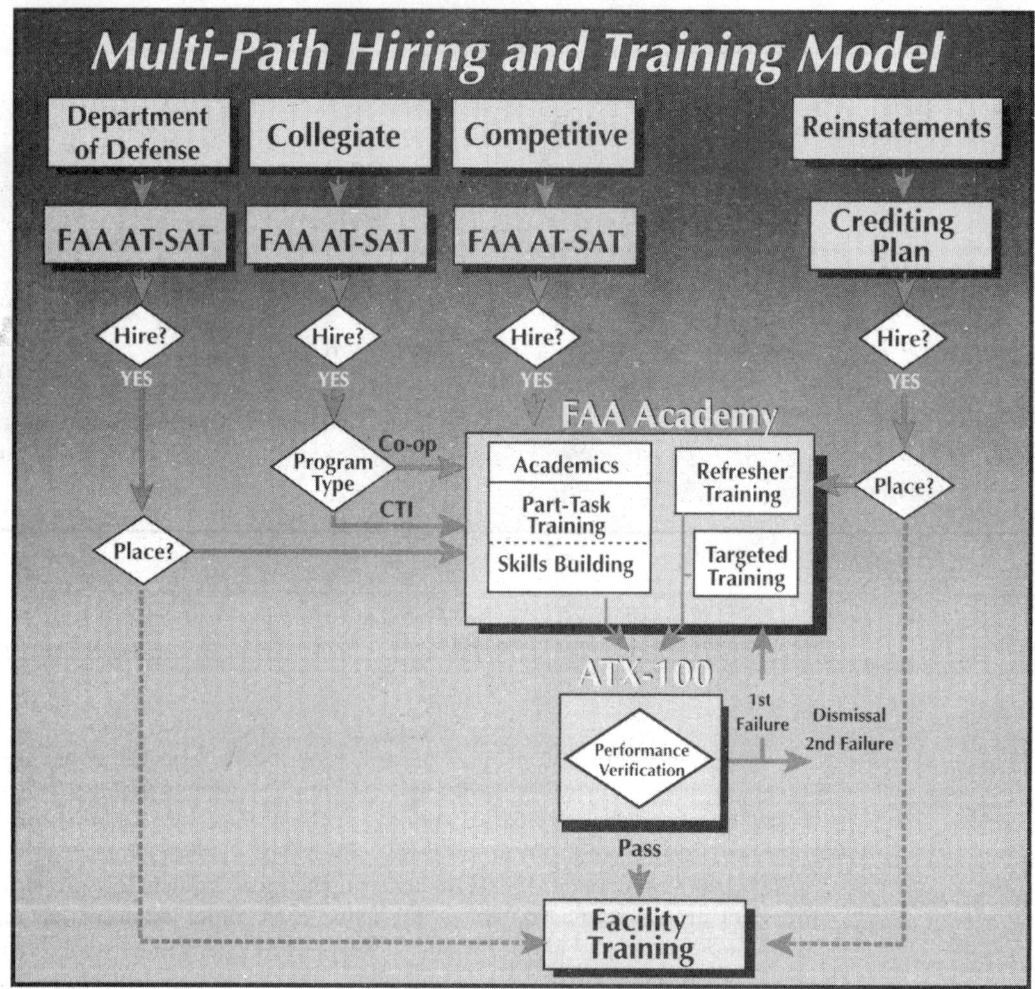

Figure 13-4. ATC hiring paths.

is unlikely, however, that this applicant pool will provide sufficient controllers for the next decade (see Figure 13-4).

In an effort to boost hiring, the FAA has selected a limited number of colleges and universities to provide initial training as part of their academic programs. This program, known as the Collegiate Training Initiative (CTI), will permit students to bypass the academic portion of the FAA training program.

Graduates of CTI programs will be given hiring priority by the FAA. The FAA hopes that the bulk of its hiring needs will be filled by graduates of CTI-approved aviation programs.

If the FAA cannot hire sufficient controllers from these two groups, competitive applications will be accepted. If selected through the competitive process, the applicant must complete the entire 15-week program at the FAA Academy. The general requirements to compete for a position with the FAA are a two- or four-year college degree (either an aviation degree from a non-CTI-approved school or a nonaviation degree) or equivalent work experience. The FAA expects to seek applicants through the competitive path only if the military and CTI schools cannot meet its employment needs. This process is highly competitive and limited to a few select individuals.

At the conclusion of the Academy training program, each student completes a series of **performance verification** exams. On successful completion of these exams, the controller is placed at an appropriate facility.

Field Training Program

Having completed the academy training program, the **developmental controller** is sent to an air traffic control facility. Depending on the complexity of the facility, it may take 1 to 4 years to become fully certified as an air traffic controller. Developmental controllers normally begin training on an operating position such as flight data. After certifying on this position, they begin to train on the other positions at the facility. At a control tower, the training sequence is normally flight data, clearance delivery, ground control, local control, and the radar control positions. Center controllers begin at flight data and progress through radar associate/nonradar controller before certifying as a radar controller.

Prior to receiving radar training at the facility, developmental controllers are sent to the **Radar Training Facility (RTF)** at the FAA Academy in Oklahoma City (see Figure 13-5). Once they have completed this training and are certified on every position at the facility, they must complete a facility rating exam. After passing this exam, they are considered facility-rated or **full performance level (FPL) controllers.**

Salaries

A controller's salary is based on the type and complexity of the facility to which he or she is assigned. ATC facilities are assigned classifications of ATC-5 though ATC-12, with ATC-12 being the most complex. Employees of more complex facilities receive correspondingly higher salaries.

Future controllers are paid while attending the training program at the FAA Academy. Upon program completion, they are sent to field facilities for on-the-job training. As they progress through their training, meeting facility-specific requirements, their pay grade is raised, step by step, from a beginning grade of ATC-3 up to the pay level assigned to the facility.

Figure 13-5. The Radar Training Facility at the FAA Academy in Oklahoma City.

Once controllers reach full performance level status at their facility, their pay will move up within the range specified for their facility. If a controller transfers to another facility, he or she normally retains the same salary, so long as it is still within the range of the pay specified for the new facility.

Controller base pay is predicated on a normal 5-day work week. Because most FAA facilities are open 24 hours a day, controllers receive additional pay for working outside the hours that most employees consider normal. If a controller works a night shift, he or she will receive additional pay known as **night differential.** Controllers also receive additional pay for working on holidays and weekends. Controllers who work overtime are paid an overtime rate that is equal to one and a half times their regular hourly rate. If they train other controllers or act in a supervisory role during any or all of their shift, they are eligible for various levels of differential pay. Controllers working in areas with higher costs of living may also be eligible for additional salary compensation. As a general rule, based on these various salary differentials, most controllers can expect to make 10 percent to 35 percent more than the base salary in any given year.

Table 13-2 lists the pay ranges assigned to the different ATC pay levels. Table 13-3 lists selected ATC facilities and their pay level classification.

Table 13-2. ATC Facility Pay Bands

ATC Grade	Min	Max
3	$37,283	$52,196
4	$39,799	$55,719
5	$45,171	$63,239
6	$49,915	$69,881
7	$55,155	$77,217
8	$60,947	$85,326
9	$67,345	$94,283
10	$77,447	$108,426
11	$81,513	$114,118
12	$85,589	$119,825
13	$89,869	$125,817
14	$94,363	$132,108

Table 13-3. ATC Facility Classifications

ID	FACILITY NAME	ATC GRADE
ZAB	ALBUQUERQUE ARTCC	10
ZAN	ANCHORAGE ARTCC	9
ANC	ANCHORAGE INTL ATCT	8
ZTL	ATLANTA ARTCC	12
ATL	ATLANTA HARTSFIELD ATCT	12
ACY	ATLANTIC CITY ATCT	6
BWI	BALTIMORE INTL ATCT/TRACON	10
BTR	BATON ROUGE METRO ATCT	7
BOI	BOISE ATCT	8
ZBW	BOSTON ARTCC	11
BOS	BOSTON LOGAN ATCT	11
A90	BOSTON TRACON	11
BUR	BURBANK ATCT	7
BTV	BURLINGTON INTL ATCT	7
APA	CENTENNIAL ATCT	9
CMI	CHAMPAIGN ATCT	7
ZAU	CHICAGO ARTCC	12

(continued)

Table 13-3. (Continued)

ID	FACILITY NAME	ATC GRADE
ORD	CHICAGO O'HARE INTL ATCT	12
C90	CHICAGO TRACON	12
ZOB	CLEVELAND ARTCC	12
DFW	DALLAS/FORT WORTH ATCT	12
D10	DALLAS/FORT WORTH TRACON	12
DAB	DAYTONA BEACH ATCT	9
ZDV	DENVER ARTCC	10
DEN	DENVER INTL ATCT	11
D01	DENVER TRACON	11
D21	DETROIT TRACON	11
IAD	DULLES INTL ATCT	11
ELP	EL PASO INTL ATCT	8
ZFW	FORT WORTH ARTCC	12
GCN	GRAND CANYON ATCT	5
GFK	GRAND FORKS ATCT	7
BUF	GREATER BUFFALO INTL ATCT	8
ILG	GREATER WILMINGTON ATCT	6
GGG	GREGG COUNTY TRACAB	7
ZHN	HONOLULU	9
HNL	HONOLULU INTL ATCT	10
ZHU	HOUSTON ARTCC	11
IAH	HOUSTON INTL ATCT	10
I90	HOUSTON TRACON/ATCT	11
HSV	HUNTSVILLE ATCT	7
ZID	INDIANAPOLIS ARTCC	12
IND	INDIANAPOLIS INTL ATCT	9
ZJX	JACKSONVILLE ARTCC	11
BPT	JEFFERSON COUNTY ATCT	6
JFK	JOHN F. KENNEDY INTL ATCT	10
ZKC	KANSAS CITY ARTCC	11
MCI	KANSAS CITY INTL ATCT	9
LGA	LA GUARDIA ATCT	10
LAF	LAFAYETTE ATCT	5
LAN	LANSING ATCT	8
LAS	LAS VEGAS INTL ATCT	10

(*continued*)

Table 13-3. (Continued)

ID	FACILITY NAME	ATC GRADE
LGB	LONG BEACH ATCT	9
ISP	LONG ISLAND MACARTHUR ATCT	7
ZLA	LOS ANGELES ARTCC	11
LAX	LOS ANGELES INTL ATCT	12
ZME	MEMPHIS ARTCC	12
MEM	MEMPHIS INTL ATCT	10
ZMA	MIAMI ARTCC	11
MIA	MIAMI INTL ATCT	12
ZMP	MINNEAPOLIS ARTCC	11
MIC	MINNEAPOLIS CRYSTAL ATCT	6
MSP	MINNEAPOLIS ST. PAUL ATCT	11
PNE	N.E. PHILADELPHIA ATCT	6
ZNY	NEW YORK ARTCC	12
N90	NEW YORK TRACON	12
EWR	NEWARK INTL ATCT	10
ZOA	OAKLAND ARTCC	11
R90	OMAHA TRACON	8
MCO	ORLANDO INTL ATCT	11
PWK	PALWAUKEE ATCT	7
P31	PENSACOLA TRACON	9
PHL	PHILADELPHIA INTL ATCT	11
PHX	PHOENIX INTL ATCT	10
P50	PHOENIX TRACON	11
PIT	PITTSBURGH INTL ATCT	11
PWM	PORTLAND INTL ATCT	7
PVD	PROVIDENCE ATCT	9
ROW	ROSWELL ATCT	7
MCC	SACRAMENTO TRACON	10
ZLC	SALT LAKE ARTCC	10
SAN	SAN DIEGO ATCT	8
MYF	SAN DIEGO MONTGOMERY ATCT	7
SFO	SAN FRANCISCO INTL ATCT	10
ZSU	SAN JUAN	9
SJU	SAN JUAN INTL ATCT	7
ZSE	SEATTLE ARTCC	10

(continued)

Table 13-3. (Continued)

ID	FACILITY NAME	ATC GRADE
SEA	SEATTLE TACOMA INTL ATCT	9
S46	SEATTLE TACOMA TRACON	10
SCT	SO. CALIFORNIA TRACON	12
GEG	SPOKANE INTL ATCT	8
T75	ST. LOUIS TRACON	11
STL	ST. LOUIS/LAMBERT INTL ATCT	11
TEB	TETERBORO ATCT	7
TVC	TRAVERSE CITY ATCT	5
TUS	TUCSON INTL ATCT	7
VNY	VAN NUYS ATCT	9
VRB	VERO BEACH MUNICIPAL ATCT	6
ZDC	WASHINGTON ARTCC	12
HPN	WESTCHESTER CNTY ATCT	7
HOU	WILLIAM P. HOBBY ATCT	8
ILM	WILMINGTON ATCT	7

KEY TERMS

Air Traffic Selection and Training (AT-SAT)
aviation academics
Department of Transportation (DOT)
developmental controller
full performance level (FPL) controller
night differential
Office of the Secretary of Transportation (OST)
part-task training
performance verification
Radar Training Facility (RTF)
regional offices (ROs)
skills building

REVIEW QUESTIONS

1. What is the basic administrative structure of the FAA?
2. What are the responsibilities of the associate administrators?
3. What are the steps in becoming an air traffic controller?
4. How are controller salaries determined?

APPENDIX A

IFR Chart Legends

99252
GENERAL INFO

GENERAL INFORMATiON

This publication includes Instrument Approach Procedures (IAPs), Departure Procedures (DPs), and Standard Terminal Arrivals (STARs) for use by both civil and military aviation and is issued every 56 days.

STANDARD TERMINAL ARRIVALS AND DEPARTURE PROCEDURES

The use of the associated codified STAR/DP and transition identifiers are requested of users when filing flight plans via teletype and are required for users filing flight plans via computer interface. It must be noted that when filing a STAR/DP with a transition, the first three coded characters of the STAR and the last three coded characters of the DP are replaced by the transition code. Examples: ACTON SIX ARRIVAL, file (AQN.AQN6); ACTON SIX ARRIVAL, EDNAS TRANSITION, file (EDNAS.AQN6). FREEHOLD THREE DEPARTURE, file (FREH3.RBV), FREEHOLD THREE DEPARTURE, ELWOOD CITY TRANSITION, file (FREH3.EWC).

PILOT CONTROLLED AIRPORT LIGHTING SYSTEMS

Available pilot controlled lighting (PCL) systems are indicated as follows:

1. Approach lighting systems that bear a system identification are symbolized using negative symbology, e.g., ●, ●, ●
2. Approach lighting systems that do not bear a system identification are indicated with a negative "●" besides the name.

A star (★) indicates non-standard PCL, consult Directory/Supplement, e.g., ●★

To activate lights use frequency indicated in the communication section of the chart with a ● or the appropriate lighting system identification e.g., UNICOM 122.8 ●, ●, ●

KEY MIKE	FUNCTION
7 times within 5 seconds	Highest intensity available
5 times within 5 seconds	Medium or lower intensity (Lower REIL or REIL-off)
3 times within 5 seconds	Lowest intensity available (Lower REIL or REIL-off)

NE-3, 03 OCT 2002

CHART CURRENCY INFORMATION

FAA procedure amendment number → Amdt 11A 99365 ← Date of latest change
Orig 00365

The Chart Date indentifies the Julian date the chart was added to the volume or last revised for any reason. The first two digits indicate the year, the last three digits indicate the day of the year (001 to 365/6) in which the latest addition or change was first published.
The Procedure Amendment Number precedes the Chart Date, and changes any time instrument information (e.g., DH, MDA, approach routing, etc.) changes. Procedure changes also cause the Chart Date to change.

MISCELLANEOUS

★ Indicates a non-continuously operating facility, see A/FD or flight supplement.
# Indicates control tower temporarily closed UFN.
"Radar required" on the chart indicates that radar vectoring is required for the approach.
Distances in nautical miles (except visibility in statute miles and Runway Visual Range in hundreds of feet). Runway Dimensions in feet. Elevations in feet. Mean Sea Level (MSL). Ceilings in feet above airport elevation. Radials/bearings/headings/courses are magnetic. Horizontal Datum: Unless otherwise noted on the chart, all coordinates are referenced to North American Datum 1983 (NAD 83), which for charting purposes is considered equivalent to World Geodetic System 1984 (WGS 84).

GENERAL INFO
99252

F1

Figure A-1. Instrument approach procedure chart legend.

99252
GENERAL INFO

GENERAL INFORMATION

This publication includes Instrument Approach Procedures (IAPs), Departure Procedures (DPs), and Standard Terminal Arrivals (STARs) for use by both civil and military aviation and is issued every 56 days.

STANDARD TERMINAL ARRIVALS AND DEPARTURE PROCEDURES

The use of the associated codified STAR/DP and transition identifiers are requested of users when filing flight plans via teletype and are required for users filing flight plans via computer interface. It must be noted that when filing a STAR/DP with a transition, the first three coded characters of the STAR and the last three coded characters of the DP are replaced by the transition code. Examples: ACTON SIX ARRIVAL, file (AQN.AQN6); ACTON SIX ARRIVAL, EDNAS TRANSITION, file (EDNAS.AQN6). FREEHOLD THREE DEPARTURE, file (FREH3.RBV), FREEHOLD THREE DEPARTURE, ELWOOD CITY TRANSITION, file (FREH3.EWC).

PILOT CONTROLLED AIRPORT LIGHTING SYSTEMS

Available pilot controlled lighting (PCL) systems are indicated as follows:

1. Approach lighting systems that bear a system identification are symbolized using negative symbology, e.g., ●, ●, ●
2. Approach lighting systems that do not bear a system identification are indicated with a negative "●" besides the name.

A star (★) indicates non-standard PCL, consult Directory/Supplement, e.g., ●★

To activate lights use frequency indicated in the communication section of the chart with a ● or the appropriate lighting system identification e.g., UNICOM 122.8 ●, ●, ●

KEY MIKE	FUNCTION
7 times within 5 seconds	Highest intensity available
5 times within 5 seconds	Medium or lower intensity (Lower REIL or REIL-off)
3 times within 5 seconds	Lowest intensity available (Lower REIL or REIL-off)

EC-2, 03 OCT 2002

EC-2, 03 OCT 2002

CHART CURRENCY INFORMATION

FAA procedure amendment number → Amdt 11A 99365 ← Date of latest change
Orig 00365

The Chart Date indentifies the Julian date the chart was added to the volume or last revised for any reason. The first two digits indicate the year, the last three digits indicate the day of the year (001 to 365/6) in which the latest addition or change was first published.

The Procedure Amendment Number precedes the Chart Date, and changes any time instrument information (e.g., DH, MDA, approach routing, etc.) changes. Procedure changes also cause the Chart Date to change.

MISCELLANEOUS

★ Indicates a non-continuously operating facility, see A/FD or flight supplement.

# Indicates control tower temporarily closed UFN.

"Radar required" on the chart indicates that radar vectoring is required for the approach.

Distances in nautical miles (except visibility in statute miles and Runway Visual Range in hundreds of feet). Runway Dimensions in feet. Elevations in feet. Mean Sea Level (MSL). Ceilings in feet above airport elevation. Radials/bearings/headings/courses are magnetic. Horizontal Datum: Unless otherwise noted on the chart, all coordinates are referenced to North American Datum 1983 (NAD 83), which for charting purposes is considered equivalent to World Geodetic System 1984 (WGS 84).

GENERAL INFO
99252

Figure A-2. Instrument approach procedure chart legend.

00279

GENERAL INFO

ABBREVIATIONS

ADF ... Automatic Direction Finder
ALS ... Approach Light System
ALSF ... Approach Light System with Sequenced Flashing Lights
APP CON ... Approach Control
ARR ... Arrival
ASOS ... Automated Surface Observing System
ASR/PAR ... Published Radar Minimums at this Airport
ATIS ... Automatic Terminal Information Service
AWOS ... Automated Weather Observing System
AZ ... Azimuth
BC ... Back Course
C ... Circling
CAT ... Category
CCW ... Counter Clockwise
Chan ... Channel
CLNC DEL ... Clearance Delivery
CNF ... Computer Navigation Fix
CTAF ... Common Traffic Advisory Frequency
CW ... Clockwise
DH ... Decision Height
DME ... Distance Measuring Equipment
DR ... Dead Reckoning
ELEV ... Elevation
FAF ... Final Approach Fix
FM ... Fan Marker
FMS ... Flight Management System
GCO ... Ground Communications Outlet
GPI ... Ground Point of Interception
GPS ... Global Positioning System
GS ... Glide Slope
HAA ... Height above Airport
HAL ... Height above Landing
HAT ... Height above Touchdown
HIRL ... High Intensity Runway Lights
IAF ... Initial Approach Fix
ICAO ... International Civil Aviation Organization
IM ... Inner Marker
Intcp ... Intercept
INT ... Intersection
LDA ... Localizer Type Directional Aid
Ldg ... Landing
LDIN ... Lead in Light System
LIRL ... Low Intensity Runway Lights
LOC ... Localizer
LR ... Lead Radial. Provides at least 2 NM (Copter 1 NM) of lead to assist in turning onto the intermediate/final course.
MALS ... Medium Intensity Approach Light System
MALSR ... Medium Intensity Approach Light System with RAIL
MAP ... Missed Approach Point
MDA ... Minimum Descent Altitude
MIRL ... Medium Intensity Runway Lights
MLS ... Microwave Landing System
MM ... Middle Marker
N/A ... Not Applicable
NA ... Not Authorized
NDB ... Non-directional Radio Beacon
NM ... Nautical Mile
NoPT ... No Procedure Turn Required (Procedure Turn shall not be executed without ATC clearance)
ODALS ... Omnidirectional Approach Light System
OM ... Outer Marker
R ... Radial
RA ... Radio Altimeter setting height
RAIL ... Runway Alignment Indicator Lights
RBn ... Radio Beacon
RCLS ... Runway Centerline Light System
REIL ... Runway End Identifier Lights
RNAV ... Area Navigation
RNP ... Required Navigation Performance
RPI ... Runway Point of Intercept(ion)
RRL ... Runway Remaining Lights
Rwy ... Runway
RVR ... Runway Visual Range
S ... Straight-in
SALS ... Short Approach Light System
SSALR ... Simplified Short Approach Light System with RAIL
SDF ... Simplified Directional Facility
TA ... Transition Altitude
TAA ... Terminal Arrival Area
TAC ... TACAN
TCH ... Threshold Crossing Height (height in feet Above Ground level)
TDZ ... Touchdown Zone
TDZE ... Touchdown Zone Elevation
TDZ/CL ... Touchdown Zone and Runway Centerline Lighting
TDZL ... Touchdown Zone Lights
TLv ... Transition Level
VASI ... Visual Approach Slope Indicator
VDP ... Visual Descent Point
VGSI ... Visual Glide Slope Indicator
WP/WPT ... Waypoint (RNAV)
X ... Radar Only Frequency

EC-2, 03 OCT 2002

EC-2, 03 OCT 2002

GENERAL INFO
00279

F2

Figure A-3. Instrument approach procedure chart legend.

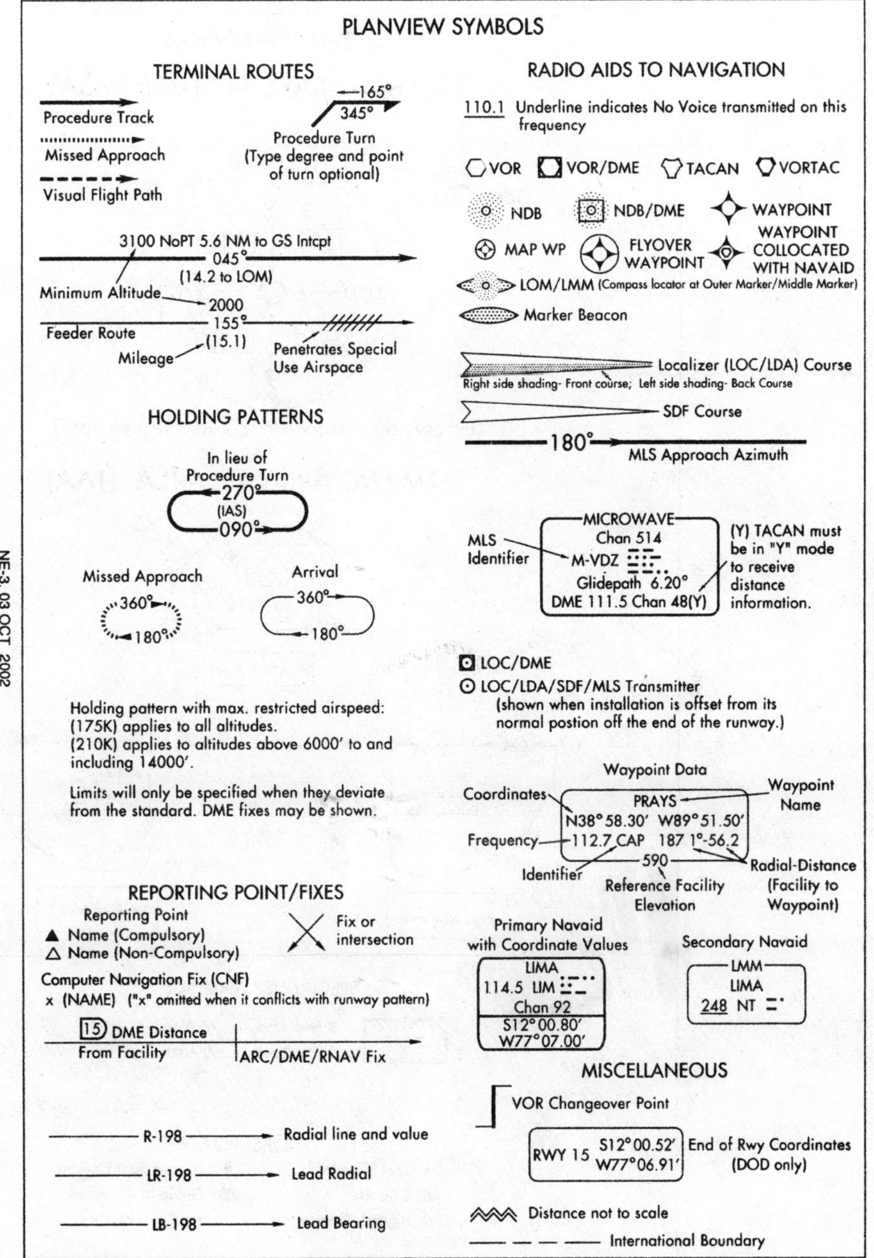

Figure A-4. Instrument approach procedure chart legend.

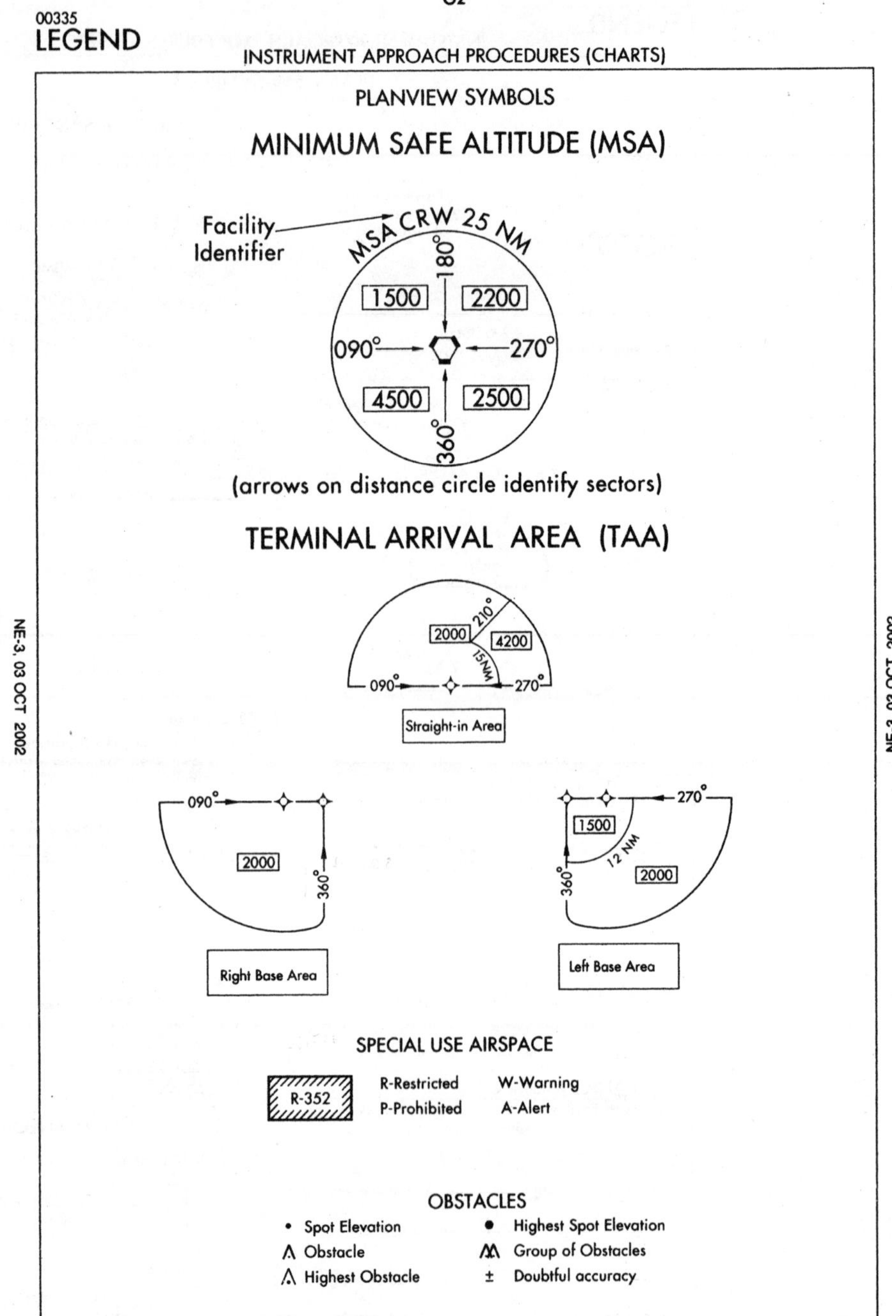

Figure A-5. Instrument approach procedure chart legend.

95145
LEGEND

INSTRUMENT APPROACH PROCEDURES (CHARTS)

PROFILE

2400 — 320° — Remain within 10 NM — 125° — Teardrop Turn

Procedure Turn — 307° — LOM

2400 — 127° — Glide Slope Altitude at Outer Marker/FAF — 2156

FAF (non-precision approaches) — ILS Glide Slope — Missed Approach Point — Missed Approach Track

Glide Slope → GS 3.00°

Threshold Crossing Height → TCH 100

Glide Slope Intercept Altitude → 2400

Airport Profiles (Primary) (Secondary)

DESCENT FROM HOLDING PATTERN

VOR — 127° — 307° — 1600 — 307° — 1300

VOR — 127° — 307° — 1600 — 307°

127° — MAP WPT — Final Approach Angle for Vertical Path Computers (RNAV Descent) — 3.02°

NE-3, 03 OCT 2002

MLS APPROACH

VOR — 360° — 3300 — 180° — MLS 00°R/L

Glidepath 3.0°
TCH 50

M-AJE 6.5 — 3250 — 3300 — Glidepath Altitude at FAF — Final Approach Fix (FAF) — M-AJE 2.2 — MLS Glidepath

NE-3, 03 OCT 2002

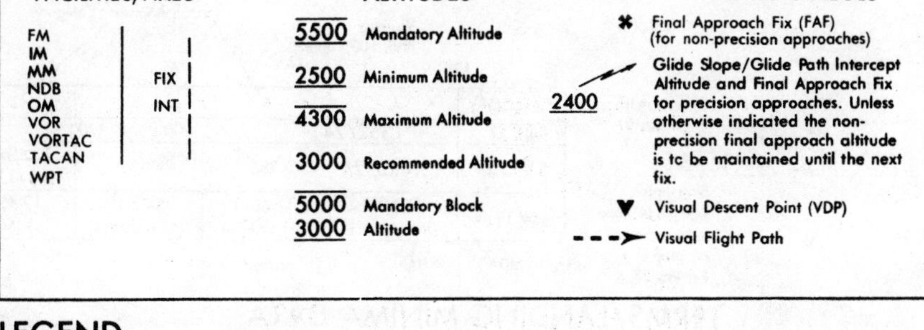

FACILITIES/FIXES

FM
IM
MM
NDB
OM
VOR
VORTAC
TACAN
WPT

FIX
INT

ALTITUDES

5500 Mandatory Altitude
2500 Minimum Altitude
4300 Maximum Altitude
3000 Recommended Altitude
5000 / 3000 Mandatory Block Altitude

PROFILE SYMBOLS

✱ Final Approach Fix (FAF) (for non-precision approaches)

2400 Glide Slope/Glide Path Intercept Altitude and Final Approach Fix for precision approaches. Unless otherwise indicated the non-precision final approach altitude is to be maintained until the next fix.

▼ Visual Descent Point (VDP)

- - -➤ Visual Flight Path

LEGEND
95145

H1

Figure A-6. Instrument approach procedure chart legend.

00111

TERMS/LANDING MINIMA DATA

IFR LANDING MINIMA

The United States Standard for Terminal Instrument Procedures (TERPS) is the approved criteria for formulating instrument approach procedures. Landing minima are established for six aircraft approach categories (ABCDE and COPTER). In the absence of COPTER MINIMA, helicopters may use the CAT A minimums of other procedures. The standard format for RNAV minima and landing minima portrayal follows:

RNAV MINIMA

CATEGORY	A	B	C	D
GLS PA DA	1382/24 200 (200-½)			
LNAV/ DA VNAV	1500/24 318 (400-½)			1500/40 318 (400-¾)
LNAV MDA	1700/24 518 (600-½)		1700/50 518 (600-1)	1700/60 518 (600-1¼)
CIRCLING	1760-1 578 (600-1)		1760-1½ 578 (600-1½)	1760-2 578 (600-2)

RNAV minimums are dependent on navigation equipment capability, as stated in the applicable AFM or AFMS and as outlined below.

GLS (GLobal Navigation System (GNSS) Landing System)

Must have WAAS (Wide Area Augmentation System) equipment approved for precise approach.
Note: "PA" indicates that the runway environment, i.e., runway markings, runway lights, parallel taxiway, etc., meets precision approach requirements. If the GLS minimums line does not contain "PA", then the runway environment does not support precision requirements.

EC-2, 03 OCT 2002

LNAV/VNAV (Lateral Navigation/Vertical Navigation)

Must have WAAS equipment approved for precision approach, or RNP-0.3 system based on GPS or DME/DME, with an IFR approach approved Baro-VNAV system. Other RNAV approach systems require special approval. Use of Baro-VNAV systems is limited by temperature, i.e., "Baro-VNAV NA below -20° C(-4° F)". (Not applicable if chart is annotated "Baro-VNAV NA".)
NOTE: DME/DME based RNP-0.3 systems may be used only when a chart note indicates DME/DME availability, for example, "DME/DME RNP-0.3 Authorized." Specific DME facilities may be required, for example: "DME/DME RNP-0.3 Authorized. ABC, XYZ required."

EC-2, 03 OCT 2002

LNAV (Lateral Navigation)

Must have IFR approach approved WAAS, GPS, GPS based FMS systems, or RNP-0.3 systems based on GPS or DME/DME. Other RNAV approach systems require special approval.
NOTE: DME/DME based RNP-0.3 systems may be used only when a chart note indicates DME/DME availability, for example, "DME/DME RNP-0.3 Authorized." Specific DME facilities may be required, for example: "DME/DME RNP-0.3 Authorized. ABC, XYZ required."

LANDING MINIMA FORMAT

In this example airport elevation is 1179, and runway touchdown zone elevation is 1152.

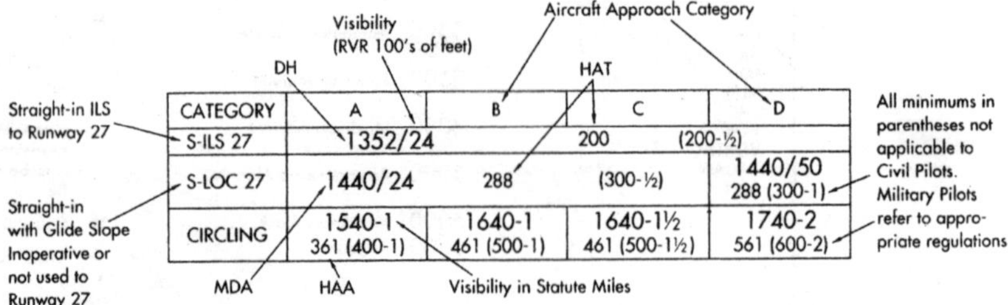

TERMS/LANDING MINIMA DATA

A1

Figure A-7. Instrument approach procedure chart legend.

A2

02052

TERMS/LANDING MINIMA DATA

COPTER MINIMA ONLY

CATEGORY	COPTER
H-176°	680-½ 363 (400-½)

Copter Approach Direction

Height of MDA/DH Above Landing Area (HAL)

No circling minimums are provided

RADAR MINIMA

PAR (c)	10 2.5°/42/1000	ABCDE	**195**/16	100	(100-¼)					
(d)	28 2.5°/48/1068	ABCDE	**187**/16	100	(100-¼)					
ASR	10	ABC	**560**/40	463	(500-¾)	D	**560**/50	463	(500-1)
		E	**580**/60	463	(500-1¼)				
	28	AB	**600**/50	513	(600-1)	C	**600**/60	513	(600-1¼)
		DE	**600**-1½	513	(600-1½)				
CIR (b)	10	AB	**560**-1¼	463	(500-1¼)	C	**560**-1½	463	(500-1½)
	28	AB	**600**-1¼	503	(600-1¼)	C	**600**-1½	503	(600-1½)
	10, 28	DE	**660**-2	563	(600-2)				

Visibility (RVR 100's of feet)

All minimums in parentheses not applicable to Civil Pilots. Military Pilots refer to appropriate regulations.

Visibility in Statute Miles

Radar Minima:

1. Minima shown are the lowest permitted by established criteria. Pilots should consult applicable directives for their category of aircraft.
2. The circling MDA and weather minima to be used are those for the runway to which the final approach is flown - not the landing runway. In the above RADAR MINIMA example, a category C aircraft flying a radar approach to runway 10, circling to land on runway 28, must use an MDA of 560 feet with weather minima of 500-1½ .

NE-3, 03 OCT 2002

NE-3, 03 OCT 2002

▲ Alternate Minimums not standard. Civil users refer to tabulation. USA/USN/USAF pilots refer to appropriate regulations.

▲ NA Alternate minimums are Not Authorized due to unmonitored facility or absence of weather reporting service.

▼ Take-off Minimums not standard and/or Departure Procedures are published. Refer to tabulation.

AIRCRAFT APPROACH CATEGORIES

Speeds are based on 1.3 times the stall speed in the landing configuration of maximum gross landing weight. An aircraft shall fit in only one category. If it is necessary to maneuver at speeds in excess of the upper limit of a speed range for a category, the minimums for the next higher category should be used. For example, an aircraft which falls in Category A, but is circling to land at a speed in excess of 91 knots, should use the approach Category B minimums when circling to land. See following category limits:

MANEUVERING TABLE

Approach Category	A	B	C	D	E
Speed (Knots)	0-90	91-120	121-140	141-165	Abv 165

RVR/ Meteorological Visibility Comparable Values

The following table shall be used for converting RVR to meteorological visibility when RVR is not reported for the runway of intended operation. Adjustments of landing minima may be required - see Inoperative Components Table.

RVR (feet)	Visibility (statute miles)	RVR (feet)	Visibility (statute miles)
1600	¼	4000	¾
1800	½	4500	⅞
2000	½	5000	1
2400	½	6000	1¼
3200	⅝		

TERMS/LANDING MINIMA DATA

Figure A-8. Instrument approach procedure chart legend.

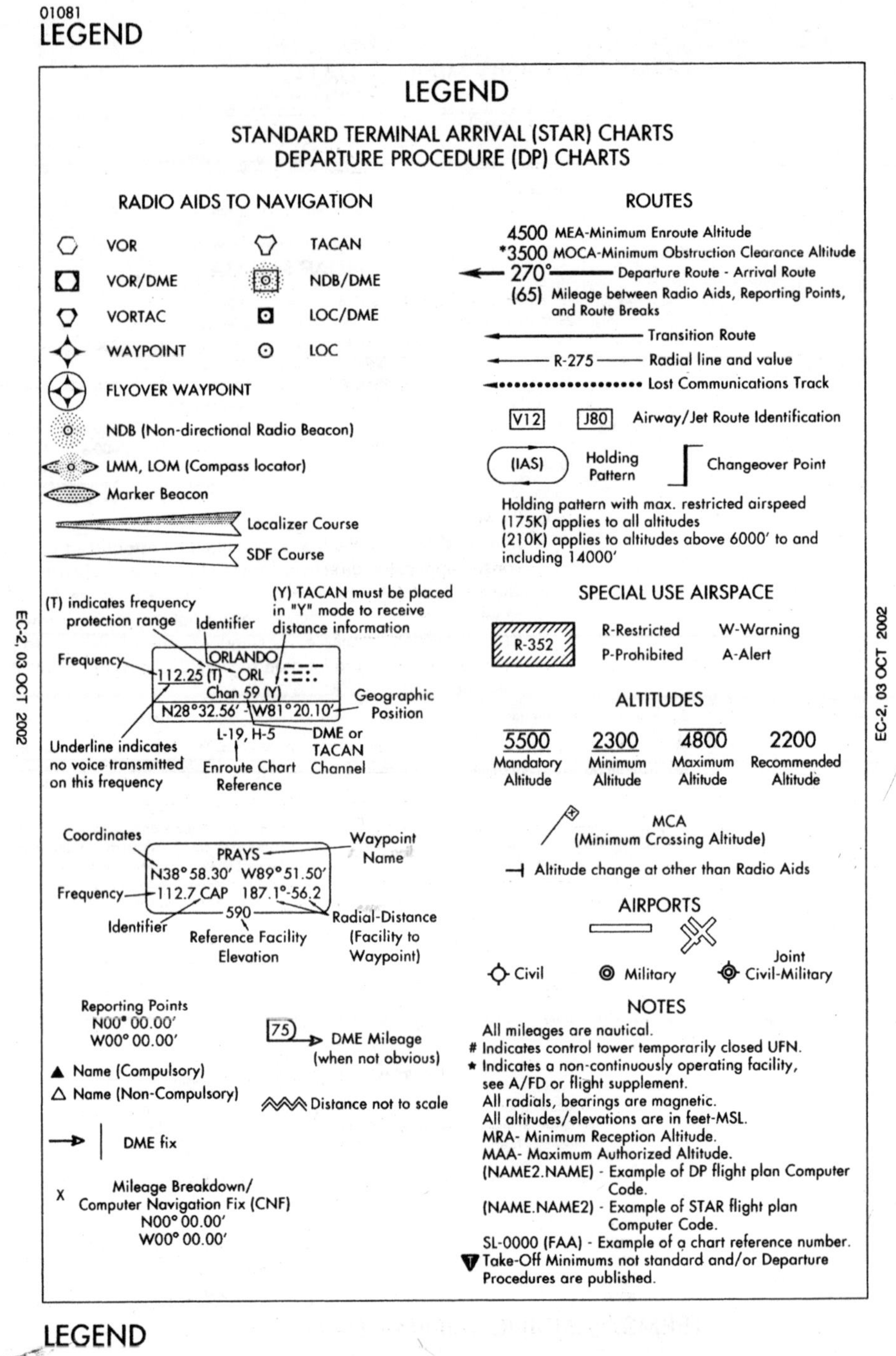

01081
LEGEND

LEGEND

STANDARD TERMINAL ARRIVAL (STAR) CHARTS
DEPARTURE PROCEDURE (DP) CHARTS

RADIO AIDS TO NAVIGATION

VOR
TACAN
VOR/DME
NDB/DME
VORTAC
LOC/DME
WAYPOINT
LOC
FLYOVER WAYPOINT
NDB (Non-directional Radio Beacon)
LMM, LOM (Compass locator)
Marker Beacon
Localizer Course
SDF Course

(T) indicates frequency protection range
Identifier
(Y) TACAN must be placed in "Y" mode to receive distance information
Frequency
ORLANDO
112.25 (T) ORL
Chan 59 (Y)
N28°32.56' - W81°20.10'
Geographic Position
L-19, H-5
Underline indicates no voice transmitted on this frequency
Enroute Chart Reference
DME or TACAN Channel

Coordinates
Waypoint Name
PRAYS
N38°58.30' W89°51.50'
Frequency
112.7 CAP 187.1°-56.2
590
Identifier
Reference Facility Elevation
Radial-Distance (Facility to Waypoint)

Reporting Points
N00° 00.00'
W00° 00.00'
▲ Name (Compulsory)
△ Name (Non-Compulsory)
75 DME Mileage (when not obvious)
Distance not to scale
DME fix
X Mileage Breakdown/ Computer Navigation Fix (CNF)
N00° 00.00'
W00° 00.00'

ROUTES

4500 MEA-Minimum Enroute Altitude
*3500 MOCA-Minimum Obstruction Clearance Altitude
270° Departure Route - Arrival Route
(65) Mileage between Radio Aids, Reporting Points, and Route Breaks
Transition Route
R-275 Radial line and value
Lost Communications Track
V12 J80 Airway/Jet Route Identification
(IAS) Holding Pattern
Changeover Point
Holding pattern with max. restricted airspeed (175K) applies to all altitudes (210K) applies to altitudes above 6000' to and including 14000'

SPECIAL USE AIRSPACE

R-352
R-Restricted W-Warning
P-Prohibited A-Alert

ALTITUDES

5500 Mandatory Altitude
2300 Minimum Altitude
4800 Maximum Altitude
2200 Recommended Altitude

MCA (Minimum Crossing Altitude)
Altitude change at other than Radio Aids

AIRPORTS

Civil
Military
Joint Civil-Military

NOTES

All mileages are nautical.
# Indicates control tower temporarily closed UFN.
* Indicates a non-continuously operating facility, see A/FD or flight supplement.
All radials, bearings are magnetic.
All altitudes/elevations are in feet-MSL.
MRA- Minimum Reception Altitude.
MAA- Maximum Authorized Altitude.
(NAME2.NAME) - Example of DP flight plan Computer Code.
(NAME.NAME2) - Example of STAR flight plan Computer Code.
SL-0000 (FAA) - Example of a chart reference number.
Take-Off Minimums not standard and/or Departure Procedures are published.

EC-2, 03 OCT 2002

EC-2, 03 OCT 2002

LEGEND

Figure A-9. Instrument approach procedure chart legend.

95201

LEGEND

INSTRUMENT APPROACH PROCEDURES (CHARTS)

AIRPORT DIAGRAM/AIRPORT SKETCH

Runways

Hard Surface — Other Than Hard Surface — Stopways, Taxiways, Parking Areas — Displaced Threshold

Closed Runway — Closed Taxiway — Under Construction — Metal Surface — Runway Centerline Lighting

ARRESTING GEAR: Specific arresting gear systems; e.g., BAK-12, MA-1A etc., shown on airport diagrams, not applicable to Civil Pilots. Military Pilots Refer to Appropriate DOD Publications.

uni-directional — bi-directional — Jet Barrier

REFERENCE FEATURES

Buildings........ ■

Tanks........ ●

Obstruction........ ⋀

Airport Beacon #........ ☆

Runway Radar Reflectors........

Control Tower #........ ▪

\# When Control Tower and Rotating Beacon are co-located, Beacon symbol will be used and further identified as TWR.

Runway length depicted is the physical length of the runway (end-to-end, including displaced thresholds if any) but excluding areas designated as stopways. Where a displaced threshold is shown and/or part of the runway is otherwise not available for landing, an annotation is added to indicate the landing length of the runway; e.g., RWY 13 ldg 5000'.

Helicopter Alighting Areas

Negative Symbols used to identify Copter Procedures landing point........

Runway TDZ elevation........TDZE 123

←0.3% DOWN

Runway Slope........0.8% UP→

(shown when runway slope exceeds 0.3%)

NOTE:
Runway Slope measured to midpoint on runways 8000 feet or longer.

■ U.S. Navy Optical Landing System (OLS) "OLS" location is shown because of its height of approximately 7 feet and proximity to edge of runway may create an obstruction for some types of aircraft.

Approach light symbols are shown in the Flight Information Handbook.

Airport diagram scales are variable.

True/magnetic North orientation may vary from diagram to diagram.

Coordinate values are shown in 1 or ½ minute increments. They are further broken down into 6 second ticks, within each 1 minute increment.

Positional accuracy within ±600 feet unless otherwise noted on the chart.

NOTE:
All new and revised airport diagrams are shown referenced to the World Geodetic System (WGS) (noted on appropriate diagram), and may not be compatible with local coordinates published in FLIP. (Foreign Only)

Runway Weight Bearing Capacity/or PCN Pavement Classification Number is shown as a codified expression.
Refer to the appropriate Supplement/Directory for applicable codes, e.g.,
RWY 14-32 S75, T185, ST175, TT325
PCN 80 F/D/X/U

SCOPE

Airport diagrams are specifically designed to assist in the movement of ground traffic at locations with complex runway/taxiway configurations and provide information for updating Computer Based Navigation Systems (I.E., INS, GPS) aboard aircraft. Airport diagrams are not intended to be used for approach and landing or departure operations. For revisions to Airport Diagrams: Consult FAA Order 7910.4B.

NE-3, 03 OCT 2002

NE-3, 03 OCT 2002

LEGEND
95201

K1

Figure A-10. *Instrument approach procedure chart legend.*

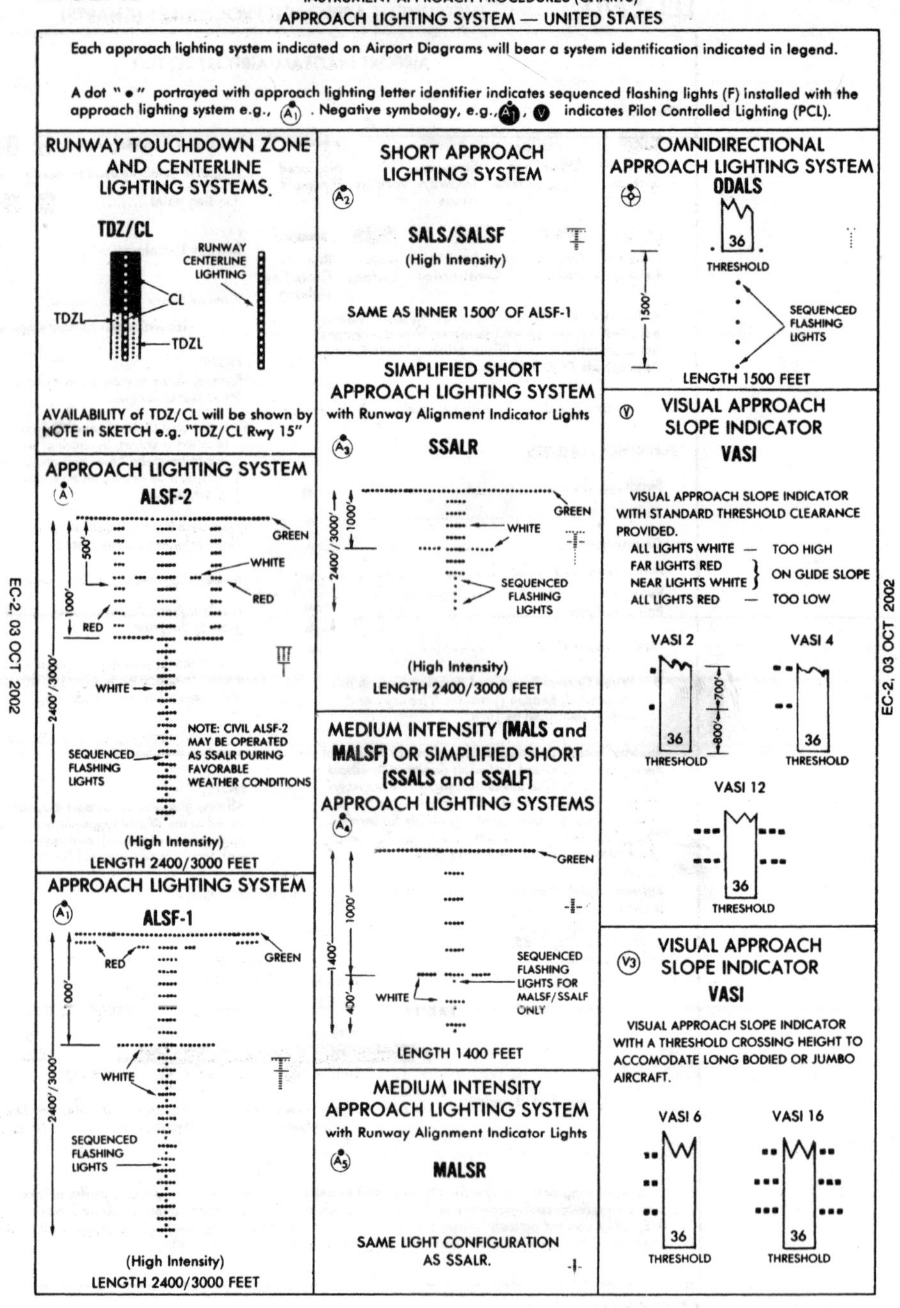

Figure A-11. Instrument approach procedure chart legend.

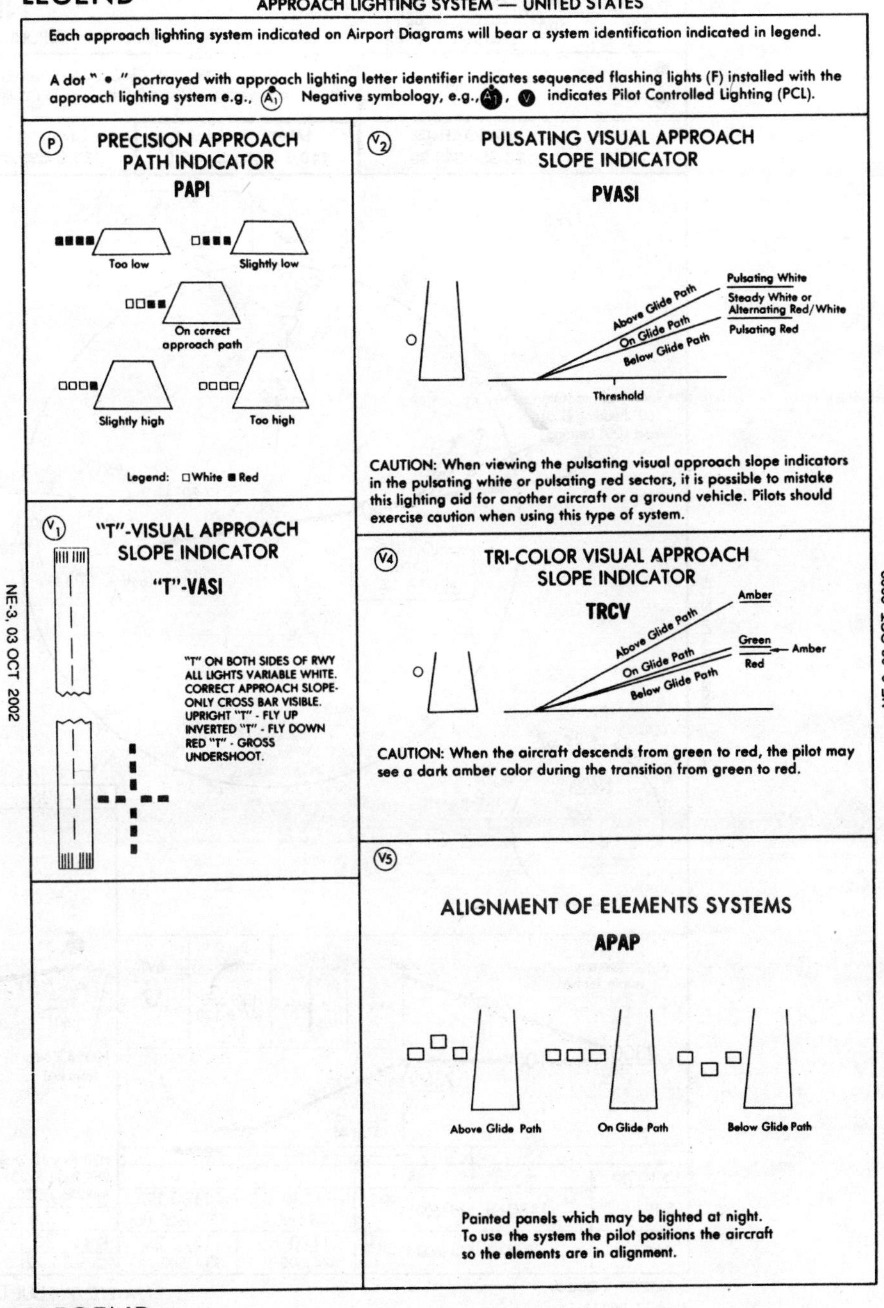

Figure A-12. Instrument approach procedure chart legend.

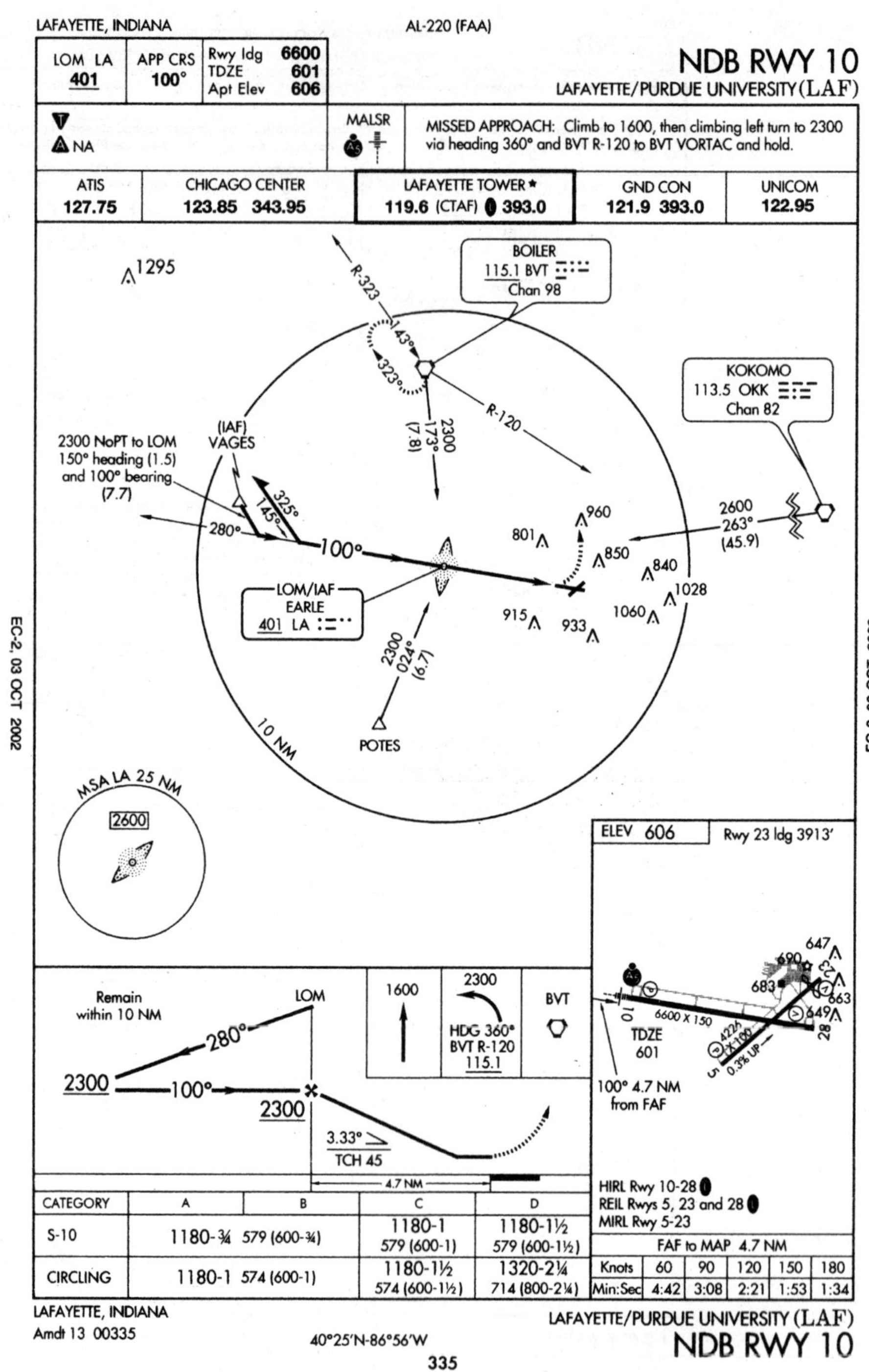

CATEGORY	A	B	C	D
S-10	1180-¾ 579 (600-¾)		1180-1 579 (600-1)	1180-1½ 579 (600-1½)
CIRCLING	1180-1 574 (600-1)		1180-1½ 574 (600-1½)	1320-2¼ 714 (800-2¼)

FAF to MAP 4.7 NM

Knots	60	90	120	150	180
Min:Sec	4:42	3:08	2:21	1:53	1:34

LAFAYETTE, INDIANA
Amdt 13 00335
40°25'N-86°56'W
LAFAYETTE/PURDUE UNIVERSITY (LAF)
NDB RWY 10
335

Figure A-13. NDB runway 10 approach at Lafayette/Purdue University airport.

AIRPORTS	
AIRPORT DATA	Airports/Seaplane Bases shown in BLUE have an approved Low Altitude Instrument Approach Procedure published. Those shown in DARK BLUE have an approved DOD Low Altitude Instrument Approach Procedure and/or DOD RADAR MINIMA published in DOD FLIPS, Alaska Supplement or Alaska Terminal. Airports/Seaplane Bases shown in BROWN do not have a published Instrument Approach Procedure.
LANDPLANE - CIVIL Refueling and repair facilities for normal traffic.	 Douglas Muni
LANDPLANE - JOINT CIVIL AND MILITARY Refueling and repair facilities for normal traffic.	 Charleston AFB/Intl
LANDPLANE- MILITARY Refueling and repair facilities for normal traffic.	MCAF Quantico
SEAPLANE-CIVIL Refueling and repair facilities for normal traffic.	 North Bay
SEAPLANE- JOINT CIVIL AND MILITARY Refueling and repair facilities for normal traffic.	 NAS Patuxent River SPB /Trapnell Naples Muni
SEAPLANE- MILITARY Refueling and repair facilities for normal traffic.	 NAS Corpus Christi SPB
HELIPORT	 Allen AHP

Figure A-14. En route low-altitude chart legend.

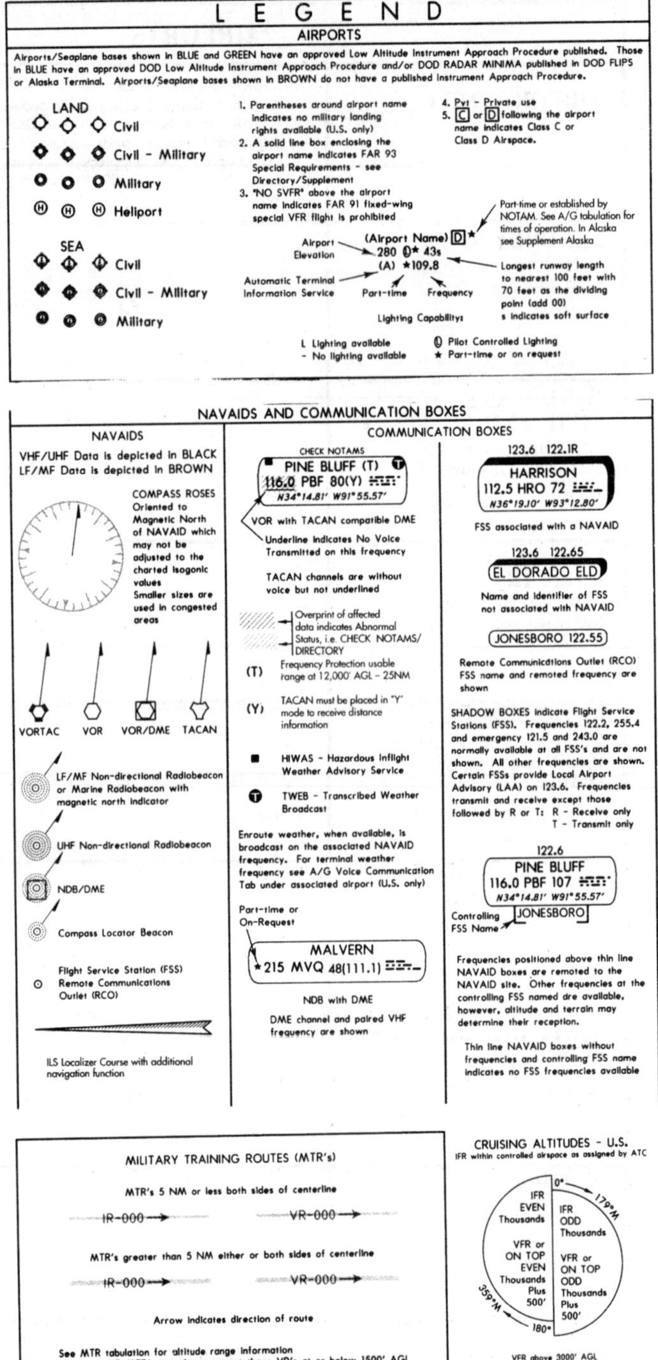

Figure A-15. Instrument navigation chart legend.

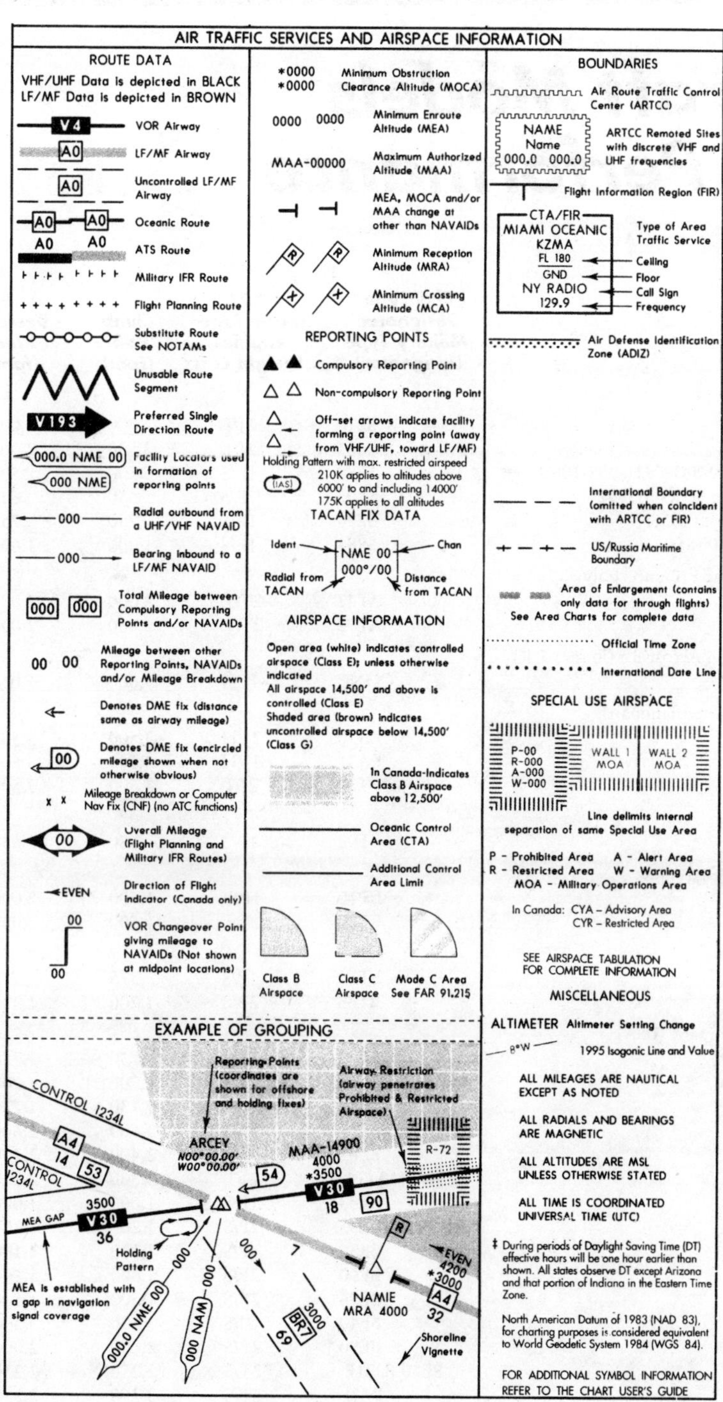

Figure A-16. Instrument navigation chart legend.

APPENDIX B

Aircraft Models and Performance

Model	Civil Type Designator (Military Type Designator)	Number and Type Engines/ Weight Class	Climb Rate (fpm)	Descent Rate (fpm)	SOIR[1] Group	SRS[2] Category
Aerospatiale (France)						
Caravelle	S210	2J/L	2,300	2,000	4	III
Corvette SN601	S601	2J/L	2,500	2,000	3	III
Rallye Club MS880/881/883/100S/100ST	S880	1P/S	500	500	3	I
Rallye Commodore Series MS892	S892	1P/S	630	630	2	I
Rallye Minerva Series MS894	S894	1P/S	750	750	2	I
Super Rallye MS885/886	S885	1P/S	1,000	1,000	2	I
Aerospatiale/Aeritalia (France/Italy)						
ATR 42-200/300	AT42	2T/L	2,000	2,000	3	III
ATR 72	AT72	2T/L	2,000	2,000	3	III
Aerospatiale/British Aerospace (France/UK)						
Concorde	CONC	4J/H	5,000	5,000	5	III
Airbus Industries (International)						
A300	EA30	2J/H	3,500	3,500	4	III
A310	EA31	2J/H	3,500	3,500	4	III
A320	EA32	2J/L	3,500	3,500	4	III
Alon Inc. (USA)						
Aircoupe A2	F02	1P/S	630	630	2	I
Avions Mudry Et Cie (France)						
Cap 10/B	CP10	1P/L	1,500	2,000	2	I
Cap 20LS-200	CP20	1P/L	1,500	2,000	2	I
Beagle Aircraft (UK)						
B.121 Pup Series	BT10	1P/S	575	750	1	I
B.206 Beagle Series	BT6S	2P/S	1,200	1,300	2	II
Beech Aircraft Company (USA)						
Airliner/Model 99	BE99	2T/S	1,750	1,750	3	II
Baron 55/Chochise	BE55 (T42)	2P/S	1,700	1,700	2	II
Baron 58	BE58	2P/S	1,730	1,730	2	II
Beech F90	BE9F	2T/S	2,600	2,600	3	II
Beech 1900/C	BE02	2T/L	2,400	2,400	4	III
Beechjet 400	BE40/BJ40 (T1)	L2J/L	3,300	2,200	3	III
Bonanza 33	BE33	1P/S	1,000	1,000	2	I
Bonanza 35	BE35	1P/S	1,200	1,200	2	I
Bonanza 36	BE36	1P/S	1,100	1,100	2	I
Duke 60	BE60	2P/S	1,600	1,600	4	II
Duchess 76	BE76	2P/S	1,500	1,500	2	II
Jet 400	BE40	2J/S	3,300	3,300	3	III
King Air C90, E90	BE90	2T/S	2,000	2,000	3	II
King Air 100/A	BE10 (U21F)	2T/S	2,250	2,250	4	II
King Air 100/B	BE1B	2T/S	2,250	2,250	4	II
Mentor	BE45 (T34)	1P/S	1,150	1,150	2	I
Queen Air 65/A65/70	BE65	2P/S	1,375	1,375	2	II

Model	Civil Type Designator (Military Type Designator)	Number and Type Engines/ Weight Class	Climb Rate (fpm)	Descent Rate (fpm)	SOIR[1] Group	SRS[2] Category
Queen Air 80	BE80	2P/S	1,275	1,275	2	II
Seminole	(U8)	2P/S	1,200	1,200	2	II
Sierra 24	BE24	1P/S	1,000	1,000	2	I
Skipper 77	BE77	1P/S	750	750	2	I
Sport 19	BE19	1P/S	680	680	2	I
Stagger Wing 17	BE17	1P/S	1,375	1,375	2	I
Starship, Model 2000	BEST	2T/L	2,650	2,650	3	III
Sundowner 23, Musketeer 23	BE23	1P/S	740	800	2	I
Super H18	BE8S	2P/S	1,400	1,400	2	II
Super King Air 200, Huron	BE20 (C12)	2T/S	2,450	2,500	4	II
Super King Air 200HDC	BE2H	2T/S	2,450	2,500	3	II
Super King Air 300	BE30	2T/L	3,000	3,000	4	III
Super King Air 300LW	BE3L	2T/S	3,000	3,000	4	II
Super King Air 350	BE3B	2T/L	3,000	3,000	4	III
Travelair 95	BE95	2P/S	1,250	1,250	2	II
Twin Beech 18	BE18 (C45)	2P/S	1,400	1,000	2	II
Twin Bonanza 50	BE50	2P/S	1,600	1,600	2	II
Ute	(U21)	2T/S	2,000	2,000	3	II
Bellanca Aircraft (Includes Aeronca, Champion, Downer, Northern) (USA)						
Aeronca Champion	AR58	1P/S	500	500	2	I
Aeronca Chief/Super Chief	AR11	1P/S	500	500	2	I
Aeronca Sedan	AR15	1P/S	500	500	2	I
Challenger	CH8	1P/S	1,150	1,000	2	I
Champion	CH5	1P/S	750	750	2	I
Champion Citabria	CH10	1P/S	1,120	1,120	2	I
Champion Citabria 7ECA	CH9	1P/S	725	725	2	I
Champion Lancer 402	CH40	2P/S	650	1000	2	II
Cruisair Sr., Cruisemaster 14-19	BL14	1P/S	1,030	1,030	2	I
Decathlon	BL30	1P/S	900	1,000	2	I
Model 8GCBC Scout	BL28	1P/S	1,080	1,080	2	I
Model 17-30A, Super Viking 300A	BL26	1P/S	1,210	1,210	2	I
Turbo Viking	BL31	1P/S	1,000	1,200	2	I
Boeing Company (USA)						
707-100/200 Series, VC-137	B707 (V137)	4J/L	3,500	3,500	5	III
707-300/400 Series	B07H	4J/H	3,500	3,500	5	III
720/B	B720	4J/L	3,000	3,000	5	III
727 (all series)	B727 (C22)	3J/L	4,500	4,500	4	III
737/200 Series	B737 (T43)	2J/L	3,000	3,000	4	III
737-300/500 Series	B73S	2J/L	5,500	3,500	4	III
737-400 Series	B73F	2J/L	6,500	3,500	5	III
747-100/200/300 Series	B747 (E4/VC25)	4J/H	3,000	3,000	5A	III
747-400 Series	B74F	4J/H	3,000	3,000	5	III
747SP/SUD	B74S	4J/H	3,000	3,000	5	III
757 (all series)	B757	2J/L	2,500	2,500	4	III
767 (all series)	B767	2J/H	3,500	3,500	5	III
E3 Sentry	EC-137 (E3A)	4J/H	3,500	4,000	5	III
Stearman	B75	1P/S	840	840	2	I
Stratofortress	(B52)	8J/H	3,000	3,000	5	III
Stratofreighter	(KC97)	4P/L	2,500	3,000	4	III
Stratolifter B717	(C135)	4J/L	2,000	2,000	5	III
Stratotanker KC-135	(KC35)	4J/L	2,500	3,000	5	III
Stratotanker KC-135E	(KE35)	4J/L	5,000	3,000	5	III
Stratotanker KC-135R	(KR35)	4J/H	5,000	3,000	5	III
British Aerospace (BAe) (UK)						
Advance Turboprop (ATP)	BATP	2T/L	3,000	3,000	3	III

Model	Civil Type Designator (Military Type Designator)	Number and Type Engines/ Weight Class	Climb Rate (fpm)	Descent Rate (fpm)	SOIR[1] Group	SRS[2] Category
BAC One-Eleven	BA11	2J/L	2,400	2,400	4	III
BAe 146-100/200 Series	BA46	4J/L	3,500	3,500	4	III
BAe HS 125 Series 400/600/700	HS25	2J/L	2,500	2,000	3	III
BAe HS 125 Series 800/1000	HS28	2J/L	3,000	4,000	3	III
BAe Harrier	(AV8)	1J/L	5,000	8,000	4	III
BAe HS 748 Series 2	HS74	2T/L	2,500	2,000	4	III
BAe Jetstream 31	B31	2T/L	2,300	2,300	3	III
Trident	HS21	3J/L	3,000	3,000	4	III
VC-10	BA10 (VC10)	4J/H	1,900	2,000	5	III
Viscount	VC7	4T/L	1,200	1,500	4	III
Vulcan	(VLCN)	4J/H	4,500	3,000	5	III
Canadair Bombardier Ltd. (Canada)						
CL600/610 Challenger	CL60	2J/L	5,000	6,000	4	III
Cessna Aircraft Company (USA)						
AGWagon/AGTruck/AGHusky 188	C188	1P/S	1,000	1,000	2	I
Bird Dog 305/321	C305 (O1)	1P/S	1,150	1,150	2	I
Caravan I 208A	C208	1T/S	1,400	1,400	2	I
Cardinal 177	C177	1P/S	850	1,000	2	I
Centurion/II	C210	1P/S	900	1,000	2	I
Cessna 120	C120	1P/S	640	640	2	I
Cessna 140	C14	1P/S	640	640	2	I
Cessna 150	C150	1P/S	670	1,000	2	I
Cessna 152	C152	1P/S	750	1,000	2	I
Cessna 170	C170	1P/S	690	1,000	2	I
Cessna 190	C190	1P/S	1,090	1,090	2	I
Cessna 195	C195	1P/S	1,200	1,200	2	I
Cessna 310	C310 (U3)	2P/S	2,800	2,000	2	II
Cessna 318	(T37*)	2J/S	3,000	3,000	2	III
Cessna 335	C335	2P/S	2,200	2,000	2	II
Cessna 340	C340	2P/S	2,900	2,000	2	II
Cessna 401	C401	2P/S	2,700	2,000	2	II
Cessna 402	C402	2P/S	2,500	2,000	2	II
Cessna 411	C411	2P/S	2,800	2,000	2	II
Chancellor 414	C414	2P/S	2,300	2,000	2	II
Citation I	C500	2J/S	3,100	3,500	3	III
Citation I-SP	C501	2J/S	4,300	3,000	3	III
Citation II/-S2	C550 (T47)	2J/L	5,300	3,000	3	III
Citation II-SP	C551	2J/S	5,300	3,000	3	III
Citation III	C650	2J/L	3,900	4,000	3	III
Citation V	C560	2J/L	6,000	3,500	4	III
Conquest/II-441	C441	2T/S	4,200	3,000	2	II
Corsair/Conquest I-425	C425	2T/S	3,500	2,500	2	II
Crusader 303	C303	2P/S	3,500	3,000	2	II
Dragonfly 318E	(A37*)	2J/S	3,370	3,000	2	III
Golden Eagle 421	C421	2P/S	3,200	2,000	2	II
Skyhawk 172/Cutlass/Mescalero	C172 (T41)	1P/S	650	1,000	2	I
Skyknight 320	C320	2P/S	2,900	2,000	2	II
Skylane 182/RG, Turbo Skylane RG	C182	1P/S	890	1,000	2	I
Skylark 175	C175	1P/S	850	1,000	2	I
Skymaster 336	C336	2P/S	1,340	1,340	2	II
Skywagon 180	C180	1P/S	1,130	1,130	2	I
Skywagon 185	C185	1P/S	1,000	1,000	2	I
Stationair 6, Turbo Stationair 6	C206	1P/S	975	1,000	2	I
Stationair/Turbo Stationair 7/8	C207	1P/S	810	1,000	2	I
Super Skymaster 337	C337 (O2)	2P/S	1,250	1,500	2	II
Super Skywagon/Super Skylane	C205	1P/S	965	1,000	2	I
Titan 404	C404	2P/S	2,600	2,000	2	II

Model	Civil Type Designator (Military Type Designator)	Number and Type Engines/ Weight Class	Climb Rate (fpm)	Descent Rate (fpm)	SOIR[1] Group	SRS[2] Category
Construcciones Aeronauticas (CASA) (Spain)						
C-212 Aviocar	CA21	2P/L	900	900	2	III
Dassault-Breguet (France)						
Falcon 10	DA01	2J/L	2,300	1,600	3	III
Falcon 20/C thru F, Fan Jet Falcon (FJF)	DA02	2J/L	2,000	2,400	3	III
Mystere Falcon 50	DA05	3J/L	1,800	1,600	4	III
Mystere Falcon 200, Falcon 20G/GF	DA20 (HU25)	2J/L	1,900	2,000	3	III
Mystere Falcon 900	DA90	3J/L	2,000	1,700	4	III
DeHavilland (Canada/UK)						
Albatross	DH91	4P/L	700	900	4	III
Beaver DHC-2	DH2 (U6)	1P/S	840	1,000	2	I
Buffalo DHC-5D/E	DH5 (C8)	2T/L	2,000	1,500	3	III
Caribou DHC-4	DH4 (C7)	2P/L	1,350	1,000	3	III
Chipmunk DHC-1	DH1	1P/S	900	1,000	2	I
Comet 2	DH62	4J/L	2,950	2,000	5	III
Comet 4	DH64	4J/L	2,850	2,000	5	III
DASH 7 DHC-7	DH7	4T/L	4,000	4,000	2	III
DASH 8 DHC-8	DH8	2T/L	1,500	1,500	3	III
Dove DH-104	DH10	2P/S	1,420	1,420	2	II
Heron DH-114	DH11	4P/S	1,075	1,075	2	III
Otter DHC-3	DH3 (U1)	1P/S	750	1,000	2	I
Turbo Beaver DHC-2T	DH2T	1T/S	1,220	1,000	2	I
Twin Otter DHC-6 (all series)	DH6 (U18)	2T/S	1,600	1,800	2	II
Dornier GmbH (FRG)						
Do 27	DO27	1P/S	700	800	2	I
Do 28/128	DO28	2P/S	1,500	1,500	3	II
Do 228-100 Series	DO81	2T/S	2,000	2,000	2	II
Do 228-200 Series	DO82	2T/L	2,000	2,000	2	III
Embraer (Brazil)						
Bandeirante EMB-110	E110	2T/S	1,500	1,500	3	II
Brasilia EMB-120	E120	2T/L	2,300	2,300	4	III
Fairchild Industries (USA—Includes Swearingen Aviation)						
Friendship F27	FA27	2T/L	3,000	3,000	3	III
Flying Box Car	FA24 (C119)	2P/L	750	750	4	III
Merlin IIA/B	SW2	2T/S	2,350	2,500	3	II
Metro III	SW3	2T/L	2,350	2,500	3	III
Metro	SW4	2T/S	2,400	2,500	3	III
Thunderbolt II	(A10A*)	2J/L	6,000	5,000	4	III
Fokker BV (Netherlands)						
Friendship F27	FK27	2T/L	3,000	3,000	3	III
Fellowship F28	FK28	2J/L	4,650	2,000	4	III
Fokker 50	FK05	2J/L	3,500	3,500	4	III
Fokker 100	FK10	2J/L	3,500	3,500	4	III
Foxjet International Inc. (USA)						
F600	FXJ	2J/S	2,500	2,500	2	III
Gates Learjet Corp. (USA)						
Learjet 23	LR23	2J/S	4,500	4,000	3	III
Learjet 24	LR24	2J/L	4,500	4,000	3	III
Learjet 25	LR25	2J/L	4,500	4,000	4	III
Learjet 28	LR28	2J/L	4,500	4,000	4	III
Learjet 29	LR29	2J/L	4,500	4,000	4	III
Learjet 31	LR31	2J/L	4,500	4,000	4	III
Learjet 35	LR35 (C21)	2J/L	4,500	4,000	4	III
Learjet 36	LR36	2J/L	4,500	4,000	4	III

Model	Civil Type Designator (Military Type Designator)	Number and Type Engines/ Weight Class	Climb Rate (fpm)	Descent Rate (fpm)	SOIR[1] Group	SRS[2] Category
Learjet 55	LR55	2J/L	5,000	4,000	4	III
Learjet 60	LR60	2J/L	—	—	5	III
General Dynamics Corp. (USA)						
Convair 240	CV24	2P/L	1,000	800	4	III
Convair 340	CV34 (C131)	2P/L	900	900	4	III
Convair 440	CV44	2P/L	900	900	4	III
Convair 580	CV58	2T/L	1,500	1,500	4	III
Convair 600	CV60	2T/L	1,500	1,500	4	III
Convair 640	CV64	2T/L	1,500	1,800	4	III
Convair 880	CV88	4J/L	2,250	2,500	5	III
Convair 990	CV99	4J/L	2,500	2,500	5	III
F-111/FB-111	(F111*)	2J/L	5,000	5,000	5	III
Fighting Falcon	(F16*)	1J/L	8,000	5,000	5	III
Government Aircraft Factories (Australia)						
N22B Nomad	N22B	2T/S	1,300	1,100	2	II
N24A Nomad	N24A	2T/S	1,300	1,100	2	II
Grumman Aerospace Corp. (USA)						
Albatross	G64 (U16)	2P/L	1,500	1,500	3	III
American TR/TS-2	AA2	1P/S	1,000	1,500	2	I
Cheetah AA-5/A/B	AA5	1P/S	660	1,000	2	I
Cougar G-93	G93 (F9)	1J/L	3,300	3,000	5	III
Cougar GA-7	GA7	2P/S	1,600	1,500	2	II
Goose/Super Goose	G21	2P/L	1,000	1,000	4	III
Greyhound	(C2)	2T/L	1,000	2,200	4	III
Hawkeye	(E2)	2T/L	2,690	3,000	5	III
Intruder	(A6)	2J/L	7,500	5,000	5	III
Mallard	G73	2P/L	1,600	1,600	3	III
Model G-164/Turbo Ag-Cat	G164	1P/S	1,500	1,500	2	I
Mohawk	G134 (OV1)	2T/S	2,100	1,300	3	I
Tomcat	(F14)	2J/L	6,000	4,000	5	III
Widgeon/Super Widgeon	G44	2P/L	1,000	1,500	3	III
Yankee AA-1B/C	AA1	1P/S	750	1,000	2	I
Gulfstream Aerospace Corp. (USA)						
GAC 159-C, Gulfstream I	G159 (VC4)	2T/L	2,000	2,000	4	III
Gulfstream II	G2 (VC11)	2J/L	5,000	4,000	4	III
Gulfstream III	G3 (C20)	2J/L	5,000	4,000	4	III
Gulfstream IV	G4	2J/L	5,000	4,000	4	III
Hamburger Flugzeubau (FRG)						
HFB-320 Hansajet	HF32	2J/L	4,500	4,500	3	III
Hamilton Aviation (USA)						
Westwind III	WSW	2T/S	2,000	2,000	2	II
Handley Page (UK)						
Jetstream HP-137	HP13	2T/S	2,100	2,100	3	II
Helio Aircraft Company (USA)						
H-250 Courier MarkII	HE25	1P/S	850	1,000	1	I
H-295 Super Courier	HE29 (U10)	1P/S	1,000	1,500	1	I
H-391 Courier	HE39	1P/S	400	500	1	I
H-395 Super Courier	HE35	1P/S	500	650	1	I
H-550/A Stallion	HE55 (AU24)	1T/S	2,200	2,200	1	I
H-580 Twin Courier	HE58	2P/S	1,250	1,500	1	II
H-634	HE63	2T/S	1,500	1,500	1	II
Ilyushin (USSR)						
IL-62	IL62	4J/H	3,500	2,500	5	III
IL-76	IL76	4J/H	3,000	2,500	5	III

Model	Civil Type Designator (Military Type Designator)	Number and Type Engines/ Weight Class	Climb Rate (fpm)	Descent Rate (fpm)	SOIR[1] Group	SRS[2] Category
Israel Aircraft Industries (Israel)						
101 Avara	RV01	2T/L	1,300	1,000	3	III
1123 Westwind	WW23	2J/L	4,000	3,500	4	III
1124 Westwind	WW24	2J/L	4,000	3,500	4	III
1124A Westwind 2	WW4A	2J/L	4,000	3,500	4	III
Israel Aircraft Industries and Astra Jet (Israel/USA)						
Astra 1125 Westwind	AJ25	2J/L	4,000	3,500	4	III
Lake Aircraft (USA)						
LA-250 Renegade/Sea Fury	LA25	1P/S	680	680	2	I
LA-250 Turbo	LA2T	1P/S	760	760	2	I
Renegade/Turbo Sea Fury LA-4/A/B	LA04	1P/S	1,000	1,000	2	I
LA-4-200 Buccaneer	LA42	1P/S	1,200	1,000	2	I
Learavia Corp. (USA)						
Lear Fan Model 2100	LRF	2J/L	2,000	1,500	2	III
Lockheed Corp. (USA)						
C-5 Galaxy	(C5A)	4J/H	2,500	2,000	5	III
C-141 Starlifter	(C141)	4J/H	3,500	3,000	5	III
Constellation 649	L649	4P/L	2,000	2,000	5	III
Constellation 749	L749	4P/L	2,000	2,000	5	III
Electra 188,	L188 (P3)	4T/L	1,850	2,000	4	III
F-104 Starflighter	(F104)	1J/L	5,000	4,000	5	III
L-1011 Tri-Star	L101	3J/H	3,500	3,000	5	III
Lodestar	L18	2P/L	1,800	2,000	3	III
Model 382 Hercules	L382 (C130)	4T/L	1,500	1,500	4	III
Model 1329 Jetstar	L329 (C140)	4J/L	4,000	3,500	4	III
Model 1329-5 Jetstar II	L295	4J/L	4,000	3,000	4	III
Super Constellation	L49	4P/L	1,620	1,620	5	III
T-33/F-80 Shooting Star	(T33/F80)	2J/L	2,000	2,000	4	III
TR-1	(TR1)	1J/L	6,000	6,000	5	III
U-2	(U2)	1J/L	6,000	6,000	5	III
Viking	(S3A)	2J/L	2,000	2,000	4	III
Martin Company (Division of Martin Marietta) (USA)						
Martin 202	M202	2P/L	1,600	1,500	3	III
Martin 404	M404	2P/L	1,600	1,500	3	III
Maule Aircraft Corp. (USA)						
M-4 Strata-Rocket	ML4	1P/S	1,000	1,000	2	I
M-5	ML5	1P/S	1,000	1,000	2	I
M-6 235 Super-Rocket	ML6	1P/S	1,500	1,000	2	I
McDonnell-Douglas Corp. (Boeing) (USA)						
DC-6/B Liftmaster	DC6 (C118)	4P/L	1,000	1,000	4	III
DC-7/B/C Seven Seas/Speed Freighter	DC7	4P/L	1,250	1,250	5	III
DC-8 (all series)	DC8	4J/L	3,000	3,000	5	III
DC-9	DC9 (C9/VC9)	2J/L	3,000	3,000	4	III
DC-9 Super/MD-80 Series	MD8	2J/L	3,500	3,000	4	III
DC-10 (all series)	DC10 (KC10)	3J/H	2,400	2,000	5	III
F-15 Eagle	(F15)	2J/L	8,000	5,000	5	III
F/A-18 Hornet	(F18/A18)	2J/L	8,000	6,000	5	III
Invader	(B26)	2P/L	1,000	1,000	4	III
Phantom II	(F4/RF4)	2J/L	8,000	6,000	4	III
Skyhawk	(A4/OA4/TA4)	1J/L	5,000	5,000	5	III
Skymaster	DC4 (C54)	4P/L	2,300	2,300	4	III
Skynight	(F10)	2J/L	5,000	5,000	5	III
Skytrain	DC3 (C47/VC47)	2P/L	1,200	1,200	3	III
Skywarrior	(A3)	2J/L	5,000	6,000	4	III

Model	Civil Type Designator (Military Type Designator)	Number and Type Engines/ Weight Class	Climb Rate (fpm)	Descent Rate (fpm)	SOIR[1] Group	SRS[2] Category
Super DC-3	DC3S (C117)	2P/L	1,330	1,330	3	III
Super DC-8 Series	DC8S	4J/H	5,500	5,500	5	III
Messerschmitt-Bolkow-Blohm (MBB) (FRG)						
BO 209 Monson	ME29	1P/S	1,100	1,100	2	I
HFB 320 Hansa Jet	ME32	2J/L	3,500	3,000	4	III
ME 108 Taifun	ME08	1P/S	400	500	1	III
Mitsubishi Aircraft International Inc. (USA/Japan)						
Mitsubishi Diamond I/MU-300	MU3	2J/L	3,500	4,000	3	III
Mitsubishi MU-2	MU2	2T/S	3,500	3,000	2	II
MU-2B-40 Solitaire	MU24	2T/S	4,000	3,000	2	II
MU-2B-60 Marquise	MU26	2T/S	4,000	3,000	2	II
Mooney Aircraft Corp. (USA)						
M-18 Mooney Mite	MO18	1P/S	750	750	2	I
Mark 10 Cadet	MO10	1P/S	800	800	2	I
Mark 20	MO20	1P/S	1,000	1,000	2	I
Mark 21/Mooney Ranger	MO21	1P/S	1,000	1,000	2	I
Mark 22	MO22	1P/S	1,300	1,300	2	I
Mooney 201/M20J	MO2J	1P/S	1,000	1,000	2	I
Turbo Mooney 231/M20K	MO2K	1P/S	1,500	1,200	2	I
Navion Rangemaster Aircraft Corp. (USA)						
Rangemaster	NA1	1P/S	1,250	1,500	2	I
Twin Navion	NA16	2P/S	1,800	1,500	3	II
Nord Aviation (Affiliate of Aerospatiale) (France)						
Super Broussard 260/262	ND26	2T/L	2,500	2,000	3	III
Transall C-160	ND16	2T/L	2,000	2,000	3	III
Northrup Corp. (USA)						
Freedom Fighter, Tiger II	(F5)	2J/L	8,000	5,000	5	III
T-38 Talon	(T38)	2J/S	8,000	5,000	5	III
Pilatus Flugzeugwerke AG (Switzerland)						
PC-6 Porter	PL6	1P/S	600	600	1	I
PC-6A/B Turbo Porter	PL6A	1T/S	1,250	1,500	1	I
Piper Aircraft Corp. (USA)						
Aero Star 600/700	PA60	2P/S	1,500	1,500	2	II
Model S-2/A/B/S	PIS2	1P/S	1,500	1,500	2	I
Apache	PA23	2P/S	1,050	1,000	2	II
Aztec	PAZT (U11)	2P/S	1,500	1,500	2	II
Brave	PA36	1P/S	800	1,000	2	I
Cherokee Archer, Dakota/Warrior	PA28 (T35)	1P/S	750	1,000	2	I
Cherokee Arrow IV	PARO	1P/S	900	1,000	2	I
Cherokee Six, Lance, Saratoga	PA32	1P/S	850	1,000	2	I
Cheyenne II	PAYE	2T/S	2,400	2,000	3	II
Cheyenne I	PA41	2T/S	2,200	2,000	3	II
Cheyenne III/IV	PA42	2T/S	2,400	2,000	3	II
Chieftan, Mohave, Navajo	PA31	2P/S	1,500	1,500	2	II
Clipper	PA16	1P/S	500	500	2	I
Comanche	PA24	1P/S	900	1,000	2	I
Cruiser	PA5	1P/S	500	500	2	I
Cub Special	PA11	1P/S	500	500	2	I
Cub Trainer	PA2	1P/S	500	500	2	I
Family Cruiser	PA14	1P/S	600	600	2	I
Malibu	PA46	1P/S	1,000	1,000	2	I
Pacer	PA20	1P/S	850	1,000	2	I
Pawnee	PA25	1P/S	650	650	2	I
Seminole	PA44	2P/S	1,100	1,000	2	II
Seneca	PA34	2P/S	1,300	1,300	2	II

Model	Civil Type Designator (Military Type Designator)	Number and Type Engines/ Weight Class	Climb Rate (fpm)	Descent Rate (fpm)	SOIR[1] Group	SRS[2] Category
Super Cruiser	PA12	1P/S	600	600	2	I
Super Cub	PA18 (U7)	1P/S	1,000	1,000	2	I
T-1040	PAT4	1P/S	1,300	1,200	2	I
Tomahawk	PA38	1P/S	750	750	2	I
Tri-Pacer	PA22	1P/S	1,000	1,000	2	I
Twin Comanche	PA30	2P/S	1,500	1,500	2	II
Vagabond	PA17	1P/S	500	500	2	I
Vagabond Trainer	PA15	1P/S	500	500	2	I
Pitts Aerobatics (Manufactured by Christen Industries Inc.) (USA)						
Model S-1S/T	PIS1	1P/S	1,500	1,500	2	I
Riley Aircraft Corp. (USA)						
M65 Rocket	RY65	2P/S	1,500	1,500	2	II
Rockwell International Corp. (USA)						
Aero Commander112	AC12	1P/S	1,000	1,200	2	I
Aero Commander 685	AC85	2P/S	1,500	1,500	3	II
Aero Commander 695	AC95	2P/S	1,500	1,500	3	II
Air-Cruiser	AC72	2P/S	1,300	1,300	3	II
Lancer	(B1)	4J/H	3,000	5,000	5	III
Bronco	(OV10)	2T/S	2,000	2,500	3	II
Buckeye	(T2)	2J/L	5,700	6,000	5	III
Commander 112A	AC2A	1P/S	1,000	1,200	2	I
Commander 112TC	AC2T	1T/S	1,200	1,500	2	I
Commander 114	AC14	1P/S	1,000	1,500	2	I
Commander 200	AC20	1P/S	1,400	1,000	2	I
Commander 500	AC50	2P/S	1,340	1,500	2	II
Commander 520	AC52	2P/S	1,340	1,500	2	II
Commander 560	AC56	2P/S	1,400	1,500	2	II
Darter 100/150	AC10	1P/S	850	850	2	I
Grand Commander 680	AC60	2P/S	1,250	1,250	2	II
Jet Commander	AC21	2J/L	5,000	4,500	3	III
Jet Commander 840/980/1000	AC84	2T/S	2,500	2,500	3	II
Jet Prop Commander	AC69	2T/S	2,500	1,500	2	II
Lark	LARK	1P/S	700	1,000	2	I
Mitchell	(B25)	2P/L	980	980	3	III
Mustang	(P51)	1P/S	2,500	2,500	3	III
Navion	N145	1P/S	750	600	2	I
Sabre	(F86)	1J/L	4,000	4,000	5	III
Sabreliner 65	N265 (T39)	2J/L	4,000	3,500	3	III
Super Commander 680S	AC68 (U4)	2P/S	1,375	1,375	2	II
Super Sabre	(F100)	1J/L	4,000	4,000	5	III
Texan	N6 (T6)	1P/S	800	800	2	I
Trojan	(T28)	1P/S	2,500	2,500	2	I
Turbo Commander 690C	AC6T	2T/S	2,500	3,000	2	III
Saab and Fairchild Industries (Sweden/USA)						
Model SF-340	SF34	2T/L	2,000	2,000	3	III
Short Brothers Ltd. (UK)						
Shorts 330	SH33	2T/L	1,380	1,380	3	III
Shorts 360	SH36	2T/L	1,400	1,400	3	III
Shorts SC.7 Skyvan	SH7	2T/S	1,500	1,500	3	II
Ted Smith Aerostar Corp. (USA)						
Aero Star	TS60	2P/S	1,800	1,500	2	II
VFW-Fokker (Zentralgesellschaft VFW-Fokker mbH) (FRG/Netherlands)						
VFW 614	VF14	2J/L	3,100	3,000	3	III

Model	Civil Type Designator (Military Type Designator)	Number and Type Engines/ Weight Class	Climb Rate (fpm)	Descent Rate (fpm)	SOIR[1] Group	SRS[2] Category
Vought Corp. (USA)						
Corsair II	(A7)	1J/L	8,000	6,000	5	III
Swift	GC1	1P/S	1,000	1,000	1	I
	Helicopters/Rotorcraft					
Aerospatiale (France)						
Lama	HR15	1T/S	1,000	1,000	1	I
Alouette II	HR30	1T/S	1,280	1,280	1	I
Alouette III	HR60	1T/S	1,500	1,500	1	I
Dauphine	HR36	1T/S	1,400	1,500	1	I
Dauphine 2	HR3C (HH65)	2T/S	1,800	1,000	1	I
Ecurevil/Astar	HR35	1T/S	1,000	1,000	1	I
Gazelle	HR34	1T/S	1,620	1,620	1	I
Puma/Super Puma	HR33	2T/L	1,250	1,500	2	I
Super Frelon	HR32	3T/L	1,200	1,500	2	I
Twin Star	HR55	2T/S	1,350	1,300	2	I
Augusta (Constuzioni Aeronautiche Giovanni Agusta SpA) (Italy)						
Model 147J-3B-1	A47J	1P/S	500	500	1	I
Model A 109/A/A-II	A109	2T/S	1,620	1,500	1	I
Model 212 ASW	A212	2T/S	1,420	1,420	1	I
Bell Helicopter Textron (USA)						
Biglifter	BH14	1T/L	1,500	1,500	2	I
Cobra	BH09 (AH1)	1T/S	1,375	1,375	1	I
Jet Ranger/Long Ranger/Sea Ranger/Kiowa/ Model 206	(OH58)	1T/S	1,200	1,000	1	I
Huey/Iroquois/Model 205 A-1	(UH1)	1T/S	1,500	1,500	1	I
Sioux/Model 47	BH47(H13)	1P/S	1,000	1,000	1	I
Twin Huey	BHTH(UH1N)	2T/S	1,420	1,420	1	I
Model 222	BH22	2T/S	1,500	1,000	1	I
Boeing Vertol Company (USA)						
Chinook Model 234	HV47 (CH47)	2T/L	1,500	1,500	2	I
Model 105	B105	2T/S	1,500	1,500	1	I
Sea Knight	HV07(CH46)	2T/L	2,130	2,130	2	45
Brantley-Hynes Helicopter Inc. (USA)						
Model B-2/A/B	HB42	1P/S	1,400	1,400	1	I
Model 305	HB43	1P/S	1,300	1,300	1	I
Enstrom Corp. (USA)						
Falcon/Model F-28/A/C/F Model 280	HF28	1P/S	800	800	1	I
Shark/Model 280FX	HF80	1P/S	1,200	1,200	1	I
Turbo Shark	HF8C	1P/S	1,500	1,500	1	I
Fairchild/Republic (includes Hiller) (USA)						
Hiller UH-12/Raven HH12	(OH23)	1P/S	1,500	1,500	1	I
Kaman Aerospace Corporation (USA)						
H-2 Seasprite	(H2)	2T/L	2,400	2,400	2	I
Huskie 600-3/5	HK60 (H43)	1T/L	2,000	2,000	2	I
Kawasaki Heavy Industries Ltd. (Japan)						
KV-107/II	KH17	2T/L	1,500	1,500	2	I
McDonnell-Douglas Helicopters (includes Hughes Helicopters) (USA)						
Model 77/Apache	HU64(AH64)	2T/L	1,500	1,500	2	I
Model 269	HU26	1P/S	1,000	1,000	1	I
Model 300/C	HU30	1P/S	1,200	1,200	1	I

Model	Civil Type Designator (Military Type Designator)	Number and Type Engines/ Weight Class	Climb Rate (fpm)	Descent Rate (fpm)	SOIR[1] Group	SRS[2] Category
Model 500C	HU50	1P/S	1,500	1,500	1	I
Osage	(H55)	1P/S	1,000	1,000	1	I
Pawnee Model 369	HU35 (OH6)	1T/S	1,500	1,500	1	I
Messerschmitt-Bolkow-Blohm (MBB) (FRG)						
Model BO 105	HM05	2T/S	1,200	1,200	1	I
MBB/Kawasaki (FRG/Japan)						
Model BK 117	HM17	2T/S	1,500	1,500	1	I
Robinson Helicopter Company Inc. (USA)						
Model R22	HR22	1P/S	800	800	1	I
Schweizer Aircraft Corp. (USA)						
Model 269C	HT26	1P/S	1,000	1,000	1	I
Sikorsky Aircraft (USA)						
Blackhawk S-70	SK70 (H60)	2T/L	2,000	2,000	2	I
Chickasaw S-55	SK55 (H19)	1P/S	800	1,000	1	I
Choctaw/Seashore/Seaboat S-58	SK58 (H34)	1P/L	1,120	1,120	2	I
LAMPS MK3	(SH60)	2T/L	2,000	2,000	2	I
Model S-51	SK51	1P/L	1,000	1,000	2	I
Model S-52	SK52	1P/L	950	1,000	2	I
Model S-59	SK59	1P/L	1,200	1,200	2	I
Model S-62	SK62 (H52)	1T/S	1,020	1,000	1	I
Model S-69	SK69 (H59)	1T/S	1,100	1,100	1	I
Model S-76	SK76	2T/S	1,300	1,300	1	I
Mojave S-56	SK56 (H37)	2P/L	970	1,000	2	I
Sea King S-61	SK61 (H3)	2T/L	1,500	1,500	2	I
Sea Stallion S-65	SK65 (H53)	2T/L	1,500	1,500	2	I
Skycrane S-64E/FSK64	(H54)	2T/L	1,300	1,300	2	I
Westland Helicopters Ltd. (UK)						
WG 30	WL30	2T/S	1,200	1,200	1	I

Engine Type Abbreviations

P = piston T = jet/turboprop J = jet

Climb and descent rates based on average en route climb/descent profiles at median weight between maximum gross takeoff and landing weights.

Homebuilt and Experimental Aircraft

Configuration diversity and the fact that airworthiness certificates are issued to aircraft builders necessitates the assignment of generic aircraft type designators based on cruise performance rather than specific manufacturer and normal descriptive/performance information.

Designator Criteria	Type Designator	Climb Rate (fpm)	Descent Rate (fpm)	SOIR Group	SRS Category
Aircraft with (indicated) cruise airspeeds of 100 knots or less	HXA	500	500	1	1
Aircraft with (indicated) cruise airspeeds of greater than 100 knots up to and including 200 knots	HXB	750	750	1	1
Aircraft with (indicated) cruise airspeeds greater than 200 knots	HXC	1,000	1,000	2	1

[1] Simultaneous operations on intersecting runways; corresponding group distance criteria can be found in FAA Order 7210.3, Chapter 12.
[2] Same runway separation.

APPENDIX C

Three-Letter Identifiers

ICAO Code	IATA Code	City	Country	Airport
AMML	MEL	Melbourne	Australia	Tullamarine International
ASSY	SYD	Sydney	Australia	Kingsford Smith International
BIKF	KEF	Keflavik	Iceland	
BIRK	REK	Reykjavik	Iceland	
CYEG	YEG	Edmonton	Canada (Alberta)	International
CYHZ	YHZ	Halifax	Canada (Nova Scotia)	International
CYMX	YMX	Montreal	Canada (Quebec)	Mirabel International
CYQM	YQM	Moncton	Canada (New Brunswick)	
CYQX	YQX	Gander	Canada (Newfoundland)	International
CYWG	YWG	Winnipeg	Canada (Manitoba)	International
CYYC	YYC	Calgary	Canada (Alberta)	International
CYYT	YYT	St. John's	Canada (Newfoundland)	
CYYZ	YYZ	Toronto	Canada (Ontario)	Lester B. Pearson
DNMM	LOS	Lagos	Nigeria	Murtalla Muhammed
EBBR	BRU	Brussels	Belgium	National
EDDF	FRA	Frankfurt	Germany	Rhein Main
EDDI	THF	Berlin	Germany	Tempelhof
EFHF	HEM	Helsinki	Finland	Malmi
EFHK	HEL	Helsinki	Finland	Vantaa
EGAA	BFS	Belfast	North Ireland	International/Aldergrove
EGCC	MAN	Manchester	England	International
EGGW	LTN	London	England	Luton
EGKK	LGW	London	England	Gatwick
EGLL	LHR	London	England	Heathrow
EGPF	GLA	Glasgow	Scotland	International
EGSS	STN	London	England	Stansted
EHAM	AMS	Amsterdam	Netherlands	Schiphol
EIDW	DUB	Dublin	Ireland	
EINN	SNN	Shannon	Ireland	
EKCH	CPH	Copenhagen	Denmark	Kastrup
ENGM	GEN	Oslo	Norway	Gardermoen
EPWA	WAW	Warsaw	Poland	Okecie
FAJS	JNB	Johannesburg	South Africa	Jan Smuts International
HECA	CAI	Cairo	Egypt	International
KABQ	ABQ	Albuquerque	USA (New Mexico)	International
KACY	AIY	Atlantic City	USA (New Jersey)	Bader Field
KATL	ATL	Atlanta	USA (Georgia)	Hartsfield International
KAUS	AUS	Austin	USA (Texas)	Bergstrom International
KBDL	BDL	Windsor Locks	USA (Connecticut)	Bradley International
KBNA	BNA	Nashville	USA (Tennessee)	Metropolitan
KBOI	BOI	Boise	USA (Idaho)	Air Terminal
KBOS	BOS	Boston	USA (Massachusetts)	Logan International
KBUF	BUF	Buffalo	USA (New York)	International

ICAO Code	IATA Code	City	Country	Airport
KBWI	BWI	Baltimore	USA (Maryland)	Washington International
KCLE	CLE	Cleveland	USA (Ohio)	Hopkins International
KCLT	CLT	Charlotte	USA (North Carolina)	Douglas Municipal
KCMH	CMH	Columbus	USA (Ohio)	Port Columbus International
KCMI	CMI	Champaign	USA (Illinois)	University of Illinois
KCOS	COS	Colorado Springs	USA (Colorado)	Peterson Field
KDAB	DAB	Daytona Beach	USA (Florida)	
KCVG	CVG	Cincinnati	USA (Kentucky)	Cincinnati/Northern Kentucky International
KDAL	DAL	Dallas	USA (Texas)	Love Field
KDAY	DAY	Dayton	USA (Ohio)	James M. Cox
KDCA	DCA	Washington	USA (D.C.)	Ronald Reagan
KDEN	DEN	Denver	USA (Colorado)	Stapleton International
KDFW	DFW	Dallas–Fort Worth	USA (Texas)	International
KDSM	DSM	Des Moines	USA (Iowa)	
KDTW	DTW	Detroit	USA (Michigan)	Metropolitan
KEWR	EWR	Newark	USA (New Jersey)	International
KFAT	FAT	Fresno	USA (California)	Air Terminal
KGFK	GFK	Grand Forks	USA (North Dakota)	
KGRR	GRR	Grand Rapids	USA (Michigan)	Country Cascade
KGTF	GTF	Great Falls	USA (Montana)	International
KHOU	HOU	Houston	USA (Texas)	William P. Hobby
KHPN	HPN	White Plains	USA (New York)	Westchester County
KIAD	IAD	Washington	USA (D.C.)	Dulles International
KIAH	IAH	Houston	USA (Texas)	Intercontinental
KICT	ICT	Wichita	USA (Kansas)	Mid-Continent
KIND	IND	Indianapolis	USA (Indiana)	International
KJAX	JAX	Jacksonville	USA (Florida)	International
KJFK	JFK	New York	USA (New York)	John F. Kennedy
KLAF	LAF	Lafayette	USA (Indiana)	Purdue University
KLAN	LAN	Lansing	USA (Michigan)	Capital City
KLAS	LAS	Las Vegas	USA (Nevada)	McCarran International
KLAX	LAX	Los Angeles	USA (California)	International
KMCI	MCI	Kansas City	USA (Missouri)	International
KMDT	MDT	Harrisburg	USA (Pennsylvania)	International
KMDW	MDW	Chicago	USA (Illinois)	Midway
KMEM	MEM	Memphis	USA (Tennessee)	International
KMIA	MIA	Miami	USA (Florida)	International
KMKE	MKE	Milwaukee	USA (Wisconsin)	General Mitchell
KMSN	MSN	Madison	USA (Wisconsin)	Truax Field
KMSP	MSP	Minneapolis	USA (Minnesota)	St. Paul International
KMSY	MSY	New Orleans	USA (Louisiana)	International
KMYR	MYR	Myrtle Beach	USA (South Carolina)	Mil. (Myrtle Beach)
KNHK		Patuxent River	USA (Maryland)	Trapnell Field
KNKT		Cherry Point	USA (North Carolina)	Cunningham Field
KNKX		Miramar	USA (California)	Mitcher Field
KNQX		Key West	USA (Florida)	Boca Chica Field
KNZY		San Diego	USA (California)	North Island NAS
KOAK	OAK	Oakland	USA (California)	International
KOKC	OKC	Oklahoma City	USA (Oklahoma)	Will Rogers World
KOMA	OMA	Omaha	USA (Nebraska)	Eppley Airfield
KONT	ONT	Ontario	USA (California)	International
KORD	ORD	Chicago	USA (Illinois)	O'Hare

ICAO Code	IATA Code	City	Country	Airport
KMCO	MCO	Orlando	USA (Florida)	International
KPHL	PHL	Philadelphia	USA (Pennsylvania)	International
KPHX	PHX	Phoenix	USA (Arizona)	Sky Harbor International
KPIT	PIT	Pittsburgh	USA (Pennsylvania)	Greater Pittsburgh
KRDU	RDU	Raleigh-Durham	USA (North Carolina)	International
KSAN	SAN	San Diego	USA (California)	Lindbergh International
KSAT	SAT	San Antonio	USA (Texas)	International
KSEA	SEA	Seattle	USA (Washington)	Tacoma International
KSFO	SFO	San Francisco	USA (California)	International
KSLC	SLC	Salt Lake City	USA (Utah)	International
KSTL	STL	St. Louis	USA (Missouri)	Lambert International
KTEB	TEB	Teterboro	USA (New Jersey)	
KTPA	TPA	Tampa	USA (Florida)	International
KTUL	TUL	Tulsa	USA (Oklahoma)	International
KTUS	TUS	Tucson	USA (Arizona)	International
KTYS	TYS	Knoxville	USA (Tennessee)	Mcgee Tyson
LEBL	BCN	Barcelona	Spain	
LEMD	MAD	Madrid	Spain	Barajas
LFPB	LBG	Paris	France	Le Bourget
LFPG	CDG	Paris	France	Charles De Gaulle
LEPO	ORY	Paris	France	Orly
LGAT	ATH	Athens	Greece	Hellenikon
LHBP	BUD	Budapest	Hungary	Ferihegy
LIRA	CIA	Rome	Italy	Ciampino
LIRF	FCO	Rome	Italy	Fiumicino
LLJR	JRS	Jerusalem	Israel	Atarot
LSZH	ZRH	Zurich	Switzerland	Kloten
MGGT	GUA	Guatemala City	Guatemala	La Aurora
MHTG	TGU	Tegucigalpa	Honduras	Toncontin International
MKJP	KIN	Kingston	Jamaica	Norman Manley International
MKJS	MBJ	Montego Bay	Jamaica	Sangster International
MMAA	ACA	Acapulco	Mexico	Juan Alvarez International
MMCS	CJS	Ciudad Juarez	Mexico	Abraham Gonzales International
MMCZ	CZM	Cozumel	Mexico	International
MMMX	MEX	Mexico City	Mexico	Benito Juarez
MMMY	MTY	Monterrey	Mexico	General M. Escobedo
MMTJ	TIJ	Tijuana	Mexico	Gen. A.L. Rodriguez
MNMG	MGA	Managua	Nicaragua	Augusto C. Sandino
MPTO	PTY	Panama City	Panama	Tocumen International
MROC	SJO	San Jose	Costa Rica	Juan Santamaria International
MTPP	PAP	Port-Au-Prince	Haiti	
MUHA	HAV	Havana	Cuba	Jose Marti International
MYGF	FPO	Freeport	Bahamas	International
MYNN	NAS	Nassau	Bahamas	International
NSTU	PPG	Pago Pago	Samoa	International (Tutuila Island)
NZCH	CHC	Christchurch	New Zealand	G. Adams
NZWN	WLG	Wellington	New Zealand	International
OBBI	BAH	Bahrain	Bahrain	Muharraq International
OEDR	DHA	Dhahran	Saudi Arabia	International
OEJN	JED	Jeddah	Saudi Arabia	King Abdul Aziz
OERK	RUH	Riyadh	Saudi Arabia	King Khaled International
OIII	THR	Teheran	Iran	Mehrabad International

ICAO Code	IATA Code	City	Country	Airport
OJAI	AMM	Amman	Jordan	Queen Alia
OKBK	KWI	Kuwait	Kuwait	International
OLBA	BEY	Beirut	Lebanon	International
OMAA	AUH	Abu Dhabi	United Arab Emirates	International
OPKC	KHI	Karachi	Pakistan	Quaid-E-Azam International
ORBS	SDA	Baghdad	Iraq	Saddam International
OSDI	DAM	Damascus	Syria	International
PAFA	FAI	Fairbanks	USA (Alaska)	International
PANC	ANC	Anchorage	USA (Alaska)	International
PGSN	SPN	Saipan	Mariana Islands	Obyan International
PHNL	HNL	Honolulu	USA (Hawaii)	International
RCTB	TPE	Taipei	Taiwan	Chiang Kai Shek
RJAA	NRT	Tokyo	Japan	Narita/New Tokyo Apt.
RJNN	NGO	Nagoya	Japan	Komaki International
RJTT	HND	Tokyo	Japan	Haneda International
RKSS	SEL	Seoul	Korea	Kimpo International
RPMM	MNL	Manila	Philippines	Nioy Aquino International
SABE	AEP	Buenos Aires	Argentina	J. Newberry International
SBBR	BSB	Brasilia	Brazil (Df)	International
SBRJ	SDU	Rio De Janeiro	Brazil (Rj)	Santos Dumont
SCEL	SCL	Santiago	Chile	A. Merino International
SEQU	UIO	Quito	Ecuador	Mariscal Sucre
SGAS	ASU	Asuncion	Paraguay	Silvio Pettirossi
SKBO	BOG	Bogota	Colombia	Eldorado
SLLP	LPB	La Paz	Bolivia	El Alto
SPIM	LIM	Lima	Peru	Jorge Chavez International
SUAA	MVD	Montevideo	Uruguay	Angel San Adami
SVMC	MAR	Maracaibo	Venezuela	La Chinita International
SVVA	VLN	Valencia	Venezuela	International
TIST	STT	St. Thomas	Virgin Islands (USA)	Cyril E. King
TJSJ	SJU	San Juan	Puerto Rico	Luis Munoz Marin
UHWW	VVO	Vladivostok	Russia	Knevichi
ULLI	LED	St. Petersburg	Russia	Pulkovo
UUWW	VKO	Moscow	Russia	Vnukovo
VABB	BOM	Bombay	India	Santa Cruz International
VHHH	HKG	Hong Kong	Hong Kong	Kai Tak
VIDP	DEL	Delhi	India	Indira Ghandi International
VTBD	BKK	Bangkok	Thailand	Don Muang International
VVTS	SGN	Ho Chi Minh City	Viet Nam	Tansonnhat
WSSS	SIN	Singapore	Singapore	Changi International
ZBAA	BJS	Beijing	China	Capital
ZKPY	FNJ	Pyongyang	Korea (North)	Sunan
ZSSS	SHA	Shanghai	China	Hongqiao

GLOSSARY

abbreviated IFR flight plans An authorization by ATC requiring pilots to submit only that information needed for ATC. It includes only a small portion of the usual IFR flight plan information. In certain instances, this may be only aircraft identification, locations, and pilot request. Other information may be requested if needed by ATC for separation/control purposes. It is frequently used by aircraft that are airborne and desire an instrument approach or by aircraft that are on the ground and desire a climb to VFR on top.

abeam An aircraft is abeam a fix, point, or object when that fix, point, or object is approximately 90° to the right or left of the aircraft track. Abeam indicates a general position rather than a precise point.

Additional Services Advisory information provided by ATC that includes but is not limited to the following: traffic advisories; vectors, when requested by the pilot, to assist aircraft receiving traffic advisories to avoid observed traffic; altitude deviation information of 300 feet or more from an assigned altitude as observed on a verified (reading correctly) automatic altitude readout (mode C); advisories that traffic is no longer a factor; weather and chaff information; weather assistance; bird activity information; and holding-pattern surveillance.

advanced automation system (AAS) A semiautomatic air traffic control system being developed by the Federal Aviation Administration.

Aeronautical Information Manual A primary FAA publication whose purpose is to instruct pilots about operating in the National Airspace System of the United States. It provides basic flight information, ATC procedures, and general instructional information concerning health, medical facts, factors affecting flight safety, accident and hazard reporting, and types of aeronautical charts and their use.

Air Commerce Act Legislation that created the Aeronautics Branch of the Department of Commerce, the first formal aviation regulatory agency of the federal government. Signed into law on May 20, 1926.

Air Coordinating Committee (ACC) Organization established on March 27, 1945, by the federal government to coordinate with the International Civil Aviation Organization and make recommendations on technical, economic, and industrial matters relating to aviation.

air defense identification zone (ADIZ) Area of airspace within which the identification, location, and control of aircraft is required for U.S. national security.

Air Navigation Conferences Conferences held by the International Civil Aviation Organization at which recommendations and changes to the ICAO Annexes are made.

Air Navigation Development Board (ANDB) Agency established in 1948 to oversee the implementation of the U.S. air traffic control system as described in the SC-31 report made by the Radio Technical Commission for Aeronautics.

air route surveillance radar (ARSR) Air traffic control radar primarily used to separate aircraft en route between terminal areas. Air route surveillance radar typically has a range of up to 250 nautical miles.

air route traffic control center (ARTCC) An air traffic control facility whose primary function is to provide separation to aircraft en route between airports.

Air Safety Board Agency established on June 23, 1938, as part of the Civil Aeronautics Act. The board was established to investigate aviation accidents and make safety recommendations.

air traffic control (ATC) A service provided by the appropriate authority to promote the safe, orderly, and expeditious flow of air traffic.

Air Traffic Control Association (ATCA) Association formed in the late 1950s to represent the interests of air traffic controllers.

Air Traffic Control Handbook **(FAAH 7110.65)** The FAA publication that delineates the procedures to be used by FAA air traffic controllers when performing their duties.

air traffic control radar beacon system (ATCRBS) A radar identification system composed of a ground-based interrogator and an airborne transponder that provides rapid and positive radar identification of an aircraft.

air traffic control tower An air traffic control facility whose primary function is to provide runway separation for aircraft landing or departing from the primary airport.

airborne delay Amount of delay to be encountered in airborne holding.

aircraft category A grouping of aircraft based on 1.3 times the aircraft's stall speed while in a landing configuration at maximum gross landing weight. Aircraft category is the primary determinant of instrument approach minima that are used by the pilot of that aircraft. The aircraft categories are:

Category A—Speed less than 91 knots.
Category B—Speed 91 knots or greater but less than 121 knots.
Category C—Speed 121 knots or greater but less than 141 knots.
Category D—Speed 141 knots or greater but less than 166 knots.
Category E—Speed 166 knots or more.

aircraft class Categorization used to determine the wake turbulence criteria that should be applied to aircraft. Aircraft class is based on maximum certificated takeoff weight. Aircraft classes are as follows:

Small—Aircraft weighing 41,000 pounds or less.
Large—Aircraft weighing more than 41,000 pounds up to and including 255,000 pounds.
Heavy—Aircraft weighing more than 255,000 pounds.

aircraft situation display (ASD) A computer system that receives radar track data from all 20 continental United States ARTCCs, organizes this data into a mosaic display, and presents it on a computer screen.

Airmail Act of 1925 Legislation authorizing the postmaster general to contract with private individuals and corporations for the purpose of transporting airmail. Also known as the Kelly Airmail Act. Signed into law on February 2, 1925.

AIRMET In-flight weather advisories issued only to amend the area forecast concerning weather phenomena that are of operational interest to all aircraft and potentially hazardous to aircraft having limited capability because of lack of equipment, instrumentation, or pilot qualifications. AIRMETs concern weather of less severity than that covered by SIGMETs or Convective SIGMETs. AIRMETs cover moderate icing, moderate turbulence, sustained winds of 30 knots or more at the surface, widespread areas of ceilings less than 1,000 feet and/or visibility less than 3 miles, and extensive mountain obscurement. (See *SIGMET, Convective SIGMET.*)

airport acceptance rate (AAR) A dynamic input parameter specifying the number of arriving aircraft that an airport or airspace can accept from the ARTCC per hour. The AAR is used to calculate the desired interval between successive arrival aircraft.

airport advisory areas The area on which a flight service station is located within 10 miles of an airport lacking a control tower. The flight service station provides airport advisory service within this area.

airport advisory service A service, provided by flight service stations, that consists of airport conditions, known traffic within the area, and weather information.

airport boundary lighting Steady-burning 40-watt white lights placed on wooden stakes every 300 feet around the perimeter of an airport.

Airport Facility Directory A publication of the Federal Aviation Administration that contains all pertinent operational information about U.S. airports. This information includes air traffic control facilities, communications frequencies, airport data, and special notices and procedures in effect.

airport radar service area (ARSA) Regulatory airspace surrounding certain designated airports where air traffic control provides full-time vectoring and sequencing for both IFR and VFR aircraft.

airport surface detection equipment (ASDE) Radar equipment specifically designed to detect moving objects on the airport surface.

airport surveillance radar (ASR) Approach control radar used to separate aircraft within the immediate vicinity of an airport. Airport surveillance radar normally has a maximum range of 60 nautical miles.

airport surveillance radar (ASR) approach An instrument approach procedure in which the air traffic controller uses airport surveillance radar to maintain the aircraft on the runway centerline while the pilot initiates a descent.

airway A formally designated control area, the centerline of which is defined by radio navigation aids.

airway traffic control centers (ATCCs) Predecessors of today's air route traffic control centers.

Air Traffic Control System Command Center (ATCSCC) An Air Traffic Operations Service facility consisting of four operational units. Central Flow Control Function (CFCF) is responsible for the coordination and approval of all major intercenter flow control restrictions on a system basis in order to obtain maximum utilization of the airspace. Central Altitude Reservation Function (CARF) is responsible for coordinating, planning, and approving special user requirements under the Altitude Reservation (ALTRV) concept. Airport Reservation Office (ARO) is responsible for approving IFR flights at designated high-density-traffic airports (John F. Kennedy, LaGuardia, O'Hare, and Ronald Reagan) during specified hours. ATC Contingency Command Post is a facility that enables the FAA to manage the ATC system when significant portions of the system's capabilities have been lost or are threatened.

airway traffic control stations (ATCSs) Predecessors of today's air route traffic control centers.

airway traffic control units (ATCUs) Predecessors of today's air route traffic control centers.

Airways Modernization Board (AMB) Independent agency formed in 1957 to coordinate civilian-military aviation electronics research and development. The AMB conducted research on air traffic control computers, transponders, and advanced radar equipment at its research and development facilities near Atlantic City, New Jersey. This complex later became the FAA's National Aviation Facilities Experimental Center (NAFEC).

alert areas Nonrestricted airspace in which a high volume of pilot training activity may be taking place.

alert notice (ALNOT) A request for an extensive communications search for an overdue, unreported, or missing aircraft.

alphanumeric display Letters and numerals used to show identification, altitude, beacon code, and other information concerning a target on a radar display.

altitude filtering A means by which the controller can control which aircraft are displayed on the radar, based on transmitted altitude.

Annexes See *ICAO Annexes.*

anticipated separation A procedure whereby the controller issues instructions to two or more aircraft based on the presumption that they will remain separated.

approach clearance Authorization by ATC for a pilot to conduct an instrument approach for which a clearance and other pertinent information is provided in the approach clearance when required.

approach control service Air traffic control service provided by an approach control facility for arriving and departing VFR/IFR aircraft and, on occasion, en route aircraft. At some airports not served by an approach control facility, the ARTCC provides limited approach control service.

approach gate An imaginary point used as the basis for vectoring aircraft to the final approach course. The approach gate is located 1 mile outside the final approach fix or 5 nautical miles from the end of the runway, whichever distance is greater.

approach lighting system (ALS) An airport lighting facility that provides visual guidance to landing aircraft by radiating light beams in a directional pattern by which the pilot aligns the aircraft with the extended centerline of the runway on the final approach for landing. Capacitor discharge sequential flashing lights/sequenced flashing lights may be installed in conjunction with the ALS at some airports. Types of approach lighting systems are:

1. ALSF-1 — approach lighting system with sequenced flashing lights in ILS Category I configuration.
2. ALSF-2 — approach lighting system with sequenced flashing lights in ILS Category II configuration. The ALSF-2 may operate as an SSALR when weather conditions permit.
3. SSALF — simplified short approach lighting system with sequenced flashing lights.
4. SSALR — simplified short approach lighting system with runway alignment indicator lights.
5. MALSF — medium-intensity approach lighting system with sequenced flashing lights.
6. MALSR — medium-intensity approach lighting system with runway alignment indicator lights.
7. LDIN — lead in light system: Consists of one or more series of flashing lights installed at or near ground level that provides positive visual guidance along an approach path, either curving or straight, where special problems exist with hazardous terrain, obstructions, or noise abatement procedures.
8. RAIL — runway alignment indicator lights (sequenced flashing lights that are installed only in combination with other light systems).
9. ODALS — omnidirectional approach lighting system that consists of seven omnidirectional flashing lights located in the approach area of a nonprecision runway. Five lights are located on the runway centerline extended with the first light located 300 feet from the threshold and extending at equal intervals up to 1,500 feet from the threshold. The other two lights are located, one on each side of the runway threshold, at a lateral distance of 40 feet from the runway edge, or 75 feet from the runway edge when installed on a runway equipped with a VASI.
10. runway lights/runway edge lights — lights having a prescribed angle of emission used to define the lateral limits of a runway. Runway lights are uniformly spaced at intervals of approximately 200 feet, and the intensity may be controlled or preset.
11. touchdown zone lighting — two rows of transverse light bars located symmetrically about the runway

centerline normally at 100-foot intervals. The basic system extends 3,000 feet along the runway.

12. runway centerline lighting—flush centerline lights spaced at 50-foot intervals beginning 75 feet from the landing threshold and extending to within 75 feet of the opposite end of the runway.
13. threshold lights—fixed green lights arranged symmetrically left and right of the runway centerline, identifying the runway threshold.
14. runway end identifier lights (REILs)—two synchronized flashing lights, one on each side of the runway threshold, which provide rapid and positive identification of the approach end of a particular runway.
15. visual approach slope indicator (VASI)—an airport lighting facility providing vertical visual approach slope guidance to aircraft during approach to landing by radiating a directional pattern of high-intensity red and white focused light beams that indicate to pilots that they are "on path" if they see red/white, "above path" if white/white, and "below path" if red/red. Some airports serving large aircraft have three-bar VASIs that provide two visual glide paths to the same runway.

approach plates A slang term used to describe instrument approach procedure charts.

approval request (APPREQ) A request by a controller to deviate from the procedures delineated in a facility directive or a letter of agreement.

area control center (ACC) An ICAO term for an air traffic control facility primarily responsible for ATC services being provided to IFR aircraft during the en route phase of flight. The U.S. equivalent facility is an air route traffic control center (ARTCC).

area navigation (RNAV) A method of navigation that permits aircraft operation on any desired course within the coverage of station referenced navigation signals or within the limits of a self-contained system capability. Random area navigation routes are direct routes, based on area navigation capability, between waypoints defined by latitude/longitude coordinates, degree/distance fixes, or offsets from published or established routes/airways at a specified distance and direction. The major types of equipment are:

1. VORTAC-referenced or course-line computer (CLC) systems, which account for the greatest number of RNAV units in use. To function, the CLC must be within the service range of a VORTAC.
2. OMEGA/VLF, although two separate systems, can be considered as one operationally. A long-range navigation system based on very low frequency radio signals transmitted from a total of 17 stations worldwide.
3. Inertial (INS) systems, which are totally self-contained and require no information from external references. They provide aircraft position and navigation information in response to signals resulting from inertial effects on components within the system.
4. MLS area navigation (MLS/RNAV), which provides area navigation with reference to an MLS ground facility.
5. LORAN-C is a long-range radio navigation system that uses ground waves transmitted at low frequency to provide user position information at ranges of up to 600 to 1,200 nautical miles at both en route and approach altitudes. The usable signal coverage areas are determined by the signal-to-noise ratio, the envelope-to-cycle difference, and the geometric relationship between the positions of the user and the transmitting stations.
6. GPS is a space-based radio positioning, navigation, and time-transfer system. The system provides highly accurate position and velocity information and precise time on a continuous global basis to an unlimited number of properly equipped users. The system is unaffected by weather and provides a worldwide common grid reference system.

arrival gates Intersections or areas used by approach control facilities primarily as inbound fixes into their areas.

arrival sequencing program (ASP) The automated program designed to assist in sequencing aircraft destined for the same airport.

ARTS-II A programmable, nontracking, computer-aided display subsystem capable of modular expansion. ARTS-II systems provide a level of automated air traffic control capability at terminals having low to medium activity. Flight identification and altitude may be associated with the display of secondary radar targets. The system has the capability of communicating with ARTCCs and other ARTS-II, -IIA, -III, and -IIIA facilities.

ARTS-IIA A programmable, radar-tracking computer subsystem capable of modular expansion. The ARTS-IIA detects, tracks, and predicts secondary radar targets. The targets are displayed by means of computer-generated symbols, ground speed, and flight plan data. Although it does not track primary radar targets, they are displayed coincident with the secondary radar as well as the symbols and alphanumerics. The system has the capability of communicating with ARTCCs and other ARTS-II, -IIA, -III, and -IIIA facilities.

ARTS-III The beacon tracking level (BTL) of the modular programmable automated radar terminal system in use at medium- to high-activity terminals. ARTS-III detects, tracks, and predicts secondary radar-derived aircraft targets. These are displayed by means of computer-generated symbols and alphanumeric characters depicting flight identification, aircraft altitude, ground speed, and flight plan data. Although it does not track primary targets, they are displayed coincident with the secondary radar as well as the symbols and alphanumerics. The system has the capability of communicating with ARTCCs and other ARTS-III facilities.

ARTS-IIIA The radar tracking and beacon tracking level (RT and BTL) of the modular, programmable, automated radar terminal system. ARTS-IIIA detects, tracks, and predicts primary as well as secondary radar-derived aircraft targets. This more sophisticated computer-driven system upgrades the existing ARTS-III system by providing improved tracking, continuous data recording, and fail-safe capabilities.

associated track Aircraft whose flight plan information has been derived from the FDP system in the ARTCC or from information entered by the controller.

association checking The correlation of an aircraft's actual three-dimensional position with its predicted position.

automated flight service station (AFSS) The final result of the Federal Aviation Administration's flight service station consolidation project: 61 automated flight service stations with advanced computer equipment will replace over 300 labor-intensive flight service stations.

automated handoff A handoff using radar data processing computer equipment.

automated radar terminal system (ARTS) A generic term that describes two computer beacon processing systems utilized in conjunction with airport surveillance radar. The two ARTS systems in use by the FAA are ARTS-II, a beacon tracking level system primarily used at low- or medium-activity airports, and ARTS-III, a radar and beacon tracking level system used at high-activity airports.

automatic direction finder (ADF) The airborne component used by pilots to navigate using nondirectional beacons.

Automatic Terminal Information Service (ATIS) The continuous broadcast of recorded noncontrol information in selected terminal areas. Its purpose is to improve controller effectiveness and to relieve frequency congestion by automating the repetitive transmission of essential but routine information.

autorotation A flight condition in which a helicopter makes a nonpowered, controlled descent to landing.

beacon data acquisition system The component of an ARTS radar system that interprets transponder replies, correlates this information with those targets detected by the primary radar system, and then sends this information in a digital format to the data processing system.

beacon slash The radar display produced by a transponder.

beacon tracking level (BTL) A radar processing system that can track only transponder-equipped aircraft.

bin A 2-mile square that contains the maximum height of any obstruction within that area.

blind speed The radial velocity at which a target will be removed from the radar display by moving target indicator equipment.

blip Slang term for target, echo, or radar return.

boresight The angular width of the transmission from a primary radar antenna.

bright radar indicator tower equipment (BRITE) A radar display system primarily designed to be used in high-ambient-light environments such as control towers.

calm wind runway The runway designated in facility directives to be used whenever the wind is less than 5 knots.

ceiliometer An optical or laser-based device that measures the height of the overlying cloud levels.

center radar ARTS presentation (CENRAP) A computer program developed to provide a backup system for airport surveillance radar in the event of a failure or malfunction. The program uses air route traffic control center radar for the processing and presentation of data on the ARTS-IIA or -IIIA displays.

central computer complex (CCC) The generic term used to describe the computers that operate both the flight data and radar processing systems at an ARTCC. Also known as the host computer.

chain One set of LORAN-C transmitting stations.

challenge pulse The collective pulses (P1, P2, and P3) transmitted by the ATCRBS ground-based transmitter.

changeover point (COP) That point on an airway where the pilot ceases to navigate from one navigation aid and begins to navigate toward the next. Unless otherwise specified, this point is normally halfway between the two navigation aids.

circular polarization (CP) A primary radar mode of operation used in an attempt to remove symmetrically shaped objects (such as precipitation) from the radar display.

Civil Aeronautical Medical Institute (CAMI) The medical branch of the Federal Aviation Administration based at the Aeronautical Center in Oklahoma City.

Civil Aeronautics Administration (CAA) Agency created in 1940 when the president restructured the Civil Aeronautics Authority. Under this reorganization, the Office of the Administrator of the Civil Aeronautics Authority was placed under the auspices of the Department of Commerce and renamed the Civil Aeronautics Administration.

Civil Aeronautics Authority (CAA) Agency created on June 23, 1938, when Congress passed the Civil Aeronautics Act, which removed the Bureau of Air Commerce from the Department of Commerce and created the Civil Aeronautics Authority. The CAA became the only independent authority established in the U.S. government at that time.

Civil Aeronautics Board (CAB) Agency created in 1940 when the president restructured the Civil Aeronautics Authority. Under this reorganization, the functions of the Air Safety Board and the five-person Civil Aeronautics Authority were combined into a new organization known as the Civil Aeronautics Board.

civil air regulations (CARs) The predecessors of today's Federal Aviation Regulations (FARs).

clearance delivery An operating position in the control tower responsible for issuing clearances to pilots.

cleared as filed (CAF) Means the aircraft is cleared to proceed in accordance with the route of flight filed in the flight plan. This clearance does not include the altitude, SID, or SID transition.

coast A condition that occurs when a computerized radar system is tracking a target but radar contact is temporarily lost. In a coast mode, the computer predicts the aircraft's location and displays it on the radar scope.

combined center—RAPCON (CERAP) An air traffic facility that combines the functions of an ARTCC and a radar approach control facility.

common digitizer (CD) A component of the NAS-A radar system that converts both primary and secondary radar information into a digital format, readying it for transmission to the central computer complex.

compass locator A nondirectional beacon (NDB) that has been co-located with a marker beacon transmitter.

composite route system An organized oceanic route structure incorporating reduced lateral spacing between routes in which composite separation is authorized.

composite separation A method of separating aircraft in a composite route system where, by management of route and altitude assignments, a combination of half the lateral minimum specified for the area concerned and half the vertical minimum is applied.

compulsory reporting points Reporting points that must be reported to ATC. They are designated on aeronautical charts by solid triangles or filed in a flight plan as fixes selected to define direct routes. These points are geographical locations that are defined by navigation aids/fixes. Pilots should discontinue position reporting over compulsory reporting points when informed by ATC that their aircraft is in "radar contact."

computer display channel (CDC) A component of the NAS-A radar system that channels digitized radar information from the central computer complex to the individual plan view displays.

computer readout device (CRD) A cathode ray tube display located next to the plan view display at an ARTCC controller's workstation. This device can be used to obtain or update flight plan information, obtain weather information, communicate with other controllers, receive generic ATC messages, and display aircraft flight plan information, ATC system status, or airport weather reports.

computer update equipment (CUE) Input/output equipment found at an ARTCC workstation that includes a computer readout device (CRD) and a quick action keyboard (QAK).

Conflict Alert A radar data processing program used at both ARTS and NAS-A sites that alerts the controllers whenever two participating aircraft are predicted to approach each other with less than the minimum separation.

Conflict Alert IFR/VFR Mode C Intruder A radar data processing program being installed at NAS-A sites that will alert the controller whenever IFR and VFR mode C–equipped aircraft are predicted to approach each other with less than the minimum separation.

conflict resolution An advanced future software function that will automatically provide the controller with resolutions to conflicts between two radar-tracked aircraft.

contact approach An approach wherein an aircraft on an IFR flight plan, having received the appropriate clearance and operating clear of clouds with at least 1-mile flight visibility and with a reasonable expectation of continuing to the destination airport under those conditions, may deviate from an instrument approach procedure and proceed visually to the airport of destination. During a contact approach, the pilot is responsible for navigation and terrain avoidance, whereas the controller is responsible for air traffic control separation. A contact approach may be initiated only by the pilot.

continuous wave (CW) radar An early form of radar that transmits constantly. Not used for air traffic control.

control sector An airspace area of defined horizontal and vertical dimensions for which a controller or group of controllers has air traffic control responsibility normally within an air route traffic control center or an approach control facility. Sectors are established based on predominant traffic flows, altitude strata, and controller workload. Pilot communications during operations within a sector are normally maintained on discrete frequencies assigned to the sector. (See *discrete frequency.*)

control slash A radar beacon slash representing the actual position of the associated aircraft. Normally, the control slash is the one closest to the interrogating radar beacon site. When ARTCC radar is operating in narrow band (digitized) mode, the control slash is converted to a target symbol.

control tower See *air traffic control tower.*

controlled airspace Airspace within which air traffic control services will be provided. IFR flights must obtain an air traffic control clearance; VFR flights may be subject to either weather or air traffic control restrictions.

controlled departure time (CDT) A program instituted by the Central Flow Control Facility to delay the departure of an aircraft attempting to depart toward a saturated airport.

controlled firing areas An area where activities, if not conducted in a controlled environment, could be hazardous to aircraft. Controlled firing areas are not charted on VFR or IFR charts because the controlling agency suspends its activities whenever nonparticipating aircraft approach the area.

controlled VFR (CVFR) An ICAO-approved type of flight in which the pilot must maintain VFR conditions but the aircraft is separated as if it is IFR. Controlled VFR is not used in the United States.

convective SIGMET A weather advisory concerning convective weather significant to the safety of all aircraft. Convective SIGMETs are issued for tornadoes, lines of thunderstorms, embedded thunderstorms of any intensity level, areas of thunderstorms greater than or equal to VIP level 4 with an area coverage of 4/10 (40 percent) or more, and hail 3/4 inch or greater.

coordinated universal time (UTC) A standardized time based on the current time at the prime meridian. Formerly called Greenwich mean time.

coordination The process that occurs when two controllers agree to a procedure or event when separating aircraft.

course-line computer (CLC) The primary component of VORTAC-based area navigation systems. The course-line computer is the device that calculates the aircraft's current position, the position of each waypoint, and the bearing and distance to each waypoint.

course scalloping A result of VOR or localizer signal reflections. Course scalloping occurs when a particular course or radial develops bends and curves within it but is still navigable.

critical area The area immediately surrounding a navigation transmitter (such as a localizer or glide slope) that must be kept clear of potentially reflective objects (such as vehicles, aircraft, or equipment).

data acquisition subsystem (DAS) A peripheral device of the ARTS radar processing system that receives raw radar data from the primary surveillance radar system in addition to beacon-derived information obtained from the secondary surveillance system. The DAS decodes this information, converts it to a digital format, and channels it to the data processing subsystem for further processing.

data block An alphanumeric display on a radar presentation that normally includes the aircraft's identity and altitude and may also include its ground speed and destination airport.

data entry and display subsystem (DEDS) Device used to display ARTS-derived information on a plan position indicator; it can also be used by the controller to input flight data into the computer. The DEDS consists of two subsystems: the data display and the data entry sets.

data entry sets (DES) Devices used to input flight data into an ARTS radar computer system. The data entry sets include an alphanumeric keyboard, a quick look selector, and a slew entry device, which is sometimes called a trackball.

data link A digital communications system that will be able to transmit data from the controller to the aircraft and vice versa. This information could include clearance and weather information, control instructions, or pilot-controller information requests. The first operational data link system, called controller pilot data link communications (CPDLC) is being tested at the Miami ARTCC.

data processing subsystem (DPS) The heart of the ARTS radar processing system. It is a high-speed, digital computer that accepts information from three sources — the data acquisition subsystem, the flight data processing system, and the data entry sets — correlates this information, and displays it to the controller on the PPI in the form of alphanumeric data blocks.

dead reckoning A method of navigation in which the pilot uses the forecast winds at the planned cruising altitude and applies trigonometry to deduce the proper heading to be flown to counteract the crosswind.

decision height (DH) The height at which, during a precision approach, the pilot must decide whether to continue the approach to land or to conduct a missed approach.

defense visual flight rules (DVFR) Rules applicable to VFR flights that will penetrate an air defense identification zone (ADIZ).

defruiter The electronic device used to remove spurious transponder replies (known as fruit) from radar displays.

Department of Transportation (DOT) A cabinet-level agency of the federal government within which the Federal Aviation Administration is located.

departure control A function of an approach control facility providing air traffic control service for departing IFR and, under certain conditions, VFR aircraft.

departure gates Intersections or areas used by departure control facilities primarily as outbound fixes.

departure sequencing program (DSP) A program designed to assist in achieving a specified interval over a common point for departures.

departure time The time an aircraft becomes airborne.

developmental controller The classification of a newly hired controller who has not yet become certified as a full performance level (FPL) controller.

digital BRITE (D-BRITE) An all-digital version of the BRITE radar display system.

discrete address radar channel (DARC) The backup system for the radar data processing system used in the ARTCCs.

discrete code As used in the air traffic control radar beacon system (ATCRBS), any one of the 4,096 selectable mode 3/A aircraft transponder codes, except those ending in zero zero; e.g., discrete codes: 0010, 1201, 2317, 7777; nondiscrete codes: 0100, 1200, 7700. Nondiscrete codes are normally reserved for radar facilities that are not equipped with discrete decoding capability and for other purposes such as emergencies (7700) or VFR aircraft (1200).

discrete frequency A separate radio frequency for use in direct pilot-controller communications in air traffic control that reduces frequency congestion by controlling the number of aircraft operating on a particular frequency at one time. Discrete frequencies are normally designated for each control sector in en route/terminal ATC facilities. Discrete frequencies are listed in the Airport/Facility Directory and the DOD FLIP IFR Enroute Supplement.

distance measuring equipment (DME) Electronic equipment, consisting of an interrogator and a transponder, that permits the pilot to accurately determine the aircraft's distance from a ground station. The ground-based DME transponder is normally co-located with either a VOR or an ILS. A precision version of DME is a functional component of the microwave landing system.

diverse vector area (DVA) In a radar environment, that area in which a prescribed departure route is not required as the only suitable route to avoid obstacles. The area in which random radar vectors below the MVA/MIA, established in accordance with the TERPS criteria for diverse departures obstacles and terrain avoidance, may be issued to departing aircraft.

domestic airspace Airspace that overlies the continental land mass of the United States plus Hawaii and U.S. possessions. Domestic airspace extends to 12 miles offshore.

Doppler radar An outdated form of area navigation that relies on the Doppler effect, or a frequency shift of reflected radar transmissions, to calculate the aircraft's ground speed and true course.

Doppler VOR (DVOR) A VOR that operates using completely different principles than a conventional VOR, although this difference in operation is transparent to the pilot. Doppler VOR is less sensitive to reflections from buildings or terrain than a conventional VOR transmitter.

duplexer A device that permits both the radar transmitter and receiver to utilize the same antenna. The duplexer ensures that the receiver is never on during pulse transmissions because a high-energy pulse would likely destroy the receiver. The duplexer also switches the transmitter off during the time that the receiver is listening for echos.

DVFR flight plan A flight plan filed for a VFR aircraft that intends to operate in airspace within which the ready identification, location, and control of aircraft are required in the interest of national security.

echo The reflection of radar energy from an object such as an aircraft, vehicle, or terrain.

emergency locator transmitter (ELT) A radio transmitter attached to an aircraft that transmits a continuous signal on 121.5 mHz and 243.0 mHz in case of an accident. The ELT is a valuable tool when searching for lost aircraft.

en route automated radar tracking system (EARTS) An automated radar and radar beacon tracking system. Its functional capabilities and design are essentially the same as the terminal ARTS IIIA system except for the EARTS' capability of employing both short-range (ASR) and long-range (ARSR) radars, use of full digital radar displays, and fail-safe design.

en route flight advisory service (EFAS) A service of selected flight service stations specifically designed to provide timely weather information to pilots en route to their destination.

en route metering (ERM) A software program resident on the ARTCC central computer complex that is able to determine an airport's acceptance rate for a 15-minute interval and then match the inbound flow of traffic to the calculated acceptance rate by issuing crossing times over specified navigational fixes known as metering fixes.

en route minimum safe altitude warning (E-MSAW) A function of the NAS-A en route computer that aids controllers by alerting them when a tracked aircraft is below or predicted by the computer to go below a predetermined minimum IFR altitude.

en route sector loading (ELOD) A software program, resident on the central computer complex, that is able to calculate every sector's current and predicted traffic load and alert personnel at the Central Flow Control Facility and at the traffic management units in the ARTCC whenever it predicts that a particular en route sector will become saturated with traffic.

en route spacing program (ESP) A program designed to assist the exit sector in achieving the required in-trail spacing.

Enhanced Conflict Alert An enhanced version of the current Conflict Alert software operational at ARTS radar sites. Enhanced Conflict Alert will not have to be desensitized near parallel approach corridors.

expect further clearance (EFC) The time that the pilot can expect to receive clearance beyond the assigned clearance limit.

expected departure clearance time (EDCT) The runway release time assigned to an aircraft by the controlled departure time software program used by the Central Flow Control Facility.

extremely high frequency The frequency band between 30 and 300 gHz.

facility directives (FD) Official documents that clarify methods and procedures used by the controllers within a particular air traffic control facility.

facility rating A certificate issued by the FAA when a controller has become certified in every assigned sector and has passed the appropriate written examinations.

false courses Courses or radials resulting from VOR or localizer signal reflections; these courses cannot be used because of their extreme inaccuracy.

false glide paths Extraneous glide paths produced by the glide slope transmitter. In every instance, the false glide paths are elevated at a greater angle than the desired glide path. There will never be a false glide path below the desired glide path.

FDC NOTAM A regulatory notice to airmen issued by the Flight Data Center.

federal airways See *airway*.

Federal Aviation Administration (FAA) The agency of the Department of Transportation charged with operating the civilian air traffic control system in the United States.

Federal Aviation Agency (FAA) The predecessor to the Federal Aviation Administration. It ceased to exist when the Department of Transportation was formed in 1967.

Federal Communications Commission (FCC) The federal authority charged with allocating, monitoring, and regulating radio communications systems.

feeder controllers Controllers whose responsibilities are to sequence arriving aircraft toward the final approach course.

feeder fix The fix depicted on instrument approach procedure charts that establishes the starting point of the feeder route.

feeder route A route depicted on instrument approach procedure charts to designate routes for aircraft to proceed from the en route structure to the initial approach fix (IAF).

feedhorn The component of the rotating radar antenna that directs the microwave radar energy toward the reflecting antenna.

field training program Air traffic control training that a developmental controller receives at his or her assigned air traffic control facility.

fields As applied to flight data processing, the individual components of a flight plan, such as aircraft type, requested altitude, and so on.

final approach fix (FAF) The fix from which the final approach segment of an instrument approach begins. The final approach fix is identified on the profile view of an instrument approach chart using the letter X.

final approach segment The segment of an instrument approach that extends from the final approach fix to the missed approach point.

final controller A controller whose responsibility is to sequence aircraft on the instrument approach.

final monitor aid (FMA) A high-resolution color display that is equipped with the controller alert system hardware/software used in the precision runway monitor (PRM) system. The display includes alert algorithms providing the target predictors, a color change alert when a target penetrates or is predicted to penetrate the no transgression zone (NTZ), a color change alert if the aircraft transponder becomes inoperative, synthesized voice alerts, digital mapping, and like features contained in the PRM system.

final monitor controller Air traffic control specialist assigned to radar monitor the flight path of aircraft during simultaneous parallel and simultaneous close parallel ILS approach operations. Each runway is assigned a final monitor controller during simultaneous parallel and simultaneous close parallel ILS approaches. Final monitor controllers utilize the precision runway monitor (PRM) system during simultaneous close parallel ILS approaches.

flight check A call sign prefix used by special FAA aircraft engaged in the flight inspection/certification of navigation aids and flight procedures.

Flight Data Center (FDC) The department in Washington, D.C., that publishes appropriate flight data defining the National Airspace System.

flight data controller An operating position within both a control tower and an ARTCC whose duties are to maintain and update relevant information concerning aircraft and the air traffic control system.

flight data input/output (FDIO) An improved communications device utilizing a video terminal and keyboard that will eventually replace the FDEP equipment.

flight data processing (FDP) The computer system in the ARTCC that provides automation capability to accept and store flight plan information, print and distribute flight plan information in the form of flight progress strips, calculate and update flight plan data, and transfer flight plan data automatically from one sector to the next within any particular ARTCC, from one ARTCC to the adjacent ARTCCs, and from ARTCCs to FDIO-equipped control towers and TRACONs.

flight information regions (FIRs) Airspace within which air traffic control services will be provided.

flight inspection field office (FIFO) The operational office where flight inspection aircraft, crews, and technicians are located.

flight levels A level of constant atmospheric pressure related to a reference datum of 29.92 inches of mercury. Every flight level is stated in hundreds of feet, with the last two zeroes being dropped.

flight management system (FMS) A computer system that uses a large database to allow routes to be preprogrammed and fed into the system by means of a data loader. The system is constantly updated with respect to position accuracy by reference to conventional navigation aids. The sophisticated program and its associated database ensure that the most appropriate aids are automatically selected during the information update cycle.

flight progress strips Standardized paper strips that contain essential flight information about aircraft participating in the National Airspace System. The flight progress strips used in the ARTCCs are of a different configuration than those used in air traffic control towers.

flight service station (FSS) An air traffic control facility that provides pilot briefings and en route communications and that conducts VFR search and rescue services. Selected flight service stations also provide en route flight advisory service.

flight trajectory modeling The procedure whereby the AERA computer projects an aircraft's position forward in four dimensions: lateral, longitudinal, vertical, and temporal.

four-course radio range An obsolete navigational aid that used an aural signal for navigation.

framing pulses The two pulses that begin and end a reply from an airborne transponder.

Fresnel lens A type of lens used on runway lights to collect and focus the light toward the approach ends of the runway.

front course The side of the ILS approach normally used for navigation. The front course is also normally equipped with a glide slope and marker beacons.

fruit Electronic radar interference caused by transponders replying to multiple interrogations from different radar systems.

fuel remaining A phrase used by either pilots or controllers when relating to the fuel remaining on board until actual fuel exhaustion. When transmitting such information in response to either a controller question or pilot-initiated cautionary advisory to air traffic control, pilots will state the approximate number of minutes the flight can continue with the fuel remaining. All reserve fuel should be included in the time stated, as should an allowance for established fuel gauge system error.

full performance level (FPL) controller A controller fully certified at every assigned operating position. Also called facility rated.

full route clearance (FRC) The procedure of verbally stating the entire route of flight to the pilot.

gain A control on a radar system that increases the intensity of displayed radar returns.

gate hold procedures Procedures at selected airports to hold aircraft at the gate or other ground location whenever departure delays exceed or are anticipated to exceed 15 minutes. The sequence for departure will be maintained in accordance with initial call up unless modified by flow control restrictions. Pilots should monitor the ground control/clearance delivery frequency for engine start/taxi advisories or new proposed start/taxi time if the delay changes.

General Schedule (GS) The universal pay schedule used to pay federal employees.

glide path See *glide slope.*

glide slope The electronic path produced by the glide slope transmitter of an instrument landing system to provide vertical guidance for the pilot.

glide slope critical area The area immediately surrounding the glide slope transmitter that must be kept clear of potentially reflective objects (such as vehicles, aircraft, or equipment).

glide slope intercept altitude The minimum altitude to intercept the glide slope/path on a precision approach. The intersection of the published intercept altitude with the glide slope path, designated on government charts by the lightning bolt symbol, is the precision FAF; however, when ATC directs a lower altitude, the resultant lower intercept position is then the FAF.

Global Positioning System (GPS) A space-based radio positioning, navigation, and time-transfer system. The system provides highly accurate position and velocity information, and precise time, on a continuous global basis, to an unlimited number of properly equipped users. The system is unaffected by weather, and provides a worldwide common grid reference system. The GPS concept is predicated upon accurate and continuous knowledge of the spatial position of each satellite in the system with respect to time and distance from a transmitting satellite to the user. The GPS receiver automatically selects appropriate signals from the satellites in view and translates these into three-dimensional position, velocity, and time. System accuracy for civil users is normally 100 meters horizontally.

gray scale The scale used to code and decode altitude transmissions from mode C transponders.

Greenwich mean time (GMT) See *coordinated universal time.*

ground clutter A pattern produced on the radar scope by ground returns which may degrade other radar returns in the affected area. The effect of ground clutter is minimized by the use of moving target indicator (MTI) circuits in the radar equipment resulting in a radar presentation which displays only targets which are in motion.

ground control The operating position in a control tower responsible for aircraft and vehicular movement about the surface of the airport, including taxiways and inactive runways. The ground controller is not responsible for the movement of aircraft on or across active runways.

ground wave The LORAN-C signal that remains close to the surface of the earth.

group form Saying several numbers as a group rather than enunciating them individually. For example, the group form pronunciation of the number 100 is "one hundred."

group repetition interval (GRI) The unique time interval between transmissions of a LORAN-C master station. Each LORAN-C chain is identified using its unique GRI.

handoff The action taken to transfer the radar identification of an aircraft from one controller to another when the aircraft will enter the receiving controller's airspace and radio communications will be transferred.

Hazardous In-Flight Weather Advisory Service (HIWAS) Continuous recorded hazardous in-flight weather forecasts broadcasted to airborne pilots over selected VOR outlets defined as an HIWAS Broadcast Area.

heavy aircraft An aircraft whose maximum certificated takeoff weight is more than 255,000 pounds.

high-altitude en route flight advisory service A service of selected flight services stations specifically designed to provide timely weather information to pilots en route at altitudes at or above FL 180.

high frequency The frequency band between 3 and 30 mHz.

high-intensity approach lighting system (ALSF) Approach lighting systems that extend 2,400 feet to 3,000 feet from the end of the runway. ALSF-1 includes sequenced flashing lights and used to be the standard for Category I ILS runways (MALSR is now the standard installation). ALSF-2 includes sequenced flashing lights and is the standard configuration for Category II ILS runways.

high-intensity runway lighting (HIRL) Runway lighting, primarily used on instrument runways, with a maximum wattage of 200 watts. High-intensity runway lights operate on one of five steps, with step one being the lowest illumination and step five the highest.

history The radar targets displayed on the plan position indicator for a number of antenna revolutions before completely disappearing. The most recent targets are the brightest, with subsequent targets becoming somewhat lower in intensity. History is what the controller uses to determine an aircraft's relative direction of flight and its velocity.

hold procedure A predetermined maneuver that keeps aircraft within a specified airspace while awaiting further clearance from air traffic control. Also used during ground operations to keep aircraft within a specified area or at a specified point while awaiting further clearance from air traffic control.

host computer The IBM 3083 computer once used in the ARTCCs for radar and data processing.

ICAO Annexes International guidelines developed by the International Civil Aviation Organization for the operation of air traffic services. These Annexes cover the following subjects: Personnel Licensing, Rules of the Air, Meteorology, Aeronautical Charts, Units of Measurement to Be Used in Air-Ground Communications, Operation of Aircraft, Aircraft Nationality and Registration Marks, Airworthiness of Aircraft, Facilitation, Aeronautical Telecommunications, Air Traffic Services, Search and Rescue, Aircraft Accident Inquiry, Aerodromes, Aeronautical Information Services, Aircraft Noise, Security, and Safe Transport of Dangerous Goods by Air.

Ident The feature of the air traffic control radar beacon system that causes the special identification pulse to be transmitted by the aircraft's transponder.

identification friend or foe (IFF) The system developed in World War II that preceded the air traffic control radar beacon system.

IFR military training routes (IR) Routes used by the Department of Defense and associated Reserve and Air Guard units for the purpose of conducting low-altitude navigation and tactical training in both IFR and VFR weather conditions below 10,000 feet MLS at airspeeds in excess of 250 knots IAS.

IFR takeoff minimums and departure procedures FAR Part 91, prescribes standard takeoff rules for certain civil users. At some airports, obstructions or other factors require the establishment of nonstandard takeoff minimums, departure procedures, or both to assist pilots in avoiding obstacles during climb to the minimum en route altitude. Those airports are listed in NOS/DOD Instrument Approach Charts (IAPs) under a section entitled "IFR Takeoff Minimums and Departure Procedures."

ILS distance measuring equipment Standard distance measuring equipment co-located with an ILS localizer transmitter.

inactive runways Runways not declared active by the local controller. Inactive runways are the responsibility of the ground controller.

inertial navigation system (INS) An area navigation system dependent on accelerometers to determine an aircraft's position and route of flight.

information request (INREQ) A request originating from a flight service station for information concerning a lost or overdue aircraft.

initial approach fix (IAF) The fixes depicted on navigation charts that identify the beginning of the initial approach segment of an instrument approach procedure.

initial approach segment The segment of an instrument approach procedure that guides the aircraft from an initial approach fix to an intermediate approach fix.

initial sector suite subsystem (ISSS) The initial improved hardware that will replace the plan view displays at area control facilities.

inner marker (IM) A marker beacon used with Category II and Category III ILS approach systems. The inner marker is approximately halfway between the middle marker and the approach end of the runway.

instrument approach procedure A series of predetermined maneuvers that permit an IFR aircraft to leave the confines of the airway structure and descend for landing at an airport.

instrument approach procedure charts A graphic depiction of the maneuvers used during an instrument approach procedure.

instrument flight rules (IFR) The rules that govern the conduct of aircraft during instrument flight.

Instrument landing system (ILS) A precision approach and landing aid that normally consists of a localizer, a glide slope, marker beacons, and an approach light system.

1. ILS Category I—An ILS approach procedure that provides for approach to a height above touchdown of not less than 200 feet and with runway visual range of not less than 1,800 feet.
2. ILS Category II—An ILS approach procedure that provides for approach to a height above touchdown of not less than 100 feet and with runway visual range of not less than 1,200 feet.
3. ILS Category III.
   a. IIIA—An ILS approach procedure that provides for approach without a decision height minimum and with runway visual range of not less than 700 feet.
   b. IIIB—An ILS approach procedure that provides for approach without a decision height minimum and with runway visual range of not less than 150 feet.
   c. IIIC—An ILS approach procedure that provides for approach without a decision height minimum and without runway visual range minimum.

Interdepartmental Air Traffic Control Board (IATCB) An organization formed on April 7, 1941, to coordinate activities between the Civil Aeronautics Administration and the military services. This board remained in existence until 1946.

intermediate approach segment The segment of an instrument approach procedure that guides the aircraft from an intermediate approach fix to the final approach fix.

International Air Transport Association (IATA) An international organization of airlines that has assisted in the definition of minimum navigation performance specification airspace.

International Civil Aviation Organization (ICAO) A specialized agency of the United Nations whose objective is to develop the principles and techniques of international air navigation and air traffic control.

***International Flight Information Manual* (IFIM)** A publication of the Federal Aviation Administration primarily designed for pilots entering foreign airspace or returning to the United States from foreign locations.

interrogator The ground-based component of ATCRBS. Also, a major component of the airborne equipment used for distance measuring.

Interstate Airway Communication Stations (INSACSs) Radio communications facilities strategically located to offer flight advisory services to aircraft operating along the federal airways. INSACS were staffed by air traffic controllers who communicated directly with pilots by radio and passed along weather information and instructions from the controllers working at the airway traffic control centers. INSACSs became flight service stations.

jet advisory areas Areas created to provide advisory services to civilian and military turbojet aircraft operating at high altitudes. The jet advisory areas extended from FL 240 to FL 410 and projected 14 nautical miles laterally on either side of every high-altitude airway. Air traffic controllers were required to use radar to constantly monitor every IFR aircraft operating on a jet route and to issue any heading change necessary to ensure that the IFR aircraft remained separated from unidentified aircraft observed on the controller's radar display. The jet advisory areas were replaced by the positive control area.

jet route A route designed to serve aircraft operations from 18,000 feet MSL up to and including FL 450. The routes are referred to as "J" routes with numbering to identify the designated route; e.g., J105.

joint surveillance systems (JSSs) Long-range radar surveillance systems jointly operated by the Federal Aviation Administration and the Department of Defense.

joint use airspace Airspace used for national defense that is released for civilian use when it is not needed by the military.

keyboard A data input device used by the flight data entry and printout system, the radar data processing system found in ARTCCs, and ARTS radar systems.

Landing Aids Experiment Station (LAES) A research center established in 1945 by the CAA, Army Air Corps, and Navy Department at the Naval Air Station at Arcata, California. It was here that most of the pioneering research in approach lighting was conducted.

landing minima The minimum visibility prescribed for landing a civil aircraft while using an instrument approach procedure. The minimum applies with other limitations set forth in FAR 91 with respect to the minimum descent altitude (MDA) or decision height (DH) prescribed in the instrument approach procedures as follows:

1. Straight-in landing minima—A statement of MDA and visibility, or DH and visibility, required for a straight-in landing on a specified runway, or
2. Circling minima—A statement of MDA and visibility required for the circle-to-land maneuver.

large aircraft Aircraft whose maximum certificated takeoff weight is more than 41,000 pounds up to and including 255,000 pounds.

lateral separation A method of separating aircraft operating at the same altitude but on different routes.

letters of agreement (LOA) Official documents that clarify the methods and procedures to be used by controllers at different air traffic control facilities.

light gun A hand-held, highly directional light-signaling device used to communicate instructions to aircraft not equipped with operable radio communications equipment.

linear polarization (LP) The normal operating mode of primary air traffic control radar.

local controller The controller whose responsibility is the sequencing and spacing of aircraft operating on the active runways of an airport.

localizer The component of the instrument landing system that provides lateral guidance to the aircraft.

localizer back course The localizer emanations as they appear to an aircraft approaching the reciprocal runway.

localizer critical area The area immediately surrounding the localizer transmitter that must be kept clear of potentially reflective objects, such as vehicles, aircraft, or equipment.

localizer directional aid (LDA) A navigation aid used for instrument approaches that operates similarly to and provides the same accuracy as an ILS localizer.

locator middle marker (LMM) A nondirectional beacon (NDB) that has been co-located with a middle marker beacon transmitter.

locator outer marker (LOM) A nondirectional beacon (NDB) that has been co-located with an outer marker beacon transmitter.

longitudinal separation A method of separating aircraft operating at the same altitude on the same route.

long-range navigation (LORAN) An area navigation system that uses multiple transmitters to plot hyperbolic lines of position.

lost communications Loss of the ability to communicate by radio. Aircraft are sometimes referred to as NORDO (No Radio). Standard pilot procedures are specified in FAR 91. Radar controllers issue procedures for pilots to follow in the event of lost communications during a radar approach when weather reports indicate that an aircraft will likely encounter IFR weather conditions during the approach.

low-altitude airway structure The network of airways serving aircraft operations up to but not including 18,000 feet MSL.

low approach An approach over a runway in which the pilot initiates a departure before making contact with the runway.

low frequency The frequency band between 30 and 300 kHz.

low-intensity runway lighting (LIRL) The most inexpensive lighting system to install, typically equipped with 15-watt bulbs that operate on one intensity level (step one).

mach The ratio of true airspeed to the speed of sound. Mach 1.0 is the speed of sound.

main bang The spot on a plan position indicator that represents the location of the rotating radar antenna.

marker beacon An electronic navigation facility that transmits a low-intensity coded signal on 75 mHz. Marker beacons are typically used as part of an instrument landing system.

maximum authorized altitude (MAA) A published altitude representing the maximum usable altitude or flight level for an airspace structure or route segment. It is the highest altitude on a federal airway, jet route, area navigation low or high route, or other direct route for which an MEA is designated in FAR 95 at which adequate reception of navigation aid signals is assured.

medium frequency The frequency band between 300 and 3000 kHz.

medium-intensity approach lighting system (MALS) An inexpensive approach lighting system that operates on three steps of intensity, with step three being equivalent in intensity to step three on an ALSF system. Using medium-intensity white lamps, MALS systems extend 1,400 feet from the runway threshold, with the light bars spaced at 200-foot intervals.

medium-intensity approach lighting system with RAIL (MALSR) An inexpensive approach lighting system similar to MALS that incorporates runway alignment indicator lights (RAIL). MALSR systems extend 2,400 feet from the runway threshold, with the light bars spaced at

200-foot intervals. MALSR approach lighting systems operate on step one through step three, with step three being equivalent in intensity to step three on an ALSF system.

medium-intensity runway lighting (MIRL) Runway edge lights equipped with 40-watt bulbs. MIRL lighting can be operated on three intensity levels. When operated on step one, medium-intensity lights produce the same light level as low-intensity lights (15 watts). When functioning on step two, they operate at about 25 watts, and on step three, they operate at the maximum allowable 40-watt level.

metering A method of time regulating arrival traffic flow into a terminal area so as not to exceed a predetermined terminal acceptance rate.

metering airports Airports adapted for metering and for which optimum flight paths are defined. A maximum of 15 airports may be adapted.

metering fix A fix along an established route from over which aircraft will be metered prior to entering terminal airspace. Normally, this fix should be established at a distance from the airport that will facilitate a profile descent 10,000 feet above airport elevation (AAE) or above.

middle marker (MM) A marker beacon normally placed approximately 1/2 nautical mile from the approach end of a runway served by an instrument landing system.

military assumes responsibility for separation of aircraft (MARSA) A condition whereby the military service involved assumes responsibility for air traffic control separation between participating military aircraft.

military operations areas (MOAs) Airspace where intensive military training operations are conducted. IFR aircraft are routed around these areas whenever the areas are active.

military training route (MTR) Airspace where military training missions are conducted at airspeeds in excess of 250 knots.

millibars Metric pressure measurement intervals; used primarily in reference to altimeter settings.

minimum descent altitude (MDA) The lowest altitude to which descent is authorized during a nonprecision instrument approach procedure.

minimum en route altitude (MEA) The lowest published altitude between navigational fixes that provides both obstacle clearance and adequate navigation radio reception.

minimum navigation performance specifications airspace (MNPSA) The airspace that extends from the northeastern United States to Great Britain, from about FL 275 to FL 420.

minimum obstruction clearance altitude (MOCA) The lowest published altitude between navigational fixes that provides obstacle clearance over the entire route and adequate navigation radio reception within 22 nautical miles of the navaid transmitter.

minimum safe altitude warning (MSAW) A function of air traffic control computer systems that alerts the controller whenever a mode C–equipped aircraft is below or is predicted to descend below a predetermined minimum safe altitude.

minimum vectoring altitude (MVA) The lowest altitude above sea level at which IFR aircraft may be vectored by the controller.

missed approach point (MAP) The point at which the missed approach procedure will be performed by the pilot if the required visual references do not exist.

missed approach procedure The maneuver performed by a pilot when an instrument approach cannot be completed to a landing.

missed approach segment The segment of an instrument approach procedure that lies between the missed approach point and a predetermined missed approach fix.

mode C altitude encoder The component on the aircraft that transmits altitude information to the ground-based radar system.

mode C intruder alert A function of certain air traffic control automated systems designed to alert radar controllers to existing or pending situations between a tracked target (known IFR or VFR aircraft) and an untracked target (unknown IFR or VFR aircraft) that requires immediate attention/action.

mode S A transponder mode that will permit individual aircraft interrogation and data link transfer.

modes The letter or number assigned to a specific pulse spacing of the radio signals transmitted by various ATCRBS components.

monitor controller A controller who continuously monitors aircraft conducting parallel ILS approaches.

moving target indicator (MTI) An electronic device that will permit radar scope presentation only from targets that are in motion. A partial remedy for ground clutter.

MTI gate The control that adjusts the range at which MTI becomes effective.

MTI/MTD video gain The control that adjusts the number of MTI- or MTD-processed radar returns that should be displayed.

narrowband The operation of NAS-A, DARC, or EARTS radar such that it creates a mosaic display.

National Air Traffic Controllers Association (NATCA) One of the current labor organizations that represent air traffic controllers.

National Airspace System The common network of airspace, airports, navigation aids, and air traffic control equipment across the United States.

National Airspace System stage A (NAS-A) The en route ATC system's radar, computers and computer programs, controller plan view displays (PVDs/radar scopes), input/output devices, and the related communications equipment that are integrated to form the heart of the automated IFR air traffic control system. This equipment performs flight data processing (FDP) and radar data processing (RDP). It interfaces with automated terminal systems and is used in the control of en route IFR aircraft.

National Beacon Code Allocation Plan (NBCAP) The national plan by which transponder codes are issued for use at individual air traffic control facilities.

National Ocean Service (NOS) The arm of the federal government that provides navigation charts to the FAA and the flying public. The National Ocean Service is part of the Department of Commerce.

National Transportation Safety Board (NTSB) The arm of the Department of Transportation charged with investigating all major transportation accidents.

National Weather Service (NWS) The arm of the Department of Commerce charged with collecting, disseminating, and forecasting weather conditions for the public.

navigation aids (navaids) Any visual or electronic device used by pilots to navigate.

next-generation radar (NEXRAD) A weather detection radar system.

night differential Extra pay that controllers receive for working night shifts.

noise As used in electronics, randomly generated electronic signals.

nondirectional radio beacon (NDB) A radio navigation beacon that transmits a uniform signal omnidirectionally using either the LF or MF radio frequency band.

nonprecision approach A standard instrument approach procedure in which no electronic glide path is provided.

normal video gain The radar control that regulates the amplification of the displayed radar signal.

North American Route A numerically coded route preplanned over existing airway and route systems to and from specific coastal fixes serving the North Atlantic.

North Atlantic Region (NAR) That area over the North Atlantic Ocean within which certain air traffic control rules apply.

North Mark A beacon data block sent by the host computer to be displayed by the ARTS on a 360° bearing at a locally selected radar azimuth and distance. The North Mark is used to ensure correct range/azimuth orientation during periods of CENRAP.

North Pacific (NOPAC) An organized route system between the Alaskan west coast and Japan.

notice to airmen (NOTAM) A notice containing information (not known sufficiently in advance to publicize by other means) concerning the establishment, condition, or change in any component (facility, service, or procedure of, or hazard in the National Airspace System) the timely knowledge of which is essential to personnel concerned with flight operations.

1. NOTAM(D)—A NOTAM given (in addition to local dissemination) distant dissemination beyond the area of responsibility of the flight service station. These NOTAMs will be stored and available until canceled.
2. NOTAM(L)—A NOTAM given local dissemination by voice and other means, such as telautograph and telephone, to satisfy local user requirements.
3. FDC NOTAM—A NOTAM regulatory in nature, transmitted by USNOF and given systemwide dissemination.

oceanic airspace Airspace over the oceans of the world, considered international airspace, where oceanic separation and procedures per the International Civil Aviation Organization are applied. Responsibility for the provisions of air traffic control service in this airspace is delegated to various countries, based generally on geographic proximity and the availability of the required resources.

Oceanic Area Control Centers (OACC) Area control centers with responsibility for oceanic airspace.

Oceanic Display and Planning System (ODAPS) An automated digital display system that provides flight data processing, conflict probe, and situation display for oceanic air traffic control.

off route vector A vector by ATC that takes an aircraft off a previously assigned route. Altitudes assigned by ATC during such vectors provide required obstacle clearance.

Office of Personnel Management (OPM) The federal office charged with conducting initial air traffic controller hiring for the FAA.

offset parallel runways Staggered runways having parallel centerlines.

offshore control area That portion of airspace between the U.S. 12-mile limit and the oceanic CTA/FIR boundary within which air traffic control is exercised. These areas are established to permit the application of domestic procedures in the provision of air traffic control services.

outer marker (OM) A marker beacon located approximately at the glide slope interception altitude of an ILS approach.

permanent echo Radar signals reflected from fixed objects on the Earth's surface; e.g., buildings, towers, terrain. Permanent echoes are distinguished from "ground clutter" by being definable locations rather than large areas. Under certain conditions, they may be used to check radar alignment.

phantom VORTAC A nonexistent VORTAC created at a predetermined point by an area navigation system. A phantom VORTAC is the same as a waypoint.

pilot-controlled lighting (PCL) systems A runway and/or approach lighting system that can be controlled by the pilot through the aircraft's VHF communications radio.

pilot reports (PIREPs) Reports made by pilots about meteorological conditions encountered while in flight.

pilotage A means of VFR navigation using navigational charts for position determination.

plan position indicator (PPI) A radar display device that can provide two-dimensional aircraft position information (azimuth and bearing).

plan view display (PVD) A radar display device used by radar mosaic systems (NAS-A, DARC, and EARTS).

point out The action taken to transfer the radar identification of an aircraft from one controller to another when the aircraft will enter the receiving controller's airspace but radio communications will not be transferred.

polar track structure (PTS) A system of organized routes between Iceland and Alaska that overlie Canadian MNPS airspace.

positive control The separation of all air traffic by air traffic control within designated airspace.

positive control area (PCA) The airspace designated by FAR 71 within which all aircraft are subject to positive control.

precision approach path indicator (PAPI) A visual navigation device that permits the pilot to judge the aircraft's position relative to the desired glide path.

precision approach radar (PAR) Radar equipment, primarily used by military air traffic controllers, that provides aircraft range, bearing, and elevation while on the approach to a runway.

preferential routes Departure or arrival routes that increase system efficiency by organizing traffic flows.

preferred arrival route (PAR) A preferential route primarily used for aircraft arriving at a terminal area.

preferred departure route (PDR) A preferential route primarily used for aircraft departing a terminal area.

PRF stagger A means of staggering a primary radar system's pulse repetition frequency to reduce the effects of aircraft blind speeds.

prime channel A slang term for the NAS-A radar processing system used in the ARTCCs.

prime meridian The 0° line of longitude that passes through Greenwich, England.

procedure turn A maneuver performed by a pilot during an instrument approach when it is necessary for the aircraft to reverse course to become established on the intermediate or final segment of the instrument approach.

Procedures for Air Navigation Services (PANS) ICAO recommendations for the establishment and operation of air traffic control systems.

Professional Air Traffic Controllers Organization (PATCO) The defunct labor organization that represented many of the controllers in the 1970s and early 1980s.

prohibited areas Designated airspace within which the flight of aircraft is absolutely prohibited.

Project Beacon The FAA task force that recommended the development and use of the ATCRBS system and ARTS and NAS-A radar processing systems.

Provisional International Civil Aviation Organization (PICAO) The predecessor of the International Civil Aviation Organization.

pulse repetition frequency (PRF) The rate at which primary radar pulses are generated.

pulse train The reply from an airborne ATCRBS transponder.

pulse-type radar Radar that transmits multiple pulses; the opposite of continuous wave radar.

QALQ message A request made by a flight service station for information concerning an overdue aircraft. Any facility that receives a QALQ must briefly check with every controller and examine recent flight strips to determine whether any contact has been made with the overdue aircraft.

quick action keyboard (QAK) A keyboard used by ARTCC controllers to extract flight information from the central computer complex.

quick look A feature of NAS-A and ARTS that provides the controller the capability to display full data blocks of tracked aircraft from other control positions.

quota flow control (QFLOW) A flow control procedure by which the Central Flow Control Facility (CFCF) restricts traffic to the ARTCC having an impacted airport, thereby avoiding sector/area saturation.

radar Radio detection and ranging equipment. Radar measures the interval between the transmission and reception of a radio pulse to determine an aircraft's bearing and distance.

radar approach An instrument approach procedure that uses either precision approach radar (PAR) to provide azimuth, distance, and elevation information or airport surveillance radar (ASR) to provide azimuth and distance information.

radar approach control (RAPCON) A terminal air traffic control facility that uses radar to provide approach control services to aircraft.

radar-assisted navigation A controller's use of radar to vector an aircraft off a published route or procedure.

radar associate/nonradar controller The controller at an ARTCC who assists the radar controller to effect aircraft separation. This controller is primarily responsible for updating the appropriate flight progress strips.

radar contact The term used by an air traffic controller to inform the pilot that the aircraft has been positively identified on the radar display.

radar contact lost The term used by an air traffic controller to inform the pilot that the aircraft is no longer identified on the radar display.

radar controller A controller that uses radar to effect aircraft separation.

radar cross section A theoretical value that describes the relative radar reflectivity of an object.

radar data acquisition subsystem (RDAS) The ARTS subsystem that digitizes primary radar information and transmits it to the data processing system for eventual display on the PPI.

radar data processing (RDP) The second phase of the ARTCC automation process that provides for radar input from multiple radar sites, radar mosaic capability, computer validation and selection of the most accurate data for display to the controller, automatic aircraft tracking, visual display of flight information, and automatic radar handoffs.

radar identification The process of ascertaining that an observed radar target is the radar return from a particular aircraft.

radar identified aircraft An aircraft whose position has been correlated with an observed target or symbol on the radar display.

radar mile The time that it takes radar signals to travel 1 nautical mile: 12.36 microseconds. A radar mile is a time measurement, not a distance measurement.

radar mosaic The type of display system used by NAS-A, DARC, or EARTS systems that use more than one radar system to generate a composite display.

radar point out An action taken by a controller to transfer the radar identification of an aircraft to another controller if the aircraft will or may enter the airspace or protected airspace of another controller and radio communications will not be transferred.

radar separation Radar spacing of aircraft in accordance with established criteria.

radar service terminated The term used by an air traffic controller to advise the pilot that radar services will no longer be provided.

radar traffic advisories Advisories issued to alert pilots to known or observed radar traffic that may affect the intended route of flight of their aircraft.

Radar Training Facility (RTF) The facility at the FAA Academy used for basic radar training of controllers.

radar vectoring Provision of navigational guidance to aircraft in the form of specific headings based on the use of radar.

radar weather echo intensity levels Existing radar systems cannot detect turbulence. However, there is a direct correlation between the degree of turbulence and other weather features associated with thunderstorms and the radar weather echo intensity. The National Weather Service has categorized radar weather echo intensity for precipitation into six levels. These levels are sometimes expressed during communications as "VIP LEVEL" 1 through 6 (derived from the component of the radar that produces the information—Video Integrator and Processor). The following list gives the "VIP LEVELS" in relation to the precipitation intensity within a thunderstorm:

Level 1. Weak
Level 2. Moderate
Level 3. Strong
Level 4. Very strong
Level 5. Intense
Level 6. Extreme

radial A magnetic bearing extending from a VOR/VORTAC/TACAN navigation facility.

Radio Technical Commission for Aeronautics (RTCA) A federal commission composed of members from the State Department, Defense Department, Coast Guard, Federal Communications Commission, and FAA.

range cells Areas of radar coverage used by the common digitizer to transmit the position of each aircraft to the central computer complex.

range mark Concentric circles displayed on a plan position indicator centered on the main bang.

range select switch The control used to select the range limits displayed on the plan position indicator.

range time The interval between the transmission of a DME interrogation signal and receipt of the reply to that interrogation.

receiver The component of a radio device that receives transmissions.

receiver gain The amplification control on a primary radar system.

regional offices The nine FAA offices located across the country that carry out the day-to-day operations of the FAA. The structure of each regional office is fairly similar to that of the FAA's Washington headquarters.

release time A departure time restriction issued to a pilot by ATC (either directly or through an authorized relay) when necessary to separate a departing aircraft from other traffic.

remote communication air/ground (RCAG) An unmanned VHF/UHF transmitter/receiver facility used to expand ARTCC air/ground communications capability.

remote communications outlet (RCO) An unmanned communications facility used by controllers at a flight service station. It is similar to a remote communication air/ground unit.

reporting point A geographical location in relation to which the position of an aircraft is reported.

Rescue Coordination Center (RCC) A facility equipped and staffed to coordinate search and rescue operations.

restricted areas Airspace designated by FAR 73 within which the flight of aircraft is not wholly prohibited but is subject to some operating restrictions.

reverse sensing The operation of a localizer indicator when receiving the localizer back course signal.

runway incursion An aircraft inadvertently taxiing onto or across an active runway without the local controller's knowledge or permission.

runway threshold lights Fixed green lights arranged symmetrically left and right of the runway centerline that identify the runway threshold. Threshold lights may be designed to appear red to aircraft approaching from the opposite direction.

runway use program A noise abatement runway selection plan designed to enhance noise abatement efforts.

runway visual range (RVR) A system that derives a value representing the horizontal distance that pilots see down the runway.

safety alert A warning issued by a controller when an aircraft may be in unsafe proximity to other aircraft, terrain, or obstructions.

satellite controllers Controllers who handle approaches and departures from low-activity airports within the approach control's area of jurisdiction.

search and rescue (SAR) A service that seeks to locate missing aircraft and aid any individual in need of assistance.

secondary surveillance radar (SSR) See *air traffic control radar beacon system.*

secretary of transportation The administrator of the Department of Transportation.

sectional charts VFR navigational charts scaled 1: 500,000, or about 8 statute miles to the inch.

sector suites See *initial sector suite subsystem.*

sectors Areas within which a single controller has responsibility for aircraft separation.

see and avoid A visual procedure wherein pilots flying in VFR conditions, regardless of the type of flight plan, are responsible for observing the presence of other aircraft and maneuvering to avoid these aircraft. Also called "see and be seen."

selective interrogation The process whereby mode S transponders will be able to interrogate individual aircraft.

semiautomated ground environment (SAGE) An air defense system developed by the U.S. Air Force in the late 1950s.

sensitivity time control (STC) Circuitry designed to provide a method by which primary radar echoes can be equalized before they are displayed on the PPI. It is an electronic means of automatically controlling the sensitivity of the receiver to equalize the display intensity of both nearby and distant targets.

separation error A loss of minimum required separation.

sequenced flashing lights (SFL) High-intensity condenser discharge strobe lights usually placed in conjunction with approach lighting systems.

service volume The area within which reliable VOR and VORTAC reception is ensured.

severe weather avoidance plan (SWAP) An approved plan to minimize the affect of severe weather on traffic flows in impacted terminal and/or ARTCC areas. SWAP is normally implemented to provide the least disruption to the ATC system when flight through portions of airspace is difficult or impossible due to severe weather.

shrimp boat A small plastic device used to mark an aircraft's position on a radar display not utilizing alphanumerics.

side lobe suppression (SLS) Electronic circuitry used by the ATCRBS to reduce replies to extraneous transmissions known as side lobes.

side lobes Unwanted transmissions from the rotating ATCRBS antenna not associated with the main transmission.

sidestep maneuver A visual maneuver accomplished by a pilot at the completion of an instrument approach to permit a straight-in landing on a parallel runway not more than 1,200 feet to either side of the runway to which the instrument approach was conducted.

SIGMET A weather advisory issued concerning weather significant to the safety of all aircraft. SIGMET advisories cover severe and extreme turbulence, severe icing, and widespread dust or sandstorms that reduce visibility to less than 3 miles.

simplified directional facility (SDF) A navigation aid used for nonprecision approaches that provides a course similar to the localizer transmitter of an ILS.

simplified short approach lighting system (SSALS) A much shorter version of the ALSF-1 approach lighting system; it is only 1,200 feet long. This system still utilizes the same high-intensity white approach lights as the ALSF-1 system, but they are spaced at 200-foot intervals.

simultaneous ILS approaches An approach system permitting simultaneous ILS/MLS approaches to airports having parallel runways separated by at least 4,300 feet between centerlines. Integral parts of a total system are ILS/MLS, radar, communications, ATC procedures, and appropriate airborne equipment.

slant range distance The actual distance between the aircraft and the ground-based DME transponder.

slew entry device (SED) A data entry device used by NAS-A, DARC, EARTS, and ARTS-III systems. Also known as a trackball.

small aircraft Aircraft of 41,000 pounds or less maximum certificated takeoff weight.

Special Committee 31 (SC-31) A special committee of the Radio Technical Commission for Aeronautics formed to try to predict the future needs of the nation's air traffic control system. The SC-31 report recommended that a common air traffic control system be developed that would serve the needs of both military and civilian pilots.

special identification pulse (SIP) The pulse transmitted by an airborne ATCRBS transmitter when the Ident feature is utilized.

special use airspace (SUA) Airspace of defined dimensions within which certain flight activities must be confined.

special VFR (SVFR) A clearance in which a VFR aircraft is provided separation and is permitted to operate within a control zone when the weather is below VFR minima.

speed adjustment An ATC procedure used to request pilots to adjust aircraft speed to a specific value for the purpose of providing desired spacing. Pilots are expected to maintain a speed of plus or minus 10 knots or 0.02 mach number of the specified speed.

squawk To activate the transponder.

standard atmospheric pressure An air pressure of 29.92 inches of mercury.

standard instrument departure (SID) A charted IFR departure procedure.

standard terminal arrival route (STAR) A charted IFR arrival procedure.

stored program alphanumeric (SPAN) An experimental alphanumeric device that preceded the development of the NAS-A system.

super high frequency The frequency band between 3 and 30 gHz.

surveillance approach An instrument approach wherein the air traffic controller issues instructions, for pilot compliance, based on aircraft position in relation to the final approach course (azimuth) and the distance (range) from the end of the runway as displayed on the controller's radar scope. The controller will provide recommended altitudes on final approach if requested by the pilot.

sweep The faint line that emanates from the main bang to the edge of the radar screen. This line corresponds with the boresight of the antenna, is synchronized with the radar antenna, and rotates in the same direction and at the same speed.

sweep decenter Two controls, one that moves the main bang in a north-south direction and one that moves the main bang in an east-west direction. The coordinated use of both controls permits the controller to move the main bang anywhere on the PPI.

tactical air navigation (TACAN) A UHF air navigation aid that provides azimuth and distance information to the pilot using a single frequency.

target The indication on a radar display resulting from the reflection of the radar transmission.

target illumination What occurs when a radar transmission reflects off of a solid object.

taxiway edge lighting Blue lights used to define the lateral limits of the taxiway surface.

taxiway turnoff lights Green lights embedded in the runway that lead the pilot to the appropriate taxiway.

taxiways Paved areas of the airport used by aircraft to proceed to or from the runways.

Technical Evaluation and Development Center The research and development facility of the federal government that was located in Indianapolis, Indiana. This facility was replaced by NAFEC in Atlantic City, New Jersey.

temporary flight restrictions Areas within which flight may be temporarily prohibited or restricted.

terminal control area (TCA) Airspace extending upward from the surface of the Earth within which all aircraft are subject to the operating rules specified in FAR 91.

terminal Doppler weather radar (TDWR) A proposed weather radar system that will be able to detect severe weather, wind shear, and microbursts around high-activity airports.

terminal instrument approach procedures (TERPS) The FAA guidelines for the development of standard instrument approach procedures.

terminal radar approach control (TRACON) A terminal air traffic control facility associated with an air traffic control tower that uses radar to provide approach control services to aircraft.

terminal radar service area (TRSA) Airspace surrounding designated airports wherein controllers provide separation to all IFR and participating VFR aircraft. TRSAs are being replaced with airport radar service areas.

terminal VOR A VOR that is designed to be used only within the terminal area for local navigation and instrument approaches.

touchdown, midpoint, and rollout RVRs Runway visual range equipment located at the approach end of the runway, midway down the runway, and at the runway end.

tower cab The glass-enclosed area of an air traffic control tower where the controllers observe and separate aircraft.

TPX-42 A numeric radar beacon decoding system.

track The computer-calculated path of an aircraft.

trackball See *slew entry device*.

traffic advisories See *radar traffic advisories*.

traffic management system (TMS) A project of the FAA designed to integrate all of the FAA's flow control functions into one fully integrated system.

traffic management unit (TMU) One component of the traffic management system. Individual TMUs will be established at each ARTCC and at many of the busier terminal facilities.

transfer of communication The action taken to transfer responsibility for communicating with an aircraft from one controller to another.

transfer of control The action taken to transfer the responsibility for the separation of an aircraft from one controller to another.

transition level The altitude at which flight levels begin.

transmissometer The component of a runway visual range that determines the runway visibility.

transmitter The component of a radio device that initiates communication and creates the electronic transmission.

transponder The airborne component of the ATCRBS that replies to ground-based interrogations.

Transport Canada The Canadian Ministry with authority and responsibility for aviation safety and regulation. It is similar to the U.S. Department of Transportation.

ultra-high frequency (UHF) The frequency band between 300 and 3,000 mHz.

unassociated track An aircraft being tracked by either the primary or the secondary radar system whose identity is unknown to the ARTS radar computer system.

uncontrolled airspace That portion of the airspace over the United States that has not been designated as controlled airspace. Within uncontrolled airspace the FAA has neither the responsibility nor the authority to exercise control over air traffic.

vector A heading issued to an aircraft by a controller using radar to provide navigation guidance.

vertical separation A method of separating aircraft operating at different altitudes while on the same route.

very high frequency (VHF) The frequency band between 30 and 300 mHz.

very low frequency (VLF) The frequency band between 3 and 30 kHz.

VFR military training routes Routes used by the Department of Defense and associated Reserve and Air Guard units for conducting low-altitude navigation and tactical training under VFR below 10,000 feet MSL at airspeeds in excess of 250 knots IAS.

VFR not recommended An advisory provided by a flight service station to a pilot during a preflight or in-flight weather briefing that flight under visual flight rules is not recommended. To be given when the current and/or forecast weather conditions are at or below VFR minimums. It does not abrogate the pilot's authority to make his or her own decision.

VFR on top ATC authorization for an IFR aircraft to operate in VFR conditions at any appropriate VFR altitude (as specified in FAR and as restricted by ATC). A pilot receiving this authorization must comply with the VFR visibility, distance from cloud criteria, and the minimum IFR altitudes specified in FAR 91. The use of this term does not relieve controllers of their responsibility to separate aircraft in Class B and Class C airspace or TRSAs as required by FAA Order 7110.65.

VHF omnidirectional range (VOR) A ground-based navigation aid that transmits a VHF navigation signal 360° in azimuth.

victor airway See *airway*.

video map An electronically displayed map on the radar display that may depict data such as airports, heliports, runway centerline extensions, hospital emergency landing areas, navaids and fixes, reporting points, airway/route centerlines, boundaries, handoff points, special-use tracks, obstructions, prominent geographic features, map alignment indicators, range accuracy marks, and minimum vectoring altitudes.

video map intensity The control that adjusts the intensity of the video map.

video map selector The control that selects which video map will be displayed.

visual approach An approach wherein an aircraft that is on an IFR flight plan, is operating in VFR weather conditions, is under the control of an air traffic control facility, and has received the appropriate clearance may proceed visually to the airport of destination. During a visual approach, the pilot is responsible for navigation and terrain avoidance. The controller is responsible for air traffic control separation. A visual approach may be initiated by either the controller or the pilot.

visual aural range (VAR) A navigation device designed in 1937 at the Bureau of Air Commerce's research center in Indianapolis, Indiana. The visual aural range operated in the VHF band around 63 mHz and was an incremental improvement over the four-course radio range.

visual flight rules (VFR) Rules that govern the procedures for conducting flight under visual conditions.

visual separation A means employed by controllers to separate aircraft in terminal areas. To utilize visual separation, either the controller or one of the pilots visually separates the involved aircraft.

VLF/OMEGA An area navigation system that uses VLF transmitters.

VOR-DME A navigational facility providing VOR azimuth and civilian distance measuring equipment at one site. VORTACs can be used only by VOR-DME–equipped aircraft. TACAN-equipped aircraft cannot utilize a VOR-DME station.

VORTAC A navigational facility providing VOR azimuth, TACAN azimuth, and TACAN distance measuring equipment at one site. VORTACs can be used by VOR-DME– or TACAN-equipped aircraft.

wake turbulence Phenomena — including vortices, thrust stream turbulence, jet blast, and propeller- and rotor-induced turbulence — resulting from the passage of an aircraft through the atmosphere.

warning areas International airspace within which special operations are conducted that may be hazardous to nonparticipating aircraft.

waveguide Hollow metallic channels that conduct radar microwave energy to and from the antenna.

waypoint A predetermined geographical point defined as a bearing/distance from a VORTAC or as a longitude/latitude coordinate.

weight class See *aircraft class*.

COMMON ABBREVIATIONS

AAS advanced automation system
ACARS Aircraft Communications Addressing and Reporting System
ACC Air Coordinating Committee
ACC area control center
ADF automatic direction finder
ADIZ air defense identification zone
AERA automated en route air traffic control
AFI ICAO African-Indian Ocean Region
AFSS automated flight service station
AIM *Aeronautical Information Manual*
AIP *Aeronautical Information Publication*
ALNOT alert notice
ALS approach lighting system
ALSF high-intensity approach lighting system
ALTRV altitude reservation
AMB Airways Modernization Board
ANDB Air Navigation Development Board
ANG Air National Guard
APPREQ approval request
ARINC Aeronautical Radio Incorporated
ARSA airport radar service area
ARSR air route surveillance radar
ARTCC air route traffic control center
ARTS automated radar terminal system
ASD aircraft situation display
ASDE airport surface detection equipment
ASR airport surveillance radar
ATA airport traffic area
ATC air traffic control
ATCA Air Traffic Control Association
ATCC airway traffic control center
ATCRBS air traffic control radar beacon system
ATCS airway traffic control station
ATCSCC Air Traffic Control System Command Center
ATCT airport traffic control tower
ATCU airway traffic control unit
ATD actual time of departure
ATIS Automatic Terminal Information Service
ATS Air Traffic Service
AZ azimuth transmitter
BAZ back azimuth
BRITE bright radar indicator tower equipment
BTL beacon tracking level
CAA Civil Aeronautics Administration
CAA Civil Aeronautics Authority
CAB Civil Aeronautics Board
CAMI Civil Aeronautical Medical Institute
CAR ICAO Caribbean Region
CARF central altitude reservation function
CARs civil air regulations
CCC central computer complex
CD common digitizer
CDC computer display channel
CDT controlled departure time
CENPAC Central North Pacific
CENRAP central radar ARTS presentation
CEP Central East Pacific
CERAP combined center/RAPCON
CFCF Central Flow Control Facility
COP changeover point
CP circular polarization
CRD computer readout device
CST controller skills test
CUE computer update equipment
CW continuous wave
DARC discrete address radar channel
DAS data acquisition subsystem
D-BRITE digital BRITE
DEDS data entry and display subsystem
DES data entry sets
DH decision height

DME	distance measuring equipment
DME/p	precision DME
DOT	Department of Transportation
DPS	data processing subsystem
DVFR	defense visual flight rules
DVOR	Doppler VOR
EARTS	en route ARTS
ECM	electronic counter measure
EDARC	enhanced discrete address radar channel
EDCT	expected departure clearance time
EFAS	en route flight advisory service
EFC	expect further clearance
EFIS	Electronic Flight Instrument System
ELOD	en route sector loading
ELT	emergency locator transmitter
ERM	en route metering
ETA	estimated time of arrival
EUR	ICAO European Region
FAA	Federal Aviation Administration
FAA	Federal Aviation Agency
FAAH7110.65	*Air Traffic Control Handbook*
FAF	final approach fix
FARs	Federal Aviation Regulations
FCC	Federal Communications Commission
FD	facility directive
FDC	Flight Data Center
FDIO	flight data input/output
FDP	flight data processing
FIFO	flight inspection field office
FIR	flight information region
FL	flight level
FLIP	Flight Information Publication
FMS	flight management system
FPL	full performance level
FRC	full route clearance
FSDO	Flight Standards District Office
FSS	flight service station
GADO	General Aviation District Office
GNSS	Global Navigation Satellite System
GPS	Global Positioning System
GRI	group repetition interval
GS	General Schedule
HIRL	high-intensity runway lighting
IAF	initial approach fix
IATA	International Air Transport Association
ICAO	International Civil Aviation Organization
IDENT	aircraft identification
IFF	identification friend or foe
IFIM	International Flight Information Manual
IFR	instrument flight rules
IFSS	international flight service station
ILS	instrument landing system
IM	inner marker
IMC	instrument meterological conditions
INREQ	information request
INS	inertial navigation system
INSACS	Interstate Airway Communication Station
IR	IFR military training route
JSS	joint surveillance system
LAES	Landing Aids Experiment Station
LDA	localizer directional aid
LDIN	lead-in lights
LIRL	low-intensity runway lighting
LLWAS	low-level wind-shear alert system
LMM	locator middle marker
LOA	letter of agreement
LOM	locator outer marker
LORAN	long-range navigation
L/MF	low/medium frequency
LP	linear polarization
MACH	mach number
MALS	medium-intensity approach lighting system
MALSR	medium-intensity approach lighting system with RAIL
MAP	missed approach point
MARSA	military assumes responsibility for separation
MCA	minimum crossing altitude
MDA	minimum descent altitude
MEA	minimum en route altitude
MIA	minimum IFR altitude
MID/ASIA	ICAO Middle East/Asia Region
MIRL	medium-intensity runway lighting
MM	middle marker
MNPSA	minimum navigation performance specifications airspace
MOA	military operations area
MOCA	minimum obstruction clearance altitude
MSAW	minimum safe altitude warning
MSL	mean sea level
MTI	moving target indicator
MTR	military training route
MVA	minimum vectoring altitudes
NAM	ICAO North American Region
NAR	North American Route
NAR	North Atlantic Region
NAS	National Airspace System
NAT	ICAO North Atlantic Region
NATCA	National Air Traffic Controllers Association
navaids	navigation aids
NBCAP	National Beacon Code Allocation Plan
NDB	nondirectional radio beacon
NEXRAD	next-generation radar
n mi	nautical mile
NOAA	National Oceanic and Atmospheric Administration
NOPAC	North Pacific
NORAD	North American Aerospace Defense Command
NOS	National Ocean Service

NOTAM	notice to airmen
NTSB	National Transportation Safety Board
NTZ	no transgression zone
NWS	National Weather Service
OACC	Oceanic Area Control Centers
ODALS	omnidirectional approach lighting system
ODAPS	Oceanic Display and Planning System
OM	outer marker
OPM	Office of Personnel Management
OTS	NAT Organized Track System
PAC	ICAO Pacific Region
PANS	Procedures for Air Navigation Services
PAPI	precision approach path indicator
PAR	precision approach radar
PAR	preferred arrival routes
PATCO	Professional Air Traffic Controllers Organization
PCL	pilot-controlled lighting
PDR	preferred departure routes
PIREP	pilot report
PPI	plan position indicator
PRF	pulse repetition frequency
PVD	plan view display
QAK	quick action keyboard
RAILs	runway alignment indicator lights
RAPCON	radar approach control
RATCF	radar air traffic control facility (USN)
RCAG	remote communication air/ground
RCC	Rescue Coordination Center
RCLS	runway centerline system
RCO	remote communications outlet
RDP	radar data processing
REILs	runway end identifier lights
RL	report leaving
RML	radar microwave link
RNAV	area navigation
RTF	Radar Training Facility
RVR	runway visual range
RVV	runway visibility value
SAGE	semiautomated ground environment
SAM	ICAO South American Region
SAR	search and rescue
SC-31	Special Committee 31
SED	slew entry device
SELCAL	selective calling system
SFL	sequenced flashing lights
SID	standard instrument departure
SIGMET	significant meteorological information
SIP	special identification pulse
SLS	side lobe suppression
SPAN	stored program alphanumeric
SSALR	simplified short approach lighting system with RAIL
SSALS	simplified short approach lighting system
SSR	secondary surveillance radar
STAR	standard terminal arrival route
STC	sensitivity time control
STMC	supervisory traffic management coordinator
STMP	sector traffic management program
STOL	short takeoff and landing
SVFR	special VFR
SWAP	severe weather avoidance plan
TACAN	tactical air navigation
TCAS	Traffic Alert and Collision Avoidance System
TDWR	terminal Doppler weather radar
TDZL	touchdown zone lighting
TERPS	terminal instrument approach procedures
TMC	traffic management coordinator
TMS	traffic management system
TMU	traffic management unit
TRACAB	terminal radar approach control in tower cab
TRACON	terminal radar approach control
TRSA	terminal radar service area
UHF	ultra-high frequency
USA	United States Army
USAF	United States Air Force
USN	United States Navy
UTC	coordinated universal time
UTM	unsuccessful transmission message
VASI	visual approach slope indicator
VFR	visual flight rules
VHF	very high frequency
VLF	very low frequency
VMC	visual meteorological conditions
VOR	VHF omnidirectional range
VOR-DME	collocated VOR and DME navigational aids (VHF course and UHF distance information)
VORTAC	collocated VOR and TACAN navigation aids (VHF and UHF course and UHF distance information)
VR	VFR military training route

A	cleared to airport (point of intended landing)
B	center clearance delivered
C	ATC clears (when clearance relayed through non-ATC facility)
CAF	cleared as field
D	cleared to depart from the fix
F	cleared to the fix
H	cleared to hold and instructions issued
L	cleared to land
N	clearance not delivered
O	cleared to the outer marker
PD	cleared to climb/descend at pilot's discretion
Q	cleared to fly specified sectors of a navaid defined in terms of courses, bearings, radials, or quadrants within a designated radius
T	cleared through (for landing and takeoff through intermediate point)

V	cleared over the fix
X	cleared to cross (airway, route, radial) at (point)
Z	tower jurisdiction
BC	back course approach
CT	contact approach
FA	final approach
FMS	Flight Management System Approach
GPS	GPS approach
I	initial approach
ILS	ILS approach
MA	missed approach
MLS	MLS approach
NDB	nondirectional radio beacon approach
OTP	VFR conditions on top
PA	precision approach
PT	procedure turn
RH	runway heading
RP	report immediately upon passing (fix/altitude)
RX	report crossing
SA	surveillance approach
SI	straight-in approach
TA	TACAN approach
TL	turn left
TR	turn right
VA	visual approach
VR	VOR approach
WATRS	West Atlantic Route System

REFERENCES

Books

Andrews, Alan. *ABC's of Radar.* Indianapolis, Ind.: Howard W. Sams & Co., 1966.

Berkowitz, Raymond S. *Modern Radar: Analysis, Evaluation, and System Design.* New York: John Wiley & Sons, 1965.

Carpentier, Michael H. *Principles of Modern Radar Systems.* Boston: Artech House, 1988.

Clausing, Donald J. *The Aviator's Guide to Modern Navigation.* Blue Ridge Summit, Penn.: Tab Books, 1989.

Jackson, William E. *The Federal Airways System.* Institute of Electrical and Electronic Engineers, 1970.

Pallett, E. H. J. *Aircraft Instruments: Principles and Applications.* London: Pitman, 1981.

Wheeler, Gershon J. *Radar Fundamentals.* Englewood Cliffs, N.J.: Prentice-Hall, 1967.

Airport System Development. Washington, D.C.: Office of Technology Assessment, 1984.

Avionics Fundamentals. Riverton, Wy.: IAP Inc., 1987.

Bonfires to Beacons. Washington, D.C.: Department of Transportation, Federal Aviation Administration, 1978.

ICAO Annexes to the Convention on International Civil Aviation. Montreal, Canada: International Civil Aviation Organization.

ICAO Bulletin. Montreal, Canada: International Civil Aviation Organization.

Safe, Separated and Soaring. Washington, D.C.: Department of Transportation, Federal Aviation Administration, 1980.

Takeoff at Mid Century. Washington, D.C.: Department of Transportation, Federal Aviation Administration, 1976.

Troubled Passage. Washington, D.C.: Department of Transportation, Federal Aviation Administration, 1987.

Turbulence Aloft. Washington, D.C.: Department of Transportation, Federal Aviation Administration, 1979.

FAA Publications

Aeronautical Information Publication, 1998.

Air Traffic Control Facility Analysis for the Microwave Landing System, 1985.

Air Traffic Handbook, 1989.

Aeronautical Information Manual, 2000.

Airport Surveillance Radar System ASR-8, 1987.

Airport Surveillance Radar System ASR-9, 1996.

ARTS-III Data Systems Personnel Course, 1982.

ASR 4-5-6, 1969.

Automated Radar Terminal System IIIA, 1987.

Aviation Weather Services, 1985.

BRITE Radar Indicator Tower Equipment, 1980.

Center Weather Service Unit, 1984.

Computer Theory and Operations for Air Traffic Control, 1973.

Data Communications, 1989.

Data Systems Specialist Course, 1980.

Direct Access Radar Channel System Instruction Book, 1985.

En Route Stage A Flight Data Processing, 1978.

En Route Stage A Radar Data Processing, 1978.

En Route, Terminal, Flight Service Navaids, 1977.

Establishment and Validation of En Route Sectors, 1984.

FAA Emergency Operations Plan, 1985.

Facility Operation and Administration, 1989.

Flight Procedures and Airspace, 1984.

Flight Services Handbook, 1989.

FSAS Overview, 1985.

Holding Pattern Criteria.
ILS Localizer, 1983.
Instrument Landing System, 1967.
International Flight Information Manual, 1989.
Introduction to Computer Operation, IBM 9020, 1981.
Introduction to Radar Techniques, 1986.
Introduction to the IBM 9020 Central Computer Complex, 1985.
Location Identifiers, 1988.
Minimum En Route IFR Altitudes over Particular Routes and Intersections, 1988.
Moving Target Indicator, 1979.
NAS Indoctrination for Engineers and Technicians, 1985.
National Airspace System Operational Equipment, 1979.
National Airspace System Plan, 1988.
Operation Rain-Check, 1988.
Planning Guide for Airport and Airway ATC Facilities and Services, 1986.
Procedures for Handling Airspace Matters, 1984.
Radar Antennas, 1984.
Rotorcraft Master Plan, 1985.
Secondary Surveillance Radar, 1967.
Siting Criteria for Instrument Landing Systems, 1985.
Studies of Poststrike Air Traffic Control Specialist Trainees, 1988.
Terminal Arrival Area Design Criteria, 1997.
Terminal Basic Air Traffic Manual, 1981.
United States Standard for Terminal Instrument Procedures, 1986.
VOR Orientation, 1971.
Wake Turbulence, 1975.

FAA Advisory Circulars

AC 20-121 *Airworthiness Approval of Airborne Loran-C Systems for Use in the U.S. National Airspace System.*
AC 73-2 *IFR Helicopter Operations in the Northeast Corridor.*
AC 90-5 *Coordination of Air Traffic Control Procedures and Criteria.*
AC 90-23 *Aircraft Wake Turbulence.*
AC 90-43 *Operations Reservations for High-Density Traffic Airports.*
AC 90-45 *Approval of Area Navigation Systems for Use in the U.S. National Airspace System.*
AC 90-50 *VHF Radiofrequency Assignment Plan for Aeronautical Operations.*
AC 90-65 *Air Traffic Fuel Economy Program.*
AC 90-67 *Light Signals from the Control Tower for Ground Vehicles, Equipment and Personnel.*
AC 90-72 *Minimum Safe Altitude Warning.*
AC 90-76 *Flight in Oceanic Airspace.*
AC 90-78 *En Route Conflict Alert.*
AC 90-79 *Recommended Practices and Procedures for the Use of Electronic Long Range Navigation Equipment.*
AC 90-82 *Random Area Navigation Routes.*
AC 90-83 *Terminal Control Areas.*
AC 90-85 *Severe Weather Avoidance Plan.*
AC 90-88 *Airport Radar Service Area.*
AC 91-14 *Altimeter Setting Sources.*
AC 91-49 *General Aviation Procedures for Flight Within the North Atlantic MNPS Airspace.*
AC 91-50 *Importance of Transponder Operation and Altitude Reporting.*
AC 97-1 *Runway Visual Range.*
AC 99-1 *Security Control of Air Traffic.*
AC 120-26 *ICAO Designator Assignment for Aircraft Operating Agencies.*
AC 120-33 *Operational Approval of Airborne Long-Range Navigational Systems for Flight Within the North Atlantic MNPS Airspace.*
AC 170-13 *Approach Lighting System Configurations and Energy Conservation.*
AC 210-5 *Military Flying Activities.*
AC 211-2 *Recommended Standards for IFR Aeronautical Charts.*

PHOTO CREDITS

Chapter 1.
5: FAA 6: FAA 7: FAA 9: Michael Nolan 11: FAA 14: FAA 19: FAA 21: Journal of Air Traffic Control 23: FAA 30: FAA 35: FAA

Chapter 2.
48: Beech Aircraft Corporation 49: Beech Aircraft Corporation 51: Michael Nolan 52, left: Beech Aircraft Corporation 52, left center: Beech Aircraft Corporation 52, right: Beech Aircraft Corporation 52, right center: Beech Aircraft Corporation 55: Beech Aircraft Corporation 58: Wadsworth, Inc. Photo by Paul Bowen Photography, Inc. 60: Bendix/ King Division of Allied Signal Aerospace Company 60: Bendix/ King Division of Allied Signal Aerospace Company 64: FAA 71: E-Systems, Montek Division 72: E-Systems, Montek Division 76: Bendix/King Division of Allied Signal Aerospace Company 79: Il Morrow Inc. 83: Garmin, Inc. 86: Delco Systems Operations, Delco Electronics Corporation 99: Michael Nolan 101: Bendix/ King Division of Allied Signal Aerospace Company 108: Michael Nolan 110, top: Bendix/ King Division of Allied Signal Aerospace Company 110, bottom: Bendix/ King Division of Allied Signal Aerospace Company 116: Michael Nolan 119: Michael Nolan 120: Michael Nolan 122: FAA 125: Michael Nolan 126: Michael Nolan

Chapter 4.
178, left: Qualimetrics, Inc. 178, right: Qualimetrics, Inc.

Chapter 5.
199: Michael Nolan 200: Michael Nolan

Chapter 8.
302: Westinghouse Corporation 308: UNISYS Corporation 310: FAA 311: FAA 321: ITT Gilfillan Division 323: ITT Gilfillan Division 324: Michael Nolan 325: Norden Systems Division of United Technologies 334: Eaton Corporation 339: UNISYS Corporation 345: FAA 348: FAA 349: FAA

Chapter 9.
357: FAA 384: FAA

Chapter 10.
392: Michael Nolan 432: FAA

Chapter 12.
461: FAA 464: FAA 473: FAA 475: FAA 478: Dimensions International 479: Dimensions International 480, top: Dimensions International 480, bottom: FAA 481, top: FAA 481, bottom: FAA 484: FAA 485: FAA 489: FAA

Chapter 13.
499: Michael Nolan 502: Michael Nolan

INDEX

D

E